HANDBOOK OF MICROWAVE AND OPTICAL COMPONENTS

VOLUME 2

HANDBOOK OF MICROWAVE AND OPTICAL COMPONENTS

Editor-in-Chief

Kai Chang
Department of Electrical Engineering
Texas A&M University
College Station, Texas

HANDBOOK OF MICROWAVE AND OPTICAL COMPONENTS

Volume 2 Microwave Solid-State Components

Edited by

Kai Chang
Department of Electrical Engineering
Texas A&M University
College Station, Texas

WILEY

A Wiley-Interscience Publication

JOHN WILEY & SONS

New York · Chichester · Brisbane · Toronto · Singapore

Library of Congress Cataloging in Publication Data:
Handbook of microwave and optical components.

   "A Wiley-Interscience publication."
   Includes bibliographies.
   Includes index.
   Contents: v. 1. Microwave passive and antenna
components v. 2. Microwave solid-state components v. 3. Optical components.
   1. Microwave devices.   2. Optical instruments.
I. Chang, Kai, 1948–
TK7876.H34   621.381'3   90.27835
ISBN 0-471-843652

Printed in the United States of America

10 9 8 7 6 5 4 3 2 1

HANDBOOK OF MICROWAVE AND OPTICAL COMPONENTS

CONTENTS

PREFACE

In the past two decades, we have witnessed a rapid development in high-frequency spectra above the microwave frequency. Components and subsystems have been built in microwave, millimeter wave, infrared, and optical spectra for applications in communications, radar, remote sensing, remote control, sensors, navigation, surveillance, electronic warfare, radiometers, medicine, plasma research, imaging, computers, signal processing, industry heating, fabrication and processing, astronomy, and other fields. Extensive use of these spectra creates an urgent need for a practical handbook to assist the practicing professionals in their daily work.

This handbook is intended to serve as a compendium of principles and design data for practicing microwave and optical engineers. Although it is expected to be most useful to engineers actively engaged in designing microwave and optical systems, it should also be of considerable value to engineers in other disciplines who have a desire to understand the capabilities and limitations of microwave and optical systems. Many handbooks for low-frequency electronics are available; a good handbook to cover microwave and optical components is nonexistent. This handbook represents the most comprehensive treatment of both microwave and optical engineering that has appeared in book form to date.

To achieve these goals, this handbook covers almost all important components in microwave, millimeter wave, submillimeter wave, infrared, and optical frequency spectra. Theoretical discussions and mathematical formulations are given only where essential. Whenever possible, design results are presented in graphic and tabular form; references are given for further study. The book provides, in practical fashion, a wealth of essential principles, methods, design information, and references to help solve problems in high-frequency spectra.

The handbook was organized into two major parts with four volumes:

Part I. Microwave Components
    Vol. 1   Microwave Passive and Antenna Components
    Vol. 2   Microwave Solid-State Components
Part II. Optical Components
    Vol. 3   Optical Components
    Vol. 4   Fiber and Electro-Optical Components

Each chapter is written as a self-contained unit. It has its own table of contents and list of references. Some overlap is inevitable among chapters, but has been kept to a minimum. It is hoped that this comprehensive handbook will offer the type of detailed information necessary for use in today's complex and rapidly changing high-frequency engineering.

The authors who have contributed chapters to this handbook have done an excellent job of condensing mountains of material into readable accounts of their respective areas. The emphasis throughout has been to provide an overview and practical information of each subject.

I would like to express my special appreciation to my friend, Mr. Algie Lance, who was writing his chapter while terminally ill. He submitted his manuscript on time and passed away just after the completion of his chapter. I would also like to thank all members of editorial board for their advice and suggestions, Dr. Felix Schwering for organizing the antenna chapters, and Mr. George Telecki, our Wiley editor, for his constant encouragement. I wish especially to thank my wife, Suh-jan, for her assistance in typing and managing this project.

KAI CHANG

College Station, Texas

CONTRIBUTORS

J. W. Archer
Division of Radio Physics
CSIRO
P.O. Box 76
Epping, New South Wales
Australia 2121

R. A. Batchelor
Division of Radio Physics
CSIRO
P.O. Box 76
Epping, New South Wales
Australia 2121

Pallab Bhattacharya
Department of Electrical Engineering
and Computer Science
The University of Michigan
Ann Arbor, MI 48109-2122

Kai Chang
Department of Electrical Engineering
Texas A&M University
College Station, TX 77843

Ho-Chung Huang
COMSAT Laboratories
22300 COMSAT Drive
Clarksburg, MD 20871-9475

Hing-Loi A. Hung
COMSAT Laboratories
22300 COMSAT Drive
Clarksburg, MD 20871-9475

Erik L. Kollberg
Department of Radio and Space Science
Chalmers University of Technology
S-412 96 Gothenburg
Sweden

H. John Kuno
Microwave Products Division
Hughes Aircraft Company
P.O. Box 2940
Torrance, CA 90509-2940

Thomas A. Midford
Microwave Products Division
Hughes Aircraft Company
P.O. Box 2940
Torrance, CA 90509-2940

Georges Salmer
Centre Hyperfréquences et
    Semiconducteurs
URA CNRS 287
Université des Sciences et Techniques de
    Lille-Flandres-Artois
59655 Villeneuve D'Ascq Cedex
France

Jasprit Singh
Department of Electrical Engineering and
    Computer Science
The University of Michigan
Ann Arbor, MI 48109-2122

Thane Smith
COMSAT Laboratories
22300 COMSAT Drive
Clarksburg, MD 20871-9475

Craig P. Snapp
Advanced Bipolar Products
Avantek, Inc.
39201 Cherry Street
Newark, CA 94560

Ajay I. Sreenivas
Aerospace Systems Division
Ball Brothers Corp.
P.O. Box 1062
Boulder, CO 80306

Ron Stockton
Aerospace Systems Division
Ball Brothers Corp.
P.O. Box 1062
Boulder, CO 80306

Cheng Sun
Department of Electrical Engineering
California Polytechnical State University
San Luis Obispo, CA 93407

Joseph F. White
Applied Microwave
Lexington, MA 02173

Jacques Zimmerman
Centre Hyperfréquences et
    Semiconducteurs
URA CNRS 287
Université des Sciences et Techniques de
    Lille-Flandres-Artois
59655 Villeneuve D'Ascq Cedex
France

HANDBOOK OF MICROWAVE AND OPTICAL COMPONENTS

VOLUME 2

1

MOLECULAR BEAM EPITAXY AND ITS APPLICATION TO MICROWAVE AND OPTICAL DEVICES

Pallab Bhattacharya and Jasprit Singh

University of Michigan
Ann Arbor, Michigan

1.1 INTRODUCTION

1.1.1 III–V Compound Devices

A new solid-state technology is required to satisfy needs in ultra high-speed computing, optical communications and other optoelectronic processing areas. Rapid advances in III–V compound semiconductors have built a convincing case that these materials may provide the base for satisfying new technological demands. There are two principal reasons that GaAs and related III–V compounds are believed to play an important role in future electronic devices and systems. First, microelectronic devices made of these new materials can operate much faster and with lower power dissipation than those made with existing silicon-based technology. Second, some optoelectronic devices, such as semiconductor lasers and detectors, can be made with these materials, which enables the integration of optical devices and electronic devices on one substrate. Especially, the recent advent of optical fibers for the 1.3- and 1.55-μm wavelength range has led to a great interest in the ternary alloy $In_{0.53}Ga_{0.47}As$ as a material for long-wavelength optoelectronic devices. In addition, high-room-temperature electron mobility and saturation velocity make this material promising for high-speed-device applications.

To build high-performance devices with these compound semiconductors, however, involves a variety of scientific and technical challenges. Thin-channel regions as well as fine-line geometries are essential to get the possible high-speed performance of field-effect transistors from these materials. Similarly, the active regions of detectors and lasers place stringent demands on thickness and doping control and materials quality. There are two different ways to form thin-channel regions: epitaxial growth such as molecular beam epitaxy (MBE) and metal-organic chemical vapor deposition (MOCVD), and ion implantation with subsequent high-temperature activation. Here we are concerned primarily with MBE as the main crystal-growth technique.

1.1.2 Molecular Beam Epitaxy

The importance of epitaxy in semiconductor device fabrication is a direct consequence of two critical needs: for thin, defect-free single-crystal films with precisely defined geometrical, electrical, and optical properties, and for heterojunction structures free of interfacial impurities and defects. The traditional techniques of liquid-phase epitaxy, which is dependent on thermodynamic phase equilibria, and vapor-phase epitaxy, which achieves growth by chemical reactions in the gas phase on a heated substrate, can certainly meet certain subsets of the foregoing requirements, but not all of them. The newer technique, molecular beam epitaxy, achieves epitaxial growth by the reaction of one or more thermal atomic or molecular beams of the constituent elements with a crystalline substrate surface held at a suitable elevated temperature under ultrahigh vacuum (UHV). The technique, as it stands today, has come a long way since the pioneering studies of Davey and Pankey [1], Arthur [2], Cho and Arthur [3], Ilegems [4], Robinson and Ilegems [5], and Gossard [6]. Work by Esaki, Chang and coworkers [7], Ploog [8], Joyce and Foxon [9], and Holloway and Walpole [10] have led to a better understanding of the growth kinetics and improved quality of materials and devices. Essentially confined to research and development until about 1976, MBE has now emerged as a feasible growth technique for the realization of stringent device requirements. The uniqueness of the technique lies mainly in the tremendous precision in controlling layer dopings, thickness, and composition; in growing modulated structures whose periods are typically less than the electron mean free path; and in achieving nearly perfect heterointerfaces and surface morphologies.

The growth of a compound semiconductor heterostructure by MBE is a many-particle, far-from-equilibrium thermodynamic problem representing an open system. In addition, numerous chemical reactions are involved in the growth process. Clearly, even for a relatively simple growth process such as MBE an extremely detailed understanding is very difficult. However, with appropriate approximations it is possible to obtain understanding of key atomistic processes in the growth. Such an understanding can throw light on the following important areas:

1. Identifying the key kinetic and energetic parameters controlling the growth process.
2. Suggesting optimum growth conditions for fabrication of an interface from a given set of semiconductors.
3. Providing insight toward the development of mathematical models for the microstructure of interfaces grown by MBE. These models could play a critical role in understanding the optical and transport properties in heterostructures as well as in identifying the effects of imperfect interfaces on device performance.
4. Identifying semiconductor combinations that are likely to be "difficult" to grow with sharp interfaces using conventional MBE technology and suggesting novel approaches for their growth.
5. Suggesting key experimental studies [such as reflection high-energy electron diffraction (RHEED) oscillations; masked growth, etc.] to understand the growth process and for the determination of growth kinetics critical for the fabrication of high-quality interfaces.

The key problems associated with any epitaxial process are the ability to grow:

1. High-quality (defect-free) epitaxial layers
2. Hyperabrupt interfaces between two different semiconductors

3. Structure with abrupt doping profiles

4. Alloys which are thermodynamically immiscible

These accomplishments should be made over a large area reliably and reproducibly. Clearly, it is critical to understand the epitaxial growth process at a microscopic level to accomplish all this.

·1.1.3   Heterostructure Technology

The last decade has seen tremendous development in the area of heterostructure research [11–13]. The new concepts involved and the potential in the heterostructure technology have drawn scientists from areas of condensed matter physics, material science, electrical engineering, and chemistry. The underlying reason for this interest, is of course, the realization that it is possible to tailor the band structure of materials using this technology. This has led to the birth not only of a number of important physics concepts, such as quantum Hall effects and physics of quasi-two dimensional systems, but has also led to the conception and realization of numerous electrical, electro-optic, and optical devices with tailored response. In fact, it is clear that heterostructure technology will be an important area of research during the next decade.

Heterostructure technology is based on the chemical modulation in real space using two or more atomic species. Thus a critical ingredient of a heterostructure is the interface between two materials. As the modulation distance becomes smaller, the role of the interface region becomes more dominant. In addition to the interface, the quality of the bulk layers are, of course, also important. Poor quality of the interface and/or the bulk regions can cause serious degradation of the potential properties of the heterostructure devices. It is thus extremely important that growth be conducted under such conditions that the heterostructure quality is as good as possible. Interest has also developed in the theoretical understanding and subsequent control of the epitaxial growth process.

In the sections to follow, we describe and discuss the theoretical and experimental aspects of MBE growth with a detailed description of the surface kinetics. The key issues in materials processing for high-speed and high-frequency microwave and optoelectronic devices will be next described. Finally, the materials and heterostructure properties, and the limitations imposed by them on device performance, are described and discussed. Most of the work at the present time have focused on the $GaAs/Al_xGa_{1-x}As$ and $In_{0.53}Ga_{0.47}As/In_{0.52}Al_{0.48}As/InP$ heterostructures grown by MBE and MOCVD, and much of our own work has centered around these systems. In this chapter we, therefore, discuss their growth and properties, and the devices made from them. These devices are modulation-doped FETs, quantum well modulators, and photodetectors.

1.2   HETEROSTRUCTURE FABRICATION BY MOLECULAR BEAM EPITAXY

1.2.1   Introduction

In contrast with liquid-phase epitaxial (LPE) and vapor-phase epitaxial (VPE) growth of semiconducting crystals under quasi-equilibrium conditions, growth by MBE is accomplished under nonequilibrium conditions and is governed principally by surface kinetic processes. Since the initial demonstration of the principle and process, MBE has come a long way and has emerged as a technique with unprecedented control over growth parameters. This has resulted from extensive research leading to a better understanding of the growth processes and great advances in system design and reliability. In the present section a very brief review of the MBE process is followed by more recent issues concerning the understanding of heterointerface growth.

1.2.2 Basic Growth Process

Molecular beam epitaxy is a controlled thermal evaporation process under ultrahigh-vacuum conditions. The process is shown schematically in Fig. 1.1 for the growth of GaAs with the possibility for dopant incorporation. The cells are designed such that realistic fluxes for crystal growth at the substrate can be realized while maintaining the Knudsen effusion condition (i.e., the cell aperture is smaller than the mean free path of the vaporized effusing species within the cell). The individual cells are provided with externally controlled mechanical shutters whose movement times are less than the time taken to grow a monolayer. Therefore, very abrupt composition and doping profiles are possible. Interfaces that are one monolayer abrupt can be obtained fairly easily. The process of crystalline growth by MBE involves absorption of the constituent atoms or molecules, their dissociation and surface migration, and finally, incorporation resulting in growth. From pulsed molecular beam experiments Arthur [2, 14] observed that below a substrate temperature of 480°C, Ga had a unity sticking coefficient on (100) GaAs. Above this temperature the coefficient is less than unity. The absorption and incorporation of As_2 or As_4 molecules is more complex. It was found that in general As sticks only when a Ga adatom plane is already established. Joyce and coworkers [9] have described the kinetic processes leading to the growth of GaAs from Ga and As_2 or As_4 molecules. The cation is in atomic form when it reaches the heated surface. It attaches randomly to a surface site and undergoes several kinetically controlled steps before it

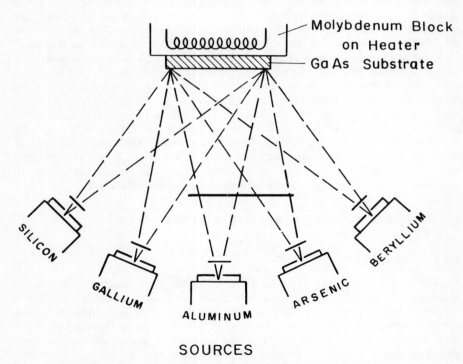

Molybdenum Block on Heater

GaAs Substrate

SILICON

BERYLLIUM

GALLIUM

ALUMINUM

ARSENIC

SOURCES

Figure 1.1 Schematic illustration of the molecular beam expitaxial process.

is finally incorporated. The As_2 or As_4 atoms is first physisorbed into a mobile, weakly bound precursor state. As this state moves on the surface, some loss occurs due to reevaporation; the rest is finally incorporated in paired Ga lattice sites by dissociative chemisorption.

The important fact which emerged from the early kinetic studies is that stoichiometric GaAs can be grown over a wide range of substrate temperatures by maintaining an excessive overpressure of As_2 or As_4 over the Ga beam pressure. Substrate temperatures during growth are usually close to and slightly higher than the congruent evaporation temperature of the growing compound. Under normal MBE growth conditions, where the incorporation rate of the cations is nearly 100%, the cation surface migration needs to be high. Otherwise, MBE growth will occur by a three-dimensional island mode rather than the step-growth mode which is necessary for producing high-quality and abrupt interfaces.

1.2.3   MBE Growth Models

The MBE growth process is an extremely complex one and is a nonequilibrium thermodynamic problem. Therefore, theoretical studies involving analytical treatment and simulation prove to be very useful in providing an insight to MBE growth and complementing experiments. They can also provide guidelines for better growth, as mentioned earlier. Singh and Bajaj [15] have subdivided the problems in three key areas: (a) energetics, (b) incorporation process, and (c) growth kinetics. More recent work by Singh and Bajaj [16] using sophisticated computer simulation models and the Monte Carlo technique have led to an understanding of the detailed atomistic processes underlying MBE growth. From these studies the importance of flux rates, cation migration, and growth rates and their relation to each other can be understood. In conclusion, it is important to realize that all these factors have to be understood and implemented to grow high-quality, defect-free materials. In Fig. 1.2 we list the important in situ and ex

	Technique	Potential Outcome
i)	Modulation Spectroscopy	- Nature of precursor state - Reaction order for chemisorption - Information on lifetimes, activation barriers
ii)	RHEED	- Growth modes - Diffusion coefficients
iii)	Masked growth studies	- Surface diffusion rates
i)	High resolution microscopy (TEM, SEM)	- Nature of interface quality - Clustering effects
ii)	High resolution photoluminescence, reflection, absorption, excitation spectroscopies	- Interface microstructure - Effects of growth interruption
iii)	Low temperature transport in MODFETs	- Interface quality - Diffusion of dopants

Figure 1.2   List of important in situ and ex situ experimental techniques and their potential role in understanding the growth process.

situ measurement techniques that have led to a better understanding of the MBE growth process.

1.2.4 System Description

A typical present-day MBE growth facility consists of the UHV growth chamber into which the growth substrate is introduced through one or two sample-exchange load-locks. The base pressure in the growth chamber is usually about 10^{-11} torr, while the other chambers are at about 10^{-10} torr. The schematic of the growth chamber is shown in Fig. 1.3. The effusion cells are made of pyrolytic BN. To obtain the dimers from the tetramer species of the group V elements, external cracking cells are incorporated. Cracking is enhanced in the presence of a loosely packed catalytic agent. Charge interlocks with auxiliary pumping is sometimes used for the more rapidly depleting group V species. The growth chamber and effusion cells are provided with liquid N_2 cryoshrouds, which are kept cold during growth.

Most growth systems are equipped with in situ surface diagnostic and analytical capabilities in the growth and auxiliary chambers. The most common facilities in the growth chambers are a quadrupole mass spectrometer (or residual gas analyzer) which gives important information regarding the ambient in the growth chamber at all times and a reflection high-energy electron diffraction (RHEED) system which gives an insight to the surface structures of crystals and the growth mechanism. Electrons from a high-energy (~ 10 keV) electron gun strike the substrate or the growing layer surface at glancing incidence, and the diffraction pattern is monitored on a fluorescent screen. In addition to the static RHEED pattern, it is possible to explore some features of the growth dynamics of MBE by monitoring temporal variations in the intensity of various features in the RHEED pattern. It has been found [17] that damped oscillations in the

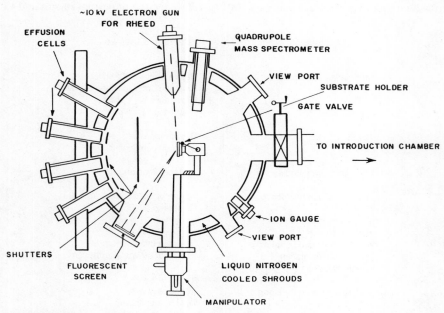

Figure 1.3 Schematic cross section of a typical MBE growth chamber equipped with in situ diagnostic tools.

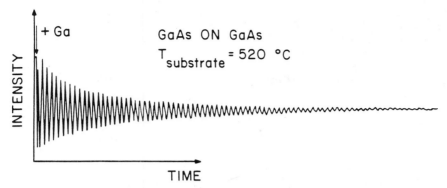

Figure 1.4   Dynamic RHEED oscillation data recorded during growth of GaAs on GaAs substrate at $T = 520°C$.

intensity of both the specular and diffracted beams occur immediately after initiation of growth, as shown in Fig. 1.4. By observation of these oscillations on misoriented surface steps, it is possible quantitatively to determine the surface diffusion rates of adatoms [17].

1.2.5   Growth of Ternary Compounds

Figure 1.5 illustrates the variation of bandgap with lattice constant in common III–V compounds, and their ternary and quaternary derivatives. $In_{0.53}Ga_{0.47}As$ and $In_{0.52}$

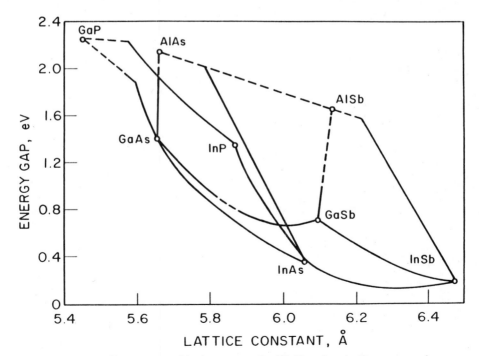

Figure 1.5   Energy gap and lattice constant for III–V semiconducting compounds.

$Al_{0.48}As$ lattice matched to InP are of tremendous technological improtance for optical communication and high-speed-device applications. For these compounds a slight deviation from the stated solid solution compositions will create a large mismatch. Despite these difficulties, these alloys are being grown with quality almost comparable to LPE material by several groups of workers [18–25]. The growth rate R_g of III–V compounds is entirely controlled by the flux densities of the group III beams, F_i:

$$R_g \propto \sum_{i=1}^{n} \alpha_i F_i \tag{1.1}$$

where n is the number of the different group III elements and α_i are their respective sticking coefficients. At normal growth temperatures, $\alpha_i \approx 1$ but decreases for elevated growth temperatures. The flux densities of the beams F_i(atoms $cm^{-2} s^{-1}$) incident on the substrate surface are controlled by the temperatures of the effusion cells provided that the cell aperture is less than the mean free path of vapor molecules within the cell (i.e., Knudsen effusion). The flux F_i of species i per unit area at the substrate surface is given by

$$F_i = \frac{A_c P_i}{\pi d_s^2 \sqrt{2\pi m_i k_B T}} \cos \phi \tag{1.2}$$

where A_c = area of the cell aperature (in square centimeters)

$\quad\quad P_i$ = equilibrium vapor pressure of species i (in torr) in the cell at the absolute temperature T

$\quad\quad d_s$ = distance of the orifice from substrate (in centimeters)

$\quad\quad m_i$ = mass of the effusion species

$\quad\quad \phi$ = angle between the beam and the normal of the substrate

From a study conducted by Wood [26] it is apparent that the best growth temperatures for the ternary compounds are based on Vegard's rule for averaging the con-

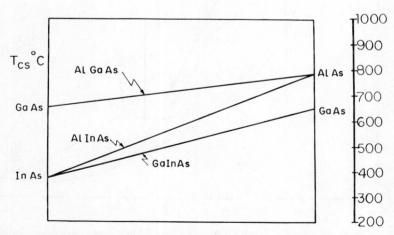

Figure 1.6 Vegard approximations for congruent sublimation temperatures for GaInAs, AlInAs, and AlGaAs. (From Ref. 26.)

gruent sublimation temperatures of the constituent binaries. The data for $In_xGa_{1-x}As$ and $Al_xGa_{1-x}As$ are shown in Fig. 1.6. Thus $In_{0.53}Ga_{0.47}As$ is usually grown in the temperature range 480 to 520°C and $Al_xGa_{1-x}As/GaAs$ is grown in the range 600 to 700°C, depending on the application. At low growth temperatures $Al_xGa_{1-x}As$ has high resistivity and its luminescence efficiency is very poor. The luminescent properties steadily improve with increase of growth temperature [27].

1.2.6 Effects of Growth Ambient

System preparation before growth has changed dramatically over the years. It is evident that all the active components and the cryoshrouds need to be baked very extensively before initiating growth of high-quality layers. By monitoring the ambient with the mass spectrometer it can be observed that concentrations of H_2O, CO, CO_2, and the hydrocarbons are reduced substantially. It is therefore improbable that any appreciable amount of impurity can be incorporated in the growing layer. Carbon is the dominant acceptor impurity found in MBE compounds and is probably produced by the reaction of background CO with the Ga adatoms [28]. It is also suspected that O from volatile Ga_2O may be incorporated [29], but the electrical and optical properties of O in MBE III–V compounds have not been clearly established. At the present time, it is believed that the most likely source of ubiquitous impurities in the films are the effusing species.

1.2.7 Electrical Doping

The common n-type dopants in III–V compounds are Si, Sn, Te, and Se, while the p-type dopants are Cd, Zn, Be, Mg, and Mn. Ge is usually amphoteric. For various reasons, as discussed by Ploog [8], most of these elements have been eliminated and Si and Be have emerged as the best n- and p-dopants, respectively. We have used these for the work reported here. However, there still exist the problems of dopant diffusion and "riding" in both cases.

1.2.8 Growth of Heterostructures and Multiquantum Wells

An increasing number of present-day electronic and optoelectronic devices make use of lattice-matched heterostructures and multilayered materials. Interesting carrier dynamics in such structures have given rise to a variety of interesting and novel concepts which are being applied to detectors, lasers, modulators, and to heterostructure electronic devices. Most of these devices have stringent requirements regarding interface quality. Although nearly perfect lattice matching at the interface can be maintained during MBE growth, the microscopic nature of the region may be far from perfect. In fact, interface roughness can extend to several monolayers, which can be of the order of the thicknesses of active regions of devices. The problem can therefore have very severe implications since it gives rise to additional carrier scattering at the interface and lower luminescence efficiency of the structure.

The magnitude of the interface roughness is related to the III–V surface bond strengths and cation migration rates under the particular growth conditions. It is therefore likely that the two layers forming a heterostructure may have different ideal growth conditions and therefore, depending on the sequence of growth, the cation adatoms may "pile up" in the layer by layer growth mode. Thus the interface, instead of being abrupt to one monolayer, can extend to several monolayers. Ideal and nonideal interface structures are schematically shown in Fig. 1.7.

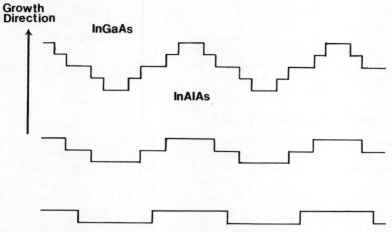

Figure 1.7   InGaAs/InAlAs interface structure, indicating various degrees of surface roughness, as obtained during MBE. The step height corresponds to a monolayer.

The problem of interface roughness has been studied theoretically by Singh and Bajaj [15, 30, 31], and these authors have suggested a number of techniques to improve the interface, such as different growth temperatures for the two layers, growth interruption, resonant laser excitation, and adding a few monolayers of a group III adatom with high surface migration rate just before the interface. Interrupted growth is being implemented by other authors [32] and we have investigated in detail growth interruption at the InGaAs/InAlAs normal and inverted interface. The probe used is the exciton transition intensity in the low-temperature photoluminescence spectra. PL studies were carried out on 120 Å InGaAs/InAlAs single-quantum-well structures and these experiments are discussed in detail in Section 1.4.

In addition to the experimental studies discussed above, detailed computer simulation experiments have been made to verify the various assumptions made as well as to predict the effect of novel growth approaches on growth modes. Computer simulations have yielded rich dividends in providing microscopic information on the MBE growth process. The general scheme followed in the computer simulations is described in Fig. 1.8. A careful coupling of experimental and theoretical information has to be established if the results of the computer simulations are to be reliable.

The output from a carefully planned computer simulation is of the following nature:

i. Partial layer coverages as a function of time
ii. Surface strain energy during growth
iii. Clustering effects in growth of alloys
iv. Interface structure parameters when interfaces are formed between different semiconductors
v. Surface hopping distances for growth under different conditions

In Fig. 1.9 we show typical results of computer simulations on the GaAs growth at two different temperatures. At low temperature, the growth front is rough, while at higher temperatures the growth front is smooth. This information can then be compared with

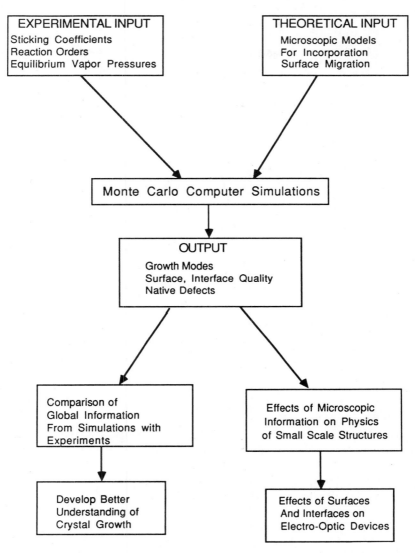

Figure 1.8 Flowchart of the coupling between experimental information and computer simulations.

specific experiments, such as RHEED, photoluminescence in quantum wells, masked growth studies, and so on, to authenticate the model.

The key results that have emerged from theoretical studies on MBE coupled with experimental studies are discussed below.

i. Cations surface kinetics are critical in controlling the growth-front (and interface) quality.

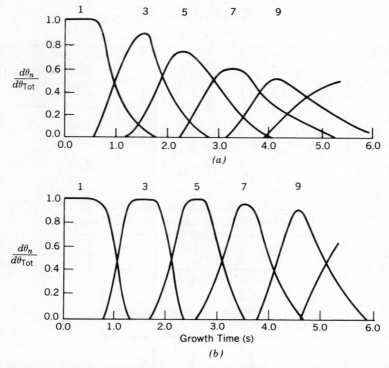

Figure 1.9 Results of computer simulations on the effect of substrate temperature on the growth modes in MBE. $d\theta_o/dt$ is the partial coverage of the nth monolayer. (a) $T_s = 700$ K (rough growth); (b) $T_s = 810$ K (layer-by-layer growth).

ii. For a fixed growth rate, a given semiconductor can grow by the layer-by-layer mode above a certain temperature where the cation hopping rate is about 10^4 hops per time interval for a monolayer growth.

iii. Since different semiconductors have different bond strengths, and hence different activation barriers for cation hopping, the ideal temperatures for growth are different.

iv. Due to the observation noted in (iii), the quality of the normal and inverted interfaces can be quite different. The term "normal" interface is used for the interface produced by the sequence in which the lower-melting-temperature component is grown first. The term "inverted" interface is used for the reversed sequence.

v. The surface roughness of a growing structure gets worse as the film thickness increases, due to statistical fluctuations of the impinging cation flux.

The results listed above have been very important in understanding a variety of observations in the MBE-grown heterostructures. The continuing computer simulation work has now acquired some predictive power which is expected to be important in heterostructure technology.

The discussion in Section 1.1 on the role of cation kinetics clearly suggest that one can expect serious difficulties in growth of high-quality heterostructures from materials with very different bond strengths. Since quite often, these are precisely the kind of heterostructures one wants to grow and study, it is important that techniques devised that will allow the growth of these structures without serious degradation of interfacial regions. These studies would all be based on the need to manipulate the kinetics of growth to ensure a layer-by-layer growth mode, at least near the region where the interface is formed. The approaches that have been examined theoretically (and experimentally in some cases) are:

1. *Growth Interruption near Interface Regions.* This approach is motivated by the recognition that a rough surface can be smoothened if growth is stopped and surface atoms are merely allowed to rearrange themselves [33, 34]. If interruption times for obtaining smooth surfaces are ≤1 to 2 min, this can be an effective technique. If fact, from the Monte Carlo studies we expect to be able to predict the interruption periods at different substrate temperatures.

2. *Pulsed Temperature Growth.* This approach is based on the recognition that if a semiconductor is grown at lower than its ideal temperature, its growth front will get rougher as the growth proceeds [35]. If at regular intervals, the substrate temperature is raised by about 100°C, for a short period (say, 1 to 5 s), it is possible to smoothen the surface through enhanced surface kinetics. This smoothening could be monitored through RHEED studies.

3. *Laser-Enhanced Processes.* Coupling of external sources of energy (e.g., via a laser system) can provide energy selectively to overcome various energy thresholds in the growth process. Among the thresholds that could conceivably be overcome are (i) the threshold for dissociative reaction for the cations and anions, and (ii) an activation barrier for surface migration of cations. Once again in situ RHEED studies would allow us to determine whether or not successful coupling of the laser energy is occurring.

4. *Use of Thin Superlattices in Improving Interface Quality.* This approach is motivated by the idea that a lower-bond-strength specie (e.g., GaAs compared to AlAs) can smoothen a rough growth front by "filling in" the surface roughness [31]. A number of groups have established the use of this technique, but its usefulness in reducing growth temperature needs to be studied carefully to utilize this concept fully in the growth of difficult heterostructures [36, 37].

1.3 DEEP-LEVEL EFFECTS IN MBE-GROWN MATERIALS AND HETEROSTRUCTURES

1.3.1 Introduction

In evaluating the quality of semiconducting materials for device applications, it has been found that centers with deep energy levels in the forbidden energy gap of large-bandgap semiconductors play an important role. Deep levels essentially act as carrier recombination or trapping centers and adversely affect device performance. The capture cross section $\sigma_{n,p}$ for electrons (or holes) of a deep-level center is a measure of how close to the center the carrier has to come to get captured. Usually, a center acts as an electron trap if $\sigma_n \gg \sigma_p$ and as a hole trap if $\sigma_p \gg \sigma_n$. Note that the charged state or the donor- or acceptor-like nature of the centers has not been specified. This is a more complex behavior and is partly decided by the absolute values of the capture cross sections.

From the principle of detailed balance the thermal emission rates for electrons or holes from their respective trapping centers are given by

$$e_n = \sigma_n \langle v_n \rangle g N_c \exp(-E_T/kT)$$
$$e_p = \sigma_p \langle v_p \rangle g^{-1} N_v \exp(-E_T/kT)$$

(1.3)

Here $\sigma_{n,p}$ are the capture cross sections of the traps, $\langle v_{n,p} \rangle$ the average thermal velocities of electrons and holes, and g the degeneracy factor for the level. The quantity $\mathscr{E}_T$ in these questions is, in fact, the free energy of ionization, or the standard Gibbs free energy of the ionization reaction, denoted by G. It relates to the ionization of an electron to the conduction band, or of a hole to the valence band. The enthalpy of ionization then is given by $H = G + ST$, where S is the entropy.

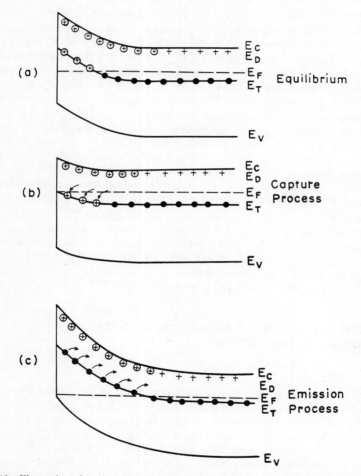

Figure 1.10 Illustration of carrier emission and capture at a deep-level electron trap located within the depletion region of a junction diode with applied (*a*) zero bias, (*b*) forward bias, and (*c*) reverse bias.

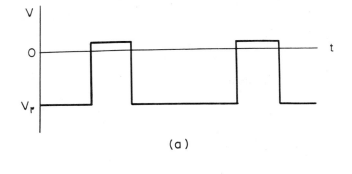

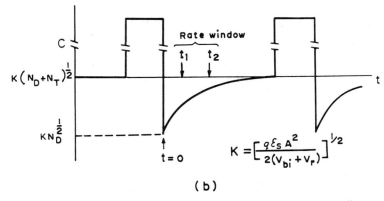

Figure 1.11   (*a*) Pulse cycling used in DLTS measurements and (*b*) the corresponding capacitance change at the semiconductor diode junction containing a deep level of density N_T. The test sample is assumed to be *n*-type with a shallow donor concentration N_D.

In this section we first describe briefly the basic principles of the most commonly used technique to identify and characterize traps—deep-level transient spectroscopy (DLTS). We then describe the measured characteristics of traps in important materials and hetero-structures grown by MBE and used for device fabrication. These are GaAs, $Al_xGa_{1-x}As$, and $Al_xGa_{1-x}As$/GaAs heterojunctions, $In_{0.53}Ga_{0.47}As$, and $In_{0.52}Al_{0.48}As$.

1.3.2   DLTS Measurement Technique

The principle of DLTS measurements is based on the physics of thermal emission and capture by traps and the associated variations in the capacitance of a junction diode. It depends on the repetitive filling and emptying of traps by use of positive and negative bias applied to a junction. The processes of carrier emission and capture in deep-level traps are illustrated in Fig. 1.10. In Fig. 1.11*a* we show the pulse cycling used in DLTS measurements, and the corresponding change in junction capacitance is depicted in Fig. 1.11*b*. The major difference between the basic capacitance transient technique and DLTS lies in the use of a "rate window" in the latter. A typical DLTS system is shown in Fig. 1.12, which can be implemented with either a boxcar averager or a lock-in amplifier for signal processing. One of the basic problems associated with DLTS or any other technique depending on capacitance transients is that the ideal condition of exponential

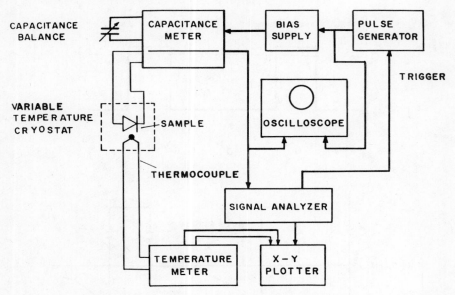

Figure 1.12   Schematic of a typical DLTS measurement system.

transients is rarely achieved under most experimental conditions. Nonexponential transients arise from large trap densities, electric field effects, nonuniform doping, nonabrupt junctions and depletion-layer edge effects.

1.3.3   Behavior of Traps in In-Doped GaAs

Improvement in the electrical and optical properties of isoelectronically doped liquid-phase epitaxial GaAs:In and InP:Ga(As) and metal-organic vapor-phase epitaxial GaAs:In has been demonstrated recently by Beneking et al. [38–40]. The effects of isoelectronic doping during molecular beam epitaxial (MBE) growth of III–V semiconductors have been studied by us both theoretically and experimentally. Recent calculations by Singh [41] show that at a growth temperature of 600°C the In adatom which arrives from the vapor and forms two In–As bonds with the growing surface has a surface migration rate 2×10^5 hops/s. The corresponding Ga migration rate under identical conditions is 10^4 hops/s. It is therefore most likely that the fast-migrating In surface atoms, with a weaker In–As bond than that for Ga–As, will occupy normally vacant Ga sites in the growing crystals and thereby lower the density of such vacancy-related defects.

The variation of the density of four dominant electron traps in GaAs with increasing amounts of In doping is shown in Fig. 1.13. The trap labeling is the same as that used originally by Lang [42]. We have confirmed that the trap densities in undoped GaAs grown under similar conditions are 1 to 4×10^{14} cm^{-3}. Our present data seem to suggest that the deep-level electron traps are probably native defects or complexes involving native defects. We believe that these defects are Ga vacancies, since according to the expected growth kinetics such vacancies can be partially filled by In atoms, thereby lowering the trap densities. It is evident that such trap-free material is ideal for fabricating microwave and optoelectronic devices.

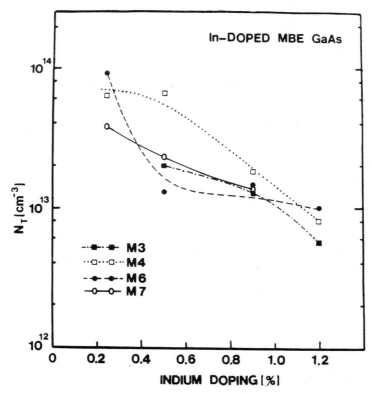

Figure 1.13 Decrease in concentration of dominant electron traps with increase of In content in MBE GaAs.

1.3.4 D-X Center in MBE $Al_xGa_{1-x}As$

This electron trap, which also behaves as a dominant deep donor in the alloys, has been studied by Hall-effect measurements in LPE [43], MOCVD [44], and MBE [45–47] crystals. Watanabe et al. [48] used the DLTS technique to demonstrate the coexistence of this deep donor with a shallow donor level in Si-doped $Al_xGa_{1-x}As$ for a certain range of x, which explained some observed anomalies in Hall data. [49].

Usual DLTS data of Si-doped $Al_xGa_{1-x}As$ with $x > 0.2$ show either a single peak or two closely spaced peaks, both believed to originate from the D-X center [50–52]. Theoretical curve fitting of DLTS data indicate the existence of three levels related to the D-X center [53]. Our own data [52] in Fig. 1.14 illustrate that by proper choice of doping it is possible to get well-resolved peaks due to the two levels separately and to characterize them experimentally. This is more apparent from Fig. 1.15, which shows the dependence of the concentration of the two levels on Si doping. Indirect evidence of the complex nature of the D-X center has also been obtained by us [52] from field-dependent transient capacitance measurements on layers grown with different orientations. Our conclusions are consistent with the original D-X center model proposed by Lang et al. [54]. However, the strong arguments presented recently in favor of a sub-stitutional donor model [55] for the D-X center cannot by fully disregarded. More work is necessary to elucidate the true nature and origin of the D-X center.

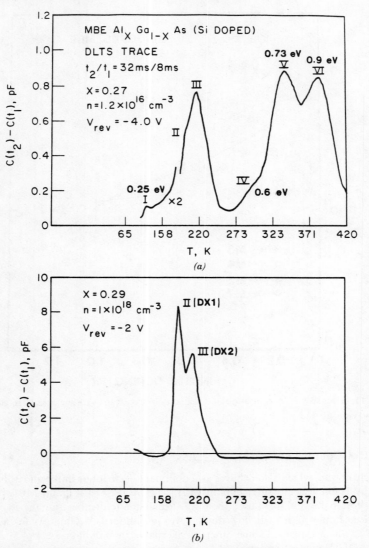

Figure 1.14   Typical DLTS data obtained from two Si-doped MBE-grown $Al_xGa_{1-x}As$ samples with different values of x and the free carrier concentration n. (a) With $x = 0.27$ and $n = 1.2 \times 10^{16}$ cm^{-3}, showing all the traps present in Si-doped $Al_xGa_{1-x}As$. (b) With $x = 0.29$ and $n = 1 \times 10^{18}$ cm^{-3}, showing DX1 and DX2 centers separately.

1.3.5 Traps in $In_{0.53}Ga_{0.47}As$

We have recently made an investigation of deep electron traps in $In_{0.53}Ga_{0.47}As$ layers grown by MBE on InP substrates with and without a buffer layer of high-purity LPE in $In_{0.53}Ga_{0.47}As$ [56]. Figure 1.16 shows typical DLTS data obtained from a regrown MBE–LPE In–GaAs hybrid p-n junction. Data at lower applied reverse bias essentially show traps detected in MBE layer, whereas the data for higher reverse bias include

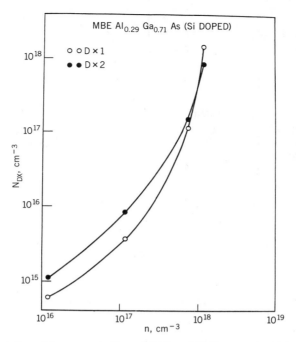

Figure 1.15 Variation of the concentrations of DX1 and DX2 centers in Si-doped MBE $Al_{0.29}Ga_{0.71}As$ as a function of free carrier concentration, n.

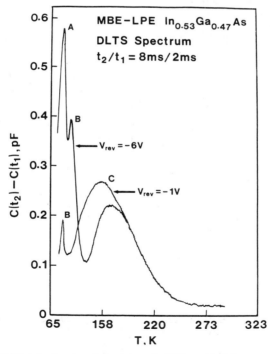

Figure 1.16 DLTS data showing electron traps in LPE- and MBE-grown regions of hybrid layers. The interface region is penetrated by increased reverse bias applied to the diode.

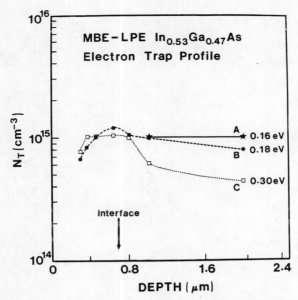

Figure 1.17 Concentration profiles of traps observed in MBE- and LPE-grown regions of hybrid layers.

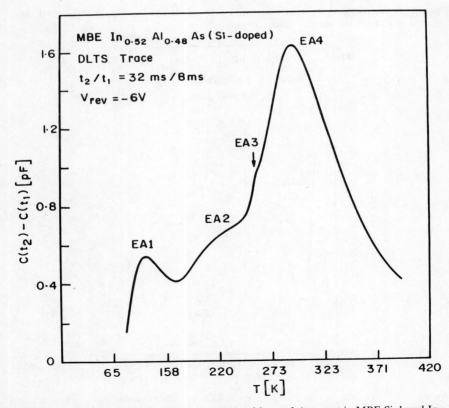

Figure 1.18 DLTS data showing electron traps (positive peaks) present in MBE Si-doped $In_{0.52}$ $Al_{0.48}As$.

traps in both MBE and LPE layers. In general, three electron traps A, B, and C, with thermal activation energies 0.16, 0.20, and 0.37 eV, were detected. Careful examination of the depth profiles (Fig. 1.17) of these traps reveals their location with respect to the MBE–LPE regrowth interface. It is apparent that traps B and C are common to both MBE- and LPE-grown layers. This conclusion cannot be made regarding trap A from our data. The characteristics of trap A are similar to a 0.16-eV level detected in LPE InGaAs [57]. This trap seems to be related to In or impurities contained therein, which is evident from our studies on In-doped GaAs described earlier.

1.3.6 Traps in $In_{0.52}Al_{0.48}As$

MBE-grown $In_{0.52}Al_{0.48}As$, lattice-matched to InP, has potential application in realizing a number of electronic and optoelectronic devices. From Hall measurement data, it is apparent that the material with low-Si doping contains a deep donor which might originate from Si [58]. Based on theoretical models, it has recently been suggested [59] that strained layers of $In_xAl_{1-x}As$ [$0.55 < (1 - x) < 0.85$] might contain a D-X-like center. The first experimental study of deep-level traps in $In_xAl_{1-x}As$ was made by us recently [60]. Measurements were made to investigate electron and hole traps present in lattice-matched InAlAs/InP, doped n- and p-type with Si and Be, respectively. Figure 1.18 shows typical DLTS data for a Si-doped InAlAs sample with $n = 5 \times 10^{16}$ cm^{-3}. Four electron traps, labeled EA1 through EA4, are present. In addition, four hole traps, HA1 to HA4, are also detected in Be-doped samples, as shown in Fig. 1.19. Trap characteristics were determined from the well-resolved peaks using the temperature dependence

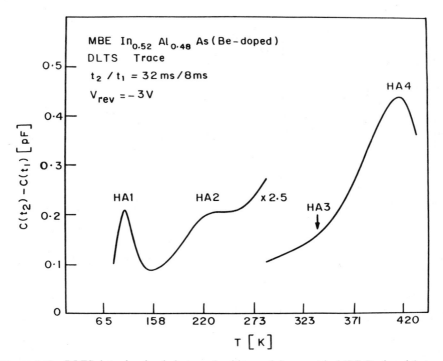

Figure 1.19 DLTS data showing hole traps (positive peaks) present in MBE Be-doped (p-type) $In_{0.52}Al_{0.48}As$. The experiments were done on n^+-p diodes with majority-carrier pulse filling.

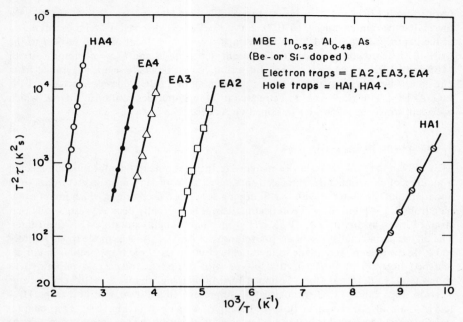

Figure 1.20 Temperature dependence of emission time constants for deep-level traps identified in Si- and Be-doped MBE $In_{0.52}Al_{0.48}As$.

of emission rates, depicted in Fig. 1.20. Table 1.1 lists the trap parameters. From these preliminary data, very little can be said about the possible origin of the trap levels. It is, however, observed that all the electron and hole traps except EA3 and HA3 show a pronounced dependence on the doping level. They can, therefore, be related to the dopants or impurities in them. From bias-dependent DLTS data it seems more likely that traps EA4 and HA4 are linked to mismatch defects or impurities at the layer–substrate interface. Traps EA1 and EA2 are detected in both Si- and Be-doped materials

TABLE 1.1· Characteristics of Traps in $In_{0.52}Al_{0.48}As$

Trap Type	Trap Label	Activation Energy ΔE_T (eV)	Capture Cross Section σ_∞^a (cm²)	Trap Concentration N_T^b (cm⁻³)
Electron	EA1	0.25[c]	—	1.3×10^{15}
Trap	EA2	0.56	9×10^{-11}	1.6×10^{15}
	EA3	0.60	1×10^{-12}	2×10^{15}
	EA4	0.71	4×10^{-12}	4×10^{15}
Hole	HA1	0.27	3×10^{-13}	6×10^{13}
trap	HA2	0.45[c]	-	7.6×10^{13}
	HA3	0.65[c]	-	2×10^{13}
	HA4	0.95	4×10^{-14}	4×10^{14}

[a] Determined from the emission rate prefactor.

[b] For a sample with $N_D - N_A = 5 \times 10^{16}$ cm⁻³.

[c] Determined approximately from the DLTS peak position.

and they can, therefore, be attributed to native defects or to Al. Traps HA3 and EA3 show invariant concentrations irrespective of the doping levels used. More detailed work is required to ascertain the origin and microscopic structure of these deep levels and then take possible steps to eliminate them.

1.3.7 Trapping Effects in Undoped and Modulation-Doped GaAs/AlGaAs Heterostructures

McAfee et al. [61, 62] used DLTS and C-V profiling techniques to investigate traps at the interface of (n)GaAs-(n)Al$_{0.25}$Ga$_{0.75}$As double heterostructures. A dominant interface trap with an activation energy of 0.66 eV was identified. From C-V measurements, the 0.66-eV trap has been found to be localized within 140°A of the GaAs layer next to the interface. The origin of this trap has been related to MBE growth conditions, and it is suspected that it plays a very important role in the operation of DH lasers. Efforts have been made [63] to improve the quality of the GaAs–AlGaAs interface by introducing thin AlGaAs prelayers with a low Al content which help to trap impurities.

The investigation of trapping and related phenomena in Al_xGa_{1-x}As-GaAs modulation-doped structures has been a matter of considerable interest. Al_xGa_{1-x}As-GaAs modulation-doped field-effect transistors (MODFET) show a number of undesirable effects, such as persistent photoconductivity, threshold voltage shift with temperature, drain I–V collapse, and persistent channel depletion by hot electrons under high electric fields. All these effects, occurring at low temperatures, have been associated directly or indirectly with the presence of the D-X center in the doped Al_xGa_{1-x}As region.

There are three principal anomalous features which we have observed when DLTS measurements are made on modulation-doped structures with the Schottky gate on the AlGaAs layer. The first is shown in Fig. 1.21a, representing DLTS data of a typical MD structure. The quiescent reverse bias applied to the diode is -2V. The data reflect traps in the Al_xGa_{1-x}As layer. The first is shown in Fig. 1.21a, representing DLTS data of a typical MD structure. The quiescent reverse bias applied to the diode is -2V. The data reflect traps in the Al_xGa_{1-x}As doping layer and a noticeable feature is the absence of the other traps observed in single ternary layers, shown in Fig. 1.21b. The second anomalous feature is the observed decrease in emission rate of carriers from the D-X centers with increasing depletion-layer electric field. This is contrary to that expected or observed for most dominant traps in single layers. Finally, the real-time capacitance transients due to electron trap emission under certain bias conditions are accompanied by a negative-going region superimposed over the usual positive-going capacitance signal. Some of the observations above were also made by Martin et al. [64] on MOCVD-grown AlGaAs–GaAs quantum well structures, which they attempted to explain by assuming emissions from the GaAs well. We have formulated a model [52] that gives a reasonable interpretation of the anomalous effects above and allows determination of the true trap characteristics. The main features of this model are:

1. The measured cpacitance is actually the sum of two capacitances in series, one due to the Schottky barrier on AlGaAs and the other due to the charged layers at both sides of the 2DEG heterointerface.

2. In the usual DLTS experiment, if the quiescent reverse bias after the filling pulse is less than the flat-band voltage at the heterointerface, some of the electrons emitted from the trap will gradually neutralize part of the accumulated positive charge at the interface. This results in a time-dependent decrease in the interface capacitance.

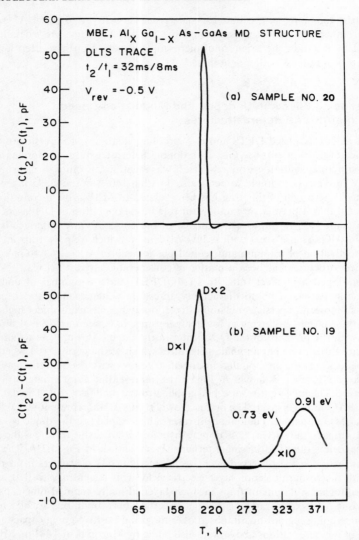

Figure 1.21 (*a*) Typical DLTS data obtained on a MBE-grown $Al_{0.3}Ga_{0.7}As$–GaAs MD structure with 350-Å Si-doped $Al_{0.3}Ga_{0.7}As$ ($n \sim 1 \times 10^{18}$ cm^{-3}) layer and a 207-Å spacer layer. The 2-DEG concentration at the heterointerface is 2.8×10^{11} cm^{-2}. The single prominent peak is due to the D-X center (*b*) DLTS data for a MD structure with 600-Å Si-doped $Al_{0.3}Ga_{0.7}As$ ($n \sim 1 \times 10^{18}$ cm^{-3}) layer and a 90-Å spacer layer. The 2-DEG concentration is 5.5×10^{11} cm^{-2}. The data clearly show DX1 and DX2 in addition to other prominent peaks observed in single-layer $Al_xGa_{1-x}As$.

1.4 EFFECTS DUE TO ALLOY AND HETEROINTERFACE QUALITY

1.4.1 Alloy Quality and Its Effect on Optical and Electrical Properties

Heterostructures are being fabricated not only from binary compound semiconductors, but also from elemental ternary and quarternary compound semiconductors. For heterostructures based on alloys, not only is the quality of the interface important, but the quality of the epitaxial alloy is also important for device performance. The use of semiconductor alloys permits a wide range of choice in tailoring the optical/electronic properties of the heterostructures. An important structural parameter that describes the quality of an alloy is the short-range order parameter, which contains information on the clustering present in the alloy. Alloy clustering can cause additional scattering effects, which, in turn, can cause broadening of optical emission lines as well as reducing the mobility of charge carriers.

Although conceptually, one can form alloys from a variety of different semiconductors, the actual realization of high-quality alloys is not easy. From thermodynamic considerations, one can determine whether or not an alloy can be produced without clustering. However, even if according to thermodynamic considerations one may not be able to grow a cluster-free alloy, one may use a far-from equilibrium growth technique to grow high-quality cluster-free alloys. MBE is a far-from-equilibrium growth technique, which, in principle, can be used to grow alloys which according to equilibrium thermodynamic considerations are immiscible at the growth temperature commonly used. However, since the growth conditions also determine the quality of the growing surface/interface, one has to ensure that improvements in alloy clustering do not come about at the expense of poorer interfaces.

Due to recent interest in the $In_{0.52}Al_{0.48}As/In_{0.53}Ga_{0.47}As$ system for high-speed devices, there is considerable interest in the quality of the MBE-grown InAlAs system. So far, experiments based on photoluminescence and transport have shown that the system is not of a high quality and may have alloy clustering present [65, 66]. Due to the large difference between In and Al related bond energies, the InAlAs system is expected to show clustering at the low temperatures employed in MBE growth [67]. We present a general theoretical formalism which relates the clustering produced in the alloy grown under different conditions to optical and transport properties of the alloy. Development of such a formalism is important if the consequences of clustering are to be understood.

To understand the effects of alloy clustering, we define an order parameter

$$C = \tfrac{1}{2} \sum_{i,i'} c_i c_i' \tag{1.4}$$

where c_i's are the occupation numbers for the cations (chosens to be $+1$ for In; -1 for Al). This order parameters represents the short-range order present in the alloy after growth. Note that for a perfectly random alloy (no clustering effect) one expects C to be given by

$$C(x) = Z\{x[x - (1 - x)] + (1 - x)[(1 - x) - x]\} = Z(4x^2 - 4x + 1) \tag{1.5}$$

where Z is the coordination ($= 12$ for tetrahedral III–V compounds) and x is the concentration of one component of the alloy. We also note that deviations from $C(x)$ given by Eq. (1.5) represent a clustering or ordered behavior. Note that for a random alloy, the value is zero.

To understand whether or not the alloy clustering has any detrimental effects, it is important to study the consequences of clustering on the optical and electrical properties of InAlAs. In this section we discuss the effects of clustering on photoluminescence line width of exitonic transitions and mobility of electrons through the semiconductor. The expression for the excitonic line width in a perfectly random alloy is given by [68]

$$\sigma = 2\left[\frac{C_A^0 C_B^0 r_c^3 (1.4)}{R_{ex}^3}\right]^{1/2} \Delta_1 (0.327) \qquad (1.6)$$

where C_A^0, C_B^0 = mean compositions of InAs and AlAs in the system
$\quad\quad r_c$ = radius associated with a cation volume
$\quad\quad R_{ex}$ = exciton radius
$\quad\quad \Delta_1$ = direct bandgap difference between InAs and AlAs ($\Delta_1 = 2.0$ eV)

For no clustering $r_c = 2.3$ Å, one finds that $\sigma = 4$ meV, assuming that $R_{ex} \simeq 200$ Å.
The mobility limited by alloy scattering can be written as [69]

$$\mu_0^{all} = \frac{32\sqrt{2}eh^4}{9\pi^{3/2}(m^*)^{5/2}V_a C_A^0 C_B^0 \Delta E^2 (kT)^{1/2}} \qquad (1.7)$$

where m^* = effective mass in the alloy
$\quad\quad V_a$ = atomic volume associated with each cation
$\quad\quad \Delta E$ = alloy scattering potential
$\quad\quad T$ = temperature

These expressions have been written for an alloy with mean composition C_A and C_B and no clustering. In case there is clustering in the system, the expressions are modified in a straightforward manner. A simple way to describe the alloy clustering is to describe the smallest scale n_c over which correlations exist ($r_c = n_c r_0$). Thus we have two types of regions each containing $(4\pi/3)n_c^3$ cations, which have concentrations different from the bulk values C_A^0 and C_B^0 (we have chosen C_A^0 and C_B^0 to be 0.5 in our work).

Let us assume that the clustering is manifested in the concentration of these clusters being different from C_A^0 and C_B^0. We assume that the concentration of A-type cations is C_A^1 in one type of cluster and C_A^2 in the other type of cluster. If $C_A^1 = 1$ and $C_A^2 = 0$, the clusters are purely A type or purely B type. But we find from our simulations that this does not occur due to the nature of MBE growth (i.e., random impingement of cation and infinite migration distances on the surface). With the simple model for clustering above, the short-range order parameter C can be calculated as follows. Within the clusters, the value of C is simply

$$C_{max} = Z(C_A^1 - C_A^2)^2 \qquad (1.8)$$

The atoms in the cluster can be divided into "core" atoms and "surface" atoms. The core atoms have an average C value equal to C_{max}, while the surface atoms on average have a C value equal to $1/2C_{max}$. Since the fraction of the core and surface atoms is approximately

$$\left(\frac{n_c - 1}{n_c}\right)^3 \quad \text{and} \quad \frac{3}{2}\left(\frac{n_c^2 - n_c}{n_c^3}\right) \qquad (1.9)$$

we get

$$C = Z(C_A^1 - C_A^2)^2 \left[\left(\frac{n_c - 1}{n_c} \right)^3 + \frac{3}{2} \left(\frac{n_c^2 - n_c}{n_c^3} \right) \right] \tag{1.10}$$

This expression is quite accurate when n_c is large (>4), but has errors when n_c is small due to the fact that the cluster is chosen to be spherical.

The expression for PL line width can now be generalized to the clustered case by the transformations $r_c \rightarrow n_c r_0$ and $\Delta_1 \rightarrow \Delta_1 |(C_A^1 - C_A^2)|$, leading to the following expression:

$$\sigma = \sigma_0 n_c^{3/2} |C_A^1 - C_A^2| \tag{1.11}$$

Similarly, the expression for mobility due to alloy scattering is given from Eq. (1.7) by the replacements

$$V_a \rightarrow n_c^3 V_a \qquad \Delta E \rightarrow \Delta E |C_A^1 - C_A^2| \tag{1.12}$$

This gives

$$\mu^{-1}(C_A^1 - C_A^2, n_c) = n_c^3 |C_A^1 - C_A^2|^2 (\mu_0^{\text{all}})^{-1} \tag{1.13}$$

Both these expressions are valid if the spatial extent of the cluster is less than the exciton size in the alloy (Ca. 200 °A) and the electron wavelength. Our simulations show that the cluster sizes obey these restrictions and

$$\frac{1}{\mu^{\text{all}}} = \left(\frac{\sigma}{\sigma_0} \right)^2 \frac{1}{\mu_0^{\text{all}}} \tag{1.14}$$

From Eqs. (1.8)–(1.11), we can see that the clustering in the alloy system can be very detrimental to both the PL line width and the electron mobility. We also note that C by itself is not able to provide information to calculate σ and μ^{all}. We have therefore estimated C_A^1 and C_A^2 for the $In_{0.5}Al_{0.5}As$ alloy grown in our simulation. These values are typically 0.7 and 0.3. We are now in a position to estimate the linewidth and mobility from Fig. 1.22 and Eqs. (1.10), (1.11), and (1.13). In Fig. 1.23 we present the results for $In_{0.5}Al_{0.5}As$ PL line width increasing from 3.8 to about 12 meV as the growth temperature is raised to 800 K. The room temperature alloy-limited mobility changes from 13,000 to 2000 cm^2 V^{-1} s^{-1}. At room temperature we have to add the effects of phonon-limited mobility to the values above according to Mathieson's rule, to get the total mobility. It is clear that a small amount of clustering can seriously hurt the electrical and optical properties of InAlAs. The best low-temperature photoluminescence line width in $In_{0.52}Al_{0.48}As$ grown in our laboratories is shown in Fig. 1.24.

We must emphasize, however, that although clustering may be suppressed at low temperatures, the poor quality of the surface growth may cause a high density of native defects, which may form nonradiative centers as well as trapping centers. Thus, ideally, one would like to reduce clustering without sacrificing the quality of the surface. This may require the use of novel growth approaches, as mentioned earlier.

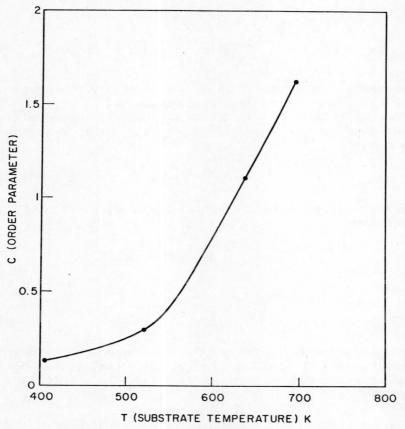

Figure 1.22   Variation of short-range cluster order parameter C (defined in the text) as a function of growth temperature as determined from computer simulations.

1.4.2   Heterointerface Quality and Its Effect on Optical and Transport Properties

In this section we discuss the effect of interface roughness on optical properties (measured by exciton line width) and electron transport properties (measured by mobility in MODFETs). In devices using heterostructures bandedge discontinuities produced at the interface of two materials, A and B are used to confine the mobile carriers near the interface. Thus the quality of the interface has an important effect on the performance of the device. In quantum well structures with well sizes of less than 200 Å, low-temperature (4 K) photoluminescence studies are widely used to estimate the quality of the interfaces forming the well.

However, little effort has been made to quantify the information obtained from the line widths in terms of the microscopic details of the interface. We present a theory correlating the line widths of the photoluminescence spectra in quantum wells to the microscopic details of the interface quality. This theory is based on the realization that the optical probe (i.e., the exciton) has a finite extent (ca. 250 Å in GaAs, for instance)

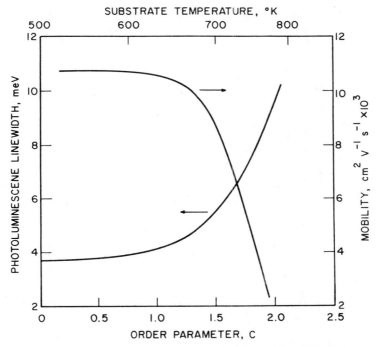

Figure 1.23 Variation of alloy-limited low-temperature PL line width and 300-K electron mobility in $In_{0.5}Al_{0.5}As$ as a function of alloy clustering or substrate temperature. Note that the improvement at low temperatures may be reduced by the intrinsic defects that may be produced due to the poor growth front quality.

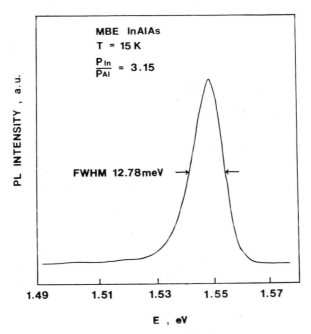

Figure 1.24 Low-temperature photoluminescence line width observed in the best MBE-grown $In_{0.52}Al_{0.48}As$ grown in our laboratory.

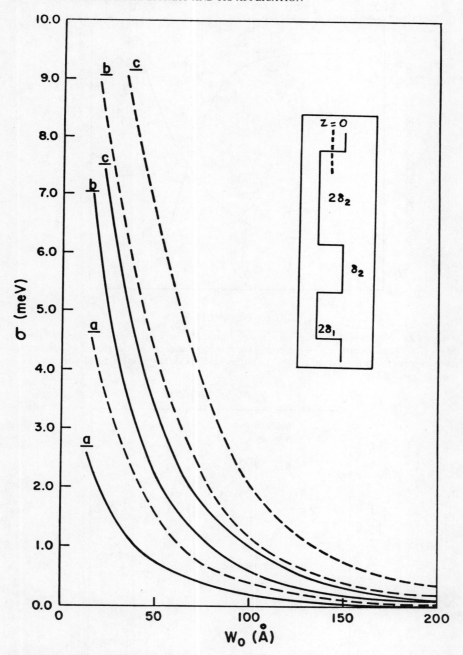

Figure 1.25 Half-width of photoluminescence line, σ, as a function of well size for value of δ_1 equal to one monolayer (solid lines) and for two monolayer (dashed lines) and δ_2 equal to (a) 20 Å, (b) 80 Å, and (c) 100 Å. The inset shows the model of the interface used.

and the energy of the emitted radiation reflect the average composition of the region seen by the exciton. We use statistical arguments similar to those invoked by Lifshitz [70] to understand the excitation spectra of disordered alloys and to find the probability distributions of the compositional fluctuations at the interface. These fluctuations are then related to the line shape of the photoluminescence spectra.

To discuss the model for the interface, let us focus on the most studied quantum well structure, $Al_xGa_{1-x}As/GaAs/Al_xGa_{1-x}As$. Depending on the growth conditions, there will be localized fluctuations in the well size around a mean value W_0. These fluctuations will arise if, for example, the growth of AlGaAs is not perfectly two-dimensional. In such a case the interface is not a step function but has a certain amount of diffusiveness. We assume that the nonideal interface can be represented by the average interface at $z = 0$ and fluctuations extending a distance $z = \pm\delta_1$ from the interface. These fluctuations arise from the presence of islands of AlGaAs in the well and islands of GaAs in the barrier regions. We assume further that the correlated lateral extent of these islands is δ_2 (i.e., the smallest size of the islands in δ_2). The inset in Fig. 1.25 shows the interface according to this model. Based on this model we view the interface as being described on a global scale by parameters C_a^0, C_b^0 and C_c^0, representing the mean concentration of the islands protruding into the wall, and out of the well and regions that are flat.

The intrinsic photoluminescence spectra arise from the formation of electrons and holes that form excitons which emit a well-defined radiation upon their collapse. If the excitons are present uniformly in a nonideal well, the energy radiated by each exciton will be determined by the spatial position of the exciton. If the exciton is present in a region where the well is narrower, it will emit a higher-energy excitation than an exciton in a region where the well is wider than the mean value. The energy of excitation is given by

$$E_{ex}(r, W) = E_e(r, W) + E_h(r, W) - E_b(r, W) + E_g \qquad (1.15)$$

where r = exciton position, E_e,

E_h = energies of the electron and hole subbands

E_g = bandgap of the well material (GaAs)

E_b = binding energy of the exciton

The binding energy E_b is not expected to change much by small fluctuations in the well size, but the values of E_e and E_h are quite sensitive to the well size changes. The exciton wave function extends over a region $\approx R_0$ so that the relevant information regarding the quantum well corresponds to the part of the well with lateral extent R_0 and centered around the position r of the exciton. The average width of the well around the point r will be given by the average microscopic nature of the interface. If we assume that the excitons are created randomly in the well, the line shape is determined by the probability distribution, $P(R_0W)$, of finding fluctuations in the well size extending over the exciton size. This distribution has recently been calculated [71].

In Fig. 1.25 we display the variation of the half-width of the photoluminescence line, σ, as a function of the well size for these three different values of δ_2. Values of δ_1, equal to one monolayer thickness (solid lines) and two monolayer thickness (dashed lines) are used. We note that for a given value of δ_1, δ_2 has an important effect on the shape. As the value of δ_2 increases, the line width increases and approaches the value it would have if the well size changed from W_0 to $W_0 + \pm 0.6\delta_1$, where $0.6\delta_1$ is the variance of the probability distribution. These results correspond to the situation where the excitons are produced uniformly in the quantum well and do not migrate to regions of lower energy. We have assumed here that as the well size changes, the quality of the interface does not change, so that the same microscopic parameters δ_1 and δ_2 describe the interfaces. This is expected to be true as long as the well size is > 10 monolayers, so that the interfaces are determined by the steady-state growth profiles of the growing system. As the well size changes, so does the size of the exciton. This effect is taken into account in obtaining the results shown in Fig. 1.25.

Note that both δ_1 and δ_2 are important in determining the excitonic line width. For high-quality heterostructures with $\delta_1 = 1$ monolayer, the exciton line widths can be quite small. This has been observed in the AlGaAs/GaAs quantum well structures. However, in material systems such as InGaAs/InAlAs, the interface quality is not as good and the exciton line widths are considerably larger. The use of growth interruption before the formation of the interfaces can improve the interface quality and reduce the exciton line width [72].

1.4.3   Electron Mobility in MODFETs

A number of workers have carried out detailed studies of the effects of interface roughness on electron mobility in MODFETs. The general consensus is that if the interface roughness is of the order of one monolayer, it does not play an important role in the mobility. However, if the interface roughness is of the order of three to four monolayers, it can play an important role, particularly at low temperatures. As mentioned in the preceding section, the latter situation does arise in the InGaAs/InAlAs system. Analysis of the photoluminescence results show that interface quality may be described by two-dimensional interface islands of height three to four monolayers and an average lateral extent of about 100 Å. We briefly describe the formalism used to calculate the effect of this roughness on the mobility of electrons.

The ground-state wave function in a MODFET structure is

$$\phi = \begin{cases} Bb^{1/2}(bz + B)\exp(-bz/2), & z > 0 \\ B'b'^{1/2}\exp(b'z/2), & z < 0 \end{cases} \tag{1.16}$$

where b, b^1, B, and B^1 are variational parameters, and the InGaAs–InAlAs interface is assumed to be located at $z = 0$. The interface model discussed above will produce a scattering potential given by

$$V_{\text{IR}} = V_0\theta(-\Delta) - V_0\theta(0) \tag{1.17}$$

where V_0 is the bandedge discontinuity between InGaAs and InAlAs, and Δ is the height of the InAlAs steps, as discussed earlier.

The interface roughness potential along the interface plane (ρ, θ) is assumed to be given by

$$V_{\text{IR}}(\rho, \theta) = \begin{cases} V, & \rho < r_0 \\ 0, & \rho > r_0 \end{cases} \tag{1.18}$$

We assume that on an average the interface structure is made up of these two-dimensional steps. To calculate the scattering of the electrons from this interface profile, one needs to evaluate the matrix element

$$M(k, k') = \int dz \int dp \, \psi^*(p, z) V_{\text{IR}}(p, \theta) \psi(p, z) \tag{1.19}$$

We use an approach outlined in Ref. 88 to express the z integration in terms of N_s, the two-dimensional electron carrier density, and the accumulation-layer charge density N_{acc}. Performing the (r, θ) integral

$$M(k, k') = \Delta r_0 \frac{8\pi^2 e^2}{\epsilon_s} \left(N_{acc} + \frac{N_s}{2} \right) \frac{J_1(q r_0)}{q} \tag{1.20}$$

where $q = k - k^1$ and J_1 is the first-order Bessel function. Assuming an equal density of two-dimensional islands and flat regions, we obtain a scattering time

$$\frac{1}{\tau(k_F)} = \frac{4\pi e^4 m^*}{\hbar^3 \epsilon_s^2} \left(N_{acc} + \frac{N_s}{2} \right)^2 \Delta^2 \int_0^\pi J_1(k_{FL} \sin \phi)^2 \, d\phi \tag{1.21}$$

In Fig. 1.26 we have plotted $1/\tau$ for the InGaAs–InAlAs interface calculated according to Eq. (1.19) for $\Delta = 1$ monolayer and for three different values of $N_s(N_{acc} \ll N_s)$ as a function of $L(2r_0)$. It is important to realize that as L increases there is initially a sharp increase in $1/\tau$ (sharp decrease in mobility), but beyond $L - 50$ Å, $1/\tau$ becomes independent of L and eventually begins to decrease (mobility starts to increase). Physically, this means that once the roughness scale increases beyond $1/k_F$, the electrons do not "see" the surface roughness as much.

As mentioned earlier, we have analyzed low-temperature photoluminescence data of InGaAs–InAlAs quantum wells to deduce the values Δ and L. In addition to interface roughness scattering discussed above, we have included the effects of acoustic phonon scattering, alloy scattering, optical phonon scattering, remote impurity scattering, and

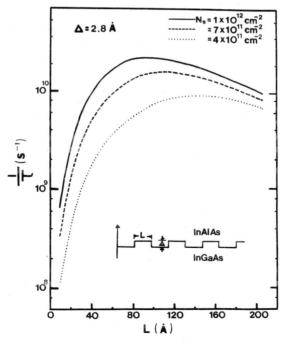

Figure 1.26 Dependence of the calculated inverse scattering time due to interface roughness on the two-dimensional island size.

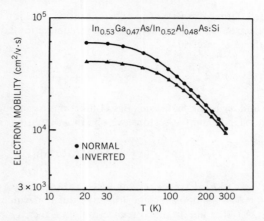

Figure 1.27 Measured temperature dependence of mobilities in normal and inverted InGaAs–InAlAs modulation-doped heterostructures. The solid lines are joins of the data points.

background impurity scattering. Normal and inverted InGaAs/InAlAs MD heterostructures were grown by MBE. In the order of growth the normal structure is made up of 3000-Å undoped InGaAs, 80-Å undoped InAlAs spacer, and 350-Å Si-doped ($2 \times 1p^{18}$ cm^{-3}) InAlAs layer. The inverted structure has a 350-Å Si-doped (2×10^{18} cm^{-3}) InAlAs layer, 80-Å undoped InAlAs spacer, and 200-Å undoped InGaAs layer. Both structures were grown on 0.4-μm undoped InAlAs buffer layers first grown on the (001) InP:Fe substrates. The substrate temperature was about 480°C during growth. The 2-DEG density for both cases is about 2.0×10^{12} cm^{-2} at 300 K and reduces slightly with lowering of temperature. The temperature-dependent mobilities measured for the normal and inverted InGaAs–InAlAs structures are shown in Fig. 1.27. The two-dimensional electron carrier concentration at 20 K is 1.5×10^{12} cm^{-2}.

Using the interface structural parameters from PL studies, we find that the following values are required to fit the mobility of the normal InGaAs–InAlAs interface at 20 K: $\Delta = 2.8$ Å (one monolayer); $L \sim 100$ Å; spacer = 80 Å; and background impurity density in InGaAs channel about 5×10^{15} cm^{-3}. On the other hand, the mobility of the inverted interface is understood on the basis of the following parameters: $\Delta \sim 12$ Å; $L \sim 100$ Å; spacer $\simeq 80$ Å; and background impurity density in InGaAs channel about 5×10^{15} cm^{-3}. Since the derived value of spacer thickness is identical to the experimental value in the case of the inverted structure, it is apparent that Si diffusion or riding effects are not prevalent at our growth temperatures.

1.5 MATERIALS PROPERTIES AND DEVICE CHARACTERISTICS

In this section we describe the properties of two devices made with InGaAs/InAlAs multilayered structures. These are the modulation-doped FET and the photodiode. We choose this material system since it presents greater challenges in terms of growth and device fabrication, and has long-range potential for high-frequency and high-speed applications.

1.5.1 InGaAs/InAlAs MODFET

As mentioned before, a high electron mobility, high electron velocity, and a large Γ–L intervalley separation are some of the favorable properties of In$_{0.53}$Ga$_{0.47}$As. Formation

n	GaAs	5E17	100 Å
n+	InAlAs	3E18	300 Å
i	InAlAs		50 Å
i	InGaAs		250 Å
i	InAlAs/InGaAs	S.L.	
i	InAlAs		4000 Å
i	InAlAs/InGaAs	S.L.	
S.I.	InP:Fe	(100)	

Figure 1.28 Schematic of $In_{0.52}Al_{0.48}AS/In_{0.53}Ga_{0.47}As$ single-quantum-well MODFET with a GaAs gate barrier layer.

of a two-dimensional electron gas (2-DEG) in $In_{0.53}Ga_{0.47}As/In_{0.52}Al_{0.48}As$ modulation-doped (MD) heterostructures lead to further improvement in the carrier transport properties. Excellent device performance has been demonstrated in FETs made with this heterostructure system [73, 74]. However, InGaAs/InAlAs MODFETs usually have two shortcomings: (a) very high gate leakage current and low drain breakdown, possibly due to significant alloy clustering in the top-doped InAlAs, and (b) high output conductance as well as poor pinch-off characteristics, due mainly to the high background conductivity $[n \sim (0.5 \text{ to } 1.0) \times 10^{16} \text{ cm}^{-3}]$ in the undoped InGaAs buffer layer. It was shown recently that an undoped InAlAs layer below the gate contact is very effective in reducing the gate leakage [75]. We describe here our recent studies on 1-μm gate devices in which features have been incorporated to circumvent the problems mentioned above. These are a quantum well channel and a thin doped GaAs layer beneath the gate contact. It was been reported that carrier confinement in the quantum well leads to smaller short-channel effects [76] and smaller energy loss of carriers under high-field conditions [77]. A high ratio of transconductance (g_m) to output conductance (g_0) for a high dc voltage gain is, therefore, expected. A schematic of the device structure is shown in Fig. 1.28.

We have also recently investigated [78] the properties of an InGaAs/InAlAs single-quantum-well (SQW) quasi-MISFET. This device consists of an inverted InGaAs/InAlAs MODFET in which a top thick (400 to 600 Å) undoped layer of InAlAs is incorporated on a relatively thin (100 to 400 Å) InGaAs layer. The structure was synthesized by molecular beam epitaxy and is shown in Fig. 1.29. The top InAlAs principally gives rise to two advantages: (a) two-dimensional electron gas (2-DEG) forming the channel is confined in a quantum well, and (b) since undoped InAlAs grown by MBE at 440 to 480°C usually exhibits high resistivity, the gate leakage current is reduced substantially. The device is, therefore, a quasi-metal-insulator semiconductor FET (MISFET). It should be remembered that present-day conventional InGaAs MISFETs have two shortcomings: current instability due to traps in the insulator [79] and reduced channel mobility and velocity due to a poor semiconductor–insulator interface [80]. The structures we report here can overcome both these limitations. Two other observations made by us are of relevance here and should be mentioned. Annealing studies show that the inverted InGaAs/InAlAs MODFET structure is thermally more stable than the normal one [81]. Second, despite differences in low-field Hall mobilities in identically grown normal and inverted InGaAs/InAlAs MD structures, due principally to interface roughness [82], the highfield mobility and the peak velocity are almost identical. Therefore, an inverted structure may be more suitable for many applications.

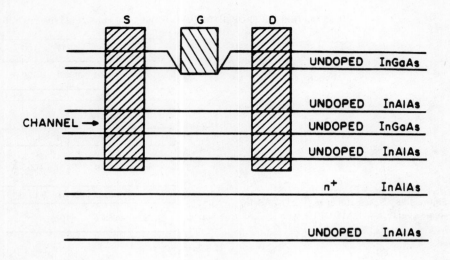

Figure 1.29 Schematic of InGaAs/InAlAs single-quantum-well quasi-MISDET grown lattice matched to InP by molecular beam epitaxy.

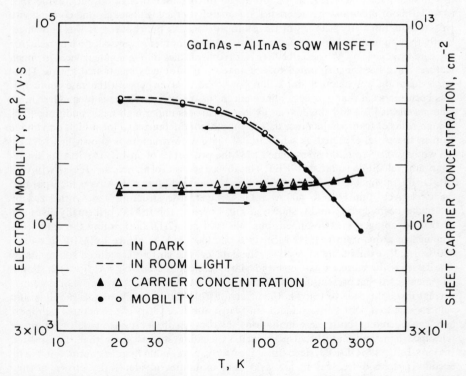

Figure 1.30 Measured temperature dependence of Hall electron mobility and sheet concentration in InGaAs–InAlAs SQW structure.

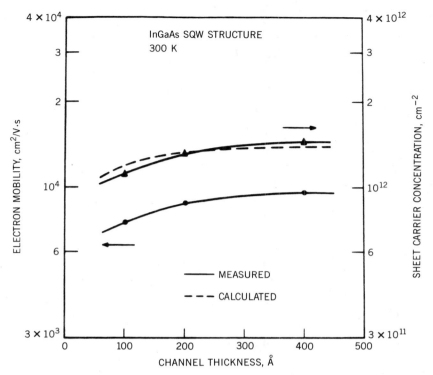

Figure 1.31 Variation of electron mobility and sheet concentration in In–GaAs/InAlAs SQW structure with quantum well thickness.

Low- and High-Field Transport Properties. Hall measurements were made with the SQW MISFET structure on photolithographically defined clover leaf samples using the van der Pauw technique. The best mobility values were obtained for a sample with a 300-Å In AlAs doping layer (1×10^{18} cm^{-2}), 80-Å undoped In AlAs spacer layer, 200-Å InGaAs SQW channel, and 400-Å InAlAs barrier layer. The variation of mobility with temperature in this sample is shown in Fig. 1.30. The 2-DEG concentration at 300 K is 2×10^{12} cm^{-2}. The low-temperature mobility is limited principally by alloy and interface roughness scattering. The variation of mobility and carrier concentration at room temperature with SQW channel thickness is shown in Fig. 1.31. It should be stressed that a fairly high mobility of the 2-DEG is maintained even for well thickness of about 100 Å. The experimental data agree fairly well with the total carrier concentrations calculated by quantum mechanical modeling of the device.

The average electron velocity-field characteristics were determined from pulsed Hall measurements and pulsed current–voltage measurements on planar H-devices [83]. The voltage pulse widths used were 1 μs for the Hall measurements and 40 ns for the velocity-field measurements. Results obtained from 300- and 77-K pulsed Hall measurements are shown in Fig. 1.32. The decrease in mobility at fields above 100 V/cm is due mainly to optical phonon scattering. The other significant features is that the sheet carrier concentration remains fairly constant at both temperatures. There is a slight lowering of the sheet concentration at higher fields. Interface trapping and defect-assisted tunneling have been cited [84] to be responsible for such behavior.

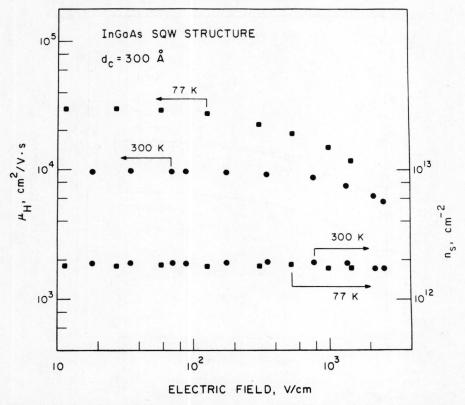

Figure 1.32 Variation of electron concentration and mobility with electric field in InGaAs SQW structure at 300 and 77 K obtained from pulsed Hall measurements.

The velocity-field characteristics determined from pulsed Hall measurements at 300 and 77 K are shown in Fig. 1.33. A current instability was observed at the highest bias data point. The reason is not clear, but intervalley and real space transfer of carriers could be contributing factors. The maximum velocity at 300 and 77 K are 1.5×10^7 and 1.7×10^7 cm/s, respectively. The small difference between the values at the two temperatures may be due to inferior transport characteristics in the InAlAs layer, in which some carriers are transferred at high fields, and the parallel conduction in the undepleted InAlAs doping layer.

Ohmic Contact Studies. The performance of lateral electron devices, such as the FET, is critically dependent on the quality of ohmic contacts, which form interfaces between the conduction channel and the external circuitry. The achievement of low contact resistances in the SQW structure becomes difficult because a thick undoped InAlAs layer exists on top of the channel region, which itself is very thin. Hence there is a need to realize contacts that have low resistances and good surface morphologies. We have investigated the performance of ohmic contacts on the SQW FET structures, using a layered metallization scheme with Ti or Cr, and with or without initial Si implantation and annealing.

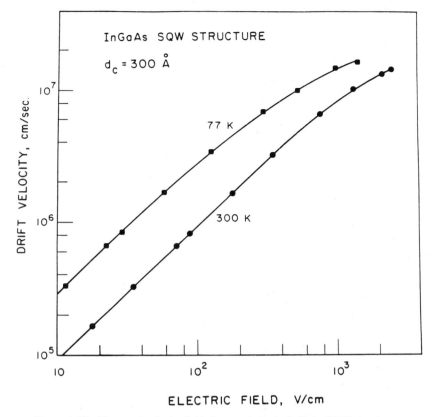

InGaAs SQW STRUCTURE

$d_c = 300\ \overset{\circ}{A}$

77 K

300 K

DRIFT VELOCITY, cm/sec.

ELECTRIC FIELD, V/cm

Figure 1.33 Measured velocity-field characteristics in InGaAs SQW structures.

The layered metals forming the contact were deposited be electron beam evaporation under a background vacuum of about 1×10^{-6} torr. The contact metals were alloyed in an AG Associates 210T halogen lamp annealing station under flowing argon. The heating rate was approximately 15°C/s and the cooling rate was between 5 and 10°C/s. We observed a degraded morphology, balling phenomenon, and an increase of the contact resistance when the heating rate was higher than 40°C/s. The samples with the ohmic pattern were placed face down on the Si wafer containing the thermocouple (chromel–alumel). The contact resistances were measured by the transmission-line measurement (TLM) method.

Contact resistance data for the InGaAs/InAlAs SQW structure are shown in Fig. 1.34, where the metallization schemes are also mentioned. The lowest contact resistance is measured for the layered metallization with Ti at an annealing temperature of 325°C. In comparison, substitution of Ti with Cr (M3) or excess Ni (M4) does not reduce the contact resistance.

A couple of facts should be stressed before comparisons are made with results reported previously [85, 86]. The SQW structure used by us has a thin-channel region and does not have a thick top InGaAs layer, which is common in conventional MODFET structures. Instead, all our structures have a 50-Å undoped InGaAs top layer. Therefore, better control over alloy depth is needed. This is achieved by reducing the total

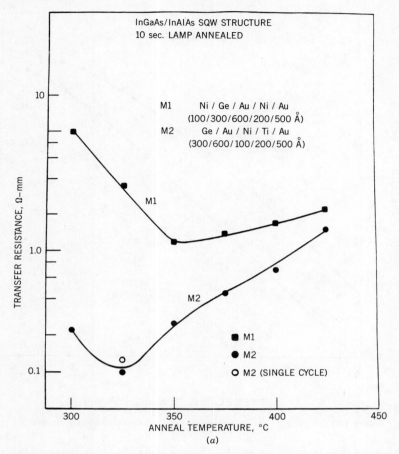

Figure 1.34   Variation of contact resistance in InGaAs SQW MISFET structure with lamp anneal temperature for layered ohmic metals: (*a*) M1 and M2; (*b*) M3 and M4.

thickness of the contact layer. There is some discrepancy in the values of the optimum annealing temperatures. Part of this discrepancy can be attributed to annealing methods (transient versus lamp annealing), the point of temperature measurement, differences in the ambient gases, and the sample structures. However, recently reported data by Kamada et al. [87] confirm that low contact resistances can be obtained at fairly low annealing temperatures.

Some insight into the role of Ti can be gained by studying the sputtered Auger data of Ito et al. [88] It is clear that even at 330°C, a considerable amount of Ti moves toward the semiconductor–metal interface and at the same time Ga outdiffuses from the semiconductor. Ti also prevents excess Au diffusion from the top layer and perhaps controls the Au–Ga/In interdiffusion.

For self-aligned gate FETs, implanted and annealed contact regions are normally used. We have, therefore, studied the characteristics of Si-implanated and annealed InGaAs/InAlAs SQW structures and the behavior of ohmic contacts made on this mate-

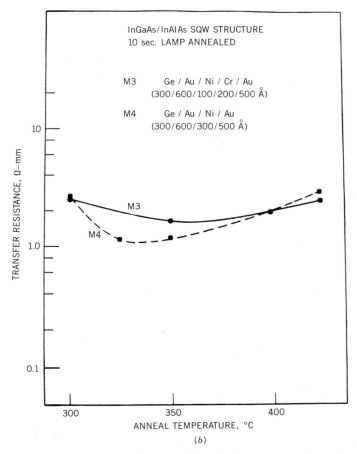

Figure 1.34 (*Continued*)

rial. Figure 1.35*a* shows the measured Hall data as a function of lamp anneal temperature. The implant energy of 50 keV was selected to ensure that the peak of the electron distribution coincides with the SQW channel. The implantation dose was 3×10^{13} or 1×10^{14} cm^{-2}. For both the doses studied, fairly small sheet resistances are obtained for an anneal temperature $> 700°$C. A recent study by us shows that the properties of the InGaAs/InAlAs QW are not degraded by lamp annealing up to $750°$C [89]. The contact resistances, using metals M1, on the implanted and annealed SQW structure are shown in Fig. 1.35*b*. In addition to the low contact resistance, the low sheet resistance of the implanted layer is useful for high-frequency performance, and therefore such implanted and annealed regions can form an integral part of the device technology with InGaAs/InAlAs.

Device Fabrication. In what follows, the typical processing steps for InGaAs/InAlAs FETs with gate lengths ≥ 1 μm are described. The SQW normal or inverted MODFETs are made by standard photolithography and lift-off techniques. The fabrication process

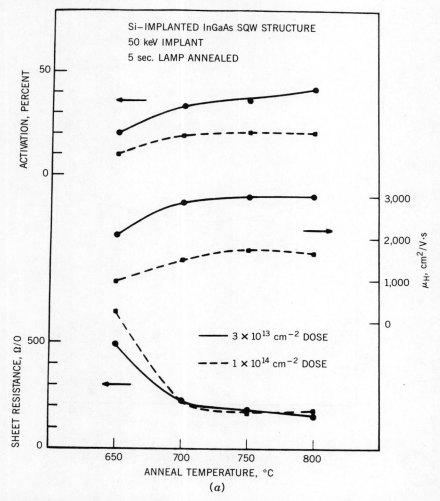

Figure 1.35   (*a*) Variation of Hall mobility, activation, and sheet resistance with anneal temperature in Si-implanted and lamp-annealed InGaAs SQW MISFET; (*b*) variation of contact resistance in these materials with post-metallization lamp-anneal temperatures.

is depicted in Fig. 1.36. Isolation between devices is first achieved by mesa etching with (1:1:38) $H_3PO_4:H_2O_2:H_2O$. The layered metals forming the source and drain ohmic contacts are then deposited by electron beam evaporation and were annealed in a halogen-lamp annealing station under flowing argon. After gate recessing with (1:1:400) $H_3PO_4:H_2O_2:H_2O$ by using the photoresist as a mask, 500 Å Ti/3500 Å Au is deposited to form the gate contact. Finally, 500 Å Ti/3500 Å is deposited to form the contact pads. Native oxides on the surface are removed just before the ohmic contact and the gate metal deposition by immersing the samples in buffered HF for 30 s.

We will first describe the dc and high-frequency characteristics of the InGaAs/InAlAs quasi-MISFET described in Fig. 1.29. Recently, we have reported the performance characteristics of a 1.8-μm gate device [90]. An extrinsic transconductance, $g_{m,\text{ext}}$, of

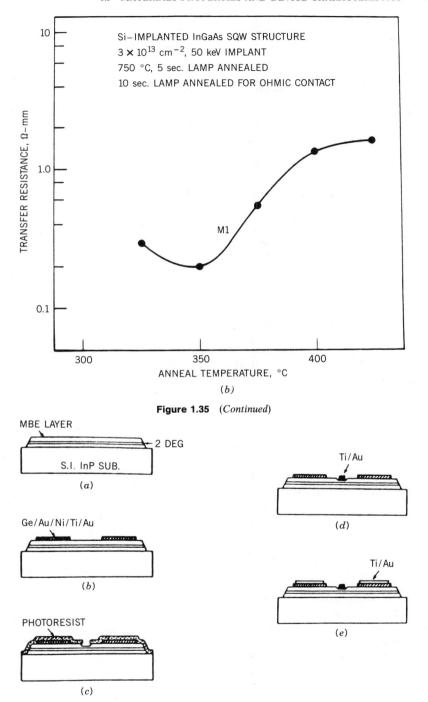

Figure 1.35 (*Continued*)

Figure 1.36 Fabrication procedure for InGaAs SQW MISFET: (*a*) mesa etching; (*b*) ohmic metal lift-off and annealing; (*c*) gate patterning and recess; (*d*) gate metal lift-off; (*e*) contact pad lift-off.

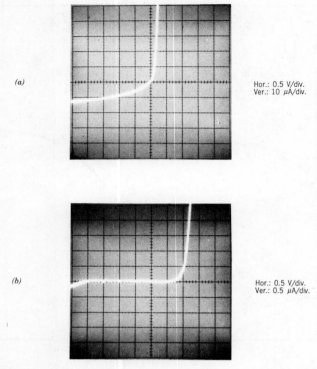

(a)

Hor.: 0.5 V/div.
Ver.: 10 μA/div.

(b)

Hor.: 0.5 V/div.
Ver.: 0.5 μA/div.

Figure 1.37 Gate-to-source current–voltage characteristics at 300 K in the quasi MISFET fabricated with mesa etching (a) before and (b) after gate metal deposition.

130 mS/mm and a cutoff frequency, f_T, of 6.8 GHz were measured in this device. It was realized that the prime reason for the rather low $g_{m,\text{ext}}$ and f_T values was a high source resistance, $R_{SG} = 4$ Ω-mm. In what follows, the dc and microwave characteristics of 1.0- and 1.3-μm gate devices with low-resistance ohmic contacts are described.

The room-temperature 300-K gate-to-source current–voltage characteristics of a 1.3-μm device are shown in Fig. 1.37. The higher leakage current in Fig. 1.37a is due to contact being made with the undepleted InAlAs layer and the InGaAs channel. Room-temperature drain current–voltage characteristics for a 1.3-μm device are depicted in Fig. 1.38a. A dc transconductance as high as 280 mS/mm is measured. We also measured transconductances of 310 mS/mm in a 1-μm device and 240 mS/mm in a 1.6-μm device. Confinement of electrons in the SQW channel is expected to provide a more linear variation of the drain current with gate bias than in conventional modulation-doped transistors. Figure 1.38b illustrates this point where a uniform transconductance over a sizable bias range and a linear variation of I_{DS} over the same range are observed. For the 1-μm devices the typical source resistance was measured to be 0.6 Ω-mm, which gives an intrinsic transconductance of 380 mS/mm.

The gate capacitances were 0.104 to 0.115 pF for the 1-μm gate devices at the peak of the transconductance. For short-channel devices, the saturation velocity v_s can be expressed as

$$v_s = \frac{1}{Z} \frac{g_m}{C_{GS}} \tag{1.22}$$

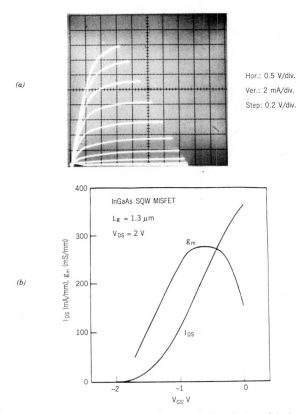

(a)

Hor.: 0.5 V/div.

Ver.: 2 mA/div.

Step: 0.2 V/div.

(b)

Figure 1.38   (a) Drain current–voltage characteristics, and (b) variation of drain current and transconductance with gate bias in a 1.3-μm gate InGaAs SQW quasi-MISFET at room temperature.

where Z is the gate width and g_m is the intrinsic transconductance. Estimated saturation velocities at 300 K are 1.7 to 1.83×10^7 cm/s, which are larger than the measured peak velocity (1.5×10^7 cm/s) mentioned in Section 1.5.1. Parallel conduction in the InAlAs doping layer is probably responsible for this discrepancy.

The microwave S-parameters in the 1-μm devices were measured in the range 0.045 to 26.50 GHz using a Cascade 26-GHz prober and a HP 8510 automatic network analyzer under various bias conditions. Figure 1.39 shows the plot of typical S-parameters of a 1- by 200-μm gate device. From these measurements and the derived equivalent circuits, values of f_T in the range 28 to 32 GHz are obtained for a 1.0-μm device.

The performance characteristics of the SQW MODFETs with GaAs gate barriers, shown in Fig. 1.28, are described next. Devices were made by standard photolithography and lift-off techniques. After the initial mesa isolation, source and drain ohmic contacts were formed by an electron beam layered deposition of Ni/Ge/Au/Ni/Ti/Au and two-step halogen lamp annealing. This was followed by Ti/Au gate formation. It should be realized that a GaAs gate barrier layer with a doping of 3×10^{17} cm^{-3} permits the gate metals to be deposited directly on the layer, maintaining a low gate–source leakage current. Thus a tedious gate recess etching can be avoided.

The measured drain current–voltage characteristics of a 1-μm gate device is shown in Fig. 1.40. The maximum extrinsic transconductance was 320 mS/mm at 300 K and

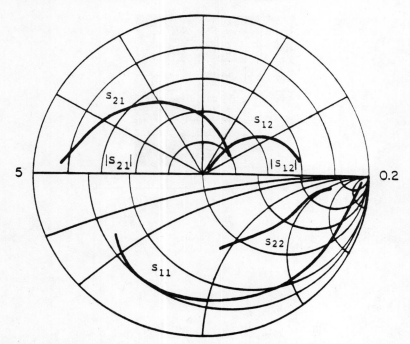

Figure 1.39   Measured S-parameters of a 1.0-μm gate InGaAs quasi-MISFET for $V_{DS} \sim 2.5$ V, $V_{DS} = -1.2$ V.

450 mS/mm at 77 K. The pinch-off characteristic also improves at the lower temperature. The microwave S-parameters, shown in Fig. 1.41, were measured in the same device, from which $f_T = 35$ GHz was derived. A maximum available power gain of 13 dB at 10 GHz is achieved. These values are among the best for 1-μm devices. Further improvements are expected by using submicron gate lengths and pseudomorphic channel regions.

Submicron Gate MODFET. For better device performance, in both the dc and high-frequency regimes, it is necessary to have submicron gate lengths. Excellent noise figures, power gains, and millimeter-wave operation have been reported for submicron gate GaAs/AlGaAs MODFETs, but the development of similar devices with InGaAs/InAlAs heterostructures is still in an early stage. It is expected that because of the more favorable properties of the system, performance characteristics superior to those in GaAs/AlGaAs devices will be achieved. We therefore conclude this section with a brief description of submicron device fabrication and their performance characteristics.

Device processing is initiated with a mesa isolation by a 0.2- to 0.3-μm $H_3PO_4:H_2O_2:H_2O$ chemical etch. Ohmic metals (Ni (50 Å)/Ge (400 Å)/Au (900 Å)/Ni (100 Å)/Ti (500 Å)/Au (1000 Å)) is then deposited on the optical-lithographically patterned source and drain regions by electron beam evaporation. The contact metals is alloyed in an AG Associates 210T Halogen lamp annealing station under flowing Ar. An alloy heating rate of $\sim 40°$C/s is used for the two-step alloying scheme, in which the sample is typically alloyed at $T_1 = 350$ to $380°$C for 10 s and then at $T_2 = 400$ to $450°$C for 7 s. Electron beam lithography is then employed after the gate definition. A simple level 4000 Å polymethyl methacrylate (PMMA) positive resist is coated for the submicrometer gate lines and the wafer is baked at 180°C for 30 min. It is then exposed to an

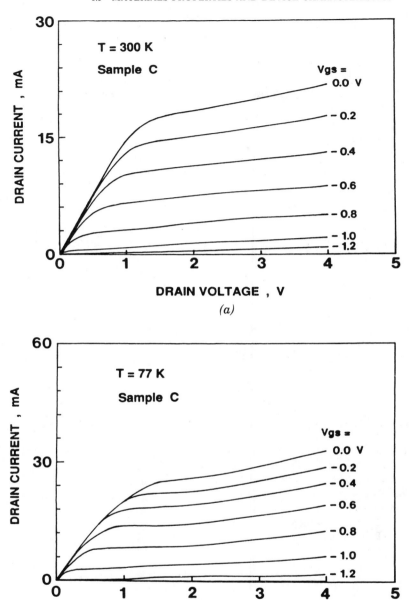

Figure 1.40 Drain current–voltage characteristics measured at (*a*) 300 K and (*b*) 77 K in 1-μm gate SQW InGaAs/InAlAs MODFET with GaAs gate barrier layer.

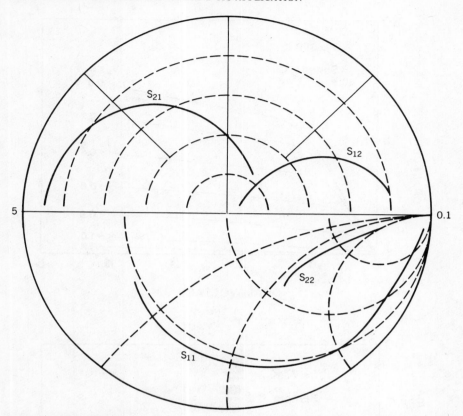

Figure 1.41   Measured S-parameters of a 1- by 300-μm gate InGaAs/InAlAs MODFET for $V_{gs} = 0$ V and $V_{ds} = 3.0$ V.

electron dose of 3.0 nC/cm in an electron beam lithography system. After the development of the gate-level resist, the exposed area is recessed using a wet chemical etch to achieve the desired channel current. Ti (500 Å)/Au (2500 Å) layers are then evaporated to form Schottky gates. For microwave characterization it is necessary to deposit very thick Ti/Au metal layers on the source, drain, and gate contact pads.

Typical current–voltage characteristics [91] of a 0.3-μm gate InGaAs/InAlAs MODFET lattice matched to InP are shown in Fig. 1.42. The device width is 25 μm

Figure 1.42   Measured drain current–voltage characteristics as a function of gate bias in a 0.3-μm gate InGaAs/InAlAs MODFET at 300 K.

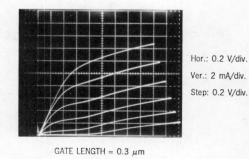

Hor.: 0.2 V/div.
Ver.: 2 mA/div.
Step: 0.2 V/div.

GATE LENGTH = 0.3 μm

and the measured $g_m = 640$ mS/mm. Even higher values of transconductance can be obtained for smaller gate dimensions. Extrinsic transconductances over 1000 mS/mm and $f_T \simeq 175$ GHz have been measured in devices with gate lengths of 0.1 μm by Mishra et al. [91].

Pseudomorphic (Strained Channel) MODFETs. The key issues controlling the intrinsic performance of a MODFET are (i) high sheet change density (controlled by a high band discontinuity); (ii) low effective mass of the carrier gas in the channel, which leads to a higher mobility; and (iii) high peak and saturation velocity of the carrier gas (controlled by high intervalley separation and high band discontinuity). In addition, one requires the material structure to have minimum light sensitivity—a problem that plagues traditional GaAs/Al$_{0.3}$Ga$_{0.7}$As MODFETs. By using a strained channel, all of the issues above can be controlled for a better device. The simplest way to introduce the strain in the GaAs/AlGaAs (or GaAs substrate) and In$_{0.53}$Ga$_{0.47}$As (or InP substrate) is to add extra In in the region 100 to 200 Å which forms the channel of the device. Both n-type [92–94] and p-type [95, 96] devices have been fabricated using this concept. Improved performance in n-type devices has occurred primarily due to an increase in sheet charge density, higher mobility, and better light sensitivity [97]. In p-MODFETs the enhanced performance arises in addition to the factors listed for n-MODFETs from the fact that hole states become quite light due to the biaxial compressive strain [98].

The ability to grow high-quality strained channel MODFETs requires considerable care to ensure that surface morphology is not destroyed. This may require changing growth temperatures during growth by MBE. This area of strained devices is expected to yield very high performance devices.

1.5.2   InGaAs/InAlAs PIN Photodiodes

In$_{0.53}$Ga$_{0.47}$As photodetectors are useful at the 1.3 range and 1.55 μm, and they can form an important front-end element in a photoreceiver for handling data at rates of 1 Gbit/s or higher. Needless to say, workers at various laboratories have investigated the properties of InGaAs photodiodes grown by different epitaxial techniques [99–106]. We will describe here the dimensional design of very high speed In$_{0.53}$Ga$_{0.47}$As photodiodes and the properties of such diodes grown by molecular beam epitaxy.

For optimum high-speed response the transit time of carrier across the absorption length should be of the same order as the circuit-limit (RC) time constant of the photodiode. Assuming that the carriers are traveling at saturation velocity (ca. 1×10^7 cm/s), the transit time can, in principle, be reduced by reducing the absorption length L. However, the quantum efficiency η, given by

$$\eta = (1 - R)(1 - e^{-\alpha L}) \tag{1.23}$$

where R is the reflectivity of the top surface and α is the absorption coefficient of the material, will be drastically reduced unless α can be enhanced at the same time.

The capacitance of the device can be reduced by reducing its active area, but again, this is achieved at the cost of the quantum efficiency. Finally, slow diffusion of carriers, which can give rise to long tails in the impulse response characteristics, can be partially eliminated by using heterojunctions. For the design of an InGaAs PIN diode we have assumed the saturation velocity of electrons to be 1×10^7 cm/s. With this parameter, a calculated junction capacitance of 80 fF, and a realistic set of parasitic elements, the equivalent circuit of Fig. 1.43 is obtained. A transient response of this equivalent circuit is obtained by taking the inverse Fourier transform of the frequency response, which is

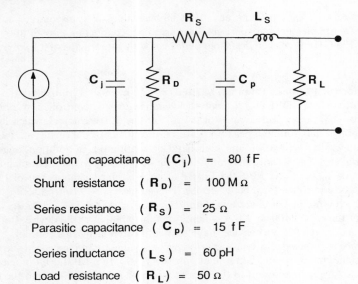

Junction capacitance (C_j) = 80 fF

Shunt resistance (R_D) = 100 MΩ

Series resistance (R_S) = 25 Ω

Parasitic capacitance (C_p) = 15 fF

Series inductance (L_S) = 60 pH

Load resistance (R_L) = 50 Ω

Figure 1.43   Equivalent circuit of InGaAs/InAlAs PIN photodiode.

found from simple circuit analysis to be

$$H(jw)\frac{R_D}{\alpha + j\omega(\beta - \delta\omega^2) - \gamma\omega^2} \tag{1.24}$$

where $\alpha = R_S + R_L + R_D$, $\beta = R_S R_L C_p + L_S + (R_S + R_L R_D C_j + R_L R_D C_p \gamma) = (R_S R_L C_p + L_S) R_D C_j + R_S L_S C_p + R_D C_p L_S$, and $\delta = R_S L_S C_p R_D C_j$. From the impulse response a rise time of 17 ps is calculated for a diode area of 5×10^{-6} cm^2 and an absorption layer thickness of 0.75 μm.

Our device design was intended to be compatible for integration with high-speed electronic devices such as field-effect transistors and is therefore realized on a semi-insulating substrate. The schematic of an InGaAs PIN photodetector and a photomicrograph are shown in Fig. 1.44a and b, respectively. The structure was grown on Fe-doped semi-insulating InP substrate and consists of an n^+ (2×10^{18} cm^{-3}) 0.5-μm InAlAs layer, an undoped n-type (1 to 5×10^{15} cm^{-3}) 0.75-μm absorption/transit layer, and a p^+ GaAs to facilitate contact, grown in that order. The top p^+ InAlAs layer with a wider bandgap also serves as a transparent window for the incident photoexcitation.

The device fabrication procedure is as follows. A square mesa (20×25 μm^2) defining the active area is formed by etching with 1 H$_3$PO$_4$:1 H$_2$O$_2$:8 H$_2$O up to the n^+ InAlAs layer. A second layer mesa (80×155 μm^2) is similarly defined to reach the semi-insulating substrate. 3000-Å SiO$_2$ is deposited on the whole structure and opening of dimension 5×5 μm^2 and 75×75 μm^2 are made by etching with HF for the p- and n-type contacts, respectively. 500/3500-Å Ti/Au and 300/400/3500-Å Ni/Ge/Au p- and n-type contact metals are deposited. The contact pads are positioned on the substrate to minimize parasitic capacitances. The zero-bias capacitance of the diode is less than 0.1 pF. Typical reverse breakdown voltages $V_{BR} \simeq 14$ V and the dark current at 10 V is less than 10 nA, which corresponds to a current density of 2×10^{-3} A/cm^2.

50 Å	p^+	GaAs	$2E18\ cm^{-3}$
0.3 μm	p^+	$In_{0.52}Al_{0.48}As$	$2E18\ cm^{-3}$
0.75 μm	n^-	$In_{0.53}Ga_{0.47}As$	$2E15\ cm^{-3}$
0.5 μm	n^+	$In_{0.52}Al_{0.48}As$	$2E18\ cm^{-3}$

Fe - doped   InP   substrate

$$\left[a\right]$$

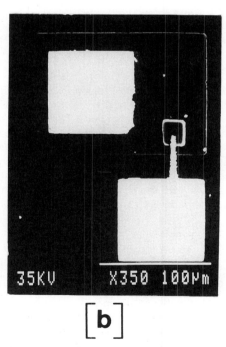

$$\left[b\right]$$

Figure 1.44   (*a*) Schematics of InGaAs/InAlAs photodiode grown by molecular beam epitaxy; (*b*) photomicrograph taken by scanning electron microscope (SEM).

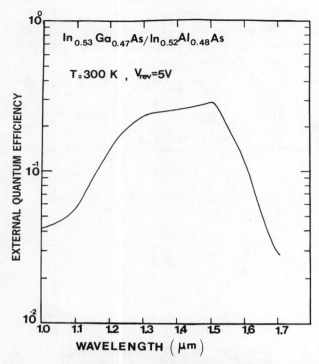

Figure 1.45   External quantum efficiency versus wavelength for InGaAs/InAlAs PIN photodiode at 5 V reverse bias.

The spectral response characteristics of the devices were recorded with a monochromatic light source and lock-in detection of the detector photocurrent. The measured photocurrents were calibrated with a Si photodiode. The external quantum efficiency of the InGaAs PIN diode at a reverse bias of 5 V, where the n-region is fully depleted is shown in Fig 1.45. A maximum value of 30% is measured at 1.5 μm, which corresponds to a measured responsivity of 0.35 A/W. This value is optimistic for a small-area diode being characterized here and is characteristic of front-illuminated devices. For higher responsivities one has to enhance the absorption artificaly by using a waveguiding structure, or by using cladding layers to effect multiple reflections in the absorption region.

For the impulse response measurements, the photodiodes were mounted on a 50-Ω microstripline package with very short bonding leads. Dc bias of 10 V to the diode was provided through a HP 11612A bias-tee with an operating range of 45 MHz to 26.5 GHz. The diodes were photoexcited on the top surface with 1 to 2-ps pulses from a Styryl 9M dye laser (810 nm) photopumped by a Nd:YAG laser. The focused incident power density on the diode was about 1 μJ/cm^2. The response of the photodiode to the optical pulses was observed on a sampling scope with an S-4 sampling head. The SMA cables used in the circuitry have an approximate rise times of 10 ps. Figure 1.46 shows the temporal response characteristics of the photodiodes. From these data the deconvolved rise time and width (FWHM) of the photodiode are calculated to be 21 and 27 ps, respectively. The weak ringing effects in the trailing edge of the response are due to mismatch effects.

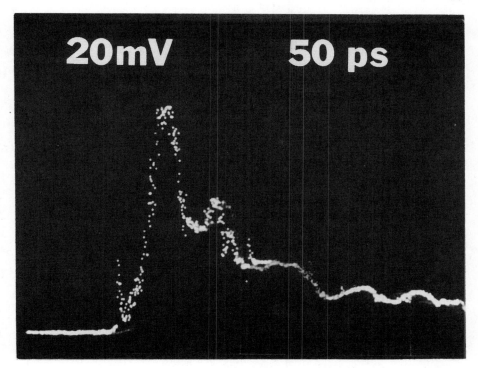

Figure 1.46 Impulse response of InGaAs/InAlAs PIN photodiode to dye laser excitation with pulse FWHM of 2 ps.

The experimental impulse response characteristics compare extremely favorably with theoretically calculated values using the same device dimensions. It can therefore by assumed that the device speed is transit-time limited. Carrier diffusion and hole "pileup" effects in the heterostructure and carrier trapping in the different regions apparently do not play a dominant role. From the response FWHM of 27 ps, an approximate 3-dB bandwidth of 18 GHz is estimated for the photodiodes. These devices will be extremely useful for microwave and millimeter wave applications. We are confident that the bandwidth in these devices can be increased even further.

Acknowledgments. The authors would like to thank Rita Szokowski for her careful typing and editing of the manuscript and several graduate students for their assistance. The work described here was supported by the Army Research Office under the URI Program and by the National Science Foundation.

References

1. J. E. Davey and T. J. Pankey, *J. Appl. Phys.*, Vol. 39, p. 1941, 1968.

2. J. R. Arthur, *Surf. Sci.*, Vol. 43, p. 449, 1974.

3. A. Y. Cho and J. R. Arthur, *Prog. Solid State Chem.*, Vol. 10, p. 157, 1975.

4. M. Ilegems, *J. Appl. Phys.*, Vol. 48, p. 1278, 1977.

5. J. Y. Robinson and M. Ilegems, *Rev. Sci. Instrum.*, Vol. 49, p. 205, 1978.

6. A. C. Gossard, *Thin Solid Films*, Vol. 104, p. 279, 1982.

7. L. L. Chang, L. Esaki, W. E. Howard, R. Ludeke, and G. Schul, *J. Vac. Sci. Technol.*, Vol. 10, p. 655, 1973.

8. K. Ploog, *Cryst. Growth Propag. Appl.*, Vol. 3, p. 73, 1980.

9. B. A. Joyce and C. T. Foxon, *J. Cryst. Growth*, Vol. 31, p. 122, 1975.

10. H. Holloway and J. N. Walpole, *Prog. Cryst. Growth Charact.*, Vol. 2, p. 49, 1979.

11. C. E. C. Wood, in G. Haas and M. H. Francombe, Eds., *Physics of Thin Films*, Vol. 2, Academic, Press, New York, 1980, p. 35.

12. A. Y. Cho, *Thin Solid Films*, Vol. 100, p. 291, 1983.

13. L. Esaki and R. Tsu, *IBM J. Res. Dev.*, Vol. 14, p. 61, 1970.

14. J. R. Arthur, *J. Appl. Phys.*, Vol. 39, p. 4032, 1968.

15. J. Singh and K. K. Bajaj, in K. Hess, Ed., *Large Scale Computational Device Modeling*, University of Illinois Press, Champaign, IL, 1986, p. 91.

16. J. Singh and K. K. Bajaj, *Superlattices Microstruct.*, Vol. 2, p. 185, 1986.

17. J. H. Neave, P. J. Dobson, B. A. Joyce, and J. Zhang, *Appl. Phys. Lett.*, Vol. 47, p. 100, 1985.

18. B. I. Miller, and J. H. McAfee, *J. Electrochem. Soc.*, Vol. 125, p. 1310, 1978.

19. H. Ohno, C. E. C. Wood, L. Rathbun, D. V. Morgan, G. W. Wicks, and L. F. Eastman, *J. Appl. Phys.*, Vol. 52, p. 4033, 1981.

20. K. Y. Cheng, A. Y. Cho, and W. R. Wagner, *Appl. Phys. Lett.*, Vol. 39, p. 607, 1981.

21. K. Y. Cheng, A. Y. Cho, T. J. Drummond, and H. Morkoc, *Appl. Phys. Lett.*, Vol. 40, p. 147, 1982.

22. Y. Kawamura, Y. Noguchi, H. Asahi, and H. Hagai, *Electron. Lett.*, Vol. 18, p. 91, 1982.

23. J. Massies, J. Rochette, P. Delescluse, P. Etienne, J. Chevrier, and N. T. Linh, *Electron. Lett.*, Vol. 18, p. 758, 1982.

24. T. Mizutani and K. Hirose, *Jpn. J. Appl. Phys.*, Vol. 24, p. L119, 1985.

25. K. S. Seo, P. K. Bhattacharya, and Y. Nasimoto, *IEEE Electron Device Lett.*, Vol. *EDL*-6, p. 64, 1985.

26. C. E. C. Wood, in T. P. Pearsall, Ed, *GaInAsP Alloy Semiconductors*, 1982.

27. V. Swaminathan W. T. Tsang, *Appl. Phys. Lett.*, Vol. 38, p. 347, 1981.

28. G. B. Stringfellow, R. A. Stall, and W. Koschel, *Appl. Phys. Lett.*, Vol. 38, p. 156, 1981.

29. P. D. Kirchner, J. M. Woodall, J. L. Freeouf, and G. D. Petit, *Appl. Phys. Lett.*, Vol. 38, p. 427, 1981.

30. J. Singh and K. K. Bajaj, *Appl. Phys. Lett.*, Vol. 46, p. 577, 1985.

31. J. Singh and K. K. Bajaj, *Appl. Phys. Lett.*, Vol. 47, p. 594, 1985.

32. T. Hayakawa, T. Suyama, K. Takahashi, M. Kando, S. Yamamoto, S. Yano, and T. Hijikata, *Appl. Phys. Lett.*, Vol. 47, p. 952, 1985.

33. H. Sakaki, M. Tanaka, and J. Yoshino, *Jpn. J. Appl. Phys.*, Vol. 24, p. L417, 1985.

34. F. Y. Juang, P. K. Bhattacharya, and J. Singh, *Appl. Phys. Lett.*, Vol. 48, p. 290, 1986.

35. J. Singh and K. K. Bajaj, *J. Vac. Sci. Technol.*, Vol. B3, No. 2, p. 520, 1985.

36. T. Ando, *J. Phys. Soc. Jpn.*, Vol. 37, p. 1233, 1974.

37. K. S. Seo, P. K. Bhattacharya, G. P. Kothiyal, and S. Hong, *Appl. Phys. Lett.*, Vol. 49, p. 966, 1986.

38. P. Narozny and H. Beneking, *Electron. Lett.*, Vol. 21, p. 1050, 1985.

39. H. Beneking and N. Emeis, *IEEE Electron Device Lett.*, Vol. EDL-7, p. 98, 1986.

40. H. Beneking, P. Narozny, P. Roentgen, and M. Yoshida, *IEEE Electron Device Lett.*, Vol. EDL-7, p. 101, 1986.

41. J. Singh, presented at the 13th Conference on the Physics and Chemistry of Semiconductor Interfaces, Pasadena, CA, 1986.

42. D. V. Lang, A. Y. Cho, A. C. Gossard, M. Ilegems, and W. Wiegmann, *J. Appl. Phys.*, Vol. 47, p. 2558, 1976.

43. A. K. Saxena, *J. Phys. C: Solid State Phys.*, Vol. 13, p. 4323, 1980.

44. P. K. Bhattacharya, U. Das, and M. J. Ludowise, *Phys. Rev.*, Vol. B29, p. 6623, 1984.

45. T. Ishikawa, J. Saito, S. Sasa, and S. Hiyamizu, *Jpn. J. Appl. Phys.*, Vol. 21, P. 2, p. L476, 1982.

46. T. Ishibashi, S. Tarucha, and H. Okamoto, *Jpn J. Appl. Phys.*, Vol. 21, p. L476, 1982.

47. R. E. Thorne, T. J. Drummond, W. G. Lyons, R. Fischer, and H. Morkoc, *Appl. Phys. Lett.*, Vol. 41, p. 189, 1982.

48. M. O. Watanabe, M. Monizuka, M. Mashita, Y. Achizawa, and Y. Zohta, *Jpn. J. Appl. Phys.*, Vol. 23, p. L103, 1984.

49. M. O. Watanabe and H. Maeda, *Jpn. J. Appl. Phys.*, Vol. 23, p. L734, 1984.

50. B. L. Zhou, K. Ploog, E. Gmelin, X. Q. Zheng, and M. Schulz, *Appl. Phys.*, Vol. A, 28, p. 223, 1982.

51. S. Subramanian, U. Schuller, and J. R. Arthur, *J. Vac. Sci. Technol.*, Vol. B3, p.650, 1985.

52. S. Dhar, W.-P. Hong, P. K. Bhattarcharya, Y. Nashimoto, and F.-Y. Juang, *IEEE Trans. Electron Devices*, Vol. ED-33, p. 698, 1986.

53. H. Ohno, Y. Akatsu, T. Hashizume, and H. Hasegawa, *J. Vac. Sci. Technol.*, Vol. B3, p. 943, 1985.

54. D. V. Lang, R. A. Logan, and M. Jaros, *Phys. Rev.*, Vol. B19, p. 1015, 19.79.

55. M. Mizuta, M. Tachikawa, H. Kukimoto, and S. Minomura, *Jpn. J. Appl. Phys.*, Vol. 24, p. L143, 1985.

56. Y. Nashimoto, S. Dhar, W.-P. Hong, A. Chin, P. R. Berger, and P. Bhattacharya, *J. Vac. Sci. Technol.*, Vol. B4, p. 540, 1986.

57. S. R. Forrest and O. K. Kim, *J. Appl. Phys.*, Vol. 53, p. 5738, 1982.

58. J. Massies, J. F. Rochette, P. Etienne, P. Delescluse, A. M. Huber, and J. Chevrier, *J. Cryst. Growth*, Vol. 64, p. 101, 1983.

59. M. Tachikawa, M. Mizuta, H. Kukimoto, and S. Minomura, *Jpn. J. Appl. Phys.*, Vol. 24, p. L821, 1985.

60. W.-P. Hong, S. Dhar, P. K. Bhattacharya, and A. Chin, *J. Electron. Mater.*, Vol. 16, p. 271, 1987.

61. S. R. McAfee, W. T. Tsang, and D. V. Lang, *J. Appl. Phys.*, Vol. 52, p. 6165, 1981.

62. S. R. McAfee, D. V. Lang, and W. T. Tsang, *Appl. Phys. Lett.*, Vol. 40, p. 520, 1982.

63. M. H. Meynadier, J. A. Brum, C. Delalande, and M. Voos, *J. Appl. Phys.*, Vol. 58, p. 4307, 1985.

64. P. A. Martin, K. Meehan, P. Gavrilovic, K. Hess, N. Holonyak, and J. J. Coleman, *J. Appl. Phys.*, Vol. 54, p. 4689, 1983.

65. W.-P. Hong, P. Bhattacharya, and J. Singh, *Appl. Phys. Lett.*, Vol. 50, p. 61B, 1987.

66. W.-P. Hong, A. Chin, N. Debbar, J. Hinckley, P. Bhattacharya, and J. Singh, *J. Vac. Sci. Technol.*, Vol. B5, p. 800, 1987.

67. K. Nakajima, T. Tanahashi, and K. Akita, *Appl. Phys. Lett.*, Vol. 41, p. 194, 1982.

68. J. Singh and K. K. Bajaj, *Appl. Phys. Lett.*, Vol. 48, p. 1077, 1986.

69. J. Singh, S. Dudley, B. Davies, and K. K. Bajaj, *J. Appl. Phys.*, Vol. 60, p. 3167, 1986.

70. I. M. Lifshitz, *Adv. Phys.*, Vol. 13, p. 483, 1985.

71. J. Singh, K. K. Bajaj, and S. Chaudhari, *Appl. Phys. Lett.*, Vol. 44, p. 805, 1984.

72. F. Y. Juang, P. K. Bhattacharya, and J. Singh, *Appl. Phys. Lett.*, Vol. 48, p. 439, 1986.

73. K. Hirose, K. Ohata, T. Mizutani, T. Itoh, and M. Ogawa, *Proc. Int. Symp. GaAs Related Compounds*, Karuizawa, Japan, pp. 529–534, 1985.

74. U. Mishra, A. Brown, L. Jelloian, L. Hackett, and M. Delaney, presented at the 45th Device Research Conference, Santa Barbara, CA. 1987.

75. C. K. Peng, M. I. Aksun, A. A. Ketterson. H. Morkoc, and K. R. Gleason, *IEEE Electron Device Lett.*, Vol. EDL-8, p. 24, 1987.

76. K. Ueno, T. Furutsuka, M. Toyoshima, M. Kanamori, and A. Higashisaka, *IEDM Tech. Dig.*, Vol. 83–85, pp. 1985.

77. M. Inoue, *Superlattices Microstruct.*, Vol. 1, p. 433, 1985.

78. K. S. Seo and P. Bhattacharya, *IEEE Trans. Electron Devices.*, Vol. ED-34, p. 2221, 1987.

79. C. C. Shen and K. P. Pande, *J. Vac. Sci. Technol.*, Vol. 2, p. 314, 1984.

80. A. S. Liao, B. Tell, R. F. Leheny, T. Y. Chang, E. A. Cardini, E. Beebe, and J. C. DeWinter, *Appl. Phys. Lett.*, Vol. 44, p. 344, 1984.

81. K. S. Seo, P. R. Berger, G. P. Kothiyal, and P. K. Bhattacharya, *IEEE Trans. Electron Devices*, Vol. ED-34, p. 235, 1987.

82. W.-P. Hong, J. Singh, and P. K. Bhattacharya, *IEEE Electron Device Lett.*, Vol. EDL-7, p. 480, 1986.

83. P. Bannerjee, P. K. Bhattacharya, M. J. Ludowise, and W. T. Dietze, *IEEE Electron Device Lett.*, Vol. EDL-4, p. 283, 1983.

84. E. F. Schubert, K. Ploog, H. Dambkes, and K. Heime, *Appl. Phys.*, Vol. A33, p. 183, 1984.

85. W.-P. Hong, K. S. Seo, P. K. Bhattacharya, and H. Lee, *IEEE Electron Device Lett.*, Vol. EDL-7, p. 320, 1986.

86. P. Zwicknagl, S. D. Mukherjee, P. M. Capani, H. Lee, H. T. Griem, L. Rathbun, J. D. Berry, W. L. Jones, and L. F. Eastman, *J. Vac. Sci. Technol.*, Vol. 4, p. 476, 1986.

87. M. Kamada, H. Ishikawa, M. Ikeda, Y. Mori, and C. Kojima, *Proc. Int. Symp. GaAs Related Compounds*, Las Vegas, NV, pp. 575–580, 1986.

88. H. Ito, T. Ishibashi, and T. Sugeta, *Jpn. J. Appl. Phys.*, Vol. 23, p. L635, 1984.

89. K. S. Seo, P. K. Bhattacharya, G. P. Kothiyal, and S. Hong, *Appl. Phys. Lett.*, Vol. 49, p. 966, 1986.

90. K. S. Seo, Y. Nashimoto, and P. K. Bhattacharya, *IEDM Tech. Dig.*, pp. 321–323, December 1985.

91. U. K. Mishra, Private communications.

92. A. A. Ketterson. W. T. Masselink, J. S. Gedymun, J. Klem, C. K. Peng, W. F. Kopp, H. Morkoc, and K. R. Gleason, *IEEE Trans. Electron Devices.*, Vol. ED-33, p. 564, 1986.

93. A. W. Swanson, *Microwaves and RF*, Vol. 25, p. 139, 1986.

94. G. I. Ng, D. Pavlidis, M. Quillec, Y. J. Chan, M. D. Jaffe, and J. Singh, *Appl. Phys. Lett.*, 1988, in press.

95. T. Drummond, T. Zipperian, I. Fritz, J. Schirbin, and T. Plut, *Appl Phys. Lett.*, Vol. 49, p. 461, 1986.

96. C. Lee, H. Wang, G. Sullivan, N. Sheng, and D. Miller, *IEEE Electron Device Lett.*, Vol. 8, p. 85, 1987.

97. T. J. Drummond, R. Fischer, W. Kopp, H. Markoc, K. Lee, and M. S. Shur, *IEEE Trans. Electron Devices*, Vol. ED-30, p. 1806 1983.

98. M. Jaffe, Y. Sekiguchi, and J. Singh, *Appl. Phys. Lett.*, Vol. 51, p. 1943, 1987.

99. T. P. Lee, C. A. Burrus, K. Ogawa, and A. G. Dentai, *Electron. Lett.*, Vol. 17, p. 413, 1981.

100. R. F. Leheny, Robert E. Nahory, M. A. Pollack, E. D. Beebe, and J. C. Dewinter, *IEEE J. Quantum Electron.*, Vol. QE-17, p. 227, 1981.

101. T. P. Pearsall, R. A. Logan, and C. G. Bethan, *Electron. Lett.*, Vol. 16, p. 611, 1983.

102. F. Capasso, K. Alavi, A. Y. Cho, P. W. Foy, and C. G. Bethan, *Appl. Phys. Lett.*, Vol. 43, p. 1040, 1983.

103. K. Li, E. Rezek, and H. D. Law, *Electron. Lett.*, Vol. 20, p. 196, 1984.

104. R. S. Sussmann, R. M. Ash, A. J. Moseley, and R. C. Goodfellow, *Electron. Lett.*, Vol. 21, p. 593, 1985.

105. J. E Bowers, C. A. Burrus, and F. Mitschke, *Electron. Lett.*, Vol. 22, p. 633, 1986.

106. V. Diadiuk and S. H. Groves, *Solid-State Electron.* Vol., 29, p. 229, 1986.

2

MIXERS AND DETECTORS

Erik L. Kollberg

Chalmers University of Technology
Gothenburg, Sweden

2.1 INTRODUCTION

In any application of microwaves, millimeter waves, or submillimeter waves, detection is a must! Detection can be performed by detector elements based on a variety of physical princles, some of which are described below. Some of these principles work over very large frequency ranges: that is why detectors that one usually associates with infrared frequencies are also mentioned in this chapter.

In some applications, straight detection is not good enough, and converting the signal to a lower frequency, using a mixer, is required. This is the case, for example, in applications when the phase of the signal must be preserved for further information processing, or when the sensitivity of the system requires extremely narrow-band filters that are impossible to realize at the signal frequency.

The major part of this chapter is devoted to mixers. Definitions, analysis, various topologies, and so on, are topics dealt with in some detail below. For general references concerning mixers and detectors, see Refs. 1–4.

2.2 SOME COMMON DETECTOR AND MIXER DEVICES

2.2.1 Introductory Remarks

In this section we discuss different types of devices that can be used as detectors and/or as nonlinear elements in mixers. The frequency range considered is from a few hundred megahertz to terahertz frequencies. Important detector and mixer devices are described briefly, with particular emphasis on sensitivity and frequency response. Important also is the ability of the detector to respond to fast modulation. Generally, two types of detectors are used for microwave and millimeter waves: rectifying diode detectors and thermal detectors. In particular, in mixer applications, the response time of the devices (nonlinear element) must be much less than the inverse of the intermediate frequency. Since thermal detectors have a slow response, they cannot be used in mixer applications.

2.2.2 *pn*-Type Diodes

There are quite a number of different ways of making diodes for detection and mixer applications out of semiconductor materials, and some of the more important methods are briefly discusses in this section. The Schottky diode, which is the most important for microwave and millimeter wave applications, is discussed separately in Section 2.2.3.

***pn*-Diodes.** A *pn*-diode is essentially made from a semiconductor (e.g., Si) with a *p*-doped region and an *n*-doped region, each with an ohmic contact (see Fig. 2.1). The doping profile can be realized using different techniques [3], and the diode properties can be tailored by proper choice of the exact doping profile. In Fig. 2.2 is shown the doping profile of the ordinary *pn*-diode, the backward diode, and the tunnel diode. The basic physics of *pn*-diodes can be studied in, for example, Sze's book on semiconductor devices [3]. The ordinary *pn*-diode is essentially a "low-frequency" diode useful up to about 1 GHz, while the tunnel and backward diodes can in principle be used up to millimeter wave frequencies.

The equivalent circuit of a *pn*-junction is a nonlinear resistance in parallel with a nonlinear capacitance (compare with Section 2.2.3 and Fig. 2.4). For the ordinary *pn*-diode, the nonlinear resistance is directly related to the dc characteristics, that is,

$$i = i_0[\exp(qV_j/kT) - 1] \tag{2.1}$$

where q = charge of the electron
k = Boltzmann constant
V_j = voltage applied over the *pn*-junction
T = physical temperature of the junction

The nonlinear capacitance for an "abrupt junction" (the doping changes abruptly from *p*- to *n*-type) is expressed as

$$C = \frac{C_0}{\sqrt{1 - \dfrac{V_j}{(V_{Bi} - 2kT/q)}}} \tag{2.2}$$

where V_{Bi} is the built-in potential, which like C_0 is dependent on the type of semiconductor and the amount of *p*- and *n*-type doping [3]. For linear graded junctions or other types of doping profiles, the capacitance voltage dependence will become slightly different from Eq. (2.2).

These formulas are a consequence of the depletion of charges, which causes the potential barrier V_{Bi} (Fig. 2.2) to hinder an excess flow of majority charges. The capacitance is essentially the capacitance between the *p*- and the *n*-doped region assuming the depleted region to be equivalent to an insulator, with a dielectric constant equal to that of the semiconductor ($\epsilon_r = 11.9$ for Si and $\epsilon_r = 13.1$ for GaAs).

However, the charge transport mechanism over the barrier is related to a flow of majority carriers, electrons and holes, respectively, that have to disappear when they have reached the *p*- and the *n*-region, respectively. The required mechanism is a recombination of charges, which is not instant: the time constants involved are of the order of nanoseconds. This phenomenon gives rise to a frequency dependence of the nonlinear resistance, and an excess capacitance in parallel with the depletion capacitance, usually referred to as the diffusion capacitance [3]. This means that at high enough frequencies,

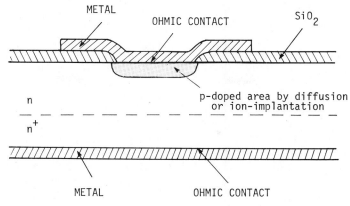

Figure 2.1   Schematic outline of a *pn*-junction diode. The semiconductor is usually silicon.

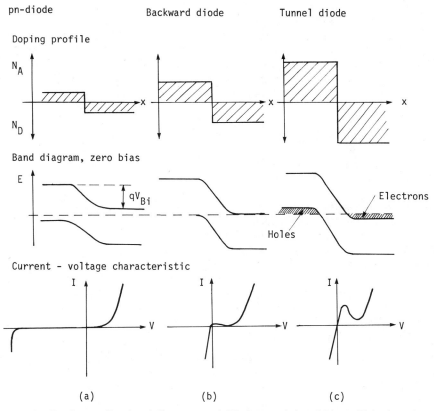

Figure 2.2   Doping profiles, band diagrams, and *IV* characteristics of (*a*) *pn*, (*b*) backward, and (*c*) tunnel diodes. Notice that the doping in the tunnel diode is high enough to make the semiconductor degenerate.

the switching capabilities of the diode are washed out. In practice, the ordinary *pn*-diode is not used above frequencies of the order of 1 GHz.

Tunnel Diodes. The tunnel diode is a *pn*-diode which is made up of such heavily doped *p*- and *n*-type material that tunneling will dominate over thermal emission and diffusion. This will cause the *IV* characteristic to show a negative resistance region (Fig. 2.2*c*) [3]. Tunneling is a majority-carrier phenomenon and is not governed by the conventional transit-time concept. Also, recombination is not a problem since the electrons can go straight into empty energy states either in the conduction band or the valence band (see Fig. 2.2). Hence tunneling devices should, in principle, work well into the millimeter wave frequency region. However, there are other limitations at high frequencies, partly related to the comparatively large depletion capacitance, which will be described in Section 2.13.2. The tunnel diode has been used as a sensitive and low-noise detector, as well as a mixer element, for microwave frequencies. The tunnel diode mixer is described in Section 2.13.2.

Backward Diodes. The backward diode is a special type of tunnel diode. The doping concentrations on the *p* and *n* sides are barely large enough to make the semiconductor degenerate (which is also the case for the tunnel diode). Therefore, a tunneling current will occur in the back-biased diode which is significantly larger than the current for the corresponding forward bias, Fig. 2.2*b*.

The backward diode, which has very low $1/f$ noise (see Section 2.2.3), can be used as a low-noise detector and as a mixer diode. Since there is no minority carrier storage effect (diffusion capacitance), it also has a good frequency response. Futhermore, since tunneling is dominant, its *IV* properties are insensitive to temperature changes and radiation effects [3]. The magnitude of the nonlinearity, defined as

$$\gamma = \frac{d^2i/dv^2}{di/dv}$$

for the backward diode can be made to exceed that of the *pn*-junction or the Schottky diode ($\gamma = q/\eta kT$) (see Section 2.2.3).

Planar Doped Barrier Diodes. The planar doped barrier diode [5] is an interesting device, since it may have a symmetrical *IV* characteristic similar to what is obtained when two ordinary diodes are coupled in antiparallel (see Fig. 2.27). Such a diode would be ideal for subharmonically pumped mixers (Section 2.6.7). Both contacts of the diode are ohmic, and an n^+-i-p^+-i-n^+ doping profile with an extremely thin, fully depleted accepter layer (p^+) forms a triangular barrier. By tailoring the widths of the different doping layers, the *IV* characteristic may be shaped, and the symmetrical *IV* characteristic is obtained when the doping profile is symmetric (see Section 2.6.7 for further comments).

2.2.3 Schottky Barrier Diode

By far the most common mixer and detector element for microwave, millimeter waves, and submillimeter waves is the Schottky barrier diode (sometimes referred to as the hot carrier diode).

Characteristics of the Schottky Barrier Diode. The nonlinear characteristics of the Schottky barrier diode are the result of the properties of metal–semiconductor interfaces [3]. In fact, the detectors often used at the beginning of this century, consisting

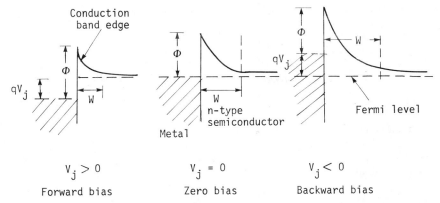

$$V_j > 0 \qquad\qquad V_j = 0 \qquad\qquad V_j < 0$$

Forward bias   Zero bias   Backward bias

Figure 2.3 Band diagram of the Schottky diode for three bias conditions. In this case the semiconductor is *n*-type, which means that forward-biased electrons will go from the semiconductor to the metal.

of a metal cat whisker contacting a silicon or a germanium crystal, was nothing but a (low-quality) Schottky diode. The difference between that diode and those we frequently use today is, of course, that modern diodes are of a much superior quality, since the properties of the metal–semiconductor interface are infintely better controlled. However, even today, the remaining problems in manufacturing high-quality diodes can be traced back to the problem of producing perfect metal–semiconductor interfaces [6–9].

The diode properties can be understood by using a band diagram model, as shown in Fig. 2.3, for a metal contact on an *n*-type semiconductor. At the metal–semiconductor surface, charges will form a dipole layer such that a potential drop ϕ, on the order of $\frac{1}{2}$ to 1 V, is created between the semiconductor and the metal. This voltage drop is called the barrier height. The exact value of the barrier height is determined by the detailed physics near (within a few tens of Angstroms) the interface, and varies depending on a number of things, such as the particular metal, the semiconductor material, the doping, the processing techique, and so on. Hence it is possible to get diodes with different barrier heights, which means that the *IV* as well as the *CV* characteristics will depend on the choice of diode.

In order to have charge neutrality, the semiconductor will be depleted of electrons (of holes in *p*-type material) up to a certain distance W from the interface. This is the reason for the parabolic form of the potential. Electrons with enough energy can pass from the semiconductor to the metal, and vice versa. For no bias, there is of course a balance (i.e., there are as many electrons passing either way, yielding a zero current). When the diode is biased, the balanced situation is upset. Referring to Fig. 2.3, it is obvious that for forward bias, the barrier is lower, as seen from the semiconductor side, letting more electrons go from the semiconductor to the metal. The current–voltage characteristic can be calculated theoretically using models that are appropriate for slightly different situations (see, e.g., Sze's book, Ref. 3). However, for all cases the same general form is obtained, that is,

$$I = I_0 \left[\exp\left(qV_j/\eta kT\right) - 1 \right] \tag{2.3}$$

where η is the ideality factor, which for a good diode at room temperature is close to 1.

The very existence of the depleted region implies that there will be a capacitance inversely proportional to the width W of the depletion region. This capacitance depends in a similar way on the bias voltage (V_j) as for the *pn*-diode [3]:

$$C = \frac{C_0}{\sqrt{1 - \dfrac{V_j}{(\phi - V_n - kT/q)}}} \qquad \left(V_j < \phi - V_n - \frac{kT}{q}\right) \tag{2.4}$$

where C_0 is the zero-bias capacitance, ϕ the barrier height, and V_n the Fermi voltage.

Equivalent Circuit of the Schottky Diode. The junction properties of a Schottky diode (and a *pn*-diode) can be modeled with a nonlinear resistance in parallel with the junction capacitance. In low-level detectors and mixers, the junction small-signal differential resistance is simply

$$r_j = \frac{\delta V_j}{\delta I} = \frac{k\eta T}{qI} \tag{2.5}$$

In practice, the diode must be modeled with an equivalent circuit comprising not only the junction itself, but also the spreading resistance from the junction area to the ohmic contact(s), and a contact resistance due to the ohmic contact itself. Figure 2.4 shows a schematic diagram of a GaAs Schottky diode chip and the corresponding equivalent circuit. Notice that the RF current is flowing around the surface of the diode. Hence the RF series resistance is slightly larger than the dc resistance. Also notice that the contribution to the series resistance from the undepleted part of the epitaxial layer is actually bias dependent, a fact that is usually neglected in device models used for analyzing detector and mixer properties. At submillimeter waves, further modifications in the device model will become necessary (see Section 2.10).

The epitaxial layer doping is chosen for optimum performance, that is, the doping must neither be too low, since then the series resistance will become high due to low conductivity in the epitaxial layer, nor must it be too high, which will cause tunneling to dominate over thermionic emission [6], making the diode noisy (high η) and the capacitance large. The doping of the substrate should be as high as possible in order to maximize the conductivity and consequently minimize the substrate contribution to the series resistance.

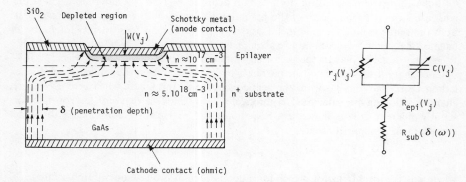

Figure 2.4 Schematic diagram of a GaAs Schottky barrier diode and its equivalent circuit. Notice that the Schottky metal is in the form of a bathtub, which will improve its performance [7, 8].

The Mott diode has an epitaxial layer thickness such that at zero bias the depleted region will just reach substrate. In this way, the contribution to the series resistance from the less conductive epilayer is minimized, and the capacitance of the diode will not vary as much for the backward-biased diode. The smaller capacitance variation will cause less parametric effects in mixer applications, a fact that may be favorable in designing low-noise mixers.

A common measure of the high-frequency response of a particular diode is the cutoff frequency, which is defined for the zero-biased diode and determined from the constraint that the applied RF voltage is divided equally over the series resistance and the junction capacitance. Hence the diode cutoff frequency is simply

$$f_c = \frac{1}{2\pi R_s C_0} \tag{2.6}$$

where R_s usually is defined at dc. It is evident that the Mott diode has an improved cutoff frequency. The cutoff frequency of GaAs diodes is in general higher than for Si diodes, due to the higher mobility of GaAs, yielding a lower R_s.

Noise in Schottky Diodes. There are several types of noise sources identified in semiconductor diodes. The most important ones are mentioned below [8–13].

Shot Noise. The shot noise is caused by the fact that the current flowing through the diode is due to the transport of individual electrons with a finite charge. The fact that their arrival at the anode can be described statistically by the Poisson distribution means that the root-mean-square (rms) fluctuation in the current δI^2 is proportional to the current:

$$\delta I^2 = 2Iq \, \Delta f \tag{2.7}$$

where Δf is an infinitesimally small frequency interval. The noise power from the diode can be calculated using Eqs. (2.5) and (2.7) as

$$P_n = \frac{\delta I^2}{4(\delta I/\delta v)} = \frac{1}{2} \eta k T \, \Delta f \tag{2.8}$$

By identifying this expression with the ordinary expression for noise power (i.e., $P_n = kT \, \Delta f$), it is seen that the equivalent noise temperature due to shot noise is

$$T_{sh} = \tfrac{1}{2}\eta T \tag{2.9}$$

1/f Noise. There usually is a noise component proportional to $1/f$, which is significant at quite low frequencies, and therefore is an important factor when detector noise is considered. For mixer applications this noise is usually less important, since the intermediate frequency is high enough that the $1/f$ noise is negligible. There are several physical phenomena that can cause the $1/f$ noise. It is most often related to something which is not perfect in the junction area [8–10]. For instance, traps at the metal–semiconductor interface, possibly located in a thin oxide, will cause $1/f$ noise. At high forward-biased currents, traps in the undepleted part of the epilayer may also cause $1/f$ noise. In this case, however, the time constants involved are sufficiently short that this noise may cause problems at frequencies of several gigahertz.

Thermal Noise. Since the series resistance is due substantially to the finite resistivity of the semiconductor substrate, it will essentially contribute thermal noise. However, the ohmic contacts may also contribute some noise related to shot and $1/f$ noise.

Hot Electron Noise and Intervalley Scattering Noise. With strong forward bias, the electrons may gain enough energy to become significantly more energetic than at thermal equilibrium. Hence the electron temperature, which of course is intimately related to the experimentally measured noise temperature, will exceed the physical temperature of the device [10–12]. This noise is referred to as hot electron noise.

In GaAs devices, energetic electrons may enter the satellite valley [3]. Since the mobility of electrons in the satellite valley is much lower than the mobility of electrons in the main valley, the transfer of electrons from the main valley into the satellite valley will cause fluctuations in the current. This noise is called intervalley scattering noise [12].

Total Diode Noise. Using the equivalent circuit of Fig. 2.4, the noise temperature of the diode (at low frequencies, where the influence of the diode capacitance can be neglected) can be expressed as follows:

$$T_{diode} \simeq \frac{r_j T_{sh} + R_{epi} T_{epi} + R_{sub} T_{sub}}{r_j + R_{epi} + R_{sub}} \tag{2.10}$$

At very high frequencies, the influence of the diode capacitance has to be taken into account [13] [i.e., Eq. (2.10) will be modified].

Cooling Schottky Barrier Diodes for Low-Noise Operation. According to the preceding section, both shot noise and thermal noise should be directly proportional to temperature. However, shot noise will, in practice, decrease with temperature, as indicated in Fig. 2.5. The reason the shot noise will not be reduced below a certain value is that at low temperatures, the tunneling of electrons through the barrier will become more important than thermionic emission. In simple words, the reason is that the tunneling current can be expressed using a formula which is essentially the same as for thermionic

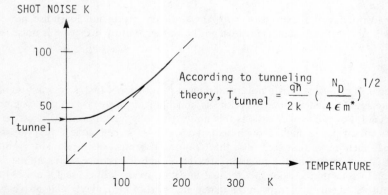

Figure 2.5 Shot noise temperature versus physical temperature for a typical GaAs diode with n-type doping of about 3×10^{16} cm^{-3}.

emission, except that temperature is exchanged for an equivalent temperature Θ, proportional to the square root of the doping concentration (see Fig. 2.5). Hence, operating the diode at 15 K, which is a typical temperature obtained in commercial closed-cycle cooling machines, will substantially decrease the diode noise. Another effect of the cooling is that the $\log[i]-v$ characteristic will become steeper, since the exponent in Eq. (2.3) includes $\eta T = \Theta$. This will improve the device properties when used in detector applications, but will not improve as much in mixer applications. Notice that for cooling only GaAs will do, since carrier freeze-out will become a serious problem in low-doped Si at low temperatures. For further information, see Section 2.9.3, and Refs. 6, 7, and 9.

Types of Diode Construction and Packaging. Many methods of constructing diodes and diode packages exist. The reader should consult the various manufacturers' catalogs for details. Here will be presented a very brief description of a few designs related to Schottky barrier detector and mixer diodes.

In Fig. 2.6 are shown schematics of different constructions. Naked diode chips are available for bonding into hybrid circuits, or whisker contacting in waveguide circuits (Fig. 2.6a and d). Figure 2.6b shows a typical outline of a beam lead diode. Also, pairs and quads of beam lead diodes are available (Fig. 2.6c) for use in balanced, double

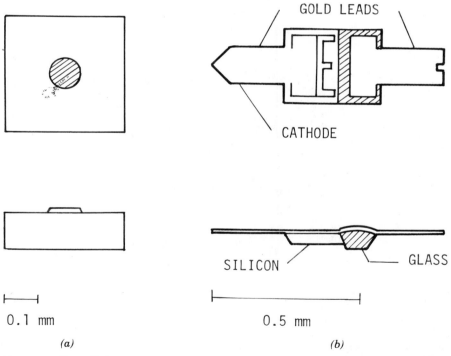

GOLD LEADS

CATHODE

SILICON   GLASS

0.1 mm

0.5 mm

(a)

(b)

Figure 2.6 Some diode constructions: (a) diode chip with one bonding pad for hybrid integrated circuits; (b) IC beam lead detector/mixer diode; (c) IC beam lead diode quad for balanced and double balanced mixers; (d) electron micrograph of a diode chip ($0.2 \times 0.2 \times 0.1$ mm) with 2-μm-diameter diodes to be whisker contacted for millimeter wave detector/mixer applications (see Fig. 2.37). (Courtesy of Farran Technology Ltd., Cork, Ireland.)

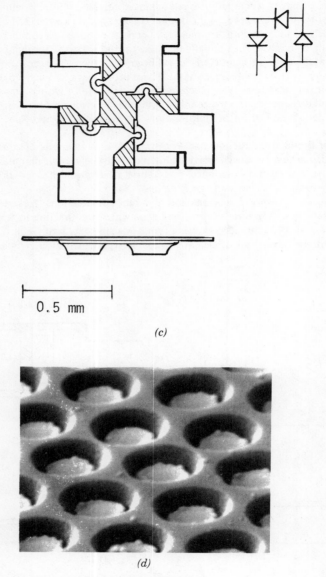

(c)

0.5 mm

(d)

Figure 2.6   (*Continued*)

balanced, and double-double balanced mixers (see Section 2.6). Each type of construction has advantages and disadvantages, depending on the application. Practical considerations are, of course, important. The performance is affected by the parasitics (parasitic capacitance, lead inductance, series resistance), which together with zero-bias capacitance, cutoff frequency, and other relevant data, are usually well described and quantified in the manufacturer's catalog. Notice that the higher mobility of GaAs makes GaAs diodes superior to Si diodes, in particular for millimeter wave applications.

TABLE 2.1   Commercially Available Monolithic Beam Lead Diode Devices

Device	Electrical Circuit	Features
Single		Ideal for use as mixers or detectors on MICs.
Series pair		Ideal for use anywhere that a closely matched pair of diodes is required.
Reverse series pair		Same as above except for polarity.
Common-cathode pair		Ideal for signal comparison detectors.
Antiparallel pair		Ideal for subharmonically pumped mixers.
Split pair		Ideal for temperature-compensated detector use.
Four-junction pair		Ideal for high-level up-converters.
Star quad		Ideal for use in star mixer circuits that do not require an IF balun.
Quad bridge		Ideal for use in termination-insensitive mixers or biased mixers.
Quad ring		Ideal for use in double balanced mixers.
Eight-junction ring		Ideal for use in double balanced mixers requiring higher compression point and/or better IM performance.
Twelve-junction ring		Ideal for use in double balanced mixers where highest compression point and/or best IM performance is required.

Source: Courtesy of Alpha Industries, Inc.

In Table 2.1 are shown different diode configurations available as monolithic beam lead devices and their respective applications. Bonding of beam lead diodes is an important art, and an overview of recommendations is given in Table 2.2. For further information, consult Ref. 14 and application notes from the various manufacturers (e.g., M/A-COM, Hewlett-Packard). *One should always follow the manufacturer's recommendations.*

Various encapsulations of diodes are available, both for single- and multiple-diode configurations (again see manufacturers' catalogs). An example of an encapsulated diode

TABLE 2.2   Some Methods for Bonding Diode Chips

Beam lead diodes
   Conducting epoxy or polyimid
   Parallel gap welding
   Thermocompression bonding
   Ultrasonic bonding
   Wobbel bonding

Chip diodes
   For chip die-down bonding techniques use, e.g.,
      Conducting epoxy or polyimid
      Eutectic or soft soldering
   For lead bonding use, e.g.,
      Ball (wire) or wedge (wire and ribbon) bonding of either thermocompression or ultrasonic type

Always follow the recommendations from the manufacturer!

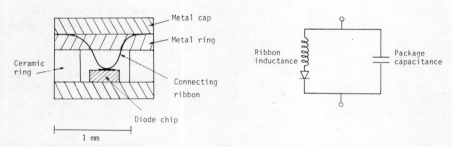

Figure 2.7   Typical diode encapsulation, ceramic pill type.

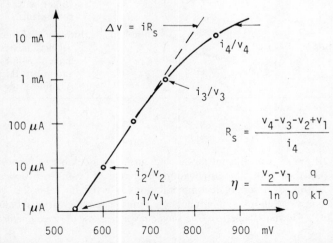

$$\Delta v = iR_s$$

$$R_s = \frac{v_4 - v_3 - v_2 + v_1}{i_4}$$

$$\eta = \frac{v_2 - v_1}{\ln 10} \frac{q}{kT_0}$$

Figure 2.8   Determination of the ideality factor η and the series resistance R_s from the measured $\log [I]$–V characteristic.

is shown in Fig. 2.7. Important parasitics are the capsule capacitance and the lead inductances, quantified by the manufacturer.

Characterization of Diodes. By "characterization" we mean the determination of the equivalent circuit of the diode (see Fig. 2.4). As was mentioned in Section 2.2.3, the series resistance varies with the bias, and is also frequency dependent. Assuming these factors can be neglected, one assumes the series resistance to be a constant circuit parameter, and the series resistance can be determined from the IV characteristic. Plotting $\log[I]$ versus voltage yields a curve as shown in Fig. 2.8. The series resistance R_s can now be calculated from the excess voltage drop at high forward currents. There are alternative techniques for determining R_s, particularly for small-area millimeter wave diodes, described in Ref. 13.

The capacitance can be determined by a suitable commercial capacitance meter, capable of accurately measuring the capacitance down to a fraction of a femtofarad if necessary. If Fig. 2.9 is shown the capacitance versus bias voltage for an ordinary Schottky diode and a Mott diode.

2.2.4 Thermal Detectors

Calorimeters. It is possible to measure power by letting the radiation be absorbed in any type of matched load and just measuring the temperature rise (e.g., with a thermal pile) of the load. Since the power absorption can be achieved quite independently of the type of temperature-measuring device, a calorimeter can be made extremely broadband.

A major problem in constructing calorimeter detectors is to avoid spurious response due to temperature drift. The temperature rise may, for example, be measured with a thermocouple, in which case the temperature reference point should be located so that the influence of ambient-temperature fluctuations will become minimal. The problem can also be tackled by using a balanced configuration, that is, measuring the response of the power-absorbing detector with respect to a calibrating twin element [15]. The amount of temperature rise, as in the bolometer case, will depend on the thermal resistance from the load to the surroundings, and the time constant will be determined by the product of the thermal mass of the load and the thermal resistance [16].

Calibration of the power meter can be accomplished by including a separate dc heating element as close to the power-absorbing load as possible [15].

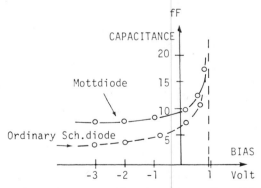

Figure 2.9 Experimental capacitance versus bias voltage for an ordinary Schottky diode [compare, e.g., Eq. (2.4)] and a Mott diode.

Bolometers. A bolometer is a type of detector that is constructed from a material whose resistance varies rapidly with temperature [16]. When radiation is absorbed, the temperature of the detector rises, causing a measurable change in the resistance. For example, a thin wire of platinum in a waveguide can be mounted to absorb the microwave power, and the wire resistance will then increase due to the temperature rise. Since the temperature rise will become larger if the thermal resistance to the surroundings is made large, it will help if the bolometer is mounted in an evacuated chamber. Doing so will cause the time constant of the bolometer to increase.

A feature of practical importance is that it is not very difficult to match the bolometer detector over a broad band. In fact, power meters based on bolometer detectors may cover a frequency band much broader than a waveguide band.

Certain types of detectors will become extremely sensitive when cooled to very low temperatures. In particular, bolometer-type cryogenically cooled detectors are common and quite important for submillimeter and infrared waves. The sensitivity is an effect not only of less temperature fluctuations and lower thermal noise, but also of access to materials with very large temperature coefficients. In particular, the resistivities of semiconductors have an extremely large temperature dependence. Semiconducting materials such as germanium or silicon with various dopants have been used as temperature-sensitive absorbers. The sensitivity of this type of detector is expressed in terms of noise equivalent power (NEP) $W/\sqrt{Hz}$, which is a common way of indicating the sensitivity of infrared detectors. The sensitivity expressed in NEP of the semiconductor bolometers can reach 6×10^{-16} $W/\sqrt{Hz}$ [17] when cooled to 0.3 K.

Golay Cell: A Pneumatic Detector. The Golay cell is a thermal type of detector [16, 17], which senses the pressure change in a gas cell heated by the incident radiation. The detector is used in particular at submillimeter wavelengths. However, the lower cutoff frequency of the detector, determined by the window opening, is typically as low as 60 GHz. The achievable sensitivity of the Golay cell at room temperature is within a factor of 10 of the theoretical limit, which is about 10^{-10} $W/\sqrt{Hz}$ [17].

Despite a slow response and severe microphony, it is a common and commercially available device for laboratory use. The problem with the microphonics is to some extent overcome by chopping the signal at the input of the detector and using a narrow-band amplifier followed by a phase detector after the detector.

Pyroelectric Detector. The pyroelectric detector [17, 18] is essentially an IR thermal detector, but can be used at millimeter and submillimeter frequencies as well. The pyroelectric effect exists in crystals exhibiting ferroelectricity (i.e., crystals showing a spontaneous electric polarization). Since this polarization is temperature dependent, the surface charge will change when the crystal is heated by absorbed radiation. The spectral range is limited in practice by the choice of material, absorber, and window size. The maximum sensitivity is typically 5×10^{-10} $W/\sqrt{Hz}$ [18].

2.2.5   Further Detector and Mixer Devices

In this section we mention briefly other, less common mixer and detector devices.

Hot Electron Detectors. Ordinary photoconductive materials can be used only for detecting radiation in the infrared; they cannot be used beyond a wavelength of about 250 μm. For submillimeter and millimeter wavelengths, a special type of photoconductivity can be used, based on the energy-dependent mobility of conduction band electrons. This "hot electron" bolometer detector also has to be cooled to liquid helium temperatures [17, 19]. The hot electron detector can reach a sensitivity of 6×10^{-13} $W/\sqrt{Hz}$ [17].

The hot electron detector can be used in mixer applications, as discussed in Section 2.13.4.

Superconducting Detectors. The most sensitive mixer systems that have been built up to now are based on superconducting elements in which two superconductors are separated with a very thin oxide (typically, 20 Å). Two types of tunneling phenomena can occur in these elements: supersonducting electrons forming Cooper pairs can tunnel through the junction, a phenomenon called the Josephson effect; and Cooper pairs can break up and tunnel as single electrons (quasi-particles), a phenomenon called quasi-particle tunneling. Both phenomena have been demostrated to be useful in detecting devices as well as in mixer devices. Superconducting tunnel elements are discussed further in Section 2.11.

2.3   OPTIMIZATION OF DIODE DETECTORS

2.3.1   Introduction

Diode detectors are rectifiers used to convert RF signals, usually modulated and of low amplitude, to (modulated) dc. Diode detectors are the most common type of detector in microwave and millimeter wave systems. Although both *pn* and Schottky diodes can be used, the latter type is much more common than the former. We discuss below how to optimize diode detectors for maximum sensitivity. For further information it is recommended that the reader consult application notes from detector diode manufacturers (e.g., Alpha, Hewlett-Packard, M/A-COM).

2.3.2   Theory of Low-Level Detection

At low levels, diode detectors act as square-law detectors, that is, the output voltage (current) is proportional to the RF input power (square of the input RF voltage). At higher signal levels, the detector will become linear, and at still higher levels the detector saturates (see Fig. 2.10).

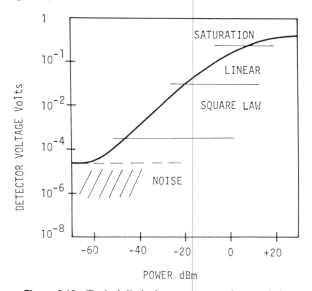

Figure 2.10   Typical diode detector output characteristic.

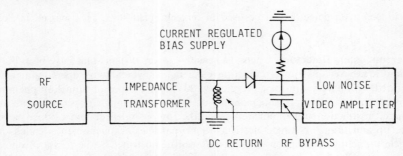

Figure 2.11 Typical detector circuit.

In Fig. 2.11 a simple detector circuit is shown. Since the bias circuit is designed to set the current through the diode equal to I_{bias} for no input power, the signal will cause the bias voltage to decrease. Let us assume that the junction voltage is V_{00} when no signal is present, and decreases by δv_0 when an RF voltage $\delta v_s \cos(\omega_s t)$ is applied. Equation 2.3 then yields ($V_0 = k\eta T/q$):

$$i = I_{\text{bias}} + \delta i = \left\langle \left(i_0 \left\{ \exp\left[\frac{V_{00} + \delta v_0 + \delta v_s \cos(\omega_s t)}{V_0} \right] - 1 \right\} \right) \right\rangle$$

$$= \left\langle \left\{ (I_{\text{bias}} + i_0) \left[1 + \frac{\delta v_0}{V_0} + \frac{\delta v_s}{V_0} \cos(\omega_s t) \right. \right. \right.$$

$$\left. \left. \left. + \frac{1}{2}\left(\frac{\delta v_s}{V_0}\right)^2 \cos^2(\omega_s t) + \cdots \right] - i_0 \right\} \right\rangle$$

$$\simeq (I_{\text{bias}} + i_0)\left[1 + \frac{\delta v_0}{V_0} + \frac{1}{4}\left(\frac{\delta v_s}{V_0}\right)^2 \right] - i_0 \qquad (2.11)$$

Hence, since the diode differential resistance is $r_j \simeq V_0/I_{\text{bias}}$ (the series resistance is neglected), we have

$$\delta i = (r_j)^{-1}\left[\delta v_0 + \frac{1}{4}\frac{(\delta v_s)^2}{V_0} \right] \qquad (2.12)$$

Consulting Fig. 2.12a and considering Eq. (2.12), it is obvious that the diode can be considered a voltage source with an internal resistance of r_j and a voltage amplitude of $\delta v_s^2/4V_0$.

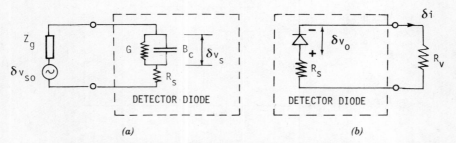

(a)   *(b)*

Figure 2.12 Detector circuit representation: *(a)* RF; *(b)* dc.

Assuming that the video amplifier input resistance R_v is infinite, the dc current remains unaltered ($\delta i = 0$) and we have

$$\delta v_0 = -\frac{1}{4}\frac{\delta v_s^2}{V_0} \tag{2.13}$$

Since δv_s^2 is proportional to the available input power times the real part of the input impedance ($\mathrm{Re}\{Z_g\}$), δv_s is maximized by choosing $\mathrm{Re}\{Z_g\}$ as high as possible, and the diode impedance (determined by the bias voltage) to be of the same order as the input impedance. A more detailed analysis, taking into account the diode capacitance and series resistance, can be done referring to Fig. 2.12. The generator is assumed to be matched to the diode, that is, the generator impedance Z_g is a complex conjugate of the diode impedance Z_d. The available signal power then can be expressed as

$$P_s = \frac{1}{4}\frac{\delta v_{s0}^2}{\mathrm{Re}\{Z_d\}} = \frac{1}{4}\delta v_{s0}^2\left(R_s + \frac{G}{G^2 + B_c^2}\right)^{-1} \tag{2.14}$$

that is,

$$\delta v_{s0}^2 = 4\left(R_s + \frac{G}{G^2 + B_c^2}\right)P_s \tag{2.15}$$

where, according to Eq. (2.4),

$$B_c = \omega C_d \simeq \frac{\omega C_0}{\sqrt{1 - V_{00}/\phi}} \tag{2.16}$$

δv_{s0}^2 is actually a function of the matching circuit and increases with $\mathrm{Re}\{Z_d\}$, and is made maximum by the proper choice of G [Eq. (2.14)]. From Fig. 2.12 we may calculate the video response as

$$\delta v_v = \frac{R_v}{r_j + R_v + R_s}\frac{\delta v_s^2}{4V_0} \tag{2.17}$$

From Fig. 2.12a, δv_s can be related to δv_{s0}, and by using Eq. (2.15), we obtain the voltage sensitivity as

$$S_v = \frac{\delta v_v}{P_s} = \frac{1}{4V_0}\frac{P_s}{G + R_s(G^2 + B_c^2)}\frac{R_v}{r_j + R_v + R_s} \quad \mu V/\mu W \tag{2.18}$$

where $G = r_j^{-1}$. One sees that it will help in reducing R_s and B_c, and by making $r_j = G^{-1}$ large and $R_v \gg r_j$. Hence for the matched case (RF), assuming that $R_v \gg r_j$ and neglecting R_s, one obtains the voltage sensitivity as

$$S_{v,\mathrm{matched}} = \frac{\delta v_v}{P_s} = \frac{r_j}{4V_0} \quad \mu V/\mu W \tag{2.19}$$

If the detector is specified for a current of 50 μA, corresponding to $G^{-1} = r_j = \eta kT/qI \simeq 560\ \Omega \gg R_s$, the voltage sensitivity for a matched detector at room temperature should be about $5 \times 10^3\ \mu V/\mu W$. In practice it may not be possible to reach this number. Degradation from this number by a few decibels should be reasonable for a practical detector.

2.3.3   Zero-Bias Detectors

For the zero-bias detector, the bias current $I_{\text{bias}} = 0$ and the diode output resistance $r_j = V_0/I_0$ (or $r_j = V_0/(I_0 + \delta i_{\text{rect}})$ if the rectified current δi_{rect} is of the same order as the saturation current I_0).

The video signal voltage is

$$\delta v_0 = \frac{1}{4} \frac{\delta v_s^2}{V_0} \frac{R_v}{R_v + r_j} \tag{2.20}$$

Maximum sensitivity is obtained if the video amplifier input resistance $R_v \gg r_j$. However, for a typical Schottky barrier diode, r_j is of the order 30 MΩ or larger [yielding a corresponding higher sensitivity according to Eq. (2.19)]. Such diodes do require a bias (e.g., $\simeq 20\ \mu$A) for achieving a fair sensitivity with video amplifiers having a reasonable input resistance. Low barrier diodes, however, are available and have a much more reasonable zero-bias resistance, typically 2 kΩ.

In diode catalogs, the bias current used for tests of the tangential sensitivity and voltage sensitivity should be given. Notice that these catalog data are given for diodes that are matched at the RF input. Since the diode RF impedance as well as the video impedance is high, this means that in practice it may be difficult to achieve high sensitivity over a large bandwidth.

Temperature effects may be important in some applications. As discussed in Section 2.2.3, both I_0 and V_0 are temperature-dependent parameters.

2.3.4   Tangential Sensitivity and Diode Figure of Merit

A term often used to define the sensitivity of a video detector is the tangential sensitivity, which is the input power in dBm required to change the dc voltage output with an amount equal to the voltage noise fluctuations. This is usually made using an oscilloscope, as shown in Fig. 2.13.

The noise fluctuations are caused by the detector diode itself and by the video detector used to amplify the output of the detector. The noise background is, in practice, negligible. The open-circuit noise of the diode is due to three sources: the thermal noise of the series resistance, the shot noise from the junction, and "low-frequency" $1/f$ noise. Forgetting for the moment the latter noise contribution, we can express the rms noise

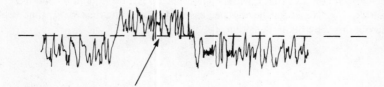

"Tangential" to noise peaks

Figure 2.13   Measurement of the tangential sensitivity.

voltage δv_n^2 at the video amplifyer input as

$$\delta v_n^2 = 4kTBR_s + 2kTBr_j + 4kTBR_a \tag{2.21}$$

where r_j is the junction differential resistance, R_a a fictitious noise resistance of the video amplifier, and B the video amplifier bandwidth. The standard number for R_a is 1200 Ω, and B is typically 10 MHz. The approximate peak-to-peak noise fluctuations are $2.8\delta v_n$. Hence the power required to change the output voltage by this amount is

$$P_{TS} = \frac{2.8\delta v_n}{S_v} \tag{2.22}$$

Using Eqs. (2.19) and (2.21), and assuming the diode bias to be 50 μA and at room temperature, a tangential sensitivity of -55 dBm is obtained for the matched detector case. Typical values seen in catalogs are 50 to 56 dBm, depending on the particular diode and bias point. Notice that matched zero-bias diodes will get a larger tangential sensitivity than was just calculated due to the larger junction resistance.

2.3.5   High-level Diode Detectors

For high input powers, the detector no longer responds as a square-law detector (Fig. 2.10). The output voltage will, rather, increase in proportion to the square root of the input power. However, this is approximate, since in this power regime the impedance of the diode will vary with the power level, making an accurate power measurement quite complicated. See also Section 2.5, where the problem of theoretically calculating the diode response of high power is dealt with in some detail.

2.4   MIXERS: SIMPLE THEORY AND BASIC DEFINITIONS

2.4.1   Introduction

In this section we discuss fundamental properties of mixers using simplified mixer models, concentrating on single-ended mixers. For further reading, consult Ref. 20, which also deals with multiple-diode mixers.

In a mixer, a usually quite weak signal, $\delta v_s \cos(\omega_s t)$, is "mixed" in a nonlinear element, such as a diode, with a strong signal $V_{LO} \cos(\omega_{LO} t)$, called the local oscillator. The resulting current will contain Fourier components of frequencies $n\omega_{LO} \pm m\omega_s$. However, if $\delta v_s \ll V_{LO}$, only the frequency components $n\omega_{LO}$ (of comparatively large amplitude) and $n\omega_{LO} \pm \omega_s$ (of small amplitude, proportional to δv_s) will be of any significance. Defining the intermediate frequency ω_{IF} as

$$\omega_{IF} = |\omega_{LO} - \omega_s| \tag{2.23}$$

(where ω_s can be either larger or smaller than ω_{LO}), the frequency components related to the signal voltage can be expressed as

$$n\omega_{LO} \pm \omega_s \equiv n'\omega_{LO} \pm \omega_{IF} \tag{2.24}$$

where $n = 1, 2, 3, \ldots, n' = 0, 1, 2, \ldots$. The frequencies $n'\omega_{LO} \pm \omega_{IF}$ are referred to as the harmonic sidebands.

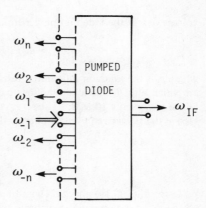

Figure 2.14 Multiport representation of the pumped mixer diode. The notations ω_{-n} and ω_n indicate the harmonic sidebands [Eq. (2.24)].

Hence the signal voltage will cause current to flow at all frequencies defined by Eq. (2.24), suggesting that signal power can be transferred to these frequencies. An equivalent multiport circuit of the pumped mixer diode based on these facts is shown in Fig. 2.14.

In the most common mixing mode, the signal frequency is close to the local oscillator frequency (i.e., $\omega_{IF} \ll \omega_{LO}$, and $\omega_s \approx \omega_{LO}$). Notice that we may have either $\omega_s < \omega_{LO}$ or $\omega_s > \omega_{LO}$ (i.e., the signal may be in either the lower sideband or the upper sideband). The signal will cause currents and/or voltages at all harmonic sidebands, and power will be transferred if the terminating loads have a resistive part.

In the harmonic mixer mode we let the IF frequency be

$$\omega_{IF} = |n\omega_{LO} - \omega_s| \tag{2.25}$$

In Fig. 2.15 the case $n = 2$ is illustrated. As for $n = 1$ (fundamental mixer), power is transferred to all harmonic sidebands with (partly) resistive terminations.

It is evident that to optimize mixer performance, it is necessary to design the mixer so that a minimum of signal power is lost to the harmonic sidebands. With a proper choice of harmonic sideband impedances, the mixer performance can in fact be considerably enhanced (see Section 2.4.5).

2.4.2 Conversion Loss

From the discussion so far, it is obvious that when analyzing mixers one has to consider the response at the harmonic sidebands. In this section we discuss the conversion efficiency of a mixer, which is usually expressed in terms of the conversion loss L, defined

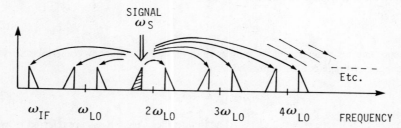

Figure 2.15 Signal transfer in a second-harmonic mixer.

as

$$L = \frac{P_s}{P_{IF}} = \frac{\text{signal power available at the input terminal(s)}}{\text{IF power delivered to the output load}} \tag{2.26}$$

In most mixers the conversion loss is less than 1. There are a few exceptions. One is when a nonlinear reactance (such as a diode which is backward biased so that the resistance is much larger than the capacitive reactance, Fig. 2.3) is used as a mixing element, and parametric effects may cause gain. Another case is when quantum effects become important, such as in superconducting quasiparticle mixers (see Section 2.11).

The conversion loss may be expressed as a product of different loss contributions, that is,

$$L = L_0 \cdot L_h \cdot L_{R_s} \cdot L_{|\Gamma_{in}|^2} \cdot L_{|\Gamma_{IF}|^2} \tag{2.27}$$

where L_0 = losses due to absorption in the (barrier) nonlinear resistance
$\quad L_h$ = losses due to power lost to the harmonic sidebands
$\quad L_{R_s}$ = losses due to absorption in the series resistance
$\quad L_{|\Gamma_{in}|^2}$ = losses due to reflection at the mixer signal
$\qquad$ input port, $L_{|\Gamma_{in}|^2} = 1/(1 - |\Gamma_{in}|^2)$
$\quad L_{|\Gamma_{IF}|^2}$ = losses due to reflection losses at the mixer
$\qquad$ output port $L_{|\Gamma_{IF}|^2} = 1/(1 - |\Gamma_{IF}|^2)$

Since the embedding impedances at the various sidebands influence all contributions listed above, the various contributions are not independent of each other. It has been shown that although a conjugate match at the IF port results in maximum power transfer, the signal input port should not, in general, be conjugate matched for minimum conversion loss. We will return to these questions in Section 2.5, where a more general and exact theory is briefly described.

Contribution from L_0. This contribution can be illustrated by evaluating the mixing properties of an idealized mixer, consisting of a purely resistive diode with an exponential relationship between the current and the voltage. We assume that the pumping voltage is purely sinusoidal (which is certainly never the case; see Section 2.5), that is,

$$I = I_0 \left(\exp \left\{ \frac{V_{LO} \cos(\omega_{LO} t)}{V_0} \right\} - 1 \right) \tag{2.28}$$

where

$$V_0 = \frac{k\eta T}{q} \qquad (\simeq 28 \text{ mV for room temperature}) \tag{2.29}$$

The next step is to calculate the small-signal current components caused by a signal, $v_s \cos(\omega_s t)$. To do this, we first derive the differential conductance of the pumped diode:

$$g(t) = \frac{dI}{dV} = \frac{I}{V_0} = \frac{I_0}{V_0} \exp \left(\frac{V_{DC}}{V_0} \right) \left[I_0 \left(\frac{V_{LO}}{V_0} \right) + 2 I_1 \left(\frac{V_{LO}}{V_0} \right) \cos(\omega_{LO} t) \right.$$

$$\left. + 2 I_2 \left(\frac{V_{LO}}{V_0} \right) \cos(2\omega_{LO} t) + 2 I_3 \left(\frac{V_{LO}}{V_0} \right) \cos(3\omega_{LO} t) + \cdots \right.$$

$$\equiv g_0 + \sum_1^\infty \cos(n\omega_{LO} t) \tag{2.30}$$

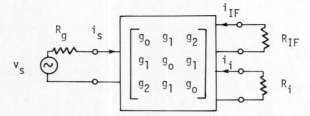

Figure 2.16 Mixer circuit analyzed in the text.

where the $I_n(V_{LO}/V_0)$ are modified Bessel functions of order n and with the argument V_{LO}/V_0.

When the signal voltage frequency components (assumed small) interact with the pumped diode, the current can be calculated as

$$i(t) = v(t)g(t) \tag{2.31}$$

where, as will be shown, $v(t) \neq v_s \cos(\omega_s t)$.

Assume that one may neglect frequency components other than ω_{IF}, ω_s, $2\omega_{LO} - \omega_s = \omega_i$ (ω_s and ω_i correspond to the two sidebands; ω_i is called the image frequency). Referring to Fig. 2.16 and assuming $R_i = R_g$, the current $i(t)$ through the diode and the voltage $v(t)$ over the diode are

$$i(t) = i_s \cos(\omega_s t) + i_i \cos(\omega_i t) + i_{if} \cos(\omega_{IF} t) \tag{2.32}$$

$$v(t) = v_s \cos(\omega_s t) - i_s R_g \cos(\omega_s t) - i_i R_g \cos(\omega_i t)$$
$$- i_{IF} R_{IF} \cos(\omega_{IF} t) \tag{2.33}$$

Hence, with Eq. (2.30), we get

$$\begin{bmatrix} i_s \\ i_{IF} \\ i_i \end{bmatrix} = \begin{bmatrix} g_0 & g_1 & g_2 \\ g_1 & g_0 & g_1 \\ g_2 & g_1 & g_0 \end{bmatrix} \begin{bmatrix} v_s - i_s R_g \\ -i_{IF} R_{IF} \\ -i_i R_g \end{bmatrix} \tag{2.34}$$

where R_g is the load impedance at the signal or the image frequencies (here assumed to be equal), and R_{IF} is the load impedance at the output of the mixer.

From this expression one can calculate the conversion loss as defined in Eq. (2.26), and one obtains

$$L = \frac{R_g(g_0 + g_2 + 1/R_g)[g_0(g_0 + g_2 + 1/R_g) - 2g_1^2]}{g_1^2} \tag{2.35}$$

where R_g is the generator impedance seen by the diode at the signal input port. By differentiating this expression with respect to R_g, the minimum conversion loss is obtained as

$$L_{min} = 2 \frac{1 + \sqrt{1 - \beta}}{1 - \sqrt{1 - \beta}} \tag{2.36}$$

TABLE 2.3 Approximate Formulas Describing Properties of the Y-Mixer[a]

	Short-Circuited Image	Broadband[b]	Open-Circuited Image
L_0 (SSB)	$1 + 2\left(\dfrac{V_0}{V_{LO}}\right)^{1/2}$	$2\left(1 + \dfrac{\sqrt{2}V_0}{V_{LO}}\right)$	$1 + \left(\dfrac{2V_0}{V_{LO}}\right)^{1/2}$
R_{input}	$\dfrac{V_0}{I_{DC}}\left(\dfrac{V_{LO}}{V_0}\right)^{1/2}$	$\dfrac{V_0}{I_{DC}}\dfrac{V_{LO}}{V_0\sqrt{2}}$	$\dfrac{V_0}{I_{DC}}\left(\dfrac{V_{LO}}{2V_0}\right)^{3/2}$
R_{IF}	$= R_{input}$	$= 2R_{input}$	$= 4R_{input}$
Pump power $\simeq I_{dc}V_{LO}$	LO impedance $\simeq \dfrac{V_{LO}}{2I_{dc}}$		

[a] The LO *voltage* waveform is assumed to be perfectly sinusoidal.
[b] Example, broadband case: $P_{LO} = 1$ mW; $I_{dc} = 3$ mA yields $V_{LO}/V_0 = 12$, $R_{input} = 80\ \Omega$, $R_{IF} = 160\ \Omega$, $L_0 = 2.3$ (3.5 dB).

where the pump parameter β can be evaluated from Eqs. (2.30)–(2.35). When the pump power increases, β also increases, approaching 1 when the pump amplitude goes toward infinity. Hence the conversion loss then approaches $L = 2$ (3 dB). The lost half of the signal power is dissipated in the image load. No power at all is dissipated in the diode, which in this case can be considered equal to a perfect switch, operated by the LO between the on and off positions.

A similar exercise can be followed for other cases of interest, such as when the image termination is reactively terminated (see Section 2.4.5 concerning image rejection and image enhancement), short circuited or open circuited, or for a harmonic mixer. Using approximate expressions for the modified Bessel equations, Saleh [20] has evaluated properties for these mixers (also considering the case when the diode is pumped with a sinusoidal voltage). The results, summarized in Table 2.3, are not at all exact but can be used for order-of-magnitude estimations, indicating relative impedance levels, comparing the various cases of image termination, and gaining an understanding of how the LO influences the various properties.

The influence of the diode parasitics $[R_s$ and $C(v_j)]$ can be accounted for approximately as indicated below.

Contribution from L_h. The loss of signal power to the harmonic sidebands can be prevented only by arranging reactive loads at those frequencies.

Contribution from L_{R_s}. The series resistance will actually cause losses at all frequencies related to the signal ($n\omega_{LO} \pm \omega_{IF}$). A first-order effect is obtained for the signal frequency itself. Assume that when the mixer is operating, we may talk about an "active" junction conductance (G), which "absorbs" signal power and transfers it to the IF. This conductance (equal to the inverse of the input resistance of Table 2.3) is in parallel with the junction capacitance, which will act as a shunt reactance ($\approx 1/\omega_s C_0$) and bypass some of the current. Hence at higher frequencies, the relative power absorbed in the series resistance will become larger and increase the conversion loss. The relative amount of power absorbed in the series resistance is readily calculated as

$$L_{R_s} = \frac{P_s}{P_G} = \frac{P_G + P_{R_s}}{P_G} = 1 + GR_s + (\omega_s C_0)^2 \frac{R_s}{G} \qquad (2.37)$$

Differentiating this expression with respect to G, one obtains the optimum case,

$$L_{R_s\min} = 1 + 2\omega_s C_0 R_s \qquad \text{for } G = \omega_s C_0 \qquad (2.38)$$

Notice that the series resistance conversion loss contribution is 3 dB for $\omega_s = 0.5\omega_{\text{cutoff}}$.

Contribution from $L_{|\Gamma_{\text{in}}|^2}$. As mentioned above, the input impedance is of importance for the mixer operation. A detailed theoretical evaluation is necessary for finding the optimum performance. However, a "reasonable" input match is of course necessary in all practical cases.

Contribution from $L_{|\Gamma_{\text{IF}}|^2}$. For the output impedance, a simple general rule is valid: The IF port has to be matched for optimum performance. Γ_{IF} can be measured with a network analyzer and subsequently made small using a suitable matching network.

2.4.3 Input and Output Impedance

Table 2.3 summarizes results that can be used for a first approximation of mixer impedances. The influence of the diode capacitance can be accounted for approximately by considering an average junction capacitance $\langle C \rangle$ (chosen to be $\simeq 1.5$ times the zero-bias capacitance) in parallel with the input and output impedances estimated from Table 2.3. The series resistance should, of course, be added in series to the resulting impedances.

2.4.4 Noise of a Mixer Receiver

In a mixer circuit there are several sources of noise that will be converted to the IF output and hence contribute to the system noise. It should be emphasized that a mixer in practice is always part of a system, and it is always necessary to analyze the system noise. The most important noise sources contributing to the mixer system noise are:

1. *Diode Noise:* noise generated within the mixer diode itself at frequencies that can be converted to the IF (see Section 2.2.3)
2. *Thermal Noise:* noise from the embedding circuit converted to the IF
3. *LO Noise:* noise from the local oscillator source, converted to the IF
4. *IF Noise:* noise from the IF amplifier

It is the ability of the mixer to respond to a number of frequencies, $n\omega_{\text{LO}} \pm \omega_{\text{IF}}$, that makes it necessary to be quite careful when the noise of the mixer is analyzed. In particular, engineers have frequently been confused when single-sideband (SSB) and double-sideband (DSB) noise is considered.

In Fig. 2.17 a mixer receiver is described schematically (omitting LO noise). Using the notations of Fig. 2.17, the equivalent noise temperature of the complete receiver can be calculated. A safe way of doing the calculation is to start calculating the noise temperature (multiplied by $k \Delta f = $ power) at the input of the circuit connected to the output of the mixer. In Fig. 2.17 this circuit is an amplifier with an equivalent input noise temperature of T_{IF}.

Assuming that the noise spectrum of the generator impedance (T_t) is flat (which is not always true; e.g., an antenna facing the sky will see a temperature that varies with

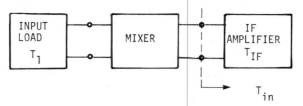

Figure 2.17   Mixer receiver configuration.

frequency, since the air and the "empty outer space" have characteristic spectra), we have at the amplifier input:

$$T_{\text{in}} = \frac{T_l}{L_s} + \frac{T_l}{L_i} + \sum_{2}^{\infty}\left(\frac{T_l}{L_{n+}} + \frac{T_l}{L_{n-}}\right) + T_{M,\text{out}} + T_{\text{IF}} \tag{2.39}$$

where L_s, L_i, L_{n+}, and L_{n-} are the conversion losses at the signal and image frequencies and at the upper and lower harmonic sidebands, respectively. $T_{M,\text{out}}$ is the noise generated in the mixer diode itself and seen at the output port of the mixer. Next we define the signal-to-noise power ratio as

$$\frac{P_s}{P_n} = \frac{P_s/L_s}{T_{\text{in}}k\,\Delta f} = \frac{P_s}{k\,\Delta f\,T_{\text{syst}}} \tag{2.40}$$

where T_{syst} here is by definition the receiver system temperature for a single-sideband system (useful signal is available at only one sideband). From Eq. (2.40) it is seen that

$$T_{\text{syst,SSB}} \quad T_{\text{in}}L_s = T_l\left(1 + \frac{L_s}{L_i} + \sum_{2}^{\infty}\left(\frac{L_s}{L_{n+}} + \frac{L_s}{L_{n-}}\right)\right) + T_{\text{MXR,SSB}} + L_sT_{\text{IF}} \tag{2.41}$$

where the equivalent temperature of the mixer itself, $T_{\text{MXR,SSB}}$, is defined as

$$T_{\text{MXR,SSB}} = T_{M,\text{out}}L_s \tag{2.42}$$

The single-sideband noise temperature of the mixer receiver, $T_{M,\text{SSB}}$, is defined as

$$T_{M,\text{SSB}} = T_{\text{MXR,SSB}} + L_sT_{\text{IF}} \tag{2.43}$$

The double-sideband (DSB) mixer noise temperature is much easier to measure than the single-sideband noise temperature. The reason is that the noise sources one uses for noise-temperature measurements are broadband, in the sense that they generate noise in both sidebands. The same is true for radiometer-type receivers: noise is received in both sidebands. We are thus interested in the case where the signal is received in both sidebands:

$$\frac{P_s}{P_n} = \frac{P_s(1/L_s + 1/L_i)}{kT_{\text{in}}\,\Delta f} \tag{2.44}$$

Hence the double-sideband system noise temperature is

$$
\begin{aligned}
T_{\text{syst,DSB}} &= \frac{T_{\text{in}}}{1/L_s + 1/L_i} \\
&= T_l \left[1 + \frac{L_s L_i}{L_s + L_i} \sum_2^\infty \left(\frac{1}{L_{n+}} + \frac{1}{L_{n-}} \right) \right] + T_{\text{MXR,DSB}} + T_{\text{IF}} \frac{L_s L_i}{L_s + L_i} \quad (2.45)
\end{aligned}
$$

where the double-sideband noise temperature of the mixer itself, $T_{\text{MXR,DSB}}$, is defined as

$$
T_{\text{MXR,DSB}} = T_{M,\text{out}} \frac{L_s L_i}{L_s + L_i} \quad (2.46)
$$

Notice that if $L_s = L_i$, both the single-sideband system noise temperature and mixer noise temperature are two times those for the double-sideband case.

For the single-ended mixer, the LO noise can be taken into account in the analyses by adding a relevant amount of noise at the input ports of the mixer. In a practical mixer, it will always help if the LO source is made inherently low noise by filtering and/or phase locking. For multidiode mixers, a more careful analysis is necessary, and we will return to this question below.

2.4.5   Image Rejection and Image Enhancement

Another important sideband is the image frequency band. The importance of controlling the image frequency becomes obvious when realizing that:

  a. A considerable amount of the signal power (about 3 dB in a broadband mixer) can be converted to the image frequency band, increasing the conversion loss.
  b. Noise from the embedding circuit at the image frequency will be converted to the IF.
  c. Unwanted signals may enter the receiver in the image band and be mistaken for proper signals in the signal band.

It will help in all three respects if the image frequency termination (see Fig. 2.14) is made reactive. Essentially, a filter may do the job, making the mixer receiver properly single sideband by image rejecting. However, the signal power that intended to escape through the image port should be reflected by the proper phase in order to minimize the conversion loss. By doing this, we realize an image-enhanced mixer.

2.4.6   Harmonic Mixers

It was mentioned in Section 2.4.1 that it is possible to construct harmonic mixers. Second-harmonic mixers are frequently constructed for millimeter wave applications; that is, a local oscillator (LO) source at approximately half the signal frequency is used. The alternative, a fundamental mixer that needs pumping at approximately the signal frequency, may be a more expensive solution due to the local oscillator costs. Harmonic mixers with quite high multiplication factors are frequently used in applications where there is no need for any particular sensitivity. Typically, such applications are in phase-locked loops and as mixers in spectrum analyzers.

The price one has to pay when using a harmonic mixer is often inferior performance. Calculating the conversion loss for a (single-diode) second harmonic mixer in a way similar to the approach described in Section 2.4.2 for L_0 reveals that for reasonable LO powers the conversion loss is typically 2 to 3 dB worse than for the fundamental mixer In practice one obtains typically 3 to 5 dB worse than for a fundamental mixer. The difference is partly due to the assumption in the simplified theory that fundamental mixing is not possible between the signal and the LO. This certainly takes place unless there are reactive terminations for these mixing products. In general, this may be difficult to arrange, since the fundamental mixing frequency is close to the LO frequency.

In mixers using two diodes in antiparallel, this problem is avoided (see Sections 2.5.6 and 2.8.8) since the fundamental modulation of $g(t)$ is actually at twice the LO frequency. In such a mixer the conversion loss can be made almost as low as for fundamental mixers. Arrangements with multiple diodes have been proposed for fourth harmonic mixers [21] (see also Section 2.8.8).

2.5 THEORETICAL MODELING OF SCHOTTKY BARRIER MIXERS

2.5.1 Introduction

In this section we give information about more accurate methods of analyzing mixer circuits. The fact that the nonlinearity of mixer diodes is quite strong makes it a nontrivial problem to analyze mixer devices theoretically. Another reason for the theory being complicated is that for accurate analyses, one must know the terminating impedances at the harmonics of the LO and at the harmonic sidebands. The number of harmonics that have to be accounted for has to be limited for different practical reasons. However, the required accuracy places a lower limit on how many harmonics one has to include. In short, analyzing a mixer involves the following steps:

Nonlinear analysis, finding the $g_j(t)$ *and* the $C_j(t)$ [compare Eq. (2.30)] of the pumped diode

Small-signal analysis, by which the conversion loss between any pair of sidebands and input (output) impedances can be calculated

Noise analysis, by which the noise of the mixer can be evaluated

2.5.2 Nonlinear Analysis

The equivalent circuit of the Schottky diode was introduced in Section 2.2.3. The analysis outlined below is based on this circuit. The series resistance R_s (which varies slightly with bias and frequency) and the ideality factor η (which may increase slightly at high forward biases) are assumed to be constant, which is an excellent approximation for most practical cases. From Eqs. (2.3) and (2.4) we obtain the following relations for the time-varying differential resistance and capacitance:

$$g(t) = \frac{dI}{dV_j} = \frac{q}{\eta k T}(I + I_0) \tag{2.47}$$

$$C(t) = C_0 \left(1 - \frac{V_j}{V_{Bi}}\right)^{-\gamma} \tag{2.48}$$

The current through the resistance is given by Eq. (2.3), and the current through the capacitance is

$$i_C = C(t)\frac{dV_j}{dt} \tag{2.49}$$

Applying LO power at ω_p means that all the voltage-dependent variables will become periodic functions that can be expressed using Fourier series. Hence we may write

$$g(t) = \sum_{k=-\infty}^{\infty} G_k \exp(jk\omega_p t) \qquad G_{-k} = G_k^* \tag{2.50}$$

$$C(t) = \sum_{k=-\infty}^{\infty} C_k \exp(jk\omega_p) \qquad C_{-k} = C_k^* \tag{2.51}$$

There are several related techniques developed for evaluating the waveforms $g(t)$ and $C(t)$:

Direct Integration in the Time Domain [22, 23]. This method is inappropriate if the mixer contains distributed elements.

The Harmonic Balance Method [24]. This method may have convergence problems when a large number of harmonics are considered. Modifications will help but not necessarily eliminate the convergence problem.

The p-Factor Method [25]. This method may be considered to be a modified harmonic balance method, with improved convergence properties.

The Multiple Reflection Method [26]. This method can be considered to be a special case of the p-factor method [25]. Experience shows that it works well in all cases tried so far, but it may converge more slowly than an optimized version of the p-factor method.

The Variables Selection Method [27]. By introducing a criterion for selecting which voltages and/or currents should be considered as unknowns, and by using a new and efficient algorithm, very fast convergence has been demonstrated, for large-signal analysis of both Schottky diodes and MESFETs.

The most practical and successful methods are the latter three. They are partly related, and the differences and convergence properties are analyzed in Ref. 25. The multiple reflection method has been found to converge in virtually all practical cases. The p-factor method seems, in general, to be faster than the multiple reflection method.

Next we describe briefly the p-factor method [25] applied to the general circuit of Fig. 2.18, which is excited by a sinusoidal source $E(t) = V_p \cos(\omega_p t)$. The iteration procedure is as follows:

i. The analysis starts by guessing an initial voltage $V^N(t)$ and calculating the resulting current $I^N(t)$ in the nonlinear element using a fourth-order Runge–Kutta method (or using an analytical expression if possible).

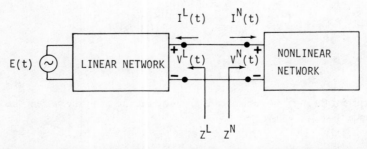

Figure 2.18   Division of the mixer circuit into one linear and one nonlinear part.

ii. By setting the current into the linear network to $I^L(t) = -I^N(t)$ and using a fast Fourier transform to get $I^L(\omega)$, one may calculate $V^L(\omega) [=Z(\omega)I^L(\omega)]$, which, like $I^L(\omega)$, can be expressed as a Fourier series with harmonics of ω_p.

iii. Using a fast inverse Fourier transform, one then calculates $V^L(t)$ and makes a comparison with the initial $V^N(t)$. If they differ by too much, $V^N(t)$ should be modified and a new iteration should take place.

iv. The iteration steps defined in steps ii and iii will continue until the difference between the V^N used in step ii and the V^L derived in step iii is sufficiently small.

Rather than using the derived V^L directly as the input voltage V^N in step i, a compromise value for V^N should be used. Hence for the next $(k + 1)$ iteration, the nth harmonic of $V^N = V^N_{k+1,n}$ of the voltage input (in step ii) is chosen according to

$$V^N_{k+1,n} = p_n V^L_{k,n} + (1 - p_n) V^N_{k,n} \qquad (2.52)$$

The way to choose p_n will not be described here; the reader is directed to Ref. 25 for further information.

When the waveform of the junction voltage $v(t)$ is known, it can be inserted into Eqs. (2.47) and (2.48) and the Fourier coefficients G_k and C_k of Eqs. (2.50) and (2.51) can be evaluated.

The outline above of the p-factor method is called the "voltage update" method. Several modifications are possible, and in some cases the "current update" method may prove to be more efficient.

Applying the algorithm introduced in the variables selection method, an optimum value for the p-factor can be found that yields much improved convergence at moderate current levels in a diode mixer [27]:

$$p_n = \left(1 - \frac{V^L_{k,n} - V^L_{k-1,n}}{V^N_{k,n} - V^N_{k-1,n}}\right)^{-1} \qquad (2.53)$$

This value of p_n will in general be a complex number, and takes different values for each iteration.

2.5.3 Small-signal Analysis

The input signal(s) are assumed to have infinitesimally small amplitudes and therefore do not affect $g(t)$ and $C(t)$. Since the junction conductance $g(t)$ and capacitance $C(t)$ are in parallel, the total small-signal current through the junction admittance is related to the corresponding small-signal voltage accordingly [28, 30, 31]:

$$\begin{pmatrix} \vdots \\ \delta I_m \\ \vdots \\ \delta I_1 \\ \delta I_0 \\ \delta I_{-1} \\ \vdots \\ \delta I_{-m} \end{pmatrix} = \left(Y_{mn} \right) \begin{pmatrix} \vdots \\ \delta V_m \\ \vdots \\ \delta V_1 \\ \delta V_0 \\ \delta V_{-1} \\ \vdots \\ \delta V_{-m} \end{pmatrix} \qquad (2.54)$$

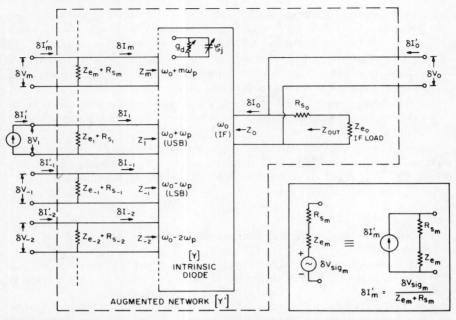

Figure 2.19 Definition of the augmented network [28]. During normal mixer operation the equivalent signal current generator $\delta I'$ is connected at port 1, the other ports being open circuited. In the noise analysis, equivalent noise sources $\delta I'_{Sm}$ and $\delta I'_{Tm}$ are connected to all ports. The inset shows the relation between the signal source $\delta V'_{\text{sig}m}$ at the mth sideband and its equivalent current source $\delta I'_m$. (From Ref. 28; copyright 1978 IEEE, reproduced by permission.)

where

$$Y_{mn} = G_{m-n} + j\omega_m C_{m-n} \tag{2.55}$$

It is convenient to form the augmented Y matrix $\mathbf{Y}'$ which is defined in Fig. 2.19. Notice that the augmented matrix includes R_s and the external embedding network within the matrix. One has

$$\mathbf{Y}' = \mathbf{Y} + \mathbf{diag}\left(\frac{1}{Z_{em} + R_{sm}}\right) \tag{2.56}$$

where Z_{em} is the embedding impedance at the mth harmonic sideband ($m\omega_p + \omega_0$), and R_{sm} is the series resistance at the same frequency.

The ports of the augmented network (Fig. 2.19) are open-circuited. The relation between voltages and currents then is

$$\delta \mathbf{I}' = \mathbf{Y}'\delta \mathbf{V} \tag{2.57}$$

Inverting (2.56) yields

$$\delta \mathbf{V} = \mathbf{Z}'\delta \mathbf{I}' \tag{2.58}$$

where

$$\mathbf{Z'} = (\mathbf{Y'})^{-1} \tag{2.59}$$

Mixer Port Impedances. The input impedance of mixer port m (Z_m) is measured by injecting an infinitesimal current δI_m into port m, keeping the input impedance $Z_{em} = \infty$ and adding the series resistance R_{sm}, yielding

$$Z_m = Z'_{mm,\infty} + R_{sm} \tag{2.60}$$

In particular, the IF output impedance is

$$Z_{\text{IF}} = Z'_{00,\infty} + R_{s0} \tag{2.61}$$

Conversion Loss. Assuming that the IF port is conjugate matched, the conversion loss between port 1 and the IF port becomes

$$L = \frac{1}{4|Z'_{01}|^2} \frac{|Z_{e0} + R_{s0}|^2}{\text{Re}\{Z_{e0}\}} \frac{|Z_{e1} + R_{s1}|^2}{\text{Re}\{Z_{e1}\}} \tag{2.62}$$

The conversion loss between any two ports i and j can be evaluated using Eq. (2.61) by replacing index 0 by j and 1 by i.

2.5.4 Mixer Noise Evaluation

According to Section 2.2.3, the junction will contribute shot noise and series resistance thermal noise (and occasionally, hot electron noise) (see also Fig. 2.19).

Shot Noise. The shot noise rms current fluctuations δI_{sh}^2 of a dc-biased diode can be calculated, according to Eq. (2.8), as $2I_{\text{sh}}q\,\Delta f$. However, when the diode is pumped with the local oscillator, the shot noise will become modulated. Assume that the pumped diode noise can be considered as consisting of amplitude-modulated pseudosinusoidal current components. Components at frequencies $\omega_m = m\omega_p + \omega_0$ ($m = 0, \pm 1, \pm 2, \dots$) can be converted to the various ports of the augmented network, including the IF port. It can be shown [28] that the shot noise rms voltage at the IF port becomes

$$\langle |\delta V_{\text{sh},o}|^2 \rangle = Z'_o \langle \delta \mathbf{I}'_{\text{sh}} \, \delta \mathbf{I}'^{\dagger}_{\text{sh}} \rangle Z'^{\dagger}_o \tag{2.63}$$

where the square matrix $\langle \delta \mathbf{I}'_{\text{sh}} \, \delta \mathbf{I}'^{\dagger}_{\text{sh}} \rangle$ is known as the noise correlation matrix, with elements

$$\langle \delta I'_{\text{sh},m} \, \delta I'^{*}_{\text{sh},n} \rangle = 2q I_{m-n} \, \Delta f \tag{2.64}$$

in which I_{m-n} is one of the Fourier coefficients of the current through the nonlinear conductance of the diode.

Thermal Noise. If hot electron noise is generated when the diode is forward biased by the LO, the noise from the series resistance may be modulated [10, 32]. Actually, the series resistance itself will also be modulated, since the contribution from the undepleted part of the epilayer varies with bias. Ignoring both effects, only thermal noise from a constant series resistance has to be considered. In this case there is no correlation of the noise at the various ports of the augmented network, and the rms thermal noise voltage at the output IF port becomes [28]

$$\langle \delta V_{T,o}^2 \rangle = Z_o' \langle \delta \mathbf{I}_T' \, \delta \mathbf{I}_T'^\dagger \rangle Z_o'^\dagger \tag{2.65}$$

where

$$\langle \delta I_{T,m}' \, \delta I_{T,m}' \rangle = \begin{cases} \dfrac{4kT_{eq}R_{sm}\,\Delta f}{|Z_{em} + R_{sm}|^2} & m \neq 0 \tag{2.66a} \\[2ex] \dfrac{4kT_{eq}R_{sm}\,\Delta f}{|Z_o|^2} & m = 0 \tag{2.66b} \end{cases}$$

and where T_{eq} is the equivalent noise temperature of the series resistance.

Total Mixer Noise. The shot noise and the thermal noise contributions defined at the output IF port are uncorrelated. Therefore, they can be combined to yield the total noise output:

$$\langle \delta V_{No}^2 \rangle = Z_o' \{ \langle \delta \mathbf{I}_{sh,o}' \, \delta \mathbf{I}_{sh,o}'^\dagger \rangle + \langle \delta \mathbf{I}_{T,o}' \, \delta \mathbf{I}_{T,o}'^\dagger \rangle \} Z_o'^\dagger \tag{2.67}$$

The equivalent input noise temperature of the mixer, that is, the temperature to which the input impedance has to be heated in order to generate a noise voltage equal to $\langle \delta V_{No}^2 \rangle$, can now be calculated in a straightforward way, yielding

$$T_{\text{MXR}} = \frac{\langle \delta V_{No}^2 \rangle}{4k\,\Delta f} \frac{|Z_{e1} + R_{s1}|^2}{|Z_{o1}'|^2 \, \text{Re}\{Z_{e1}\}} \tag{2.68}$$

2.5.5 Noise of a Purely Resistive Mixer

It is illuminating to consider the noise of a purely resistive exponential diode mixer (i.e., a mixer with a diode having the series resistance equal to zero), a constant capacitance, and a current–voltage characteristic as determined by Eq. (2.3). Using the theoretical approach outlined in this section, the available noise power from the mixer diode at the IF output terminal is derived as [33]

$$P_{o,\text{avail}} = k\,\Delta f \left[\frac{\eta T}{2} \left(1 - \sum_{m \neq 0} \frac{1}{L_m} \right) + T_e \sum_{m \neq 0} \frac{1}{L_m} \right] \tag{2.69}$$

where T is the physical temperature of the diode and T_e the physical temperature of the embedding network. It is of particular interest to consider the following two special cases.

Single-Sideband Mixer. Assuming that all ports of the augmented network, except for the signal and the IF ports, are reactively terminated,

$$T_{\text{MXR,SSB}} = \frac{\eta T}{2} [L_1 - 1] \tag{2.70}$$

This result is identical to the input noise of an ordinary attenuator with the physical temperature equal to $\eta T/2$ and an attenuation of $10 \log[L_1]$ decibels.

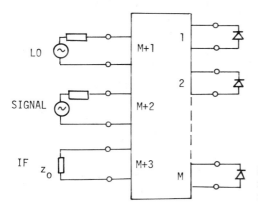

Figure 2.20 General equivalent circuit of a mixer with M diodes. (From Ref. 34; copyright 1980 IEEE, reproduced by permission.)

Double-Sideband Mixer. Assuming that all ports of the augmented network are terminated by reactive impedances, except for the signal (1), image (2), and IF (0) ports, then

$$T_{\mathrm{MXR,DSB}} = \frac{\eta T}{2}\left(1 - \frac{1}{L_1} - \frac{1}{L_2}\right)\left(\frac{1}{L_1} + \frac{1}{L_2}\right)^{-1} \tag{2.71}$$

2.5.6 Multiple-Diode Mixers

There are several types of mixers using more than one diode (see Section 2.6). Several theoretical papers have been published on multiple-diode mixers. Extensive analysis has been done on two-diode balanced mixers [31, 34]. Two-diode subharmonically pumped mixers have also been analyzed [31, 35].

Consider a mixer with M diodes, attached to a linear network having $M + 3$ ports, as shown in Fig. 2.20. The ports $M + 1$, $M + 2$, and $M + 3$ are connected to the LO pump, the signal input, and the IF output, respectively. Since the only case considered is that when the signal is much lower in amplitude than the LO pump, the problem of analyzing the mixer can again be divided into one nonlinear part, one linear part, and one part dealing with the noise. Faber and Gwarek [34] have analyzed this multidiode mixer by essentially using the techniques outlined above for single-diode mixers. In the nonlinear analysis, they used essentially the method described by Held and Kerr for the single-diode case [28]. Subharmonically pumped mixers have been analyzed in a similar way by Kerr [31] and Hicks and Kahn [35].

2.5.7 Some Further Comments on Mixer Analysis

In an interesting paper by Hines it is pointed out that the actual model chosen for analyzing the mixer can affect the result. In the analysis above it is assumed that the bias of the diode is constant in voltage and that the bias supply has zero output impedance. Hence, if the signal changes the dc current somewhat, it will not cause any power transfer from the signal to the dc load.

Hines [36], using a time-domain method, analyzed the situation when the diode was modeled as a perfect switch which is switched between infinite and zero impedance, and the bias was obtained using a "battery" with a certain open-circuited output voltage and an output resistance equal to the IF load. He then found that irrespective of the "diode" being assumed lossless (an on–off switch), and all ports, except for the signal and the IF ports, being terminated by reactive loads, the mixer will have a significant

TABLE 2.4 Mixer Comparison Guide

Mixer Type	Number of Diodes[a]	Conversion Loss[b]	VSWR			Port-to-Port Isolation		LO AM-Noise Rejection	Intermodulation Products Generated	
			RF	LO	IF	RF-IF LO-IF	LO-RF		Signal	LO
Single diode	1	Good[b]	Poor[c]	Good[c]	Poor[c]	Moderate[d]	Moderate[d]	None	All	All
Single balanced, 180°	2	Good	Fair[c]	Fair[c]	Fair[c]	Fair	Very good	Good	Odd[e]	All[e]
Single balanced, 90°	2	Good	Good	Good	Fair[c]	Fair	Poor	Good	Odd[e]	All[e]
Doubled balanced	4	Very good	Poor	Poor	Good	Good	Very good	Good	Odd	Odd
Double balanced, image rejection	4	Very good	Good	Good	Good	Good	Good	Good	Odd	Odd
Image recovering	4	Excellent	Good	Good	Good	Good	Very good	Good	Odd	Odd
Double-double balanced mixers	8	Good	Good	Good	Good	Good	Very good	Good	Odd	Odd
Subharmonically pumped	2	Good	Fair[c]	Fair[c]	Fair[c]	Depends on filters	Depends on[f] filters	Good	All	Even

[a] LO power and saturation power is proportional to number of diodes.
[b] Image enhancement possible.
[c] Matching circuit recommended.
[d] Depends on filters.
[e] If signal fed in phase and LO out of phase (compare Fig. 2.22c); if the other way around, reverse the columns (see Section 2.7).
[f] For RF → LO isolation, filters are necessary.

conversion loss which cannot be explained by any mismatch at the input or the output port. A subsequent analysis of the same problem using the frequency-domain analysis as discussed above fortunately gave the same result. The explanation of the phenomenon is slightly different depending on the method of analysis. The time-domain analysis suggests that the signal energy is converted to dc and then dissipated in the dc load. A frequency-domain analysis suggests that the signal energy is converted to an infinite number of high-order modulation products, being dissipated in the infintesimal resistance of the switch in its low-impedance "on"-state, resulting in a finite loss.

The conclusion from this brief discussion is a recommendation that extra thought be given as to how to model the actual mixer being analyzed.

2.6 SINGLE- AND MULTIPLE-DIODE MIXERS

2.6.1 Introduction

Below we compare the properties of different mixer circuits using one or several mixer diodes (see Table 2.4). Basically, four types of fundamental frequency mixer circuits have found practical use:

- The single-ended mixer (SE) with one diode
- The single balanced mixer (SB) with two diodes
- The double balanced mixer (DB) with four diodes
- The double-double balanced mixer (DDB) with eight diodes

It should be kept in mind that there are alternative circuits possible for most of the mixer configurations discussed below (i.e., the circuits given in the figures should just be considered as examples). Besides fundamental frequency mixers, subharmonically pumped multiple-diode mixers will also be introduced. The discussion in the previous sections essentially applies also to multiple-diode mixers. In fact, multiple-diode mixers can be understood in terms of interconnected single-diode mixers. For further information, consult major suppliers of commercial mixers (e.g., Anaren, Alpha, Hewlett-Packard) and Refs. 37–40. Notice that different multiple-diode configurations are commercially available as monolithic beam lead devices in various types of packages (see Table 2.1).

2.6.2 The Single-Ended Mixer

The principal advantage of the single-ended mixer is its simplicity (Fig. 2.21). The signal, local oscillator, and intermediate-frequency ports can be isolated from each other by

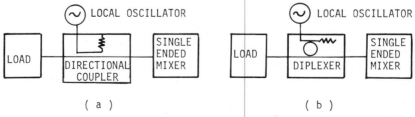

Figure 2.21 Examples of single-ended mixer configurations: (*a*) local oscillator injected through a broadband directional coupler; (*b*) local oscillator injected via a narrow-band diplexer (e.g., a ring filter).

means of filter circuits. In several common and simple circuit solutions, however, there is poor isolation, if any, between the LO and signal ports (see Section 2.8). Therefore, to achieve broadband coupling of the LO and the signal to the diode requires a broadband coupler (Fig. 2.21a), which will cause increased conversion loss, a worse noise figure, and a multiplied LO power requirement (e.g., a 6-dB directional coupler will increase the conversion loss and the noise figure with 1.25 dB, and require four times more LO power). A narrowband LO injection diplexer (Fig. 2.21b) avoids this problem, but may make tunability difficult.

Another important fact is that LO noise is more difficult to avoid, in particular if the IF is a low frequency. In multiple-diode mixers this problem is considerably alleviated, as discussed below. Since the required local oscillator power is proportional to the number of diodes of the mixer, the SE mixer, from that point of view, requires less power than do the other mixer types. However, this also means that the dynamic range of the single-ended mixer is smaller, since the 1-dB compression point is typically 5 to 10 dB below the LO power level.

SE mixers are used in microwave systems where a simple solution is desirable and/or lower performance levels can be tolerated. SE low-noise mixers are frequently used for the millimeter wave frequency range (see Sections 2.9 and 2.10).

2.6.3 Single Balanced Mixers

Some of the disadvantages of the SE mixer can be overcome using single balanced (SB) mixers. Such mixers can be constructed using various types of hybrid circuits and baluns for coupling the LO and the signal to the diodes. Either 90° or 180° 3-dB hybrids can be used, each having certain advantages and disadvantages, but in both cases offering better performance than the SE mixer. Thus they offer reduced spurious response, cancellation of the dc component, suppression of local oscillator noise, and isolation between the various ports.

Schematic diagrams of 90° and 180° hybrid SB mixer structures are shown in Fig. 2.22b. Working out the relative phases of the signal and the LO at the terminals of the diode pair (see Fig. 2.22) reveals that in both cases the LO has a 180° phase shift from the diode pair, while the signal has zero phase shift. Therefore, the origin of the IF signal is the same for both cases, and can be combined as indicated in Fig. 2.22b.

The 180° hybrid balanced mixer can be described readily using the equivalent circuit of Fig. 2.22c. Assuming the diodes (D_1 and D_2) to be exactly equal, they form a voltage divider, causing a virtual ground (zero phase relative to ground) at point A. The way that signal current i_s and LO current i_{LO} is fed to the circuit means that i_s and i_{LO} add in one diode and subtract in the other. This will cause an imbalance at A, which will slowly cycle with a frequency equal to the IF. Consequently, the IF signal can be extracted between A and ground. Although ideally, there is infinite isolation between the various ports, the practical solutions available may, for various reasons, not be good enough, and therefore filter circuits for improving the isolation may be necessary. Notice that noise from the LO will not cause an imbalance at A, and will consequently not add noise in the IF circuit.

Using the 180° hybrid is equal to combining two SE mixers in parallel, 180° out of phase. Terminating the two output arms of the 180° hybrid with identical impedances will cause reflected power to go back to the input port. Consequently, there will in this case be perfect isolation between the signal and the LO port. In practice, with well-matched diodes, the isolation is typically 20 dB or more. However, as in the SE mixer, the VSWR at the RF ports will depend on the match between the diodes and the circuit, typically yielding a VSWR of 2:1. For optimum performance, filters should be used to separate RF and IF frequencies.

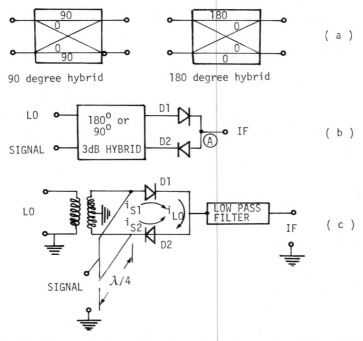

Figure 2.22 Single balanced mixer configurations: (*a*) the phase shifts of 90° and 180° hybrids; (*b*) schematic balanced mixer configurations; (*c*) equivalent circuit of the 180° hybrid mixer.

With the 180° hybrid, the mixer can be designed to suppress the even harmonics of one of the input signals, usually the LO signal. The amount of suppression depends on how well the diodes are matched and on the balance of the hybrid.

Using 90° hybrids yields significantly different properties. The VSWR is excellent over the full performance range of the hybrid. Feeding a signal into one of the RF ports, similar diodes will cause the reflected waves to combine at the other RF port. Hence, either input will have a low VSWR (typically, less than 1.5:1). The isolation between the RF ports will, in this case, depend on the match between the circuit and the diodes, and is typically not better than 7 dB.

To obtain optimum performance, filters to separate RF and IF ports have to be included. Since the 90° hybrid is as relatively easy to design and fabricate as either a microstrip, stripline, or coaxial-line circuit, it has been used widely in broadband (octave-band) mixers. Examples of 90° hybrids are (see also Refs. 1 and 41):

- Properly designed 3-dB stripline or microstrip couplers [42]
- The Lange coupler [43]
- The branch-line hybrid [44]

Examples of 180° hybrids are:

- The waveguide magic T [45]
- The ring or "rat race" hybrid [41]
- Various combinations of microstrip, coplanar, slotline, finline, and so on, junctions (see Section 2.8)

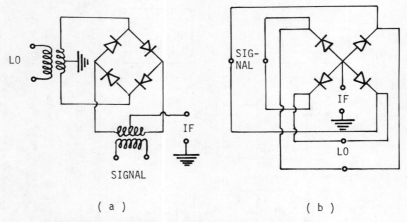

Figure 2.23   Double balanced mixers: (*a*) ring mixer; (*b*) star mixer.

Baluns [46] (balanced-to-unbalanced line transformer; the LO transformer in Fig. 2.22*c* is a balun) are also used in balanced mixer circuits. A balun evidently has the exact properties required and can be designed with decade bandwidth ratios [47]. In Section 2.8, practical balanced mixer circuits are described in some detail.

2.6.4   Double Balanced Mixers

Double balanced (DB) mixers are composed of two SB mixers coupled in parallel and 180° out of phase (Fig. 2.23). The diodes can be arranged in either a star or a ring configuration. The ring modulator can be obtained as a very compact monolithic circuit, and is the more commonly used configuration in practical mixers. The symmetry of the circuit ensures complete isolation between the LO and the signal port if the diodes are perfectly matched. Furthermore, the topology of the circuit now yields the suppression of even harmonics of both the signal and the LO frequencies. This fact also means that intermodulation is reduced compared to SE and SB mixers.

The DB mixers use two baluns rather than one as in the SB mixer. At low frequencies (a few gigahertz) the baluns can be fabricated from bifilar-wound transformers with central taps for the IF output. The isolation between the three ports becomes independent of frequency and quite large (more than 35 dB). At higher frequencies this type of transformer will not work, and other solutions must be found. Mixing different types of transmission-line circuits and using their symmetry properties is one way (see Section 2.8). Another way is through use of distributed circuits. These types of structures will not achieve the same isolation as will the transformer type; an isolation of about 25 dB may be expected. The RF input ports have the same properties as the 180° hybrid SB mixer (i.e., typically the SVWR is about 2.5:1).

2.6.5   Double-Double Balanced Mixers

Double-double balanced (DDB) mixers are composed of two DB mixers (an example is shown in Fig. 2.24). Since all three ports (signal, LO, and IF) are balanced, they could also be called "triple balanced." Eight diodes are used, and therefore twice as much LO

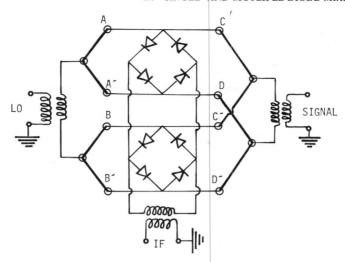

Figure 2.24 Example of a double-double balanced mixer circuit. The bold lines connected to *A*, *A'*, *B*, *B'*, *C*, *C'*, *D*, *D'* indicate, for example, $\lambda/4$ lines, necessary not to short circuit the diode quads. (From Ref. 37; reproduced from *Microwave Systems News*, October 1981, by permission of the publisher, EW Communications, Inc.)

power is needed as for the DB mixer, and eight times the LO power as for the SE mixer. This means that the dynamic range is increased eightfold compared to the SE mixer, and so on, when compared to the other mixer types. In addition to the larger dynamic range, the intermodulation properties are superior. The drawbacks of the DDB mixer are the requirement for greater LO power and the increased cost due to more diodes being used.

In all mixers there must exist ground return paths for both RF and IF currents. This fact results in loss, adding to the conversion loss of the mixer. DDB mixers do not require a return path for the IF, eliminating an additional signal loss, which can be on the order of 1 dB.

2.6.6 Quadrature IF Mixers and Image-Rejection Mixers

The mixer circuits discussed above can be used in various combinations for applications other than simply converting an incoming signal to another frequency. The quadrature IF (QIF) mixer is an example of such a circuit, and can be used, for example, for Doppler systems and network analyzers. The QIF mixer can also be converted to form an image rejection (IR) mixer.

The QIF mixer consists basically of two balanced mixers, one 90° hybrid for injecting the signal (LO) and one in-phase power divider for injecting the LO (signal) (Fig. 2.25). Thus a 90° phase difference is introduced between the RF and the LO signals. This will result in two IF output signals (IF_1 and IF_2) of equal amplitude but with a phase difference of $\pm 90°$. The sign of this phase difference will depend on which frequency is higher, the LO or the RF signal. If the LO is constant, the QIF mixer can distinguish between a signal higher or lower than the LO frequency. Thus the QIF mixer can be used in Doppler systems and in network analyzers as a vector voltmeter.

Knowing that the QIF mixer can distinguish between the upper and the lower sidebands, we may combine the two IF outputs in a 90° hybrid, directing the lower sideband

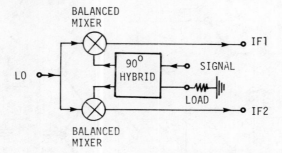

Figure 2.25 Quadrature IF mixer outline.

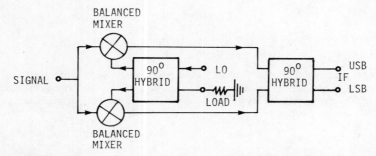

Figure 2.26 Single-sideband mixer using two balanced mixers and two 90° hybrids.

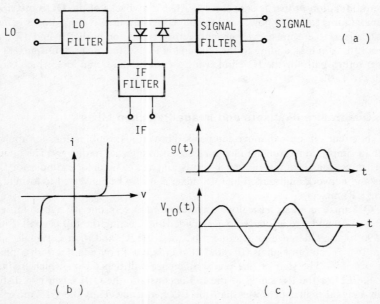

Figure 2.27 Subharmonically pumped mixer using an antiparallel diode pair: (*a*) mixer circuit; (*b*) dc *iv* characteristic; (*c*) time dependence of the local oscillator voltage and the differential conductance.

into one of the output ports of the IF hybrid, and the upper sideband into the other (Fig. 2.26).

2.6.7 Subharmonically Pumped Two-Diode Mixers

Figure 2.27 shows how two diodes coupled in an antiparallel configuration can be used for frequency conversion. Since the bias voltage is zero volts, the LO voltage will swing over the iv characteristic so as to produce a modulated small-signal conductance $g(t)$ with a modulation rate which is twice the LO frequency. Hence frequency conversion will occur only for frequencies close to twice the LO frequency, and no fundamental mixing at all will occur near the LO frequency.

The concept of using two antiparallel diodes for mixing is of practical importance particularly for millimeter wave mixers, when it is difficult to realize enough LO power near the signal frequency. However, there are also several advantages compared to the one-diode harmonic mixer. One is that fundamental mixing is avoided in a two-diode mixer, while in a one-diode harmonic mixer, fundamental mixing will take place unless there are reactive terminations for these mixing products. This is obviously difficult to arrange unless a considerable LO power loss (mismatch) is accepted, since the fundamental mixing frequency is usually close to the LO frequency. Also, noise from the LO circuit will be converted to the IF in the one-diode mixer.

In summary, the advantages of the subharmonically pumped mixer with antiparallel diodes are:

- Reduced conversion loss due to suppressed fundamental mixing
- Lower noise through suppression of LO noise
- Suppression of direct video detection
- Inherent self-protection against large peak inverse voltage burnout

It is possible to use the planar doped barrier diode, which has a symmetrical iv characteristic described briefly in Section 2.2.2) instead of two diodes in antiparallel. In Section 2.8.8, some practical circuit solutions are described.

2.7 INTERMODULATION IN MICROWAVE MIXERS

2.7.1 Introduction

In many applications, the intermodulation properties are important. The receiver designer should know in advance what spurious responses can occur and what can be done to master intermodulation problems. Intermodulation is discussed briefly here. For further information the reader may consult Refs. 38 and 4.

2.7.2 One-Signal Intermodulation

In a fundamental mixer ($f_{\text{IF}} = |f_{\text{LO}} - f_s|$), when the signal power becomes of the same order of magnitude as the LO, undesired intermodulation (IM) products $|mf_s \pm nf_{\text{LO}}| = f'_{\text{IF}}$ (where m and n are integers) will become significant. For single-ended mixers, this type of intermodulation response has been analyzed with the assumption that it uses a purely resistive exponential diode [48] (see Fig. 2.28). The theoretical intermodulation output power is, in this approximate model, proportional to the power m of the signal power (P_s^m).

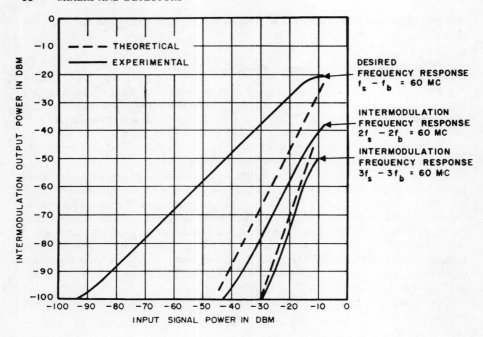

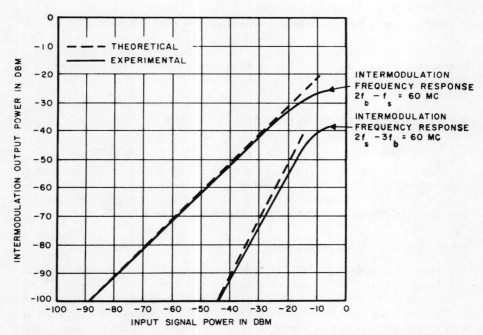

Figure 2.28   Intermodulation output power versus signal power according to Ref. 48. (From Ref. 48; copyright 1964 IEEE, reproduced by permission.)

2.7.3 Intermodulation with More Than One Signal Present

When more than one signal is mixed with the LO, IM products at the following frequencies will be produced:

$$f_{IM} = (\pm m_1 f_{s1} \pm m_2 f_{s2} \pm \cdots) \pm n f_{LO} \tag{2.72}$$

Some insight into the relative importance of the various IM products can be gained by considering the following expression for the diode current [37]:

$$i = i_0 \exp\left(\frac{v_{s1} + v_{s2} + \cdots + V_{LO}}{V_0}\right) = i_0 \sum_{n=0}^{\infty} \frac{v_{s1} + v_{s2} + \cdots + V_{LO})^n}{V_0^n n!} \tag{2.73}$$

where the voltages are sinusoidal and $V_0 = \eta T k/q$.

Let us first discuss briefly single-tone intermodulation, the case when only one signal frequency component and the LO are involved ($v = v_{s1} + V_{LO}$). For this case the term $n = 0$ corresponds to the dc current, $n = 1$ to the fundamentals, and $n = 2$ to the second harmonics and the frequency sums and differences (including the IF). The term $n = 3$ yields not only fundamental frequency components, but also the third harmonic components $2f_{LO} \pm f_s$, $2f_s \pm f_{LO}$. From Eq. (2.73), it can be seen that the IM products will become attenuated by $n!$ appearing in the denominator of (2.73), where n equals the order.

Now consider the case when there are two signals (f_{s1} and f_{s2}) mixing with the LO. The two-tone third-order IM products are defined as

$$f_{IM1} = (\pm 2f_{s1} \pm f_{s2}) \pm f_{LO}; f_{IM2} = (\pm f_{s1} \pm 2f_{s2}) \pm f_{LO} \tag{2.74}$$

Notice that these IM products are obtained for $n = 4$, but are named "third-order IM" since the coefficients in front of f_{s1} and f_{s2} add up to 3. In fact, the third-order IM products are generated not only for $n = 4$, but also for $n = 6$, and so on.

A much more accurate analysis of two-tone intermodulation in a one-diode mixer can be made using the approach outlined in Section 2.5. Such an analysis shows [49] that intermodulation distortion is minimized by using a low diode junction capacitance, a small series resistance, and low embedding impedances. It is also demonstrated [49] that a high LO drive is advantageous.

Fairly sparse theoretical information is available concerning IM in balanced, double, balanced, and double-double balanced mixers [50, 51]. Simple logic indicates, the experience shows, that the IM properties are better when more diodes are involved in the mixing process. Hence the double-double balanced mixer should have considerably better IM performance than that of the single-ended mixer. Moreover, for symmetry reasons, certain IM products will cancel in (perfectly) balanced, DB, and DDB mixers:

- In balanced mixers, only half of the possible IM products are generated. Referring to Fig. 2.22b, only those involving odd harmonics of the signal frequency are generated. Notice: If the LO is connected to the in-phase port and the signal to the out-of-phase port, even harmonics of the signal can be generated, which is usually not desired.
- In both DB and DDB mixers (star or ring configuration), only one-fourth of the possible IM products are generated: those involving odd harmonics of the signal and the LO.

Broadband mixers, require special attention. In any mixer, the intermodulation products will exit through the three mixer ports. They may subsequently be reflected back into the mixer, remix, and generate further IM products. Since the phase of the IM products certainly will vary with frequency, this phenomenon will cause ripple on the conversion loss versus frequency characteristic, and variations in the intermodulation suppression upon frequency. It is therefore important to match the three ports as closely as possible by using attenuators and isolators. When selecting a mixer it is also important to choose a bandwidth that is not too wide, in order to limit the number of IM products that can leave and reenter the mixer.

2.7.4 The Intercept Point and Classes of Mixers

The intercept point is a figure of merit for IM product suppression, and should be as high as possible. It is defined as the power level at which the power of the undesired IM products at the IF port becomes equal to the desired IF power (i.e., these two power levels intercept each other). However, in practice, when the input power is increased over a certain level, the mixer becomes saturated and the intercept will not take place. Therefore, the intercept point is theoretically evaluated from low-input-power-level IM data by means of extrapolation. The intercept point can be specified at either the input or the output; that is, the input intercept point is the signal power at the input port to reach the intercept point, and the output intercept point is the intercept power measured at the output port. A word of warning: Some manufacturers specify the intercept point as the power required for obtaining actual interception, a definition that will yield higher ("better") numbers. For further discussion, see Ref. 37.

As mentioned above, the more diodes used in the mixer, the better the intermodulation properties. Different classes of mixers are defined in such a way that the higher the class, the more LO power is required and the better the intermodulation properties. The normal DB mixer, which has a single diode in each leg, is a class 1 mixer. A class 2 DB mixer has an extra series diode in each leg. In Fig. 2.29 are listed the various classes and types of mixers. Further information may be found in Ref. [37].

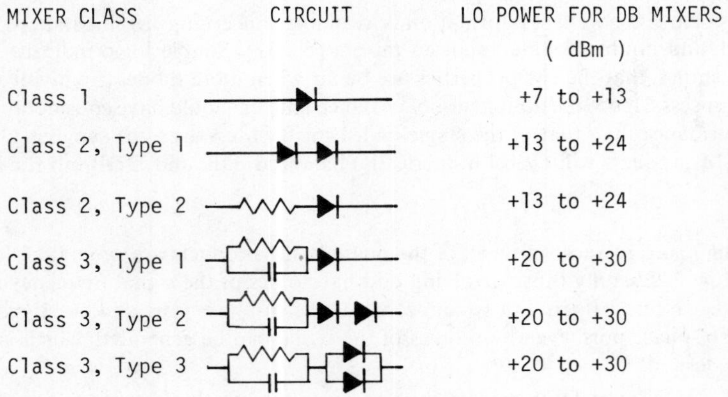

MIXER CLASS	CIRCUIT	LO POWER FOR DB MIXERS (dBm)
Class 1		+7 to +13
Class 2, Type 1		+13 to +24
Class 2, Type 2		+13 to +24
Class 3, Type 1		+20 to +30
Class 3, Type 2		+20 to +30
Class 3, Type 3		+20 to +30

Figure 2.29 Classes of mixers. (From Ref. 37; reproduced from *Microwave Systems News*, October 1981, by permission of the publisher, EW Communications, Inc.)

2.8   PRACTICAL IMPLEMENTATION OF MICROWAVE MIXERS

2.8.1   Introduction

In this section, circuits used for the practical implementation of microwave mixers are presented. Although the structures discussed below were developed originally for microwave frequencies, modern technology has made it possible to use some of them at millimeter frequencies as well.

Early mixer circuits were made using waveguide technology or coaxial-line technology. Modern designs are often made in microstrip technology, or using a combination of technologies, such as waveguide, coaxial-line, stripline, and microstrip technology. We do not discuss the very simplest circuits, where only simple filtering (or none at all) is used to separate the various frequency components, but concentrate on more sophisticated circuits featuring special properties, such as image rejection, interport isolation, and LO-noise reduction. There are quite a number of mixer circuits described in the literature (see, e.g., Ref. 4), and those described below should be considered as educational examples rather than suggestions of optimum circuits.

2.8.2   Single-Ended Image-Enhanced MIC Mixer

A simple and straightforward image-rejecting and image-enhanced single-ended C-band mixer is described in Fig. 2.30 [52, 53]. By varying the position of the resonant stub responsible for the image rejection, the minimum noise figure was found to occur for a distance L, causing a short-circuited image at the diode. Part of the reason for this optimum was that the mixer became well matched to the IF amplifier. Other single-ended image rejection and/or image-enhanced mixers are described in Refs. 23 and 54. A theoretical analysis using a simple model is presented in Ref. 54.

2.8.3   Image-Enhanced Balanced MIC Mixer

There are a great number of imaginative schemes for designing image-enhanced balanced mixers to be found in the literature, and some of them are discussed in Ref. 4. Here we show a mixer designed using microwave integrated-circuit (MIC) techniques on alumina substrate [55]. As indicated in Fig. 2.31 both sides of the substrate are used. The substrate is viewed from the ground-plane side. The RF signal (9.3 GHz) enters the substrate at the right edge (signal input) on a microstrip line at the opposite side of the substrate. The signal is then coupled to the diodes via a microstrip-slotline transition

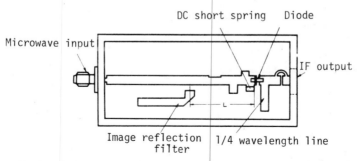

Figure 2.30   Single-ended image-enhanced MIC mixer. (From Ref. 52; copyright 1971 IEEE, reproduced by permission.)

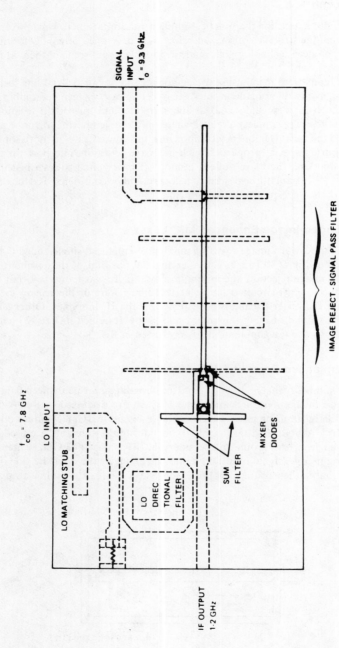

Figure 2.31 X-band image-enhanced balanced mixer. (From Ref. 55; copyright 1975 IEEE, reproduced by permission.)

and through a bandpass, image-rejecting, and impedance-matching filter consisting of microstrip lines coupled to the slotline.

The LO at 7.8 GHz is injected via the "LO input" microstrip terminal. The LO power then passes through the directional ring filter to the IF output microstrip line, which is connected to the coplanar line at the ground-plane side via a pin through the substrate. Slotline stubs at the end of the coplanar-line section present a short circuit to the diodes at the sum frequency. The wide frequency separation between the IF and the LO allow very simple diplexing. The two diodes are in parallel to the IF and the LO ports, but in series to the signal port. The mixer has a maximum of 3 dB conversion loss over approximately 10% bandwidth at 10 GHz.

2.8.4   Phasing-Type Image Recovery Double Balanced Mixers

In Section 2.6.4, double balanced mixers were discussed briefly. Again, there are several papers discussing various ways of designing double balanced mixers [4, 56]. Oxley [57] has carefully investigated some design schemes using hybrids to achieve proper phasing between the two balanced mixers constituting the double balanced one. In Fig. 2.32 is shown one of the X-band designs he has investigated. The best configuration showed a maximum conversion loss of 2.9 dB over 20% bandwidth. The image rejection was better than 20 dB.

The image recovery can be seen as the result of the image power generated in mixer 1 being converted to IF power by mixer 2, and vice versa. These IF signals, resulting from frequency conversion of the image powers, are added in phase, with the IF generated by the primary signal power.

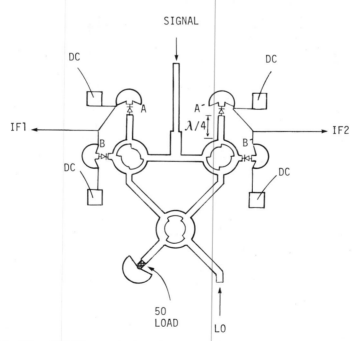

Figure 2.32   X-band MIC image rejection mixers. (From Ref. 57; copyright 1980 IEEE, reproduced by permission.)

There are several other designs reported in the literature similar to the one described above. Mixers also using 90° hybrids, but with a quite different circuit solution, are described, for example, in Refs. 58 and 59.

2.8.5   Crossbar Mixers

Several designs of microwave crossbar mixers have been presented in the literature [4]. As an example of a design, a 75- to 110-GHz mixer is shown in Fig. 2.33. The crossbar mixer is a balanced mixer. The signal enters via the waveguide while the LO is connected to the diodes via the crossbar, and the IF is also retrieved from the crossbar. The mixer diodes are in series with respect to the RF signal, while the diodes are in parallel to the LO. The particular mixer shown in Fig. 2.33 is constructed on a suspended substrate. The conversion loss of this mixer is between 6 and 7 dB over the waveguide band.

It is possible to make a more elaborate design, for example by adding filters to arrange for image rejection and for proper filtering of the second and third harmonics [61]. An octave-bandwidth mixer can be constructed by using a ridged waveguide rather than an ordinary waveguide [62].

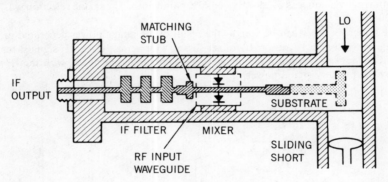

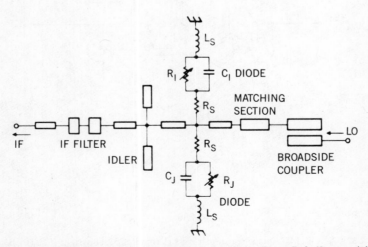

Figure 2.33   Crossbar stripline mixer and its equivalent circuit. (From Ref. 60; copyright 1983 IEEE, reproduced by permission.)

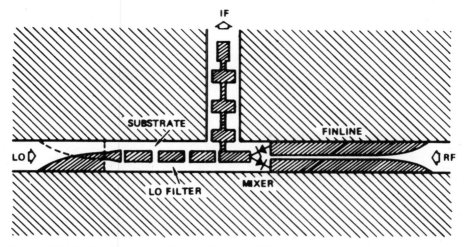

Figure 2.34 Finline mixer circuit layout. (From Ref. 60; copyright 1983 IEEE, reproduced by permission.)

2.8.6 Finline Balanced and Double Balanced Mixers

For millimeter wave frequencies up to 100 GHz, finline circuits have been adopted successfully for mixer design. The finline has a simple geometry and is compatible with waveguides. Several designs of balanced mixers have been presented [4]. A fairly straightforward design is shown in Fig. 2.34. The signal is coupled to the mixer via a broadband waveguide-to-finline transition, and the LO via a waveguide-to-finline-to-microstrip transition. This mixer was constructed for the waveguide band 75 to 110 GHz, and had an insertion loss of about 9 dB. This particular circuit was designed to have a very broad IF bandwidth, and 32 GHz was obtained.

There are several other, more-or-less similar designs reported in the literature [63–65]. An interesting design of a double balanced mixer based on the finline concept has been reported by Blaisdell et al. [66]. A conversion loss between 5 and 6 dB was obtained over the full waveguide band (18 to 26 GHz and 26 to 40 GHz), and between 5 and 7 dB for a version using a ridged waveguide over 18 to 40 GHz.

2.8.7 Mixers Using Baluns for Double and Double-Double Balanced Configurations

Extremely broadband mixers have been designed using baluns rather than ordinary hybrids for coupling the signal and the LO, respectively, to the mixer diodes in balanced, double balanced, and double double balanced mixers. Baluns based on an original coaxial design described by Marchand have been realized in planar designs, yielding as large bandwidths as 10:1 or more [47].

Several mixer designs based on the double-double balanced mixer concept using orthogonal structures and baluns have been described in the literature, and frequency coverage from about 1 to 25 GHz have been reported [67–69]. Typical conversion loss numbers obtained are between 6 and 9 dB. Although balun-type balanced mixers are thought to offer the potential of very low conversion loss, experience seems to indicate difficulties in getting below about 5 dB [70]. Further research in this area might be rewarding.

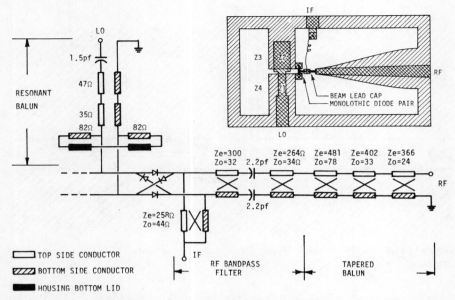

Figure 2.35 Balun-coupled mixer. (From Ref. 68; copyright 1982 IEEE, reproduced by permission.)

Figure 2.35 describes the design of a double balanced balun-coupled mixer for 2 to 18 GHz [68]. The diode ring is constructed with a monolithic beam lead pair on the top side of the suspended substrate, connected to another pair at the other side of the substrate via two plated-through holes. This design ensures good isolation between the signal and the LO port since the LO voltage fed via two edge-coupled strips is orthogonal to the signal, which is present on the two broad-side coupled strips. The RF balun is a nonresonant tapered balun about a quarter-wavelength long at the lowest frequency. The resonant balun for the LO power is designed accurately using available theoretical knowledge of the Marchand balun [47].

2.8.8 Harmonic Mixers

There are two types of harmonic mixers of practical importance:

i. Harmonic mixers with a large harmonic multiplication number, for applications where sensitivity is of minor importance. Typical applications are as mixers for spectrum analyzers and as harmonic mixers for phase-locked circuits.

ii. Harmonic mixers in applications where the sensitivity is important. Such mixers are particularly important for millimeter waves, since simpler and cheaper local oscillators can be used.

In Fig. 2.36 is shown a design useful in millimeter applications. Special so-called notch-front diodes are used [71], that is, honeycomb diode chips with metallization on adjacent sides. These chips can be soldered into a microstrip circuit and contacted by a whisker. In a 55 GHz mixer, a conversion loss as low as 3 dB with beam-lead diodes has been demonstrated. At 100 GHz with notch-front diodes, a conversion loss of 6.8 dB was obtained.

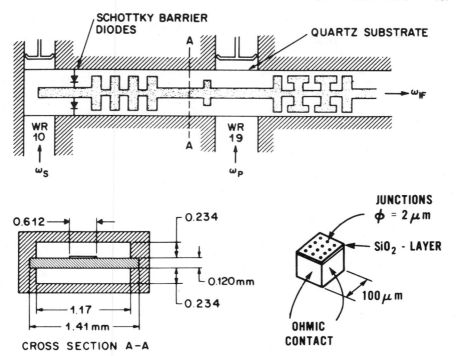

Figure 2.36 Subharmonic mixer using notch-front diodes. (From Ref. 71; copyright 1975 IEEE, reproduced by permission.)

Also, MIC designs have been developed [72, 73]. In Ref. 73 are discussed various designs of subharmonically pumped multiple-diode mixers, including a four-diode fourth-harmonic mixer. The planar doped diode should be remembered in connection with subharmonically pumped mixers since it is a device that can be designed to have a symmetrical *iv* characteristic [5].

2.9 PRACTICAL IMPLEMENTATION OF MILLIMETER WAVE SCHOTTKY BARRIER MIXERS

2.9.1 Introduction

In Section 2.8 we described mixers constructed using various design techniques that, at least in some cases, can be used for millimeter wave mixers. However, those designs do require beam lead or quad-type diodes, which for shorter millimeter wavelengths (>100 GHz) are not yet available. Therefore, the waveguide design is important, and in this section it is described in some detail. Since low noise amplifiers are not yet available for millimeter waves, cryogenic (cooled) mixer receivers, described in Section 2.9.3, are still the alternative for low-noise receivers.

2.9.2 Single-Ended Waveguide Mixers

In a single-ended waveguide mixer, the LO and the signal enter the mixer mount through the same waveguide (see Fig. 2.37). A diplexer, which may be constructed either with

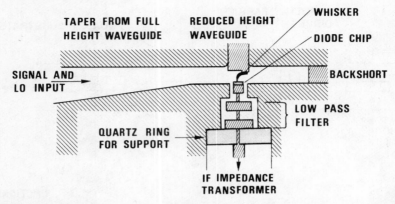

Figure 2.37 Schematic diagram of a waveguide mixer mount. (From Ref. 78; copyright 1983 IEEE, reproduced by permission.)

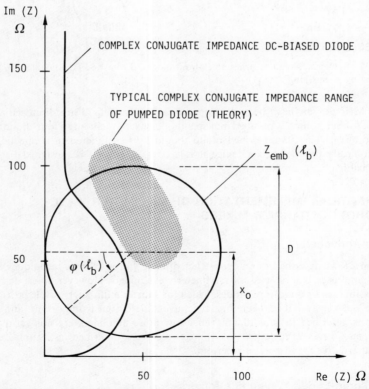

Figure 2.38 Locus of the embedding impedance as seen from the diode, and the complex conjugate impedance of the dc- and LO-biased diode. (From Ref. 78; copyright 1983 IEEE, reproduced by permission.)

waveguide techniques using a narrow-band injection filter [74] or with quasi-optical techniques using interferometers [75], is used for combining the LO and the signal.

The diode mount has to be designed for an appropriate match to the diode. In Fig. 2.38 the situation is described in an approximate way. The solid line describes the complex conjugate of the diode impedance for a dc-biased diode [see Fig. 2.4 and Eqs. (2.3)–(2.5)]:

$$Z_d^* = R_s + \left[\frac{qi}{k\eta T} - j\omega C(v_j) \right]^{-1} \tag{2.75}$$

The shaded area in Fig. 2.38 shows where in the impedance diagram one typically has the corresponding impedance of the pumped diode. The exact impedance of the pumped diode can only be found using the exact theory outlined in Section 2.5. It can be concluded that the diode capacitance affects the maximum real part of the diode impedance.

It is quite difficult to determine accurately the impedance of the diode mount. Theories exist for an approximate evaluation of the mount impedance [76, 77]. In Fig. 2.39 a simple model is given for the mount impedance. The capacitance X_c from the whisker to the diode chip may be neglected if the waveguide height is made low enough to make the ratio $X_w/X_c \ll 1$. For this case, the diode mount impedance circle has a diameter equal to the waveguide impedance:

$$Z_{wg} = \frac{754b}{a} \left[1 - \left(\frac{f}{f_c} \right)^2 \right]^{-1/2} \quad \Omega \tag{2.76}$$

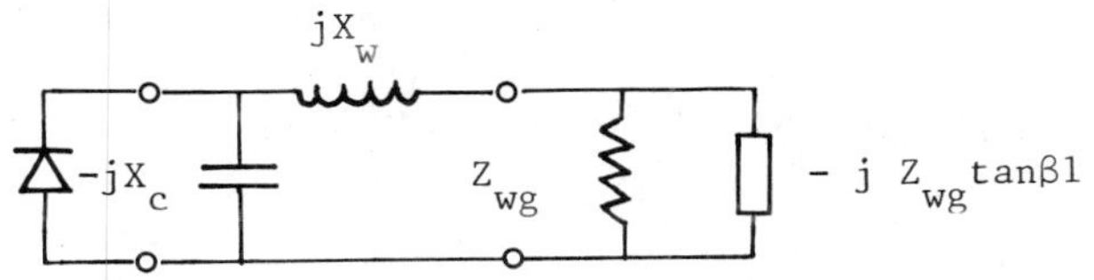

$$D = \frac{Z_{wg}}{(1-X_w/X_c)^2} \qquad\qquad X_o = \frac{X_w}{1-X_w/X_c}$$

$$\phi = -2\arctan \left[\frac{X_s'/Z_{wg}'}{1 - \dfrac{X_s}{X_c - X_w}} \right] \qquad\qquad \frac{1}{X_s'} = \frac{1}{Z_{wg}\tan\beta 1} - \frac{1}{X_c}$$

Figure 2.39   Simple equivalent circuit for the waveguide mixer mount. The parameters describing the embedding impedance circle shown in Figure 2.38 are described in terms of the circuit impedances, where X_c is the whisker-diode chip capacitance, X_w the whisker inductance, and Z_{wg} the waveguide characteristic impedance. (From Ref. 80; copyright 1980 IEEE, reproduced by permission.)

For a given diode, an increase in frequency will lower the diode impedance, forcing one to decrease the ratio b/a. In particular, the waveguide height may not be decreased much below, say, 150 μm, a fact that forces one to use lower-capacitance diodes. In practice, it is found that the whisker inductance usually has to be made as low as possible by making the length of the whisker short. The whisker can be made of, for example, 12.5-μm-diameter phosphor-bronze wire. For more details, see Refs. 78 and 79.

The embedding impedance seen by the diode at the fundamental frequencies can be evaluated by a technique using the dc-biased diode impedance as a reference, and simply measuring the reflected power versus the backshort setting [80]. However, the embedding impedance at harmonic frequencies cannot be obtained in this way. These impedances can be obtained as well by performing measurements using a microwave network analyzer on a scaled model of the mixer mount [29, 30].

Of importance for optimum performance is also the design of the adjustable backshort, the IF filter, and IF impedance transformer. It is recommended that a backshort be designed that will work well for the fundamental *and* for the first harmonic frequencies. A design that fulfills this requirement is given in Ref. 81. The IF filter should be designed to stop the escape of the fundamental frequency band as well as the first harmonic frequency band. The IF transformer should typically match a diode impedance of 150 to 200 Ω to the 50-Ω IF amplifier input. For further details, see Ref. 79.

The choice of the intermediate frequency is based on performance considerations, primarily the receiver noise temperature and instantaneous bandwidth. FET amplifiers will show lower noise the lower the frequency [82]. However, if the IF becomes too low, the amplifier cannot be made very broadband, and LO noise may become a problem (see Section 2.9.4). IF frequencies between 1 and 5 GHz are typical. At 1.5 GHz, a room-temperature noise temperature of about 40 K and a bandwidth of about 400 MHz can be obtained, while at 4 GHz, 70 K and 600 GHz are feasible.

Choosing a higher IF (e.g., 4 GHz) has the advantage in facilitating the construction of a single-sideband receiver. Since the conversion loss and mixer noise temperature usually are not rapid functions of the backshort setting, it is possible to set the backshort in a position where it will short-circuit the image sideband without a significant deterioration of the conversion loss and noise temperature. A low IF will make this trick more difficult, which is easily understood if one considers the case when the IF is zero Hz. Reference 78 describes a receiver utilizing this technique for constructing a true 100-GHz single-sideband receiver.

2.9.3   Cryogenic Mixers

Since the mixer at present is the only practical type of receiver for frequencies above about 50 GHz, a straightforward way of realizing better sensitivity is to construct a cooled (typically 15 K) version (see Section 2.2.3). By cooling the diode, the mixer noise temperature (T_{MXR}) typically drops by a factor of 3 while the conversion loss remains approximately the same as at room temperature. The noise does not drop further because the shot noise at low temperatures is limited by tunneling, and the hot-electron-noise phenomenon is approximately temperature independent (see Section 2.2.3). Another advantage of cooling is that the noise temperature of the IF amplifier becomes considerably lower. The noise temperature of a 4-GHz (1.5-GHz) FET amplifier will typically drop from about 50 K (40 K) at room temperature to about 13 K (8 K) at 15 K ambient temperature [82, 83]. Hence a receiver temperature drop of at least a factor of 3 is typical when the entire receiver is cooled from room temperature to say 15 K. Recently, it has been shown that HEMTs at low temperatures offer still lower noise temperatures, typically 1 K/GHz [84].

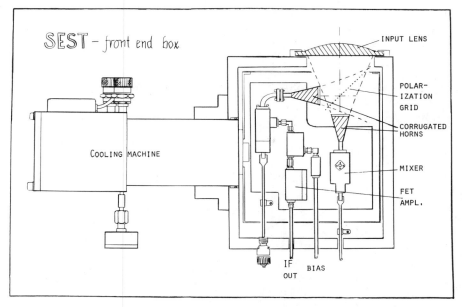

Figure 2.40   Schematic outline of a cryogenic millimeter wave mixer receiver. For thermal reasons, the mixer receiver has to be housed in a vacuum, and the lens acts as a vacuum window.

The design of the mixer mount follows essentially the same general rules as outlined in Section 2.9.2. The fact that (part of) the receiver is cooled will, of course, cause special design constraints. In the design shown in Fig. 2.37, the IF filter is mounted in the mixer block using a quartz or Macor dielectric ring [78, 79], an arrangement that will avoid differential expansion problems upon cycling the mixer in temperature. It is also essential to take precautions to avoid the jamming of adjustment screws in the cryogenic environment (i.e., avoiding any grease) [85].

In Fig. 2.40 is shown schematically the outline of a cryogenic mixer receiver. A closed-cycle cooling machine will cool the receiver parts to about 15 K. Vibrations from the machine may be avoided by using a flexible copper strip for the thermal contact between the coldhead and the mixer-FET amplifier. The Dewar window has to be either thin enough (Mylar film can be used) not to cause any severe reflections, or have an anti-reflection coating. The latter can be arranged by using quartz ($\epsilon = 3.85$) as the window material and a quarter-wavelength-thick Teflon ($\epsilon = 2.0$) sheet as the coating. The window of the receiver shown in Fig. 2.40 is also a lens for adapting the feed-horn antenna pattern to the a quasi-optical interferometer for injection of local oscillator power.

2.9.4   Millimeter Wave Local Oscillators

Essentially, the four types of LO sources listed in Table 2.5 can be considered. In Fig. 2.41, output power versus frequency for the various LO sources is shown. Important also is that the LO noise at $f_{LO} \pm f_{IF}$ is low enough so as not to contribute to the receiver noise. Since the millimeter wave local oscillators are all quite expensive, a short lifetime will severely affect the running costs of the receiver.

TABLE 2.5

	Typical Lifetime (h)	Noise	Tunability	Reference
Klystron	$\simeq 1,000$	Low	Medium	86
Gunn oscillator	$> 10,000$	Low	Large	87
IMPATT oscillator	$> 10,000$	High	Small	88
Low-frequency oscillator + multiplier	$> 10,000$	Low	Medium	89

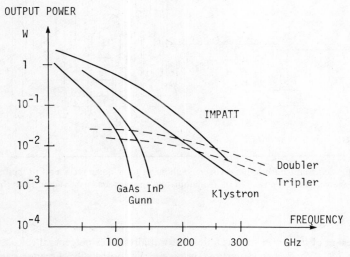

Figure 2.41   Power versus frequency for LO sources.

In Fig. 2.42 information is given about the LO noise for some LO sources. The noise properties of a fundamental oscillator followed by a multiplier deteriorate with the multiplication factor squared. In Fig. 2.42 only AM noise is shown. The FM noise can be reduced considerably using phase-locking techniques.

2.10   SUBMILLIMETER WAVE SCHOTTKY DIODE MIXERS

2.10.1   Introduction

The Schottky diode mixer has been demonstrated to work up to infrared frequencies [90–94]. For frequencies up to about 2.5 THz, Schottky diode mixers have been used in practical systems [91]. In Fig. 2.43 is shown a diagram of mixer receiver noise temperatures as obtained up to now (1986).

2.10.2   Diode Considerations

For submillimeter mixers it is, of course, imperative to have diodes with a very high cutoff frequency. This implies that the product of diode capacitance and diode series resistance must be made as small as possible. Some further insight into the problem of optimizing the diode can be gained by realizing that the pumped mixer diode can be

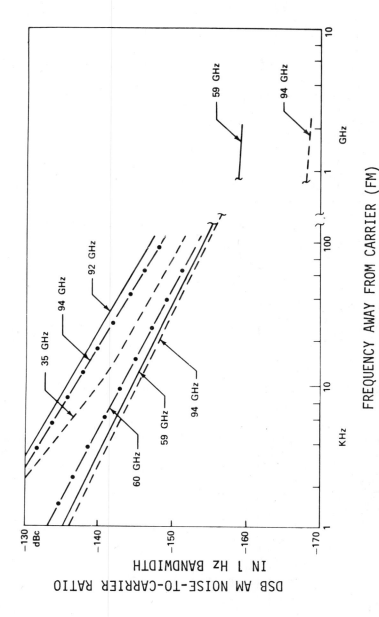

Figure 2.42   AM noise characteristics of millimeter wave oscillators. (From Ref. 88; copyright 1979 Academic Press, Inc., reproduced by permission.) Solid line, IMPATT; dashed line, Gunn; dashed-dotted line, klystron.

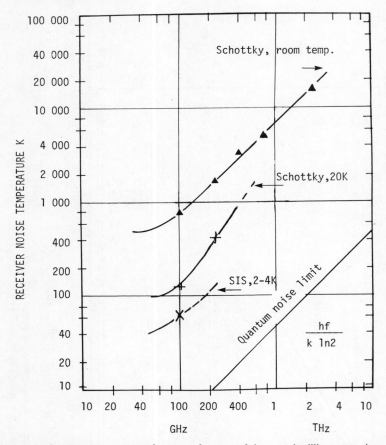

Figure 2.43 Noise temperature versus frequency for state-of-the-art submillimeter receivers (1986).

compared to a switch which, ideally, is switched on and off. The epilayer thickness of a mixer diode is, as mentioned in Section 2.2, chosen in such a way that it is completely depleted at zero-bias voltage (Mott diode). When the diode is pumped by a high-level local oscillator signal, Fig. 2.4 suggests that the diode is switched between a high-impedance state at minimum bias equal to $R_{\text{spread}} - j(1/\omega C_D)$, and a low-impedance state at maximum forward bias equal to $R_{\text{spread}} + R_{\text{epi}}/(1 + j\omega C_D R_{\text{epi}})$. To achieve efficient mixing, these high and low impedance levels should be made large (by making C_D small) and small (by making $R_{S\text{epi}} + R_{\text{spread}}$ small), respectively, compared to the embedding resistance (R_A). Notice that the diode area S and the epilayer thickness t_e affect R_{epi} and C_D in such a way that the RC time constant $\tau = 2\pi R_{\text{epi}} C_D$ is independent of S and t_{epi}. The epilayer doping concentration, however, can be optimized by minimizing R_{epi} proportional to $1/[\mu(N_d)N_D]$, suggesting that the doping concentration should be as large as possible. However, too large a doping concentration will cause two less desirable effects: One is that the capacitance will become large since the epilayer thickness becomes thin; and another is that tunneling will become dominant at room temperature, making the mixer noisier due to the increase in shot noise. Further information can be obtained from, for example, Refs. 13, 32, and 93.

However, the diode model introduced in Section 2.2 has some shortcomings when the frequency becomes of the order of 1 THz or larger [93]:

i. The displacement current density ($=\omega\epsilon E$) and the conductive current density ($=\sigma E$) in the semiconductor will become of the same magnitude at the dielectric relaxation frequency $\omega_c = \sigma/\epsilon$, which at room temperature is 1 THz for $N_d = 1 \times 10^{16}$ cm^{-3} and 7 THz for $N_d = 1 \times 10^{17}$.

ii. The carrier inertia will introduce an equivalent inductive current, which becomes equal to the conductive current at a frequency $\omega_L = q/m^*\mu$, where m^* and μ are the effective mass and the mobility of the carrier.

iii. There will be a considerable increase in the series impedance near the plasma frequency $\omega_p = \sqrt{\omega_L\omega_c}$.

iv. The relaxation properties of the carriers become important since the momentum relaxation time is of the order 0.1 ps, corresponding to a frequency of 1.6 THz, and the energy relaxation time is of the order 1 ps, corresponding to a frequency of 160 GHz!

These facts have not yet been taken into account in mixer modeling (1986), and the relative importance of the phenomena (i) to (iv) is not known in any greater detail. Experiments on harmonic mixing between a 0.4-THz oscillator and FIR lasers at around 5 THz, and direct detection experiments, seem to indicate that the Schottky diode up to 5 THz still behaves essentially as a diode [94].

2.10.3   Submillimeter Wave Mixer Circuits

The resistive part of the junction impedance $r_j/[1 + (\omega C_D r_j)^2]$ decreases with increasing frequency due to the influence of the depletion-layer capacitance. For a waveguide mixer, it is not satisfactory to scale equally all the dimensions of a mixer block in order to increase its operating frequency. Assuming the diode to be the same, it is also necessary to reduce the height of the waveguide further relative to its width, and shorten the whisker further in order to match the diode. This obviously cannot be done beyond a certain physical limit. The highest-frequency waveguide mixers operate in the frequency band 600 to 750 GHz, and the waveguide dimensions then are as small as 0.29 × 0.09 mm [92]. The waveguide mount in this case has been fabricated using electroforming techniques, and the mixer has been operated cooled to 77 K, yielding a SSB receiver noise temperature of approximately 4300 K at 700 GHz. Since the waveguide is reduced only slightly, a very low capacitance diode is required. For this particular mixer, a diode having a zero-bias capacitance of 2 fF (corresponding to an impedance of 114 Ω at 700 GHz) and a dc series resistance of 18 Ω was used.

Another approach is to use an open structure mixer mount. The most frequently used structure, shown in Fig. 2.44, has a long (4λ) wire antenna located near the junction of two faces of a 90° corner reflector. The diode is located just outside the region enclosed by the reflecting surfaces, and sometimes in a conducting ground plane. The advantages of this structure are that it can be made to operate at far higher frequencies than can the waveguide structure, and that it has a relatively low loss. A particular mixer of this type can operate over a fairly large instantaneous frequency range. The main disadvantage is that the antenna beam (Fig. 2.45) has a large fraction of the power in sidelobes [96], and typically only about 60% of the power is coupled into a fundamental Gaussian beam. Theoretical work, however, indicates that up to 80% efficiency

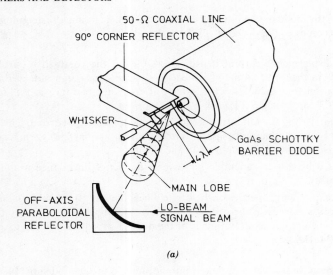

50-Ω COAXIAL LINE
90° CORNER REFLECTOR
WHISKER
GaAs SCHOTTKY BARRIER DIODE
4λ
MAIN LOBE
OFF-AXIS PARABOLOIDAL REFLECTOR
LO-BEAM
SIGNAL BEAM

(a)

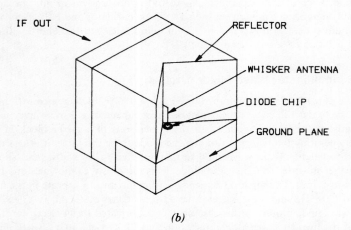

IF OUT
REFLECTOR
WHISKER ANTENNA
DIODE CHIP
GROUND PLANE

(b)

Figure 2.44 Corner reflector mixer mount: (*a*) Mixer with no ground plane (from Ref. 90; copyright 1984 Plenum Publishing Corporation, reproduced by permission); (*b*) mixer with ground plane, "corner cube" mixer (from Ref. 95; copyright 1985 SPIE, reproduced by permission.) There is no essential difference in performance between (*a*) and (*b*).

should be possible to obtain [95] (see Fig. 2.46). Also, the direction of the beam (see Fig. 2.44*a*) changes slightly with frequency. Another disadvantage is that it cannot readily be modified for optimizing the embedding impedance.

Anyhow, excellent results have been obtained [90, 91]. Using diodes with a zero-bias capacitance of only 1.2 fF and a series resistance of 12 Ω (corresponding to a cutoff frequency of 11 THz), a noise temperature as low as 5400 K was obtained at 803 GHz for an uncooled receiver. With a 7-THz cutoff frequency diode, a 17,000-K SSB receiver temperature was measured at 2.5 THz. Other mixer structures, such as one using a conical antenna [97], have been tried experimentally, but virtually no information is available on the performance.

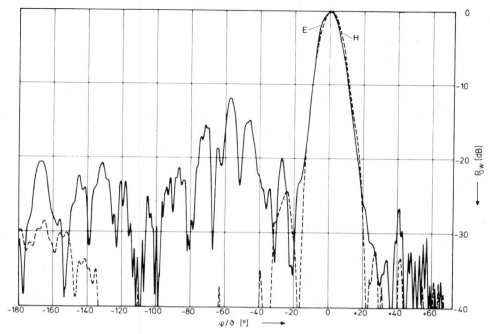

Figure 2.45 Experimental antenna diagram for a scaled model of a 90° corner reflector (Fig. 2.44*a*) with a 4λ-long whisker antenna. The distance (d) between the whisker and the apex of the 90° corner is 1.4λ. (From Ref. 96; copyright 1984 Plenum Publishing Corporation, reproduced by permission.)

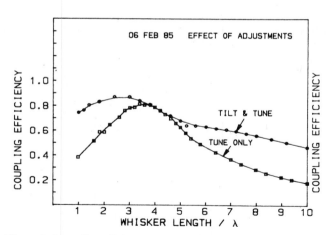

Figure 2.46 Theoretical coupling efficiency (defined as the maximum amount of radiated power coupled into a Gaussian beam) for a corner cube (see Fig. 2.44*b*) mixer mount. The tilt and tune curve refers to the fact that the mixer has to be optimized when operated off the $L/\lambda = 4$ operating point by tilting the mount and tuning the distance (d) between the whisker and the apex of the corner reflector. The tune-only curve is for fixed tilt and d, optimized for $L/\lambda = 4$. (From Ref. 95; copyright 1986 SPIE, reproduced by permission.)

2.11   SUPERCONDUCTING MILLIMETER WAVE MIXERS

2.11.1   Introduction

Superconducting tunneling elements can be used in two different modes as nonlinear mixing elements. One of these modes, where the Josephson effect is used, was the first tried in mixer applications. Today the other, called the quasi-particle tunneling mode, is considered the most promising. The Josephson effect, caused by tunneling of the Cooper pairs responsible for the supercurrent, basically creates a nonlinear reactance which can be used for parametric frequency down-conversion. Quasi-particle tunneling, on the other hand, is obtained when Cooper pairs first break up into single electrons which tunnel through the barrier and then recombine into Cooper pairs. Electrically, the device in this case will behave as a nonlinear resistance. In both cases, quantum effects play an important role; classical mixer theory as outlined in Section 2.5 or for parametric devices in Chapter 3, is not able to explain the experimental results fully. Both effects can be observed in a tunneling element in which two superconductors are separated by a very thin isolator (on the order 20 Å), as shown in Fig. 2.47. The Josephson effect may also occur in "weak links" such as microbridges or point contacts (i.e., where two superconductors are separated by a very thin, small, superconducting contact). The contact should be small enough to prevent any Cooper pairs from passing. At present, the quasi-particle tunneling mixers have a clear lead in low-noise applications. For detailed information on both Josephson effect mixers and quasi-particle mixers, the review paper by Tucker and Feldman [98] is recommended.

The research on superconducting mixers and detectors is still quite intensive, and important developments can be foreseen in the near future. In particular, superconducting technology is expected to be important for low-noise millimeter and submillimeter wave detectors and mixers. For further reading, see Ref. 98.

2.11.2   Superconducting Tunnel Junctions

The most common type of superconducting–insulator–superconducting (SIS) tunnel elements, based on lead alloys, may be fabricated in several different ways. Often, the base electrode is made from lead with a small percentage of indium, resulting in a transition temperature $\simeq 6.5$ K. The top electrode is also a lead alloy, often containing bismuth and sometimes a small fraction of gold. Researchers at IBM have developed a technique for reliably fabricating these lead-type junctions [99]. However, there is a serious drawback with the lead alloy technology since these elements are sensitive to

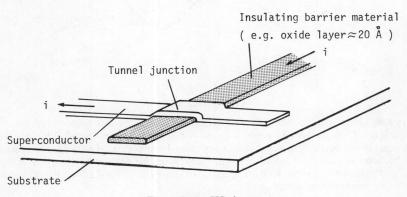

Figure 2.47   SIS element.

water and temperature cycling, a problem that is very difficult to avoid when cryogenic cooling is involved: If the elements are not made properly, they may be destroyed after being cycled in temperature a few times.

At present, other electrode and barrier materials offering potentially much more reliable SIS elements are being tested. In some of the most promising of these techniques, niobium (transition temperature $\simeq 9.3$ K) is used as the superconducting electrode. Niobium is a metal with excellent mechanical and chemical properties. Niobium elements can also benefit from using artificial barrier materials rather than the native oxide. Such elements seem to be excellent for SIS mixer applications [100], although at the time of writing, no startling results have been obtained. More advanced materials with higher transition temperatures than those of niobium will probably soon turn out to be even better than niobium. In particular, niobium nitride (transition temperature $\simeq 16$ K) is a very promising electrode material [101]. If this material or some of the hard superconducting materials such as niobium–tin (transition temperature $\simeq 19$ K) can be used in SIS junctions, cooling will become much less of a problem, since cooling machines for temperatures down to about 12 K are available for moderate prices. Even more spectacular is the possibility of using the very recently announced superconducting materials with transition temperatures above the boiling point of liquid nitrogen (i.e., $T_c > 77$ K) [102].

Since the insulating layer in a tunnel junction is very thin, the capacitance of the element becomes quite large even if the surface area is only a few square micrometers. Techniques have thus been developed to make the area very small, either by defining the area directly by using photolithographic techniques, by using certain tricks with masks and oblique evaporation [103] of the superconductor, or by creating the junction at the edge of a superconducting film which will make one dimension typically 1000 Å [104]. The effective capacitance can also be made small by arranging several equal elements in series, a technique that also increases the power level at which the mixer saturates [98, 105].

2.11.3 Quasi-Particle Mixers

The current–voltage characteristic of an SIS element is shown schematically in Fig. 2.48. The physics that will explain this behavior can be found, for example, in Ref. 98. The fact that the *iv* curve bends considerably within a voltage interval hf/q (where h is Planck's constant and q is the charge of the electron) leads to quantum phenomena when the junction is exposed to radiation of frequency f. The effect, called photon-assisted tunneling, is illustrated in Fig. 2.48. The formal complexity inherent in the analysis of the photon-assisted tunneling effect is considerable [98]. It is found that although the mixing occurs due to a "resistive" nonlinear effect, the conversion gain can, in fact, be larger than 1 [98]! From the classical point of view this is an unexpected result, and can only be explained as due to quantum phenomena.

The noise of the quasi-particle mixer is very low. Essentially, shot noise yields the dominating contribution. Applying Eq. (2.8) to the *iv* characteristic reveals that the shot noise is at least one order of magnitude lower than for the cooled Schottky mixer. Moreover, there is no series resistance, which for the diode mixer causes further noise and conversion loss. Since the SIS mixer may have very low conversion loss or even conversion gain, the mixer noise may, in fact, be comparable to the quantum noise, which is often represented as an equivalent temperature of [98]:

$$T_{qn} = \frac{hf}{k \ln 2} \simeq 0.069 f_{GHz} \qquad K \qquad (2.77)$$

At 115 GHz, where the quantum noise is approximately 7 K, receiver noise temperatures as low as 70 K have been reported, and very careful measurements at 36 GHz

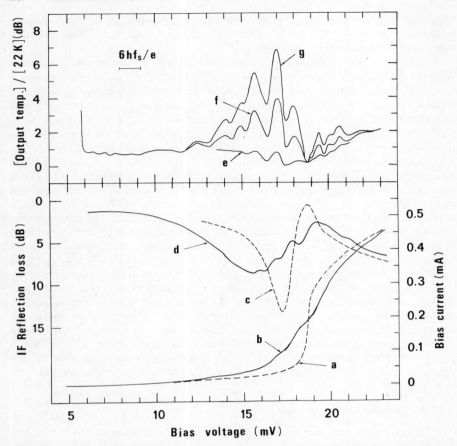

Figure 2.48 DC IV curves for SIS elements (an array of six elements in series) without (*a*) and with (*b*) LO power applied. In curve (*c*) and (*d*) are shown the IF reflection loss without LO and with LO, respectively. Curves (*e*), (*f*) and (*g*) show the IF output with LO and with loads of 2, 80, and 297 K, respectively. (From Ref. 107; copyright 1983, Plenum Publishing Corporation, reproduced by permission.)

show that $T_{MXR} \simeq 5.6 \pm 2.5$ K, a noise temperature that is very close to the quantum limit [106].

2.11.4 Practical Implementation of Quasi-Particle Mixers

Since the quantum effects that improve the conversion efficiency become more prominent at millimeter wave frequencies, and since at microwave frequencies, cooled FET (HEMT) amplifiers have very low noise temperatures, the quasi-particle mixer is important primarily in the millimeter and submillimeter wave region. Receiver systems for radio astronomical applications have been constructed and used from about 35 GHz [107] up to a few hundred gigahertz [108]. Figure 2.43 shows some results reported by early 1986.

The SIS element is a true planar device, fabricated directly on a substrate material. It is therefore natural to make part of the coupling structure on the same substrate. In Fig. 2.49 some examples of different SIS mixer designs are shown. Of particular

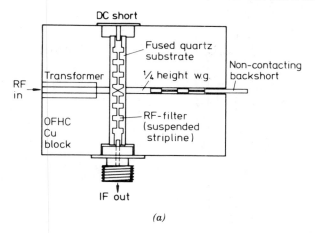

(a)

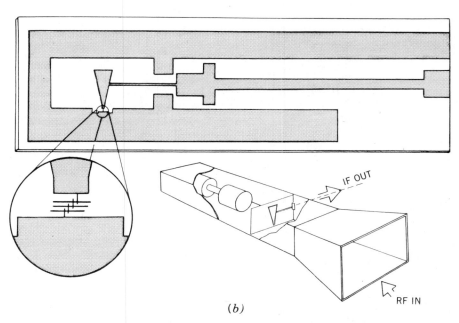

(b)

Figure 2.49   (*a*) Schematic drawing of an 80- to 115 GHz mixer mount [110]. A channel waveguide transformer [111] reduces the waveguide height by a factor of 4. The RF matching is based on scaled model measurements, and optimized by designing the filter structure on the quartz substrate for optimum performance. (From Ref. 110 copyright 1986 Int. J. IR & MM Waves) (*b*) Alternative SIS mixer waveguide mount. An input and output coupling structure and the SIS junctions are fabricated on one piece of glass or quartz substrate. By proper design of the triangular waveguide antenna probe, proper coupling can be arranged to the array of, in this case, six SIS junctions, formed at the crossings of superconducting strips a few micrometers wide [107]. (From Ref. 107 copyright 1983 Int. J. IR & MM Waves).

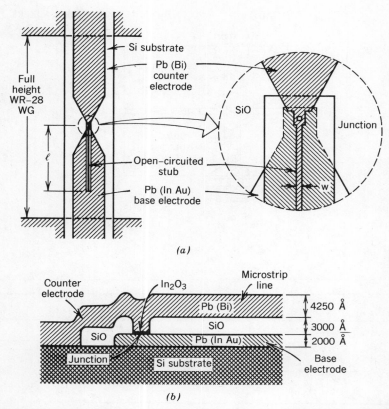

(a)

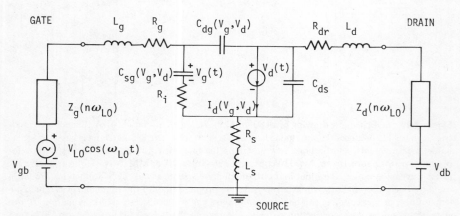

(b)

Figure 2.50 (a) Open-circuited stub slightly longer than $\lambda/4$ used for creating a resonating inductance in parallel with the junction capacitance (from Ref. 113; copyright 1985 IEEE, reproduced by permission); (b) fabricational details.

Figure 2.51 Somewhat simplified large-signal equivalent circuit of an FET mixer, indicating the dominating nonlinear circuit elements. R_g, R_s, and R_{dr} are the parasitic resistances, and L_g, L_s, and L_d are the parasitic inductances in the gate, source, and drain leads, respectively. $Z_g(n\omega_{LO})$ and $Z_d(n\omega_{LO})$ are the embedding impedances at the gate and the drain terminals, respectively. The LO could be applied to the drain terminals (drain mixer) instead of the gate (gate mixer).

interest are the attempts to integrate the SIS mixing element with planar antennas [108, 109].

A problem in the matching of the SIS element to the input circuit is the large capacitance of the element. Typically, an element with a normal resistance of approximately 50 Ω is required. The capacitance for such an element may cause a shunt reactance of less than 10 Ω. A natural way to improve the matching is to incorporate a shunt inductance on the substrate which tunes out the junction capacitance [112, 113]. It is not possible to use a simple inductor, since it would prevent voltage bias of the junction. An open-circuited stub slightly longer than $\lambda/4$ is a natural way to solve this problem [113]. An elegant way of introducing the stub required is shown in Fig. 2.50. Excellent results have been obtained using this approach. A conversion gain (loss) of about 0 dB and mixer receiver noise temperatures below 50 K have been demonstrated at 100 GHz.

2.12   FET MIXERS

2.12.1   Introduction

The FET mixer is an active device that combines frequency down-conversion and amplification in one device: the field-effect transistor (FET). This is its principal advantage over the Schottky mixer. The interest in FET mixers has been relatively low, mainly because most experimental work found that the noise of the mixer is inferior to the noise of the Schottky diode mixer–FET amplifier combination. However, this may be due to the lack of theoretical knowledge concerning how to optimize the embedding circuit.

In this section we describe different FET mixer circuits of practical interest and give an introduction to the analysis of mixer properties. Single- and dual-gate devices can be used in mixer applications. It is possible to apply the LO either between the gate and the source terminals (gate mixer), or between the drain and the source terminals (drain mixer). Also, balanced and double balanced mixer configurations are possible. For further information, see Refs. 1 and 114.

2.12.2   Mixing in a FET

Several of the circuit elements of a FET are bias dependent. Hence, when a low-level signal is applied to a FET pumped with a strong LO signal, the modulated circuit elements will cause signal power to be converted to other frequencies.

Figure 2.51 shows an equivalent circuit of a FET, indicating which circuit parameters are voltage dependent. Most important for the mixing is the pumped transconductance $g_m(t)$. The other bias-dependent circuit parameters play a minor role in the mixing process, but cannot be ignored if an accurate analysis is required. The drain resistance R_{dr} should always be included, while the gate–source capacitance C_{sg} and the charging resistance R_i are of less importance. The equivalent transconductance of an FET for different bias conditions is shown in Fig. 2.52.

2.12.3   Single-gate FET Mixers

More work has been carried out on gate mixers than on drain mixers, although the latter may have some advantages [115]. In this section we discuss some features and design rules for the single-gate FET "gate mixer".

Large-Signal Analysis.   For a more accurate analysis of the FET mixer, as for the diode mixer, first a large-signal analysis has to be performed to determine the Fourier

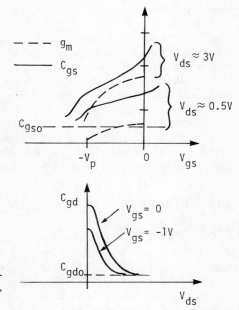

Figure 2.52 Typical behavior of the bias-dependent circuit parameters, $g_m(V_{gs})$, $C_{gs}(V_{gs}, V_{ds})$, and $C_{gd}(V_{gs}, V_{ds})$.

components of all the voltage-dependent circuit parameters. As for the diode mixer, the behavior of the device itself may be described by a few nonlinear differential equations, which can be integrated numerically, while the response of the embedding circuit may be described in the frequency domain. The two sets of equations can be solved in a way analogous to the diode mixer. However, since the device is active, the large-signal algorithm may not converge if the circuit is unstable. Such a case is, of course, of no practical interest. However, one must keep in mind that failures of the large-signal algorithm to converge may also be due to numerical instabilities in the computer algorithm.

The three basic differential equations for analyzing the FET mixer are identified in Fig. 2.53. By analyzing three current loops, one obtains

$$-2V_{ii} + R_1(I_f + I_{g1}) + V_f + R_2(I_f - I_{d1}) + 2V_{oi} = 0 \quad (2.78)$$

$$-2V_{ii} + R_1(I_f + I_{g1}) + V_g + I_{g1}R_i + R_s(I_{g1} + I_{d1}) + L\frac{d(I_{g1} + I_{d1})}{dt} = 0 \quad (2.79)$$

$$-2V_{oi} + R_2(I_{d1} - I_f) + R_s(I_{d1} + I_{g1}) + L\frac{d(I_{g1} + I_{d1})}{dt} = 0 \quad (2.80)$$

In these equations, relations exist between I_{g1}, and V_g, between V_f and I_f, and between I_{d1} and V_g and V_d. The embedding source and load netweeorks are treated in the frequency plane (see Fig. 2.51), that is,

$$V_{gn} - Z_{gn}I_{gn} - V_{gsn} = 0 \quad (n = 0, 1, 2, \ldots) \quad (2.81)$$

$$V_{dn} - Z_{dn}I_{dn} - V_{dsn} = 0 \quad (n = 0, 1, 2, \ldots) \quad (2.82)$$

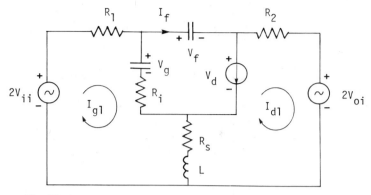

Figure 2.53 Simplified FET equivalent circuit illustrating the voltage and current loops used in the time-domain analysis.

where $n = n$th Fourier component of the different time-dependent currents
                and voltages
V_{gsn} and V_{dsn} = gate–source and the drain–source Fourier components, respectively
   V_{gn} and V_{dn} = source voltages at the gate and drain terminals, respectively

That is, for a gate mixer (Fig. 2.51), $V_{g0} = V_{gb}$, $V_{d0} = V_{db}$, and $V_{g1} = V_{LO}$, while all the other components are zero. For further details, see Refs. 1, 27, and 114.

Small-Signal Analysis. Knowing the time dependence of the various circuit elements from the large-signal nonlinear analysis, the small-signal behavior can be calculated in a way that is analogous to how it was made in the diode case, described in Section 2.5.3. The small-signal linear and time-varying equivalent circuit is shown in Fig. 2.54. Details concerning the mathematical approach of the analysis can be found in Ref. 1. The analysis allows for calculation of the conversion gain and the input and output impedances.
    It is illuminating to make a simplified analysis: By assuming that when the signal and

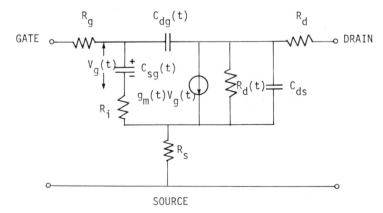

Figure 2.54 Small signal linear time-varying equivalent circuit. R_g, R_s, and R_{dr} are the parasitic resistances in the gate, source, and drain leads, respectively, and R_i is the charging resistance for the source–gate capacitance C_{sg}.

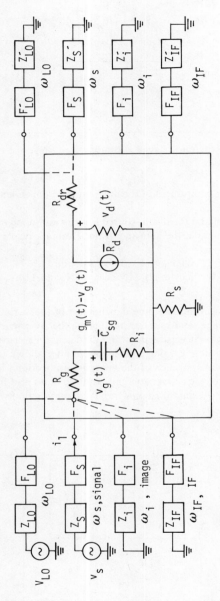

Figure 2.55 Simplified schematic of FET mixer, including only signal, image, and IF circuits ($R_{dr} \equiv R_d$ of Fig. 2.54). (From Ref. 116; copyright 1976 IEEE, reproduced by permission.)

the LO are applied to the gate–source terminals, only ω_{LO}, ω_s, $\omega_{IF} = \omega_s - \omega_{LO}$, and $\omega_i = 2\omega_{LO} - \omega_s$ have to be accounted for (see Fig. 2.55), and $g_m(t)$ is the dominating nonlinear circuit element (the time dependence of C_{gs} and R_{ds} is neglected). It follows that the available mixer conversion gain simply is [116]

$$G_c = \frac{g_1^2}{4\omega_i^2 \bar{C}^2} \frac{\bar{R}_d}{R_{in}} \tag{2.83}$$

where g_1 = Fourier component at frequency ω_{LO} of the transconductance
$\bar{C}$ = time-averaged value of the source–gate capacitance
$\bar{R}_d$ = time-averaged value of the drain resistance (R_{ds})
R_{in} = input resistance ($R_{in} = R_g + R_i + R_s$, where R_g is the gate (parasitic) resistance, R_i the charging resistance, and R_s the source (parasitic) resistance)

From Eq. (2.83) it is obvious that the conversion gain can be larger than 1.

It is an interesting fact that formula (2.83) is identical with the expression for the FET amplifier gain, if g_1, $\bar{C}$, and $\bar{R}_d$ are replaced by the corresponding values for the amplifier. In fact, the mixer conversion gain can exceed the amplifier gain.

Noise Analysis. Concerning the FET mixer noise, very little has been published until now [117]. Thermal noise emanates from the gate, source, and drain leads, related to R_g, R_s, and R_d. Noise is also caused by fluctuations in the drain current and induced-gate current noise. The latter two contributions are partially correlated [117].

No detailed information exists yet on how to minimize the FET mixer noise. Experimental results yield noise figures at X-band of about 4 dB. Further work using the more elaborate theoretical analysis outlined briefly above, and proper experimental implementation, may lead to significantly improved FET mixer performance.

Some Design Rules. The optimum performance of an FET mixer is a compromise between minimum noise figure and conversion gain. The IF amplifier noise contributes to the total receiver noise in the same way as for the diode mixer.

It seems to be advantageous to make the transconductance dominant in the mixing process and the influence from the other nonlinear impedances negligible. Essentially, this can be achieved by keeping the FET in the current-saturated region throughout the LO cycle. It also seems advantageous to short-circuit the gate to the source at all mixing products other than the LO and the signal frequency [1]. The output impedance of a FET mixer is normally quite high, on the order of 1 kΩ, which may cause some problems. Another problem with the single-gate mixer is that the LO is amplified and will leak into the IF output port, which implies that the IF circuit must include a low-pass filter to prevent the LO from causing problems in the IF circuit. The schematic of a FET gate mixer design is shown in Fig. 2.56.

2.12.4 Drain Mixers

As mentioned above, it is also possible to construct a FET mixer such that the LO is applied between the drain and the source terminals. The principal nonlinearities are again the transconductance and the drain resistance. Thus the voltage amplification factor $\mu = g_m R_d$ becomes a time-varying function. Very few papers have been published describing drain mixers. Experimental evidence indicates that the performance may at least equal the performance of the gate mixer [115]. A simple analysis yields the

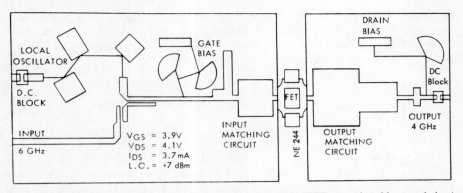

Figure 2.56 Gate mixer design. (From Ref. 115; copyright 1976 IEEE, reproduced by permission.)

following formula for the conversion gain [118]:

$$G = \frac{|\mu_1|^2}{4\omega_s^2 C_{sg}^2 (R_{gm} + R_i + R_s)(R_{dr} + R_s + R_{d0})} \tag{2.84}$$

where μ_1 is the Fourier component of the voltage amplification factor with the frequency equal to ω_{LO}, and R_{d0} is the time average of the drain resistance R_d. Figure 2.57 shows a schematic of a drain mixer design.

2.12.5 Dual-Gate Mixers

The dual-gate FET has two gate electrodes in parallel, between the source and the drain electrodes. This allows for separate circuits for the LO and the signal, yielding a very good LO/signal separation. This is the primary advantage of this mixer. The dual-gate FET can be modeled as two single-gate FETs in series, as shown in Fig. 2.58. Notice, however, that the connection point between the drain of the lower FET and the source

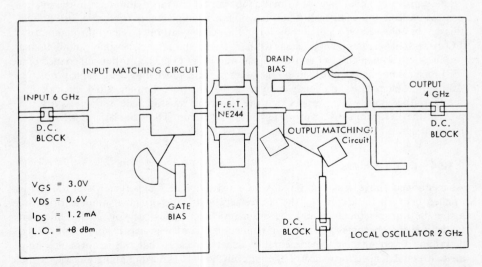

Figure 2.57 Drain mixer design. (From Ref. 115; copyright 1976 IEEE, reproduced by permission.)

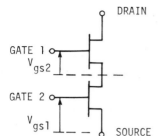

Figure 2.58 Two FETs in series is equivalent to a dual-gate FET.

of the upper one is not physically accessible. Also note that the operating voltages of the two individual FETs are not applied directly to the dual-gate FET terminals.

As for the single-gate mixer, the best operation is obtained when the mixer is operated as a transconductance mixer (i.e., the unwanted LO and mixing products should be short-circuited at the gate and drain terminals). Since the LO for this mixer is amplified, it is possible to realize a self-oscillating mixer (i.e., no external LO source is necessary) [119].

2.12.6 Balanced and Double Balanced FET Mixers

As for diode mixers, balanced and double balanced mixer configurations can be considered. Either a 90° or 180° hybrid may be used. Essentially, the properties are the same as for the diode mixers, as discussed in Section 2.6. Hence, improved spurious response properties, LO noise rejection, and LO/signal isolation are obtained. As indicated in Fig. 2.59, both the 90° and the 180° hybrid mixer require a 180° output hybrid for combining the IFs of the two individual mixers.

In balanced configurations, single-gate FETs may be more straightforward to use than dual-gate FETs since the latter requires separate hybrids for the LO and the signal.

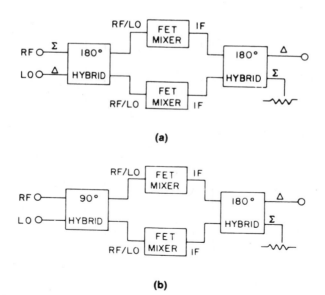

Figure 2.59 Balanced FET mixers. (From Stephan A. Maas, *Microwave Mixers*, p. 306; copyright 1986 Artech House, Inc., reproduced by permission.)

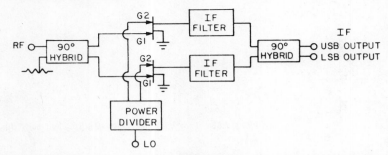

Figure 2.60 Dual-gate–FET image–rejection mixer (From Stephan A. Maas, *Microwave Mixers*, p. 308; copyright 1986 Artech House, Inc., reproduced by permission.)

However, as discussed in Section 2.12.5, the dual-gate FET (single-device) mixer has excellent inherent LO/signal isolation. The choice among the available configurations will be a matter of performance required.

Image rejection mixers and other devices may be realized using a pair of single-gate balanced mixers. However, using two dual-gate FET mixers in a balanced configuration (Fig. 2.60) yields a much simpler circuit with only two hybrids and one power divider. Double-double balanced mixers should also be constructed using dual-gate mixers, to keep the complication of the circuit within reasonable limits. Since the nonlinearities of the FET devices are relatively weak compared to diodes, the balance will usually become very good, resulting in excellent spurious signal rejection and port-to-port isolation.

2.13   FURTHER ASPECTS ON MICROWAVE AND MILLIMETER WAVE MIXERS

2.13.1   Introduction

In this section we discuss mixers using types of nonlinear devices not discussed above, and mixers made using monolithic techniques on GaAs substrates.

2.13.2   Tunel Diode Mixers

Considerable effort in designing tunnel diode frequency converters was put forth in the 1960s [120]. However, the tunnel diode mixer never turned out to be of great practical use, mainly because the noise properties were inferior to the Schottky diode mixers. It is, however, wise to learn the basics about this mixer, and keep in mind that a diode which exhibits a negative resistance, like the tunnel diode, has certain very interesting properties, such as available conversion gain. The tunnel diode is discussed briefly in Section 2.2.2.

In Fig. 2.61 is shown a small-signal equivalent circuit of the tunnel diode. From this circuit it is evident that above a certain frequency f_{max}, the device will not show any negative resistance. For f_{max} we get

$$f_{max} = \frac{1}{2\pi} \frac{1}{2R_{min}C_j} \sqrt{\frac{R_{min}}{R_s} - 1} \tag{2.85}$$

where R_{min} is the smallest value of the differential negative resistance.

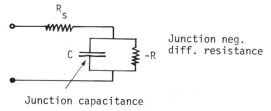

Figure 2.61 Small-signal impedance of the tunnel diode.

The iv characteristic of a tunnel diode and the differential conductance versus bias voltage is shown in Figs. 2.2c and 2.62, respectively [120]. In Fig. 2.62 is also shown the time-dependent differential conductances for four bias points, A to D. To prevent oscillations regardless of terminations, g_0 must be positive, a condition that is fullfilled for case A. However, if g_0 is negative, LO power can be produced by means of self-oscillations, an arrangement that may cause problems due to instabilities.

For the bias point A, an external LO is required, and the operational principle is similar to that of the ordinary mixer, as discussed in Section 2.4. The main differences are that the conversion gain of the mixer can be made larger than unity, a feature which results from the fact that contrary to the ordinary diode mixer, it is possible to get $|g_1| > g_0$ [compare Eq. (2.35)]. Bias point C is useful for second-harmonic mixer applications.

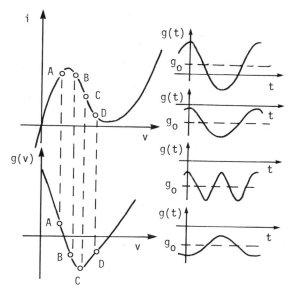

Figure 2.62 Various bias points for the tunnel diode mixer. [From Ref. 120; copyright 1961 IRE (now IEEE), reproduced by permission.]

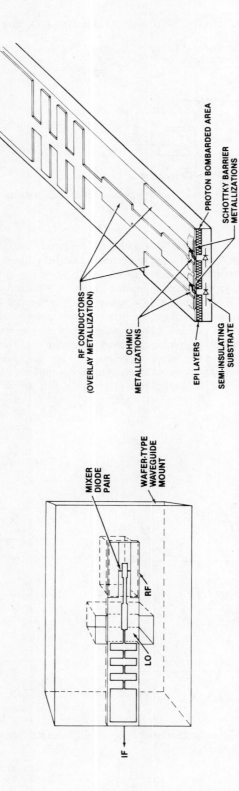

Figure 2.63 Millimeter wave Ka-band GaAs monolithic mixer chip for a crossbar-type waveguide mixer. (From Ref. 124; copyright 1983 IEEE, reproduced by permission.)

The noise of the tunnel diode mixer can be shown to be approximately equal to the noise of the tunnel diode amplifier [120]. Typical noise figures for Ge tunnel diode amplifiers are 5 dB at 6 GHz and 6 dB at 14 GHz [3], and similar numbers should be valid for mixers [121].

2.13.3   Monolithic Mixers

Microwave monolithic integrated circuits (MMIC) fabricated in GaAs have become rather mature, and several designs have been reported concerning both various types of diode mixers and FET mixers. The advantage of monolithic circuits is, in the first place, that they are small, in particular when designed for millimeter waves, and second, when fabricated in large amounts, they will become quite cheap.

Monolithic diode mixers have been successfully designed for frequencies up to 100 GHz. Today (1986), at ordinary microwave frequencies it is possible to buy balanced and double balanced MMIC mixers up to at least 15 GHz. Some GaAs foundries offer balanced and double balanced mixers as macrocells from a cell library, to be included in designs of larger circuits consisting of, for example, low-noise input amplifier stages, followed by a double balanced mixer and further amplification at the IF [122].

Due to the diode parasitics, the cutoff frequency of the monolithic diodes cannot easily be made as high as for the mixers using whiskered honeycomb diode chips. Special design features such as air bridges or proton bombardment for making the GaAs semi-insulating will isolate the diodes from the circuit parasitics and improve the cutoff frequency. Typical results show that it is possible to fabricate monolithic millimeter wave balanced mixers with diodes having a cutoff frequency of about 640 GHz, and a single-sideband conversion loss of about 6 dB at Ka band [123]. Two Ka band designs are shown in Figs. 2.63 and 2.64. In Fig. 2.64, notice that the mixer is integrated with a "bow-tie" dipole antenna.

MMIC FET mixers have also been constructed, and as for diode mixers, it is now possible to include such mixers in larger MMICs. A high-volume, low-cost 3- to 6-GHz receiver front-end MMIC on one chip, with three stages of low-noise amplification in front of a double balanced FET mixer followed by three stages of IF amplification, has been described [125].

2.13.4   Other Types of Mixers

In this chapter we have described most of the common mixers used in various, mainly low-noise applications. In this section we mention only some less commonly used mixers, both low-noise and "high-noise" types.

Hot electron bolometer mixers are very low noise mixers that are used for submillimeter applications. The most commonly used detector of this type is the indium-antimonide (InSb) detector, developed initially as a microwave and far-infrared detector. To work properly, the detector has to be cooled to a few kelvin. The detection is achieved when the conduction band electrons are heated by the absorbed radiation, causing a change in their mobility. The fact that the electrons have to transfer their excess energy to the lattice through relatively weak coupling mechanisms means that the detector is relatively slow. Hence the maximum IF frequency is only of the order of a few megahertz.

The InSb detector can be used with or without a magnetic field. Typically, double-sideband receiver noise temperatures obtained without a magnetic field at 1.6 K are about 180 K at 350 GHz, 350 K at 492 GHz, and 650 K at 625 GHz [126, 127]. With a magnetic field of about 6 kG, the detector will operate in a cyclotron resonance mode, and improved performance is possible. When operated at 1.6 K, the best receiver

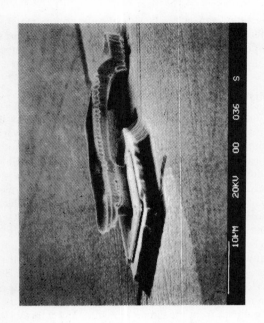

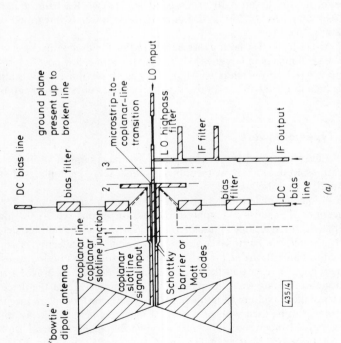

Figure 2.64 A 30 GHz GaAs monolithic single balanced mixer incorporated with a bowtie antenna: (*a*): circuit; (*b*): SEM micrograph of one Mott diode. (From Ref. 123; published previously by the Institution of Electrical Engineers in *Handbook of Microwave and Optical Components*.)

The circuit (a) includes the following labeled elements:

- DC bias line
- ground plane present up to broken line
- bias filter
- microstrip-to-coplanar-line transition
- LO input
- LO highpass filter
- IF filter
- "bowtie" dipole antenna
- coplanar line
- coplanar slotline junction
- coplanar slotline signal input
- Schottky barrier or Mott diodes
- bias filter
- DC bias line
- IF output

435/4

(a)

(b)

noise temperature for the cyclotron resonance detector is 250 K at 495 GHz, 350 K at 625 GHz, and 510 K at 812 GHz [127].

Self-mixing oscillators have applications in low-cost front ends such as doppler radar or burglar alarms. Self-mixing Gunn oscillators have consequently been developed for low-cost applications where sensitivity is of minor importance. The Gunn diode will serve both as a local oscillator, and because nonlinearities are always present, as a mixing element. A 60-GHz system based on an InP Gunn diode self-mixing oscillator is able to detect signals levels of -80 dBm (using a 100-MHz bandwidth), a result that is consistent with information reported for lower frequencies [128].

Self-mixing BARITT oscillators: The BARITT diode is a transit-time device and a relative to the IMPATT diode [3]. However, since the mechanism responsible for the microwave oscillation is thermoionic emission and diffusion, it will have a much better signal-to-noise performance than that of the IMPATT diode. The BARITT diode has also been operated in a self-mixing mode, and experiments at Ka-band indicate the sensitivity to be similar to that of the Gunn diode self-mixing oscillator [129].

REFERENCES

1. S. A. Maas, *Microwave Mixers*, Artech House, Dedham, MA, 1986.

2. M. J. Howes and D. V. Morgan, Eds., *Variable Impedance Devices*, Wiley, New York, 1978.

3. S. M. Sze, *Physics of Semiconductor Devices*, 2nd ed., Wiley, New York, 1981.

4. E. L. Kollberg, Ed., *Microwave and Millimeter-Wave Mixers*, IEEE Press, New York, 1984.

5. R. J. Malik, "A Subharmonic Mixer Using a Planar Doped Barrier Diode with Symmetric Conductance," *IEEE Electron Device Lett.*, Vol. EDL-3, pp. 205–207, 1982.

6. M. V. Schneider, "Metal–Semiconductor Junction as Frequency Converters," in K. J. Button, Ed., *Infrared and Millimeter Waves*, Vol. 6, *Systems and Components*, Academic Press, New York, 1982.

7. R. A. Linke, M. V. Schneider, and A. Y. Cho, "Cryogenic Millimeter-Wave Receiver Using Molecular Beam Epitaxy Diodes," *IEEE Trans. Microwave Theory Tech.*, Vol. MTT-26, No. 12, pp. 935–938, 1978.

8. G. K. Sherrill, R. J. Mattauch, and T. W. Crowe, "Interfacial Stress and Excess Noise in Schottky-Barrier Mixer Diodes," *IEEE Trans. Microwave Theory Tech.*, Vol. MTT-34, No. 3, pp. 342–345, 1986.

9. E. L. Kollberg, H. Zirath, and A. Jelenski, "Temperature-Variable Characteristics and Noise in Metal–Semiconductor Junctions," *IEEE Trans. Microwave Theory Tech.*, Vol. MTT-34, No. 9, pp. 913–922, 1986.

10. H. Zirath, "High-Frequency Noise and Current–Voltage Characteristics of MM-Wave Platinum n-n^+ GaAs Schottky Barrier Diodes," *J. Appl. Phys.*, Vol. 60, pp. 1399–1408, August 15, 1986.

11. P. J. Price, "Fluctuations of Hot Electrons," in R. E. Burgess, Ed., *Fluctuation Phenomena in Solids*, Academic Press, New York, 1965, Chap. 8.

12. W. Baechtold, "Noise Behavior of GaAs Field-Effect Transistors with Short Gate Lengths," *IEEE Trans. Electron Devices*, Vol. ED-19, No. 5, pp. 674–680, 1972.

13. A. Jelenski, E. Kollberg, and H. Zirath, "Broadband Noise Mechanisms and Noise Measurements of Metal–Semiconductor Junctions," *IEEE Trans. Microwave Theory Tech.*, Vol. MTT-34, pp. 1193–1201, 1986.

14. T. S. Laverghetta, *Microwaves Materials and Fabrication Techniques*, Artech House, Dedham, MA, 1984.

15. A. E. Fantom, "Millimeter Wave Power Standard," *IEE Colloquium on Measurement at Millimetre and Near Millimetre Wavelengths*, Institution of Electrical Engineers, Stevenage, Hertfordshire, England, 1981.

16. F. L. Warner, "Detection of Millimetre and Submillimetre Waves," in F. A. Benson, Ed., *Millimetre and Submillimetre Waves*, Iliffe Books, London, 1969.

17. T. G. Blaney, "Detection Techniques at Short Millimeter and Submillimeter Wavelengths: An Overview," in K. J. Button, Ed., *Infrared and Millimeter Waves*, Vol. 3, *Submillimeter Techniques*, Academic Press, New York, 1980.

18. A. Hadni, "Pyroelectricity and Pyroelectric Detectors," in K. J. Button, Ed., *Infrared and Millimeter Waves*, Vol. 3, *Submillimeter Techniques*, Academic Press, New York, 1980.

19. P. L. Richards and L. T. Greenberg, "Infrared Detectors for Low-Background Astronomy: Incoherent and Coherent Devices from One Micrometer to One Millimeter," in K. J. Button, Ed., *Infrared and Millimeter Waves*, Vol. 6, *Systems and Components*, Academic Press, New York, 1982.

20. A. A. M. Saleh, *Theory of Resistive Mixers*, MIT Press, Cambridge, MA, 1971.

21. J.-D. Buchs and G. Begemann, "Frequency Conversion Using Harmonic Mixers with Resistive Diodes," *IEE J. Microwave Opt. Acoust.*, Vol. 2, No. 1, pp. 71–76, 1978.

22. D. A. Fleri and L. D. Cohen, "Nonlinear Analysis of the Schottky-Barrier Mixer Diode," *IEEE Trans. Microwave Theory Tech.*, Vol. MTT-21, No. 1, pp. 39–43, 1973.

23. S. Egami, "Nonlinear, Linear Analysis and Computer-Aided Design of Resistive Mixers," *IEEE Trans. Microwave Theory Tech.*, Vol. MTT-22, No. 3, pp. 270–275, 1974.

24. W. K. Gwarek, "Nonlinear Analysis of Microwave Mixers," M.S. thesis, Massachusetts Institute of Technology, Cambridge, 1974.

25. R. G. Hicks and P. J. Kahn, "Numerical Analysis of Nonlinear Solid-State Device Excitation in Microwave Circuits," *IEEE Trans. Microwave Theory Tech.*, Vol. MTT-30, No. 3, pp. 251–259, 1982.

26. A. R. Kerr, "A Technique for Determining the Local Oscillator Waveforms in a Microwave Mixer," *IEEE Trans. Microwave Theory Tech.*, Vol. MTT-23, No. 10, pp. 828–831, 1975.

27. C. Camacho-Penalosa, "Numerical Steady-State Analysis of Nonlinear Microwave Circuits with Periodic Excitation," *IEEE Trans. Microwave Theory Tech.*, Vol. MTT-31, No. 9, pp. 724–730, 1983.

28. D. N. Held and A. R. Kerr, "Conversion Loss and Noise of Microwave and Millimeter-Wave Mixers. Part 1. Theory," *IEEE Trans. Microwave Theory Tech.*, Vol. MTT-26, No. 2, pp. 49–55, 1978.

29. D. N. Held and A. R. Kerr, "Conversion Loss and Noise of Microwave and Millimeter-Wave Mixers. Part 2. Experiment," *IEEE Trans. Microwave Theory Tech.*, Vol. MTT-26, No. 2, pp. 55–61, 1978.

30. P. H. Siegel and A. R. Kerr, *A User-Oriented Computer Program for the Analysis of Microwave Mixers, and a Study of the Effects of the Series Inductance and Diode Capacitance on the Performance of Some Simple Mixers*, NASA Technical Memorandum, Goddard Institute for Space Studies, New York, N.Y. 1979.

31. A. R. Kerr, "Noise and Loss in Balanced and Subharmonically Pumped Mixers. Part I. Theory," *IEEE Trans. Microwave Theory Tech.*, Vol. MTT-27, No. 12, pp. 938–943, 1979.

32. T. W. Crowe and R. J. Mattauch, "Analysis and Optimization of Millimeter and Submillimeter-Wavelength Mixer Diodes," *IEEE Trans. Microwave Theory Tech.*, Vol. MTT-35, No. 2, pp. 159–168, 1987.

33. A. R. Kerr, "Shot-Noise in Resistive-Diode Mixers and the Attenuator Noise Model," *IEEE Trans. Microwave Theory Tech.*, Vol. MTT-27, No. 2, pp. 135–140, 1979.

34. M. T. Faber and W. K. Gwarek, " Nonlinear-Linear Analysis of Microwave Mixer with Any Number of Diodes," *IEEE Trans. Microwave Theory Tech.*, Vol. MTT-28, No. 11, pp. 1174–1181, 1980.

35. R. G. Hicks and P. J. Kahn, "Numerical Analysis of Subharmonic Mixers Using Accurate and Approximate Models," *IEEE Trans. Microwave Theory Tech.*, Vol. MTT-30, No. 11, pp. 2113–2119, 1982.

36. M. E. Hines, "Inherent Signal Losses in Resistive-Diode Mixers," *IEEE Trans. Microwave Theory Tech.*, Vol. MTT-29, No. 4, pp. 281–292, 1981.

37. B. Henderson, "Mixer Design Considerations Improve Performance," *MSN Commun. Technol.*, Vol. 11, No. 10, pp. 103–118, 1981.

38. C. W. Gerst, "New Mixer Designs Boost D/F Performance," *Microwaves*, Vol. 12, No. 10, pp. 60–69, 1973.

39. J. F. Reynolds and M. R. Rosenzweig, "Learn the Language of Mixer Specification," *Microwaves*, Vol. 17, No. 6, pp. 72–80, 1978.

40. R. B. Mouw, "A Broad-Band Hybrid Junction and Application to the Star Modulator," *IEEE Trans. Microwave Theory Tech.*, Vol. MTT-16, No. 11, pp. 911–918, 1968.

41. G. L. Matthai, L. Young, and E. M. T. Jones, *Microwave Filters, Impedance Matching Networks, and Coupling Structures*, Artech House, Dedham, MA, 1970.

42. S. Rehnmark, "High Directivity CTL-Couplers and a New Technique for the Measurement of CTL-Coupler Parameters," *IEEE Trans. Microwave Theory Tech.*, Vol. MTT-25, pp. 1116–1121, 1977.

43. Wen Pin Ou, "Design Equations for an Interdigital Directional Coupler," *IEEE Trans. Microwave Theory Tech.*, Vol. MTT-23, pp. 253–255, 1975.

44. V. K. Tripathi, H. B. Lunden, and J. P. Starski, "Analysis and Design of Branch-Line Hybrids with Coupled Lines," *IEEE Trans, Microwave Theory Tech.*, Vol. MTT-32, pp. 427–432, 1984.

45. N. Marcuwitz, *Waveguide Handbook*, Dover, New York, 1951.

46. B. R. Hallford, "A Designer's Guide to Planar Mixer Baluns." *Microwaves*, Vol. 18, No. 12, pp. 52–57, 1979.

47. J. H. Clotte, "Exact Design of the Marchand Balun," *Proceedings of the 9th European Microwave Conference*, Microwave Exhibitions and Publishers, Turnbridge Wells, Kent, England, 1979, pp. 480–484.

48. L. M. Orloff, "Intermodulation Analysis of Crystal Mixer," *Proc. IEEE*, Vol. 52, No. 2, pp. 173–179, 1964.

49. S. A. Maas, "Two-Tune Intermodulation in Diode Mixers," *IEEE Trans. Microwave Theory Tech.*, Vol. MTT-35, pp. 307–314, 1987.

50. J. G. Gardiner, "An Intermodulation Phenomenon in the Ring Modulator," *Radio Electron. Eng.*, Vol. 39, No. 4, pp. 193–197, 1970.

51. M. A. Maiuzzo and S. H. Cameron, "Response Coefficients of a Double-Balanced Diode Mixer," *IEEE Trans. Electromagn. Compat.*, Vol. EMC-21, No. 4, pp. 316–319, 1979.

52. M. Katoh and Y. Akaiwa, "4-GHz Integrated-Circuit Mixer," *IEEE Trans. Microwave Theory Tech.*, Vol. MTT-19, No. 7, pp. 634–637, 1971.

53. M. Akaike and S. Okamura, "Semiconductor Diode Mixer for Millimeter Wave Region," *Trans. Inst. Electron. Commun. Eng. Jpn.*, Vol. 52-B, No. 10, pp. 601–609, 1969.

54. C. J. Burkley and R. S. O'Brien, "Optimization of an 11 GHz Mixer Circuit Using Image Recovery," *Int. J. Electron.*, Vol. 38, No. 6, pp. 777–787, 1975.

55. L. E. Dickens and D. W. Maki, "An Integrated-Circuit Balanced Mixer, Image and Sum Enhanced," *IEEE Trans. Microwave Theory Tech.*, Vol. MTT-23, No. 3, pp. 276–281, 1975.

56. H. Ogawa, M. Aikawa, and K. Morita, "K-Band Integrated Double-Balanced Mixer," *IEEE Trans. Microwave Theory Tech.*, Vol. MTT-28, No. 3, pp. 180–185, 1980.

57. T. H. Oxley, "Phasing Type Image Recovery Mixer," *IEEE MTT-S Int. Microwave Symp. Dig.*, pp. 270–273, 1980.

58. L. E. Dickens and D. W. Maki, "A New 'Phased-Type' Image Enhanced Mixer," *IEEE MTT-S Int. Microwave Symp. Dig.*, pp. 149–151, 1975.

59. G. P. Kurpis and J. J. Taub, "Wide-Band X-Band Microstrip Image Rejection Balanced Mixer," *IEEE Trans. Microwave Theory Tech.*, Vol. MTT-8, No. 12, pp. 1181–1182, 1970.

60. R. S. Tahim, G. M. Hayashibara, and K. Chang, "Design and Performance of W-Band Broad-Band Integrated Circuit Mixer," *IEEE Trans. Microwave Theory Tech.*, Vol. MTT-31, pp. 277–283, 1983.

61. G. B. Stracca, F. Aspesi, and T. D'Arcangelo, "Low-Noise Microwave Down-Converter with Optimum Matching at Idle Frequencies," *IEEE Trans. Microwave Theory Tech.*, Vol. MTT-21, No. 9, pp. 544–547, 1973.

62. L. T. Yuan, "Design and Performance Analysis of an Octave Bandwidth Waveguide Mixer," *IEEE Trans. Microwave Theory Tech.*, Vol. MTT-25, No. 12, pp. 1048–1054, 1977.

63. P. J. Meier, "*E*-Plane Components for a 94-GHz Printed-Circuit Balanced Mixer," *IEEE MTT-S Int. Microwave Symp. Dig.*, pp. 267–269, 1980.

64. R. N. Bates, "Millimeter-Wave Low Noise *E*-Plane Balanced Mixers Incoporating Planar MBE GaAs Mixer Diodes," *IEEE MTT-S Int. Microwave Symp. Dig.*, pp. 13–15, 1982.

65. L. Bui and D. Ball, "Broadband Planar Balanced Mixers for Millimeter Wave Applications," *IEEE MTT-S Int. Microwave Symp. Dig.*, pp. 204–205, 1982.

66. A. Blaisdell, R. Geoffroy, and H. Howe, "A Novel Broadband Double Balanced Mixer for the 18–40 GHz Range," *IEEE MTT-S Int. Microwave Symp. Dig.*, pp. 33–35, 1982.

67. B. Henderson, "Full-Range Orthogonal Circuit Mixers Reach 2 to 26 GHz," *MSN Commun. Technol.*, Vol. 12, No. 9, pp. 122–126, 1982.

68. R. B. Culbertson and A. M. Pavio, "An Analytic Design Approach for 2–18 GHz Planar Mixer Circuit," *IEEE MTT-S Int. Microwave Symp. Dig.*, pp. 425–427, 1982.

69. B. R. Hallford, "Single Sideband Mixers for Communications Systems," *IEEE MTT-S Int. Microwave Symp. Dig.*, pp. 30–32, 1982.

70. B. R. Hallford, "Investigation of a Single-Sideband Mixer Anomaly," *IEEE Trans. Microwave Theory Tech.*, Vol. MTT-31, No. 12, pp. 1030–1038, 1983.

71. E. R. Carlson, M. V. Schneider, and T. F. McMaster, "Subharmonically Pumped Millimeter-Wave Mixers," *IEEE Trans. Microwave Theory Tech.*, Vol. MTT-26, No. 10, pp. 706–715, 1978.

72. M. V. Schneider and W. W. Snell, Jr., "Harmonically Pumped Stripline Down-Converter," *IEEE Trans. Microwave Theory Tech.*, Vol. MTT-23, No. 3, pp. 271–275, 1975.

73. J.-D. Buchs and G. Begemann, "Frequency Conversion Using Harmonic Mixers with Resistive Diodes," *IEE J. Microwave Opt. Acoust.*, Vol. 2, pp. 71–76, 1978.

74. Hong-Ih Cong, A. R. Kerr, and R. J. Mattauch, "The Low-Noise 115-GHz Receiver on the Columbia GISS 4-ft Radio Telescope," *IEEE Trans. Microwave Theory Tech.*, Vol. MTT-27, No 3, pp. 245–248, 1979.

75. P. F. Goldsmith, "Quasi-optical Techniques at Millimeter and Submillimeter Wavelengths," in K. J. Button, Ed., *Infrared and Millimeter Waves*, Vol. 6, *Systems and Components*, Academic Press, New York, 1982.

76. R. L. Eisenhart and P. J. Kahn, "Theroretical and Experimental Analysis of a Waveguide Mounting Structure," *IEEE Trans. Microwave Theory Tech.*, Vol. MTT-19, No. 8, pp. 706–719, 1971.

77. A. G. Williamson, "Analysis and Modelling of a Single-Post Waveguide Mounting Structure," *IEE Proc.*, Pt. H, Vol. 129, No. 5, pp. 271–277, 1982.

78. E. L. Kollberg and H. Zirath, "A Cryogenic Millimeter-Wave Schottky-Diode Mixer," *IEEE Trans. Microwave Theory Tech.*, Vol. MTT-31, No. 2, pp. 230–235, 1983.

79. C. R. Predmore, A. V. Räisänen, N. R. Erickson, P. F. Goldsmith, and J. L. R. Marrero, "A Broad-Band, Ultra-Low-Noise Schottky Diode Mixer Receiver from 80 to 115 GHz," *IEEE Trans. Microwave Theory Tech.*, Vol. MTT-32, No. 5, pp. 498–507, 1984.

80. C. E. Hagström and E. L. Kollberg, "Measurements of Embedding Impedance of Millimeter-Wave Diode Mounts," *IEEE Trans. Microwave Theory Tech.*, Vol. MTT-28, No. 8, pp. 899–904, 1984.

81. M. K. Brewer and A. V. Räisänen, "Dual-Harmonic Noncontacting Millimeter Waveguide Backshorts: Theory, Design, and Test," *IEEE Trans. Microwave Theory Tech.*, Vol. MTT-30, No. 5, pp. 708–714, 1982.

82. S. Weinreb, "Low-Noise Cooled GaAs FET Amplifiers," *IEEE Trans. Microwave Theory Tech.*, Vol. MTT-28, No. 10, pp. 1041–1054, 1980.

84. M. Pospieszalskie and S. Weinreb, "FET's and HEMT's at Cryogenic Temperatures—Their Properties and Use in Low-Noise Amplifiers," *IEEE MTT-S Int. Microwave Symp. Dig.*, pp. 955–958, 1987.

85. G. K. White, *Experimental Techniques in Low-Temperature Physics*, Clarendon Press, Oxford, 1979.

86. A. F. Pearce and D. J. Wootton, "Reflex Klystrons," in F. A. Benson, Ed., *Millimeter and Submillimeter Waves*, Iliffe Books, London, 1969.

87. I. G. Eddison, "Indium Phosphide and Gallium Arsenide Transferred-Electron Devices," in K. J. Button, Ed., *Infrared and Millimeter Waves*, Vol. 11, *Millimeter Components and Techniques*, Part 3, Academic Press, New York, 1984.

88. H. J. Kuno, "IMPATT Devices for Generation of Millimeter Waves," in K. J. Button, Ed., *Infrared and Millimeter Waves*, Vol. 1, *Sources of Radiation*, Academic Press, New York, 1979.

89. J. W. Archer, "Low-Noise Receiver Technology for Near-Millimeter Wavelengths," in K. J. Button, Ed., *Infrared and Millimeter Waves*, Vol. 15, *Millimeter Components and Techniques*, Part 6, Academic Press, New York, 1986.

90. H. P. Röser, E. J. Durwen, R. Wattenbach, and G. V. Schultz, "Investigation of a Heterodyne Receiver with Open Structure Mixer at 324 and 693 GHz," *Int. J. Infrared Millim. Waves*, Vol. 5, No. 3, pp. 301–314, 1984.

91. H. P. Roeser, R. Wattenbach, E. J. Durwen, and G. V. Schultz, "A High Resolution Heterodyne Spectrometer from 100 μm to 1000 μm and the Detection of CO (J = 7 − 6), CO (J = 6 − 5) and ^{13}CO (J = 3 − 2)," *Astron. Astrophys.*, Vol. 165, No. 1/2, pp. 287–289, 1986.

92. P. F. Goldsmith and N. R. Erickson, "Waveguide Submillimeter Mixers," in E. Kollberg, Ed., *Instrumentation for Submillimeter Spectroscopy*, Proc. SPIE 598, pp. 52–59, 1986.

93. W. M. Kelly and G. T. Wrixon, "Optimization of Schottky-Barrier Diodes for Low-noise, Low-Conversion Loss Operation at Near-Millimeter Wavelengths," in K. J. Button, Ed., *Infrared and Millimeter Waves*, Vol. 3, *Submillimeter Techniques*, Academic Press, New York, 1980.

94. C. O. Weiss and A. Godone, "Harmonic Mixing and Detection with Schottky Diodes Up to the 5 THz Range," *IEEE J. Quantum Electron.* Vol. QE-3, No. 2, pp. 67–99, 1984.

95. W. M. Kelly, M. J. Gans, and J. G. Eivers, "Modelling the Response of Quasi-optical Corner Cube Mixers," in E. Kollberg, Ed., *Instrumentation for Submillimeter Spectroscopy*, Proc. SPIE 598, pp. 72–78, 1986.

96. E. Sauter, G. V. Schultz, and R. Wohlleben, "Antenna Patterns of an Open Structure Mixer at a Submillimeter Wavelength and of Its Scaled Model," *Int. J. Infrared Millim. Waves*, Vol. 5, No. 4, pp. 451–463, 1984.

97. J. J. Gustincic, "A Quasi-optical Receiver Design," *IEEE MTT-S Int. Microwave Symp. Dig. Tech. Pap.*, pp. 99–102, 1977.

98. J. R. Tucker and M. J. Feldman, " Quantum Detection at Millimeter Wavelengths," *Rev. Mod. Phys.*, Vol. 57, No. 4, pp. 1055–1115, 1985.

99. M. R. Beasly and C. J. Kircher, "Josephson Junction Electronics: Materials Issues and Fabrication Techniques," in S. Fonerand and B. B. Schwartz, Eds., *Superconductor Materials Science*, Plenum Press, New York, 1981, pp. 605–684.

100. J. M. Lumley, R. E. Somekh, J. E. Evetts, and J. H. James, "High Quality All Refractory Josephson Tunnel Junctions for Squid Applications," *IEEE Trans. Magn.*, Vol. MAG-21, No. 2, pp. 539–542, 1985.

101. E. J. Cucauskas, M. Nisenoff, H. Kroger, D. W. Jillie, and L. R. Smith, "All Refractory, High T_c Josephson Device Technology," in A. F. Clark and R. P. Reed, Eds., *Advances in Cryogenic Engineering*, Vol. 30, Plenum Press, New York, 1984, pp. 547–558.

102. M. K. Wu, J. R. Ashburn, C. J. Torng, P. H. Hor, R. L. Meng, L. Gao, Z. J. Huang, Y. Q. Wang, and C. W. Chu, " Superconductivity at 93 K in a New Mixed-phase Y-Ba-Cu-O Compound System at Ambient Pressure," *Phys. Rev. Lett.*, Vol. 58, No. 9, pp. 908–910, 1987.

103. G. J. Dolan, T. G. Phillips, and D. P. Woody, "Low Noise 115 GHz Mixing in Superconducting Oxide-Barrier Tunnel Junctions," *Appl. Phys. Lett.*, Vol. 34, No 5, pp. 347–349, 1979.

104. A. W. Kleinsasser and R. A. Burman, "High-Quality Submicron Niobium Tunnel Junctions with Reactive-Ion-Beam Oxidation," *Appl. Phys. Lett.*, Vol. 37, No.9, pp. 841–843, 1980.

105. M. J. Feldman and S. Rudner, "Mixing with SIS Arrays," in K. J. Button, Ed., *Reviews in Infrared And Millimeter Waves*, Vol. 3, Plenum Press, New York, 1983.

106. W. R. McGrath, A. V. Räisänen, and P. L. Richards, "Variable-Temperature Loads for Use in Accurate Noise Measurements of Cryogenically Cooled Microwave Amplifiers and Mixers," *Int. J. Infrared Millim. Waves*, Vol. 7, No. 4, pp. 543–554, 1986.

107. L. Olsson, S. Rudner, E. Kollberg, and C. O. Lindström, "A Low-Noise SIS Array Receiver for Radio Astronomical Applications in the 35–50 GHz Band," *Int. J. Infrared Millim. Waves*, Vol. 4, No. 6, pp. 847–858, 1983.

108. M. J. Wengler, D. P. Woody, R. E. Miller, and T. G. Phillips, "A Low Noise Receiver for Millimeter and Submillimeter Wavelengths," *Int. J. Infrared Millim. Waves*, Vol. 6, No. 8, pp 697–706, 1985.

109. D. Winkler, W. R. McGrath, B. Nilsson, T. Claesson, J. Johansson, E. Kollberg, and S. Rudner, "A Submillimeter Wave Quasiparticle Receiver for 750 GHz—Progress Report," in Kollberg Ed., *Instrumentation for Submillimeter Spectroscopy*, Proc. SPIE 598, pp. 33–38, 1985.

110. A. V. Räisänen, D. G. Crete, P. L. Richards, and F. L. Lloyd, "Wide Band, Low Noise MM-Wave SIS Mixers with Single Tuning Element," *Int. J. Infrared Millim. Waves*, Vol. 7, No. 12, pp. 1834–1852, 1986.

111. P. H. Siegel, D. W. Peterson, and A. R. Kerr, "Design and Analysis of the Channel Waveguide Transformer," *IEEE Trans. Microwave Theory Tech.*, Vol. MTT-31, No. 6, pp. 473–484, 1983.

112. L. R. D'Addario, "An SIS Mixer for 90–120 GHz with Gain and Wide Bandwidth," *Int. J. Infrared Millim. Waves*, Vol. 5, No. 11, pp. 1419–1442, 1984.

113. A. V. Räisänen, W. R. McGrath, P. L. Richards, and F. L. Lloyd, "Broad-Band RF Match to a Millimeter-Wave SIS-Quasiparticle Mixer," *IEEE Trans. Microwave Theory Tech.*, Vol. MTT-33, No. 12, pp. 1495–1500, 1985.

114. R. S. Pengelly, *Microwave Field-Effect Transistors—Theory, Design and Applications*, 2nd ed., Research Studies Press (Wiley), New York, 1986.

115. P. Bura and R. Dikshit, "FET Mixers for Communication Satellite Transponders," *IEEE MTT-S Int. Microwave Symp. Dig.*, pp. 90–92, 1976.

116. R. A. Pucel, D. Masse, and R. Bera, "Performance of GaAs MESFET Mixers at X-Band," *IEEE Trans. Microwave Theory Tech.*, Vol. MTT-24, No. 6, pp, 351–360, 1976.

117. G. K. Tie and C. S. Aitchison, "Noise Figure and Associated Conversion Gain of a Microwave MESFET Gate Mixer," *Proceedings of the 13th European Microwave Conference*, Microwave Exhibitions and Publishers, Turnbridge Wells, Kent, England, 1983, pp. 579–584.

118. G. Begemann and A. Jacob, "Conversion Gain of MESFET Drain Mixers," *Electron. Lett.*, Vol. 15, No. 18, pp 567–568, 1979.

119. C. Tsironis, R. Stahlman, and F. Ponse, "A Self-Oscillating Dual Gate MESFET X-band Mixer with 12 dB Conversion Gain," *Proceedings of the 9th European Microwave Conference*, Microwave Exhibitions and Publishers, Turnbridge Wells, Kent, England, 1979, pp. 321–325.

120. C. S. Kim, "Tunnel Diode Converter Analysis," *IRE Trans. Electron Devices*, Vol. ED-8, No. 9, pp. 394–405, 1961.

121. F. Sterzer and A. Presser, "Stable Low-Noise Tunnel Diode Frequency Converters," *RCA Rev.*, Vol. 23, No. 1, pp. 3–28, 1962.

122. D. Lockie, A. Podell, and S. Moghe, "Cell Libraries Provide Stepping-Stone into Monolithic Integration," *MSN Commun. Technol.*, Vol. 16, No. 8, pp. 74–85, 1986.

123. U. K. Mishra, S. C. Palmateer, S. J. Nightingale, M. A. G. Upton, and P. M. Smith, "Surface-Oriented Low-Parasitic Mott Diode for EHF Mixer Applications, *Electron. Lett.*, Vol. 21, No. 15, pp. 652–653, 1985.

124. C. Chao, A. Contulatis, S. A. Jamison, and P. E. Baudhahn, "Ka-Band Monolithic GaAs Balanced Mixer," *IEEE Trans. Microwave Theory Tech.*, Vol. MTT-31, No. 1, pp. 11–15, 1983.

125. A. Podell and W. W. Nelson, "High Volume, Low Cost, MMIC Receiver Front End," *IEEE Microwave Millim.-Wave Monolithic Circuits Symp.*, 1986, pp. 57–59.

126. T. G. Phillips and K. B. Jefferts, "A Low Temperature Bolometer Heterodyne Receiver for Millimeter Wave Astronomy," *Rev. Sci. Instrum.*, Vol. 44, pp. 1009–1014, 1973.

127. E. R. Brown, J. Keene, and T. G. Phillips, "A Heterodyne Receiver for the Submillimeter Wavelength Region Based on Cyclotron Resonance in InSb at Low Temperatures," *Int. J. Infrared Millim. Waves*, Vol. 6, No. 11, pp. 1121–1138, 1985.

128. S. Dixon and H. Jacobs, "Millimeter-Wave InP Image Line Self-Mixing Gunn Oscillator," *IEEE Trans. Microwave Theory Tech.*, Vol. MTT-29, No. 9, pp. 958–961, 1981.

129. P. N. Förg and J. Freyer, "Ka-Band Self-Oscillating Mixer with Schottky BARITT Diodes," *Electron. Lett.*, Vol. 16, No. 22, pp. 827–829, 1980.

3

MULTIPLIERS AND PARAMETRIC DEVICES

J. W. Archer and R. A. Batchelor

Division of Radiophysics
CSIRO, Sydney, Australia

3.1 INTRODUCTION

This chapter deals with the theory and practical design of microwave circuits containing nonlinear reactance devices. Many of the microwave circuits and components discussed incorporate a semiconductor device known as a varactor diode. A varactor diode is but one example of a variable-reactance circuit element. The varying reactance is provided by the diode junction capacitance, which changes nonlinearly as a function of the applied voltage. The capacitance variation can be used to produce a number of different effects in a microwave circuit. Use of the varactor diode to generate harmonics of an applied microwave signal is one of the most important applications of such devices in modern microwave technology. Very efficient microwave and millimeter wave frequency multipliers can be built using varactors. At lower microwave frequencies a variant of the varactor diode, the step-recovery diode, has become popular because the conversion efficiencies attainable are substantially higher than those of conventional varactor-based circuits. Until recently, a most important use of varactor devices was in parametric amplifiers and up/down-converters. However, the advent of the high-performance field-effect transistor has now displaced the parametric amplifier in low-noise amplifier applications. Other applications of the varactor are: (a) the switching and modulation of a microwave signal through variation of the reactance by means of an externally applied bias, (b) the electronic tuning of resonant structures, and (c) microwave power limiting.

In Section 3.2 we discuss theoretical relationships necessary for an understanding of nonlinear reactance devices. Section 3.3 presents a description of practical variable reactance devices, with emphasis on the widely used varactor diode. Methods for evaluating the quality of a varactor diode are summarized. Sections 3.4 and 3.5 deal with the design of practical microwave frequency multipliers using varactor diodes. Design examples are given for a millimeter wave varactor frequency tripler and a 2.7 GHz step-recovery diode frequency sextupler. In Section 3.6 we discuss parametric amplifiers, up-converters, and down-converters, including a design example of a 3.25 GHz nondegenerate parametric amplifier.

3.2 MANLEY–ROWE RELATIONSHIPS

Manley and Rowe [1] have derived a set of general equations that relate power flow into and out of a nonlinear reactance. These equations are very useful for understanding the behavior of harmonic generators, parametric amplifiers, and frequency up/down-converters. They may be used to predict the ultimate power gain and conversion efficiency that can be obtained. The Manley–Rowe relationships for a nonlinear lossless reactance, which is excited so that the current and voltage have frequency components of the form $mf_1 + nf_2$, where m and n are integers, are given by the following two independent equations: [2]

$$\sum_n \sum_m \frac{mP_{m,n}}{mf_1 + nf_2} = 0 \tag{3.1}$$

$$\sum_n \sum_m \frac{nP_{m,n}}{mf_1 + nf_2} = 0 \tag{3.2}$$

where $P_{m,n}$ is the average power flowing into the nonlinear reactance at the frequencies $mf_1 + nf_2$. These equations are a result only of the nonlinear variation of the reactance and are independent of the shape of its characteristic and of the driving power levels. Since the device is considered lossless, the sum of all inward power flow at the different frequencies must be zero.

By placing constraints on the impedance presented to the reactance at frequencies $mf_1 + nf_2$, it is possible to obtain a variety of useful circuits. If the nonlinear device is excited at f_1 and f_2 and is presented with an open circuit at all frequencies other than $f_3 = f_1 + f_2$, then, provided that $f_1 \ll f_1 + f_2$, the Manley–Rowe relationships reduce to

$$\frac{P_1}{f_1} + \frac{P_3}{f_3} = 0 \tag{3.3}$$

$$\frac{P_2}{f_2} + \frac{P_3}{f_3} = 0 \tag{3.4}$$

The source at frequency f_2 is usually called the pump source and is primarily responsible for driving the reactance. Since pump power (P_2) is supplied to the capacitor $(P_2 > 0)$, then $P_3 < 0$ and hence P_3 is supplied by the reactance—therefore $P_1 > 0$. It follows that the device is absolutely stable and has power gain equal to f_3/f_1. This type of amplifier is called an upper-sideband up-converter.

If a small amount of power at the signal frequency (f_3) is applied to the varactor, P_1 and P_2 are negative. The input power P_3 is split between P_1 and P_2, and if $f_3 \gg f_1$, most of the output power is delivered at f_2. Such a configuration behaves as an upper-sideband down-converter. However, since P_2 is negative, a negative resistance is presented to the pump circuit and the device is potentially unstable.

If $f_3 = f_2 - f_1$, where f_2 is again the pump frequency, the Manley–Rowe relations become

$$\frac{P_1}{f_1} - \frac{P_3}{f_3} = 0 \tag{3.5}$$

$$\frac{P_2}{f_2} + \frac{P_3}{f_3} = 0 \tag{3.6}$$

Since $P_2 > 0$, then $P_3 < 0$ and hence $P_1 < 0$. That is, the reactor emits more power than is supplied by the generator at f_1. Thus, with this embedding configuration, the signal power can be amplified at the input frequency, in contrast to the previous case. The output powers P_1 and P_3 are dependent on the pump power and the external impedance levels, but they are always related by the equations above. When the input and output frequencies are the same, power at f_3 is simply dissipated in the circuit and is unused. For this reason the third frequency is usually called the idler frequency. The idler signal is an unavoidable result of this type of amplification, and suppressing it would suppress the amplification of the input signal. The separation of idler and signal frequencies is an important parameter in the design of an amplifier—the closer the idler to the signal, the more difficult it will be to separate them by filtering. If the signal and idler frequencies are separated sufficiently for the signal circuit to reject the idler, the amplifier is called a nondegenerate amplifier. Conversely, if the signal circuit passes both signal and idler bands and the input termination is common to both of them, the amplifier is called degenerate.

If the output power is withdrawn at f_3 and $f_3 \gg f_1$, the device is called a lower-sideband up-converter. When $f_3 < f_1$, the circuit performs the function of a lower-sideband down-converter.

If the nonlinear reactor is excited only at f_1, then $n = 0$. The power flow relationships then become

$$\sum_m - P_m = P_1 \tag{3.7}$$

describing a nonlinear reactor harmonic generator. The total harmonic output power is equal to the fundamental input power. If the circuit is adjusted to reactively terminate all harmonics other than the required output, then for an ideal nonlinear reactor, the conversion efficiency for that harmonic is 100%.

3.3 PRACTICAL NONLINEAR REACTANCE DEVICES

3.3.1 Varactor Diodes

The varactor diode [2, 3] is the most convenient nonlinear reactance element available to practical circuit designers. As stated earlier, the varying reactance is provided by the diode junction capacitance, which changes nonlinearly as a function of the applied (reverse) voltage. Varactors may be classified into two broad groups, depending on the method of fabrication: the junction varactor, widely used at microwave frequencies, and the Schottky diode devices, usually used for millimeter wavelength applications. Figure 3.1 shows a cutaway view of typical varactor wafers of each type, as well as giving a typical equivalent circuit for a varactor diode, which applies equally to the two cases. The equivalent circuit for the device includes the following:

a. C_j—the junction capacitance, which is a function of the applied voltage.
b. R_j—the junction resistance, in shunt with C_j and also a function of bias.
c. R_s—the series resistance, which may also be a function of bias. This parasitic resistance includes the resistance of the bulk semiconductor material external to the junction, the resistance of the undepleted epitaxial material, and the resistance of the ohmic contacts to the device.

A typical junction varactor is made on an n-type silicon or GaAs wafer, with a heavily doped conducting substrate upon which is grown a lightly doped epitaxial layer.

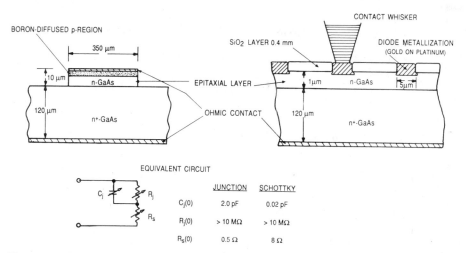

Figure 3.1 Cutaway views of typical varactors, showing approximate dimensions. An equivalent circuit applicable to either Schottky or junction varactors is also shown, along with typical element values.

A suitable *p*-type dopant is then diffused into the epitaxial layer to form the *pn*-junction. Ohmic contacts are made to a small circular area on the top of the wafer and to the back of the wafer. Most of the epitaxial layer is then etched away, except in the area of the top contact, forming a mesa of the desired diameter.

A Schottky barrier varactor diode consists of a circular metallic contact pad deposited on a lightly doped, epitaxially grown, *n*-type layer. Once again the substrate is heavily doped GaAs material. The epitaxial material is conductive except in the vicinity of the metal, where an insulating depletion zone is formed.

Figure 3.1 shows typical values for the physical dimensions and equivalent circuit parameters for both Schottky and junction GaAs varactors, with a substrate resistivity of 0.004 Ω-cm and a 1-μm thick epitaxial layer with a donor doping density of 6×10^{16} cm^{-3}. The high electron mobility and the inherently low spreading resistances in GaAs diodes make GaAs the material of choice for most microwave and millimeter wave applications.

Varactor diodes are usually operated under reverse-bias conditions, where the junction resistance is negligible by comparison with the capacitive reactance of the junction at the operating frequency. Therefore, at microwave frequencies the equivalent circuit of the reverse-biased varactor diode (either junction or Schottky) may be considered to be simply a (bias-dependent) capacitor and resistor in series. The equivalent circuit of the forward-biased junction varactor at microwave frequencies is generally more complex, since it must incorporate the diffusion capacitance of the injected minority carriers as well as the effect of these carriers on the conductance of the semiconductor material. Minority carrier injection does not occur for Schottky barrier varactors. Hence if forward-bias operation is employed, the junction varactor will exhibit greater capacitance variation than its Schottky equivalent, because of the absence of charge storage of minority carriers in the latter device.

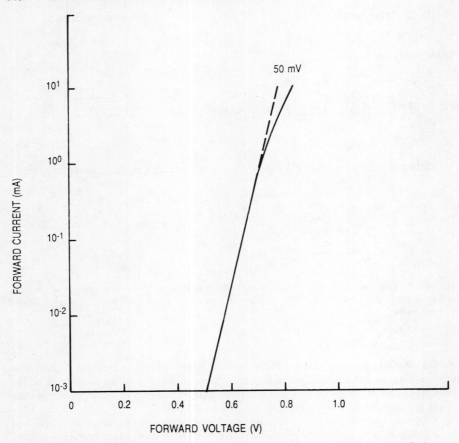

Figure 3.2   Typical forward current–voltage characteristics for a GaAs varactor diode.

At low frequencies the varactor exhibits straightfoward behavior in both the forward- and reverse-biased conditions. The forward-bias current increases exponentially with the applied voltage, and for reverse bias a small saturation current flows. The current–voltage characteristic is given by (see Fig. 3.2)

$$i_d = I_s \left[\exp\left(\frac{qV_d}{\gamma kT}\right) - 1 \right] \tag{3.8}$$

where k = Boltzmann's constant
$\quad T$ = device physical temperature
$\quad I_s$ = saturation current
$\quad q$ = electronic charge
$\quad V_d$ = applied voltage
$\quad \gamma$ = measure of the ideality of the device ($\gamma > 1$)

The ideality factor (γ) is normally in the range 1 to 2 at 300 K.

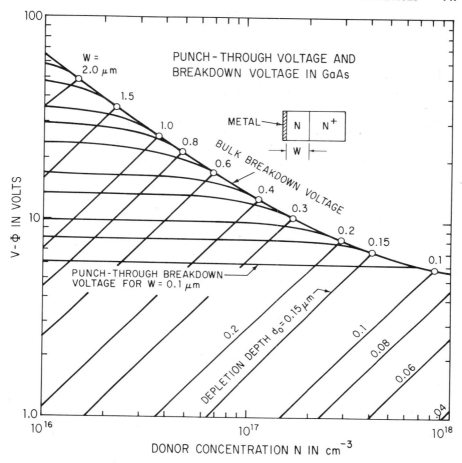

Figure 3.3 Variation of the onset of bulk and punch-through breakdown in GaAs as functions of doping concentration and thickness for an ideal, uniformly doped sample. (From Ref. 5; reproduced by permission).

As the reverse bias is increased, a point is reached where avalanche breakdown occurs in the epitaxial material and the diode current increases very rapidly, since it is limited only by the diode resistance and the external circuit resistance. In general, avalanche breakdown occurs as a result of impact ionization in which a high-energy electron or hole collides with a lattice site and generates a hole–electron pair—hence the reverse current rapidly increases [4–6]. Avalanche breakdown may occur in one of two ways: a bulk-type breakdown or a premature breakdown, commonly called punch-through. Breakdown voltage as a function of doping for GaAs is given in Fig. 3.3, for both the bulk and punch-through mechanisms [5]. The magnitude of the punch-through breakdown is dependent on the epitaxial layer thickness as well as its doping. For GaAs material doped higher than 2×10^{17} cm^{-3}, tunneling also becomes an important mechanism in junction breakdown, but this will not be considered here.

GaAs exhibits a higher defect density and lower obtainable donor density than silicon. For these reasons it is difficult to obtain high breakdown voltages with GaAs

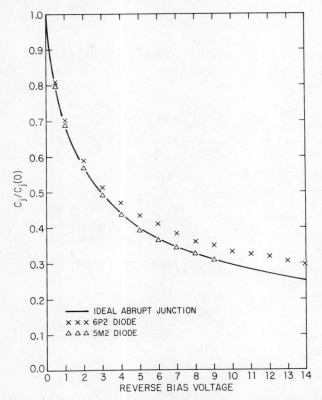

Figure 3.4 Normalized junction capacitance versus reverse voltage for an ideal abrupt junction varactor and for typical Schottky GaAs varactors fabricated by R. J. Mattauch at the University of Virginia. (From Ref. 18; reproduced by permission.)

diodes, especially for thin (ca. 1 μm) epitaxial layers. In general, the reverse breakdown in Schottky barrier devices is not as sharp as that of a diffused junction of the same material.

The dependence of the junction capacitance on the applied reverse voltage for an ideal abrupt junction varactor is given by [6] (see Fig. 3.4)

$$C_j = C_j(0)\left(1 - \frac{V_d}{\phi}\right)^{-\zeta} \tag{3.9}$$

where ϕ is the barrier potential and $C_j(0)$ is the zero-bias capacitance given by (neglecting fringing effects)

$$C_j(0) = A\sqrt{qN\varepsilon/2\phi} \tag{3.10}$$

where $N = N_d$ for Schottky diodes, $(N_a N_d)/(N_a + N_d)$ for junction diodes

A = diode area

ε = permittivity of the epi material

N_d = donor doping density

N_a = acceptor doping density

The capacitance law exponent, ζ, depends on the doping profile of the epitaxial layer of the device and usually lies in the range 0.1 to 0.5. For Schottky diode varactors with uniform epitaxial layer doping it is usually close to 0.5. Ideally, the diode junction capacitance should increase with reverse bias until the breakdown voltage of the diode is exceeded. However, in many practical devices the junction capacitance reaches a minimum value at a voltage significantly below the limit imposed by breakdown. This arises because, in practice it is difficult to achieve a discontinuous step in doping concentration between the epitaxial material and the heavily doped substrate. Real diodes usually exhibit a transition region through which the doping grades continuously from that of the epitaxial layer to that of the substrate. In addition to the reduced capacitance variation resulting from this effect, a decrease in the diode reverse breakdown voltage also occurs relative to that expected for a uniformly doped layer with abrupt interface.

In a Schottky diode, the substrate resistance (i.e., one component of the series resistance) decreases approximately in proportion to the device diameter. For the epilayer component, the resistance decreases in proportion to the diode area [7, 8]. For a given diode it also increases with frequency, owing to the skin effect and plasma resonance phenomena [9]. The skin effect arises at high frequencies because the current becomes confined to a thin surface layer in the semiconductor as a result of a decrease in skin depth. Plasma resonance effects are a consequence of interaction between the displacement current and the charge carrier inertia, leading to a large increase in diode series resistance at frequencies close to the plasma resonance frequency, f_p, for the epilayer or substrate material. The plasma resonance frequency is given by

$$f_p = 2\pi \left(\frac{N_d q^2}{m^* \varepsilon} \right)^{1/2} \tag{3.11}$$

where m^* is the effective mass of the charge carriers. For n-type GaAs, with $N_d = 2 \times 10^{16}$ cm^{-3}, the plasma resonance frequency is approximately 1000 GHz.

For Schottky devices with diameters of a few micrometers, the series resistance is dominated by the contribution from the undepleted epitaxial layer. Thus, to minimize the series resistance associated with a given device, the doping concentration should be as high as possible and the epitaxial layer as thin as possible, consistent with the desired breakdown voltage [10].

Additional sources of parasitic resistance exist in the diffused junction diode [3]. These are the resistance of the diffused p-region and the resistance of the ohmic top contact. Because of the greater ionization energies of acceptor impurities compared with those of donors, as well as the dominance of impurity scattering in a diffused layer, the p-region resistance becomes large at cryogenic temperatures, causing serious degradation of diode performance. Hence GaAs Schottky varactors have been the best choice for cooled parametric amplifier applications because of their low, relatively temperature-insensitive resistance. Figures 3.2 and 3.4 show typical dc IV characteristics and 1 MHz CV characteristics for both junction and Schottky barrier GaAs varactor diodes.

Although the variation in junction capacitance is the most important feature of varactor diode behavior, parasitic capacitances, inductances, and resistances associated with the necessary packaging and/or mounting structures can significantly affect device performance. Figure 3.5 shows a typical package for a microwave GaAs junction varactor, illustrating the effects of the package parasitics on the device equivalent circuit. The package parasitics are of particular consequence in the design of nondegenerate parametric amplifiers, since they are often used to resonate the device at the idler frequency (see Section 3.6.2).

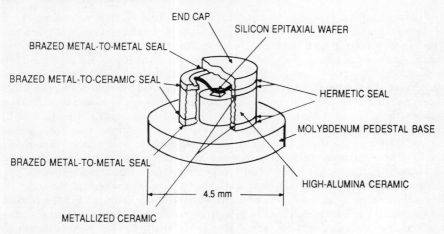

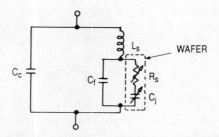

Figure 3.5 Sketch of a packaged varactor and a suitable equivalent circuit for this structure.

3.3.2 Figures of Merit

For efficient varactor operation it is necessary for the reactance of the junction capacitance of the varactor to be much larger than the device series resistance. This constraint places an upper frequency limit on the usefulness of a given varactor, and figures of merit for varactors have been developed to quantify this limit [2]. Four of the most commonly employed figures of merit are listed below.

Static Figures of Merit

a. The cutoff frequency at a specified bias voltage (v), given by

$$f_c(v) = \frac{1}{2\pi R_s C_j(v)} \tag{3.12}$$

b. The quality factor at a specified bias voltage and frequency (f_0), given by

$$Q(v) = \frac{f_c(v)}{f_0} \tag{3.13}$$

Dynamic Figures of Merit

c. The dynamic cutoff frequency, given by

$$f_{cd} = \frac{1}{2\pi R_s} \left(\frac{1}{C_{j,\min}} - \frac{1}{C_{j,\max}} \right) \tag{3.14}$$

Usually $C_{j,\min}$ is taken to be the capacitance at the reverse breakdown voltage and $C_{j,\max}$ is taken to be the capacitance at a small forward bias of about 10 μA.

d. The dynamic quality factor, given by

$$Q_d = \frac{S_1}{2\pi f_0 R_s} \tag{3.15}$$

where S_1 is the first Fourier coefficient of the time-varying elastance (reciprocal of capacitance) of the pumped varactor. Q_d is related to $f_c(0)$ approximately as

$$Q_d = \alpha \left(\frac{C_j(0)}{C_j(v)} \right) \left(\frac{f_c(0)}{f_0} \right) \tag{3.16}$$

where $C_j(v)$ is the junction capacitance at the operating bias and

$$\alpha = \frac{C_1}{C_0} \tag{3.17}$$

is a function of the Fourier coefficients of the time-varying capacitance, where

$$C_j(t) = \sum_n C_n \exp(jn\omega t) \tag{3.18}$$

For abrupt junctions, $\gamma \simeq 0.25$.

Clearly, the first two figures of merit listed above do not take into account the degree or nature of the capacitance variation of the varactor. Since the dynamic figures of merit partly take this into account, they give a better description of the usefulness of the varactor at a given operating frequency.

However, these quality indicators still have failings in that they result in an incomplete description of the device behavior. For example, the details of the capacitance variation between extremes is ignored in calculating the dynamic cutoff frequency. Since the dynamic quality factor depends on the Fourier coefficients of the capacitance time dependence, it takes the nature of the nonlinear response into account more completely. However, because the capacitance time variation is dependent on the driving waveform, as determined by the impedance presented to the device by the microwave embedding circuit, this figure of merit is still not uniquely dependent on the varactor alone.

Several well-known techniques are available for the practical determination of these quality factors [3, 11–14]. These characterization techniques will not be discussed here, since they have been described extensively in the literature. Clearly, the figures of merit must be used only as a guide to device usefulness at a given frequency. A more complete analysis of the behavior of the varactor response, taking into account the characteristics of the microwave embedding circuit, is necessary if the potential of a device for a given circuit application is to be fully evaluated.

3.3.3 Step-Recovery Diodes

Another important class of device, the step-recovery diode, is widely used as the active element in frequency multipliers for output frequencies below about 30 GHz [15, 16]. So far in this chapter we have emphasized the varactor diode, a device exhibiting a dielectric capacitance variation as a consequence of the varying width of the diode depletion region. The step-recovery or charge-storage diode is a *pn*-junction device which has been designed to enhance the charge storage capacitance associated with minority carrier injection during forward conduction in a junction varactor and its subsequent withdrawal under reverse-biased conditions. In conventional junction varactor diodes the charge storage capacitance is often used to enhance the nonlinearity, and it can result in a higher power-handling ability and higher efficiency. In the step-recovery diode the nonlinearity is derived exclusively from the charge-storage capacitance.

A prerequisite for the existence of charge-storage capacitance is that the junction must be capable of minority carrier injection (i.e., the injection of holes from a *p*-type region into an *n*-type region); thus Schottky barrier varactors are excluded. Furthermore, the injected charge must be recoverable during the reverse-biased portion of the pump cycle. For this to occur, two conditions must be met:

a. The carriers must not recombine before they are withdrawn.
b. The carriers must not diffuse so far from the junction that they cannot be retrieved.

Although the first condition can easily be met using silicon material (minority carrier lifetimes between 10^{-8} and 10^{-6} s), the minority carrier lifetime in GaAs is so short ($<10^{-10}$ s) that it renders the use of that material impractical, except possibly at millimeter wave frequencies.

Therefore, all step-recovery diodes are currently made from silicon, and device optimization consists of satisfying the second condition above without unduly affecting other performance criteria, such as breakdown voltage and cutoff frequency. Steep impurity profiles are required to produce the large electric fields on either side of the junction that are necessary to prevent the charge carriers from diffusing away too rapidly. Futhermore, because of the finite time required for the carriers to cross the depletion region

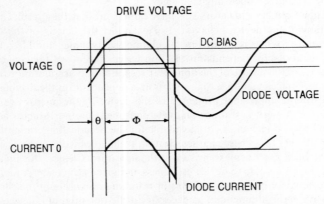

Figure 3.6 Typical current and voltage waveforms in a step-recovery diode driven by an RF signal.

and the current-limiting effects of the fields produced by the minority carriers themselves, the depletion region must not be too wide. These constraints result in an inevitable trade-off between increased frequency response and decreased breakdown voltage.

A typical device waveform for several pump cycles is shown in Fig. 3.6. An important characteristic of the response of the device is the very abrupt transition, during the reverse cycle, from reverse charge storage conduction to cutoff. With optimum design of the impurity profile the duration of this transition can be made less than 100 ps, resulting in an harmonic-rich transient that can be used for harmonic generation. The frequency multiplication process for these devices is thus fundamentally different in concept from the usual varactor operation. Frequency multipliers incorporating step-recovery diodes are capable of high-order harmonic generation with efficiencies substantially greater than can be achieved with single-varactor devices, and they are commonly used where single-step multiplication of order $\geqslant 5$ is required [17].

3.4   MULTIPLIER DESIGN USING VARACTOR DIODES

3.4.1   Design Procedures

Any nonlinear device driven by a sinusoidal signal generates power at all the harmonics of the input frequency. The Manley–Rowe [1] relations for power flowing into and out of a lossless nonlinear device state that the sum of all the powers flowing at the harmonic frequencies must be zero:

$$\sum_i P_i = 0 \tag{3.19}$$

Assuming that the nonlinear device is driven by a signal of frequency f_1 with a power P_1, we have

$$P_1 = -\sum_i P_i \tag{3.20}$$

Theoretically, therefore, power at any harmonic frequency can be extracted from such a device and delivered to a load. This is the fundamental principle of operation for the varactor diode harmonic generator; a nonlinear device is driven by a source at frequency f_1 and power is extracted at a frequency multiple, nf_1, and delivered to a load.

Clearly, a varactor diode is capable of harmonic generation through the nature of its voltage-variable capacitance. If the circuitry to which the diode is connected is designed such that power is delivered to a load only at the desired output harmonic, while the power at all other harmonics is reactively terminated, conversion efficiencies of 100% can, in theory, be achieved.

A practical varactor exhibits a series resistance in addition to the nonlinear capacitance and so is not lossless. Practical conversion efficiencies are therefore somewhat less than 100%. Furthermore, at microwave and millimeter wave frequencies the losses in the embedding circuitry will be nonzero, further degrading the attainable conversion efficiency. At a given output frequency, f_0, high conversion efficiencies can be achieved only with a high dynamic cutoff frequency ($f_{cd} \gg f_0$) varactor diode that has been optimally impedance matched to its embedding circuit at input and output frequencies. Furthermore, all other harmonics must be terminated optimally and with low loss by the microwave circuit.

For a complete analysis of varactor multiplier performance, the nature of the microwave circuit to which the diode is connected must be considered [18]. Furthermore, the

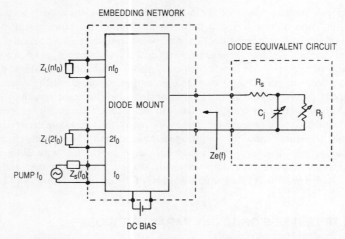

Figure 3.7 Conceptual equivalent circuit for a varactor multiplier showing the interconnection of the embedding network and the varactor diode.

series resistance must be taken into account when calculating the voltage drop across the diode. An equivalent-circuit model, which may be used to predict the performance of a varactor diode as a frequency multiplier when mounted in a given microwave circuit, is shown in Fig. 3.7. Two interconnected networks represent (a) the lossless, nonlinear junction capacitor, the dynamic diode resistance, and the series resistance, and (b) a multiport embedding network that models the impedance presented to the diode by the microwave mounting circuit at all harmonics of the pump. The two networks are generally optimized to obtain maximum power transfer between the embedding network and the time-varying reactance at the input and output harmonics.

The conversion efficiency of a given multiplier depends on the large-signal diode current and voltage waveforms induced by the pump signal. These waveforms are, in turn, determined by the impedance presented to the diode by the embedding network at the various harmonics of the pump frequency. For multipliers operating in the millimeter wavelength range it is difficult to predict the embedding impedances at frequencies substantially greater than the pump frequency, because the mounting structures (such as waveguide, stripline, or coaxial lines) become multimoded. One approach to this problem is to measure these impedances on a low-frequency scale model of the diode mount. The measured impedance data and the characteristics of the diode can then be analyzed to determine the diode voltage and current waveforms and hence the conversion efficiency of the multiplier for the chosen output harmonic [19].

In general, the steady-state, large-signal response of the multiplier circuit can be described in terms of the Fourier coefficients of the voltage, v_j, and current, i_c [19]:

$$v_j(t) = \sum_k V_k e^{jk2\pi f_1 t} \qquad V_k = (V_{-k})^* \tag{3.21}$$

$$i_c(t) = \sum_k I_k e^{jk2\pi f_1 t} \qquad I_k = (I_{-k})^* \tag{3.22}$$

where f_1 is the pump frequency. Two sets of boundary conditions must be satisfied simultaneously by these quantities. The first, imposed by the diode, is most easily expressed in the time domain, while the second set, imposed by the embedding network (the embedding impedance Z_e), is more conveniently considered in the frequency domain. Assuming that the diode does not conduct in the forward direction during the pump cycle (i.e., $v < \phi$),

$$i_c = C_j(v_j)\frac{dv_j}{dt} \tag{3.23}$$

where $C_j(v_j)$ is given by Eq. (3.9).

The embedding network requires that

$$
\begin{aligned}
V_k &= -I_k[Z_e(kf_1) + R_s(kf_1)] \qquad k = \pm2, \pm3, \ldots \\
V_{\pm1} &= V_p - I_{\pm1}[Z_e(\pm f_1) + R_s(\pm f_1)] \\
V_0 &= V_{dc} - I_0[Z_e(0) + R_s(0)]
\end{aligned}
\tag{3.24}
$$

where V_p and V_{dc} are the pump signal voltage and dc-bias voltage, respectively. The frequency dependence of R_s was discussed in a previous section.

Except in the case of the doubler, if currents flow only at the input and output frequencies, it is impossible to generate harmonics with the abrupt junction varactor. To generate higher-order harmonics, intermediate harmonic (or idler) currents must flow in the diode [2, 20].

Methods of solving Eqs. (3.21) and (3.22) to obtain the nonlinear, large-signal behavior of specific multipliers have been described by other authors [19, 21]. Figures 3.8 and 3.9 show the results of calculations using idealized models for a frequency doubler

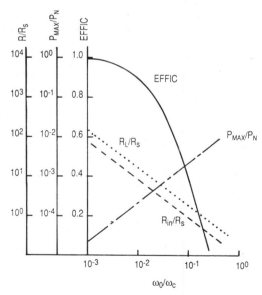

Figure 3.8   Predicted performance of an idealized varactor doubler as a function of frequency (normalized to the dynamic cutoff frequency of the diode). The doubler equivalent circuit used for this analysis is shown in Fig. 3.10a. P_N is given by $(V_B - \phi)^2/R_S$, R_L is the load resistance, and R_{in} is the diode input resistance.

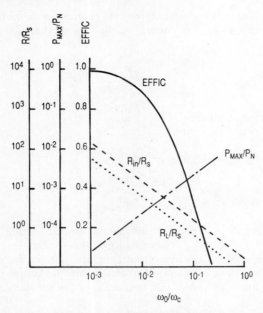

Figure 3.9 Predicted performance of an idealized frequency tripler as a function of frequency (normalized to the dynamic cutoff frequency of the diode). The tripler equivalent circuit used for this analysis is shown in Fig. 3.10b.

and a frequency tripler. These results were obtained using a computer-aided technique developed by Gwarek [22] and refined by Kerr and Siegel [21]. Current flow in the diode was permitted only at the pump and output frequencies, and at the second harmonic idler frequency in the case of the tripler. In the latter case the termination was assumed to be an inductive reactance in shunt, and resonant with, the average diode capacitance at the second harmonic. Conceptual equivalent circuits are shown in Fig. 3.10. The pump signal was assumed to be sufficiently large to swing the diode voltage, v_d, between its reverse breakdown limit, V_{br}, and the onset of forward conduction ($V_{br} < v_d < \phi$). The plots show how multiplier efficiency, power handling, and input and load resistances vary with frequency (normalized to the diode dynamic cutoff frequency). The results demonstrate that the dynamic cutoff frequency should be much higher than the output frequency if efficient multiplication is to be obtained. The analysis of more realistic model structures, outlined in a subsequent section of this chapter, will demonstrate how, for a given diode, practical, nonideal mounting structures result in lower conversion efficiencies.

A number of general guidelines should be followed in the design of any practical multiplier.

a. The mount should be designed to enable a good impedance match between the input circuit and the varactor device over the desired pump frequency range. The same conditions should apply between the varactor and the output circuit at the desired output harmonic. Often power flow is required at a harmonic between the input and output frequencies—an idler frequency. The mount design should allow these components to be reactively terminated with low loss.

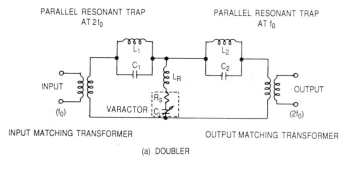

(a) DOUBLER

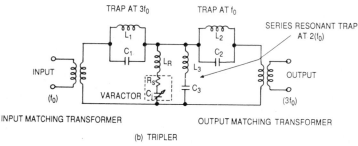

(b) TRIPLER

Figure 3.10 Equivalent circuits for the idealized multiplier analyses presented in Figs. 3.8 and 3.9.

b. At frequencies other than the input, output, and idler harmonics, the mount should be poorly matched to the diode to minimize power loss to these unwanted output components.

c. The input and output circuits should be physically and electrically isolated.

Although varactor-diode frequency multipliers have long been used at microwave frequencies, they have now been largely superseded in this role by the step-recovery diode. However, the recent development of highly efficient harmonic generators for millimeter wavelengths has revitalized interest in the technology [23]. Schottky diodes with diameters of a few micrometers and new mounting structures have been the major advances that have led to this renewed interest. For the most part, therefore, in the remainder of this section we consider only millimeter wave varactor multiplier design. The design techniques for varactor multipliers at microwave frequencies have much in common with those used at millimeter wavelengths. A number of other excellent reviews of multiplier design for microwave frequencies already exist in the literature (see, e.g., Refs. 20 and 24).

At millimeter wavelengths the best multiplier performance has been obtained with crossed-waveguide mounts [25, 26], similar to that shown in Fig. 3.11. The whisker-contacted varactor diode chip is mounted in the output guide, which is cut off at the input frequency and usually of reduced height. Pump power is coupled from the input waveguide to the diode via a low-pass filter, which is cut off at the output frequency. The input circuit usually incorporates a means of dc biasing the diode and tuning the mount to optimally couple the pump power to the varactor. Similarly, a movable back-short in the output guide is used to aid in matching the varactor to the output circuit.

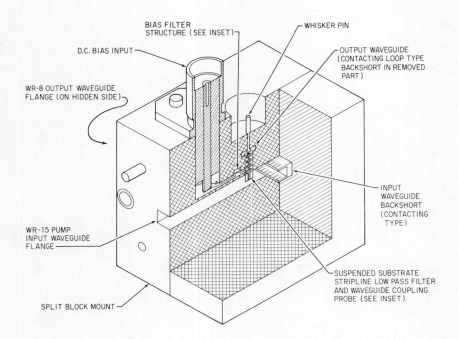

BIAS FILTER
STRUCTURE (SEE INSET)

WHISKER PIN

D.C. BIAS INPUT

OUTPUT WAVEGUIDE
(CONTACTING LOOP TYPE
BACKSHORT IN REMOVED
PART)

WR-8 OUTPUT WAVEGUIDE
FLANGE (ON HIDDEN SIDE)

INPUT
WAVEGUIDE
BACKSHORT
(CONTACTING
TYPE)

WR-15 PUMP
INPUT WAVEGUIDE
FLANGE

SUSPENDED SUBSTRATE
STRIPLINE LOW PASS FILTER
AND WAVEGUIDE COUPLING
PROBE (SEE INSET)

SPLIT BLOCK MOUNT

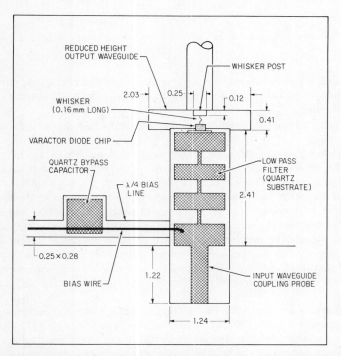

REDUCED HEIGHT
OUTPUT WAVEGUIDE

WHISKER POST

2.03   0.25   0.12

WHISKER
(0.16 mm LONG)

0.41

VARACTOR DIODE CHIP

QUARTZ BYPASS
CAPACITOR

λ/4 BIAS
LINE

LOW PASS
FILTER
(QUARTZ
SUBSTRATE)

2.41

0.25 × 0.28

BIAS WIRE

1.22

INPUT WAVEGUIDE
COUPLING PROBE

1.24

Figure 3.11 Isometric sketch of a typical crossed waveguide frequency doubler, which shows some of the main features important to the design of a milimeter wavelength harmonic generator. The inset shows details of the stripline low-pass filter, the diode mounting arrangement, and the dc bias input connection. (Dimensions are in millimeters.) (From Ref. 28; copyright 1985 IEEE, reproduced by permission.)

For generation of harmonics of third order and higher with high efficiency, low-loss idler terminations must be provided. Optimum idler terminations have been incorporated in millimeter-wave multiplier mounts in a number of ways, resulting essentially in the diode being short circuited or reactively terminated at the idler frequency.

Theoretical studies of the performance of crossed-waveguide multipliers require that the diode embedding circuit be determined. The embedding circuit impedances at the harmonics of the pump frequency can be determined by low-frequency scale modeling, or calculated approximately by analytical techniques such as described by Eisenhart and Kahn [27]. Such procedures usually show that it is difficult to choose a unique diode capacitance that will enable a simultaneous pump and output frequency impedance match to be obtained over a wide range of frequencies (greater than, say, 5% of a given center frequency). The choice of diode capacitance (i.e., the mean value of the time-varying junction capacitance) must represent a trade-off between several, possibly competing requirements. In general, broadband impedance matching at the input frequency is more easily achieved if the diode reactance at the pump frequency is near resonance with the mount impedance (whisker inductance). Since the output guide is cut off at the pump frequency, the embedding impedance seen by the varactor at this frequency is not greatly influenced by the tuning of the output backshort, except when this backshort is positioned very close to the diode. The whisker length and diode capacitance should also be chosen so that, with the aid of the backshort reactance, a reasonable impedance transformation can be achieved in the output circuit between the low dynamic resistance of the diode (50 to 100 Ω) and the guide wave impedance. For mechanical reasons, it is difficult to achieve output guide impedances of much less than 200 Ω in the millimeter frequency range, so that the impedance transformation required to achieve good match at the output frequency is significant.

For a selected mount and diode configuration at a predetermined diode bias, the dynamic, large-signal analysis techniques described by Siegel and Kerr [21] can be used to determine the instantaneous voltage and current waveforms in the mounted varactor diode for one pump cycle. From these waveforms the minimum reverse breakdown voltage required of the diode may be determined, in order for it to handle a given pump power level. Furthermore, by Fourier analysis of the diode current and voltage variation, the conversion efficiency of the multiplier may be estimated.

The results of analyses of simplified models of millimeter wave varactor doubler and tripler mounts of the type illustrated in Fig. 3.11 are shown in Tables 3.1 to 3.7. These analyses provide a useful illustration of the way in which diode and mount design can affect the diode voltage waveform and the conversion efficiency of millimeter wave Schottky varactor multipliers. In deriving the data in the tables, the diode capacitance was modeled by Eq. (3.9), the forward IV characteristic of the diode by Eq. (3.8), and the frequency dependence of the series resistance was modeled by

$$R_s(f) = R_s(0) + R_{skin}\sqrt{f} \qquad (3.25)$$

The following constant values were also assumed: $\phi = 0.9$ V, $q/\gamma kT = 35.0$ V^{-1}, $\zeta = 0.5$, $I_{sat} = 1 \times 10^{-15}$ A, $R_{skin} = 1 \times 10^{-6}$ Ω Hz$^{-1/2}$, and $V_{dc} = -10$ V. The abbreviated column headings are defined as follows: $C_j(0)$, diode zero-bias capacitance; $R_s(0)$, diode dc series resistance; ϵ_p, pump signal-to-diode coupling efficiency; ϵ_n, nth harmonic conversion efficiency; Z_s^{opt}, optimum pump source impedance; Z_1^{nth}, mount embedding impedance at the nth harmonic; x_{short}, backshort spacing relative to the diode; V_r^{max}, maximum instantaneous diode voltage; p_{pump}^{max}, maximum pump power that can be handled without forward conduction.

TABLE 3.1   Predicted Performance of Millimeter Wave Frequency Doublers for Various Diode Zero-Bias Capacitances and Series Resistances at a Pump Frequency of 50 GHz[a]

$C_j(0)$ (fF)	$R_s(0)$ (Ω)	ϵ_p	ϵ_2	Z_s^{opt} R (Ω)	X (Ω)	Z_1^{2nd} R (Ω)	X (Ω)	x_{short} (mm)	ϵ_3/ϵ_2	V_r^{max} (V)	P_{pump}^{max} (mW)
10	1	0.96	0.88	34	1010	451	163	1.40	0.054	21.2	17.0
	5	0.91	0.80	58	1012	359	347	1.35	0.073	21.7	29.3
	10	0.84	0.73	65	1000	359	347	1.35	0.086	22.1	34.7
	15	0.78	0.67	69	1000	359	347	1.35	0.092	22.1	37.5
20	1	0.95	0.82	26	479	359	347	1.35	0.039	26.2	10.8
	5	0.82	0.71	30	479	359	347	1.35	0.045	26.2	12.4
	10	0.70	0.60	35	479	359	347	1.35	0.053	26.1	14.4
	15	0.61	0.52	39	479	359	347	1.35	0.061	26.1	16.4
40	1	0.82	0.79	7	239	210	−60	1.60	0.026	23.9	9.1
	5	0.53	0.50	11	235	253	−59	1.55	0.054	24.8	17.6
	10	0.37	0.35	16	235	253	−59	1.55	0.056	24.8	25.4
	15	0.27	0.25	21	236	314	−42	1.50	0.066	24.9	32.5

[a] The geometry of the waveguide mount is illustrated in Figs. 3.11 and 3.12; for this example, $A = 2.03$ mm, $B = 0.51$ mm, $C = 0.26$ mm.

TABLE 3.2   Predicted Performance of Millimeter Wave Frequency Doublers for Various Pump Frequencies, $C_j(0) = 20$ fF, $R_s(0) = 5$ Ω[a]

f_{pump} (GHz)	ϵ_p	ϵ_2	Z_s^{opt} R (Ω)	X (Ω)	Z_1^{2nd} R (Ω)	X (Ω)	x_{short} (mm)	ϵ_3/ϵ_2	V_r^{max} (V)	P_{pump}^{max} (mW)
40	0.87	0.81	41	674	114	368	2.70	0.071	20.6	31.5
50	0.82	0.71	30	479	359	347	1.35	0.045	26.2	12.4
60	0.70	0.68	18	387	80	18	1.40	0.029	22.0	7.7

[a] The geometry of the waveguide mount is illustrated in Figs. 3.11 and 3.12; for this example, $A = 2.03$ mm, $B = 0.51$ mm, $C = 0.26$ mm.

TABLE 3.3   Predicted Performance of Millimeter Wave Frequency Doublers for Various Diode Zero-Bias Capacitances with $R_s(0) = 5$ Ω, at a Pump Frequency of 50 GHz[a]

$C_j(0)$ (fF)	ϵ_p	ϵ_2	Z_s^{opt} R (Ω)	X (Ω)	Z_1^{2nd} R (Ω)	X (Ω)	x_{short} (mm)	ϵ_3/ϵ_2	V_r^{max} (V)	P_{pump}^{max} (mW)
10	0.79	0.68	25	965	206	5	1.80	0.107	20.8	16.5
20	0.75	0.72	21	477	206	5	1.80	0.027	22.6	6.3
40	0.62	0.61	14	236	206	5	1.80	0.007	25.1	20.7

[a] The geometry of the waveguide mount is illustrated in Figs. 3.11 and 3.12; for this example, $A = 2.03$ mm, $B = 0.51$ mm, $C = 0.34$ mm.

TABLE 3.4 Predicted Performance of Millimeter Wave Frequency Doublers for Various Diode Zero-Bias Capacitances with $R_s(0) = 5 \, \Omega$, at a Pump Frequency of 50 GHz[a]

$C_j(0)$ (fF)	ϵ_p	ϵ_2	Z_s^{opt}		Z_1^{2nd}		x_{short} (mm)	ϵ_3/ϵ_2	V_r^{max} (V)	P_{pump}^{max} (mW)
			$R\,(\Omega)$	$X\,(\Omega)$	$R\,(\Omega)$	$X\,(\Omega)$				
10	0.90	0.56	53	946	94	211	1.50	0.198	19.6	27.1
20	0.82	0.68	29	472	150	90	1.90	0.191	22.5	9.0
40	0.71	0.65	18	239	119	61	2.10	0.094	24.0	24.2

[a] The geometry of the waveguide mount is illustrated in Figs. 3.11 and 3.12; for this example, $A = 2.03$ mm, $B = 0.26$ mm, $C = 0.24$ mm.

TABLE 3.5 Predicted Performance of Millimeter Wave Frequency Doublers for Various Pump Frequencies, $C_j(0) = 20$ fF, $R_s(0) = 5 \, \Omega$[a]

f_{pump} (GHz)	ϵ_p	ϵ_2	Z_s^{opt}		Z_1^{2nd}		x_{short} (mm)	ϵ_3/ϵ_2	V_r^{max} (V)	P_{pump}^{max} (mW)
			$R\,(\Omega)$	$X\,(\Omega)$	$R\,(\Omega)$	$X\,(\Omega)$				
40	0.91	0.68	60	604	173	226	3.20	0.164	22.9	11.0
50	0.82	0.68	29	472	150	90	1.90	0.191	22.5	9.0
60	0.82	0.80	29	390	120	104	1.40	0.012	24.1	13.2

[a] The geometry of the waveguide mount is illustrated in Figs. 3.11 and 3.12; for this example, $A = 2.03$ mm, $B = 0.26$ mm, $C = 0.24$ mm.

TABLE 3.6 Predicted Performance of Millimeter Wave Frequency Triplers for Various Diode Zero-Bias Capacitances with $R_s(0) = 5 \, \Omega$, at a Pump Frequency of 50 GHz[a]

$C_j(0)$ (fF)	ϵ_p	ϵ_3	Z_s^{opt}		Z_1^{3rd}		X_1^{2nd} (Ω)	x_{short} (mm)	ϵ_4/ϵ_3	V_r^{max} (V)	P_{pump}^{max} (mW)
			$R\,(\Omega)$	$X\,(\Omega)$	$R\,(\Omega)$	$X\,(\Omega)$					
10	0.71	0.67	18	962	34	281	135	2.40	0.001	21.5	1.0
20	0.84	0.62	34	456	68	123	135	1.80	0.229	18.1	7.5
40	0.93	0.49	75	276	58	126	135	3.00	0.136	32.3	34.1

[a] The geometry of the waveguide mount is illustrated in Figs. 3.11 and 3.12; for this example, $A = 2.03$ mm, $B = 0.26$ mm, $C = 0.24$ mm. The cutoff filter for the second harmonic was spaced 2.23 mm from the diode.

TABLE 3.7 Predicted Performance of Millimeter Wave Frequency Triplers for Various Pump Frequencies, $C_j(0) = 20$ fF, $R_s(0) = 5 \, \Omega$[a]

f_{pump} (GHz)	ϵ_p	ϵ_3	Z_s^{opt}		Z_1^{3rd}		X_1^{2nd} (Ω)	x_{short} (mm)	ϵ_3/ϵ_2	V_r^{max} (V)	P_{pump}^{max} (mW)
			$R\,(\Omega)$	$X\,(\Omega)$	$R\,(\Omega)$	$X\,(\Omega)$					
40	0.44	0.27	9	584	87	85	-185	2.40	0.368	23.2	2.0
45	0.79	0.63	25	501	162	179	147	3.00	0.154	18.1	5.2
50	0.84	0.62	34	456	68	123	135	1.80	0.229	18.1	7.5
55	0.77	0.36	23	520	130	134	280	2.40	0.592	20.0	2.8
60	0.81	0.72	28	375	77	152	113	1.40	0.040	20.2	10.7

[a] The geometry of the waveguide mount is illustrated in Figs. 3.11 and 3.12; for this example, $A = 2.03$ mm, $B = 0.26$ mm, $C = 0.24$ mm. The filter spacing was 2.23 mm.

The following simplifying assumptions were made.

1. The pump circuit can be conjugately matched to the diode impedance at the pump frequency.
2. The losses associated with waveguide, coaxial, and stripline structures incorporated in the mount are negligible.
3. The output circuit is perfectly decoupled from the pump circuit by a low-pass filter and the filter presents a short circuit to the diode at second and higher harmonics.
4. The diode can be represented by the simple equivalent circuit shown in Fig. 3.7. The series resistance is assumed bias independent but has a half-power frequency dependence due to skin effect.
5. The lossless output waveguide tuning short is adjusted for maximum second- or third-harmonic conversion efficiency. This is achieved when the varactor reactance is resonated as well as possible by the mount at the output and idler frequencies.
6. The reduced height output waveguide is coupled to a broadband termination by a perfect, broadband impedance transformer. In the case of the tripler, the reduced height guide in the vicinity of the diode is not cut off at the second harmonic. However, the transformer is designed to be cut off at the idler frequency; it couples all other harmonics losslessly to the termination and is positioned a half guide wavelength from the diode plane at a frequency near the desired mount center frequency.
7. The dc bias on the diode is fixed at 10 V reverse bias. The pump power level is adjusted so that the maximum instantaneous forward voltage across the diode is equal to ϕ (i.e., no foward current flows in the diode, but the voltage swing is maximized).

The mount impedances seen by the diode at the first five pump harmonics were determined using analytical techniques. The current and voltage waveforms were then derived by a nonlinear analysis technique. The analysis was carried out for half-height, third-height, and quarter-height mounts. As stated in assumption 7, the pump drive at this bias voltage is sufficient to maximize the effect of the capacitance nonlinearity of the diode, without causing forward current to flow in the device. For each of these mounts the device zero-bias capacitance was set at 10, 20, and 40 fF, and for the half-height mount only, the series resistance was varied between 1 and 15 Ω. The analysis was carried out for both the doubler and tripler configurations at several frequencies in the output waveguide band (80, 100, and 120 GHz for the doubler; 120, 135, 150, 165, and 180 GHz for the tripler).

The conversion efficiencies, peak reverse diode voltage, pump power level, optimized backshort position, and source and load impedances are given in Tables 3.1 to 3.7. Figure 3.12 illustrates the resultant, theoretical diode voltage waveforms for a number of selected cases. The theoretical efficiency values are not attainable in a practical mount because of the unavoidable ohmic losses that occur in waveguide and stripline structures. These losses can be on the order of 1 to 3 dB in a millimeter wave multiplier block. In addition, where the diode voltage swing is large, additional losses will be incurred, owing to forward current flow in the varactor. However, with these limitations in mind, the results can still provide useful guidelines when designing and optimizing practical millimeter wavelength frequency multipliers.

The conversion efficiencies predicted by these analyses of more realistic millimeter wavelength mounting structures can be seen to be significantly lower than would be

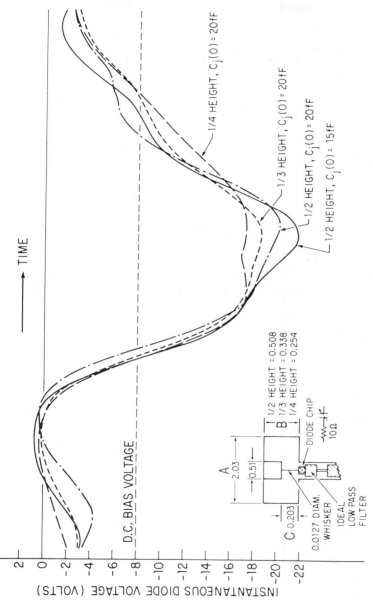

Figure 3.12 Theoretical diode voltage waveforms (one period) for an idealized doubler mount at a pump frequency of 50 GHz and a pump power of 10 mW. The analysis was carried out for the three different waveguide structures indicated, using the simplifying assumptions noted in the text. For the half-height mount, the curves also show the effect of decreasing the zero-bias capacitance of the diode. (Dimensions modify as shown are in millimeters.) (From **Ref. 18**; reproduced by permission.)

163

TABLE 3.8 Summary of State-of-the-Art Performance for Millimeter Wave Frequency Multipliers

Mount Type	Tunable Output Operating Band (GHz)	Minimum Output		Maximum Output			Maximum Pump Power (mW)	Notes[a]	Reference
		Effic. (%)	Power (mW)	Effic. (%)	Power (mW)	Freq. (GHz)			
Doubler	80–120	9.5	18	14.0	26.6	88 and 105	190	2, 3, 9	28
	80–120	10.7	16	15.5	23.2	100	150	1, 2, 3	28
	80–120	10	7	16	11	104	70	1, 4, 3	25
	100			25	20	100	80	6, 4	29
	110–170	10	8	15	12.0	120	80	1, 2, 3	30
	140–150	10	8	22	17.6	145	80	1, 2, 3, 5	30
	190–260	10	8	27	21.5	215	80	1, 2, 3	30, 31
	200			19	18	200	150	6, 4	29
	400			8.5	0.44	300	5.1	1, 2, 3, 7	32
	500–600	7	0.7			—	10	1, 2, 8	33
Tripler	85–115	4	1.2	8	2.4	106	28	1, 2, 8	26
	96–120	1.8	1.8	8.2	8.2	110	100	1, 2, 3	34
	105			25	18	105	72	6, 4	29
	200–290	2.5	2.0	7.5	6	225	80	1, 2, 3	35, 36
	190–240	1	0.3	10	3	230	30	1, 2, 8	26
	260–350	1.8	1.5	3.75	3.0	340	80	2, 3, 6	37
	300			2	2	300	100	6, 4	29
	450			1	0.079	450	6.3	1, 2, 3, 7	32
×6 balanced doubler/tripler	310–350	0.3	0.6	0.4	0.75	345	190	1, 2, 3, 6, 9	28

[a] 1, Crossed waveguide mount; 2, tuning and bias optimized at each operating frequency; 3, microstrip low-pass filter; 4, fixed tuning and bias; 5, narrow-banded version of NRAO 110- to 170-GHz doubler; 6, quasi-optical mount; 7, limited pump power available; 8, coaxial low-pass filter; 9, two diode balanced cross guide mounts.

expected from the simplified approach discussed earlier, the results of which were presented in Figs. 3.8 and 3.9. This is because power is lost to the output load at unwanted harmonics and because the waveguide mounting structure cannot always attain the optimum load and source impedances at the input and output frequencies. The choice of diode capacitance and mount geometry for a given application can, to a certain extent, be traded off one against the other. For example, in the case of a doubler with a pump frequency of 50 GHz and target conversion efficiency of 0.7, similar conversion performance can be obtained with the following configurations:

(i) $B = 0.510$ mm, $C = 0.255$ mm, $C_j(0) = 10$ fF,

$R_s = 10\ \Omega$, $P_{pump} = 34.7$ mW

(ii) $B = 0.510$ mm, $C = 0.255$ mm, $C_j(0) = 20$ fF,

$R_s = 5\ \Omega$, $P_{pump} = 12.4$ mW

(iii) $B = 0.255$ mm, $C = 0.242$ mm, $C_j(0) = 40$ fF,

$R_s = 5\ \Omega$, $P_{pump} = 24.2$ mW

It is clear from these data that power-handling ability and hence output power are strongly dependent on the choice of diode and mount. Furthermore, the tables show that for a fixed junction capacitance, the diode series resistance has a strong effect on the multiplier performance. It should also be noted from the tables that if the diode capacitance is made too small, it becomes increasingly difficult to conjugately match the pump circuit because of the large inductive source reactance required.

Table 3.8 shows the best performance measured with broadly tunable frequency doublers and triplers. Clearly, present multiplier designs are capable of providing adequate power for local oscillator applications in receivers in the millimeter wavelength range. The varactor diodes available today are of sufficiently high quality that higher efficiencies than indicated in Table 3.8 (and multiplication to higher frequencies) should, in theory, be possible. However, increased harmonic efficiency and the extension of multiplier technology to higher output frequencies awaits further improvement in mount design.

3.4.2 Noise Characteristics of Varactor Multipliers

The output of any microwave signal source contains unwanted, random frequency-modulation (FM) and amplitude-modulation (AM) of the desired spectral component. These noise components may be characterized in terms of their power spectra. It is usually preferable to minimize the energy contained in these unwanted sidebands of the wanted output spectral line.

Noise at the output of a varactor frequency multiplier arises from two sources. The primary factor determining the spectral purity of the multiplier output is the output spectrum of the pump source. It is particularly important that the pump source have low FM noise, as the frequency multiplication process results in a degradation of the ratio of the signal to FM noise by n^{-2}, where n is the multiplication order [20]. An ideal varactor multiplier, such as discussed in relation to the data given in Figs. 3.8 and 3.9, does not affect the ratio of the signal to AM noise. However, in a real multiplier the conversion efficiency may vary rapidly with frequency, causing significant FM-to-AM and AM-to-FM noise conversion at frequencies near these excursions. Furthermore, if the bias circuit time constant for the multiplier is too short, the average diode elastance will vary with pump signal amplitude, resulting in a phase modulation of the output signal. Clearly, care is necessary in choice of pump source and in multiplier design to minimize these effects.

3.4.3 Specific Design Example—A 200 to 290 GHz Frequency Tripler

The recent development of efficient frequency triplers for the 1 mm band has made it possible to construct simple, portable all-solid-state receivers in this wavelength range. A 200 to 290 GHz frequency tripler described recently in the literature [36] illustrates one practical approach to the design of an efficient, broadly tunable harmonic generator.

Mount Description.   The tripler employs a split-block construction. Figures 3.13 and 3.14 illustrate the mount design. Power incident in the full-height, 3.1 by 1.5 mm input guide is fed to the varactor diode via a tunable waveguide to stripline transition and a seven-section, low-pass filter, which passes the pump frequency with low loss but is cut off for higher harmonics. The varactor chip, a 0.1 mm sided cube, is mounted on the filter substrate adjacent to the reduced-height, 1.14 by 0.93 mm output waveguide. One of the many diodes on the chip is contacted and coupled to the output guide with a post-mounted, 0.0125 mm diameter by 0.15 mm long, gold-plated, phosphor-bronze whisker, which has been suitably pointed and pre-bent. Output tuning is accomplished with the aid of an adjustable backshort in this guide. The bias circuit comprises a 140 Ω transmission line, consisting of a 0.025 mm diameter gold wire center conductor bonded at one end to a low-impedance section of the low-pass filter and at the other end to a 100 fF quartz dielectric bypass capacitor. The outer shield is a slot of rectangular cross section milled into the face of one of the blocks. At 95 GHz the bias line approximates a quarter-wave short-circuited stub, thus minimizing its effect on the performance of the low-pass filter near cutoff.

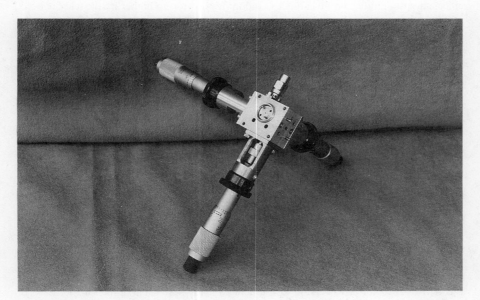

Figure 3.13   Photograph of the 200 to 290 GHz frequency tripler designed by Archer [36].

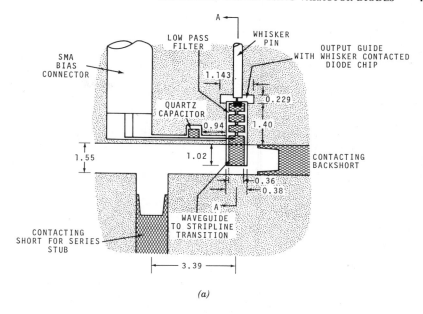

(a)

SECTION AA

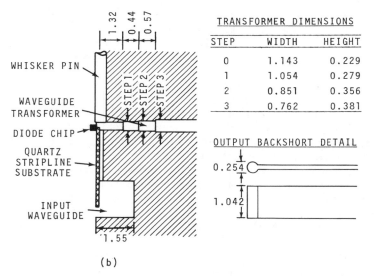

TRANSFORMER DIMENSIONS

STEP	WIDTH	HEIGHT
0	1.143	0.229
1	1.054	0.279
2	0.851	0.356
3	0.762	0.381

OUTPUT BACKSHORT DETAIL

(b)

Figure 3.14 (*a*) View of the 200 to 290 GHz tripler block split along the partition between blocks, showing the input guide, stripline filter, bias circuit, and output waveguide. (Dimensions are in millimeters.) (*b*) Section through the block detailing the waveguide transformer and diode mounting arrangement. The output backshort design is also shown (From Ref. 36; copyright 1984 IEEE, reproduced by permission.)

A quarter-wave, two-section impedance transformer, with dimensions shown in Fig. 3.14b, couples the 1.14 by 0.23 mm reduced height guide to the 0.76 by 0.38-mm output guide. Power can flow in the wider guide at the second harmonic, whereas the output guide is cut off at this frequency. The transformer is thus used to implement a reactive second-harmonic idler termination in the manner described in the preceding section. The transformer is spaced 0.352 mm from the plane of the diode (approximately $\lambda_g/2$ at the second harmonic, where the guide wavelength equals λ_g).

The varactor diode used in this multiplier (type 5M2) was a Schottky diode device fabricated under the supervision of R. J. Mattauch at the University of Virginia. The zero-bias capacitance was 21 fF, the dc series resistance was 8.5 Ω, and the reverse breakdown voltage was 14 V at 1 µA. These devices have a highly nonlinear capacitance versus voltage law, which approximates the inverse half-power behavior of an ideal abrupt junction varactor to within about 2 V of the breakdown limit (see Fig. 3.4).

The length of the contact whisker is chosen so that its inductance approximately series resonates the average capacitance of the pumped diode at the input frequency. Furthermore, this choice of whisker length provides, with the aid of the tuning short, a convenient transformation between the diode impedance and the waveguide impedance at the output frequency. At the pump frequency the low-pass filter, which is about a half wavelength in total length, transforms the approximately real-valued impedance of the whisker/varactor combination (of the order of 20 to 50 Ω) to a similar real-valued impedance at the plane of the waveguide-to-stripline transition. Pump circuit impedance matching is achieved using two adjustable waveguide stubs with sliding contacting shorts. One stub acts as a backshort for the probe type waveguide-to-stripline transition and a second as an E-plane series stub located $\lambda_g'/2$ (at the pump wavelength λ_g') toward the source from the plane of the transition. This tuning configuration, with two degrees of freedom, facilitates the matching of the guide impedance to a wide range of impedances at the input to the low-pass filter. Mechanical adjustment of these tuners typically enables the input to be matched to the diode impedance with a VSWR of 2:1 or less at any frequency within the operating bandwidth of the WR-12 pump waveguide.

Low-Pass Filter Design. A special feature of this multiplier mount was the novel stripline structure used to implement the low-pass filter. The seven-section low-pass filter was a quasi-lumped element, 0.2-dB ripple Chebycheff design, implemented using high/low-impedance stripline sections on a crystalline quartz substrate. The stripline geometry and design data are given in the original paper [36]. A significant advantage of the stripline structure used in the design is that it allows the channel to be milled in only one of the pair of split blocks, while maintaining a large ratio between high and low impedances.

For broadband multiplier performance, the low-pass filter should present, at the very minimum, a short circuit in its stopband to all expected second- and third-harmonic frequencies. In this design the channel and substrate dimensions were carefully chosen so that the moding cutoff frequency of the channel was above the stopband limit imposed by spurious resonances in the transmission-line sections, thus maximizing the useful stopband [36]. Figure 3.15 shows the line dimensions for the filter used in the multiplier and compares its predicted frequency response to an earlier filter design using conventional suspended substrate technology. The input frequency band is 67 to 97 GHz and the maximum −20 dB stopband width achieved is 130 to 350 GHz, providing a reactive termination for the diode at second-, third-, and most fourth-harmonic frequencies. A computer analysis predicts that the filter should appear as short circuit to the diode at 265 GHz and that at 200 GHz it should exhibit a capacitive reactance of 10 Ω. When used in the multiplier mount the length of the low-impedance section to which the varactor diode is mounted is shortened by about 0.25 mm to compensate for the stray capacitance between the diode and the channel walls.

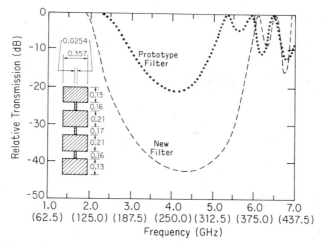

Figure 3.15 Measured transmission response of 62.5 × scale models of the filter and an earlier design that used suspended-substrate stripline. The frequencies in parentheses are the corresponding millimeter wave frequencies. The inset shows the metallization pattern for the millimeter wave version of the filter. (From Ref. 36; copyright 1984 IEEE, reproduced by permission.)

Multiplier Performance. The frequency tripler exhibits significantly wider tuning bandwidth than do previously reported designs [26, 35, 29]. As shown in Fig. 3.16, between 200 and 290 GHz the device provides more than 2 mW output power for 80 mW in. Backshort tuning and dc bias were optimized at each measurement frequency. The typical reverse dc bias voltage was 5 V, with forward currents between 0.1 and 0.5 mA. The peak power output of 4.6 mW occurs at 220 GHz, with a corresponding conversion efficiency of 5.7%. As shown in Fig. 3.16, higher conversion efficiencies may be obtained at lower pump powers. At 35 mW input power, the maximum efficiency obtained was 8% at 222 GHz.

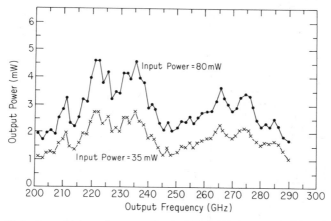

Figure 3.16 Output power versus frequency for the 200 to 290 GHz tripler. Bias and tuning were optimized at each measurement frequency. The points have been arbitrarily interconnected with straight-line segments for clarity. (From Ref. 36; copyright 1984 IEEE, reproduced by permission.)

3.5   MULTIPLIERS USING STEP-RECOVERY DIODES

3.5.1   Multiplier Design

A conceptual block diagram of a step-recovery diode multiplier is shown in Fig. 3.17. A signal source at the input frequency delivers power to the step-recovery diode, which is used as an impulse generator, converting the energy in each input cycle into a narrow, large-amplitude voltage pulse. The pulse excites a resonant output circuit, converting the impulse into a damped ringing waveform. This signal is then filtered to select the desired CW output component.

To form the impulse generator the diode is embedded in the circuit shown in Fig. 3.18. The high-amplitude, short impulse is formed by storing energy in the drive inductance just prior to the transition of the diode from forward to reverse bias. Since the step-recovery diode is driven hard into forward and then into reverse bias, its equivalent circuit may simply be represented as a small resistor, approximately equal to the series resistance, in the forward-biased state, and as a capacitor, C_r, in the reverse-biased state. The stored energy appears across C_r, after switching, as a negative, half-sine pulse with a peak voltage given by [17]

$$V_p = I_p \sqrt{\frac{L}{C_r}} \exp\left(\frac{-\pi\xi}{2\sqrt{1-\xi^2}}\right) \tag{3.26}$$

where I_p is the peak current flowing in the diode. The impulse width is determined by the resonant frequency of the LC combination and is given by

$$t_p = \pi\sqrt{LC_r}/\sqrt{1-\xi^2} \tag{3.27}$$

where

$$\xi = \frac{1}{2R_1}\sqrt{\frac{L}{C_r}} \tag{3.28}$$

is a damping factor determined by the loaded Q of the circuit. The total power in the impulse train is given by

$$P_{\text{im}} = \frac{\pi\xi V_p^2 f_{\text{in}} C_r}{\sqrt{1-\xi^2}} \tag{3.29}$$

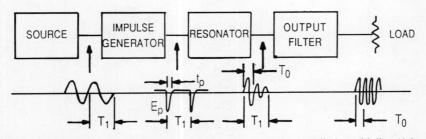

Figure 3.17   Simplified conceptual block diagram of a step-recovery diode multiplier. (Adapted from Ref. 38, courtesy of Hewlett-Packard.)

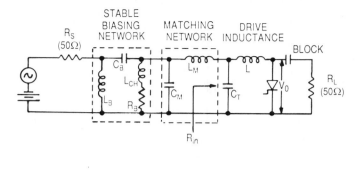

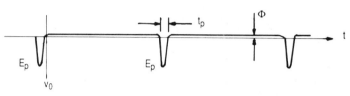

Figure 3.18 Circuit diagram for a practical impulse generator in a step-recovery multiplier and its output voltage waveform. (Adapted from Ref. 38, courtesy of Hewlett-Packard.)

The input circuit of the impulse generator must include a means for dc biasing the diode and a reactive network to match the impedance of the step-recovery diode to the source impedance of the generator. A typical (narrow-band) L-C network that can be used to achieve these functions is shown in Fig. 3.18. The tuning capacitor resonates the drive inductance at the input frequency. Since this capacitor carries the RF current at all harmonics of the input signal (at least up to $f = 1.5/t_p$) it must be a very high quality capacitor with a high self-resonant frequency (certainly greater than $1/2t_p$). The input impedance at the terminals of C_T is resistive (at f_{in}) and is given approximately by $R_{IN} = 2\alpha\pi f_{in}L$, where $\alpha \simeq 1 + 1.6\xi$. For $R_S/R_{IN} < 10$, where R_S is the source resistance, the matching circuit element values are given by [38]

$$L_M = \frac{\sqrt{R_S R_{IN}}}{2\pi f_{in}} \tag{3.30}$$

$$C_M = \frac{1}{2\pi f_{in}\sqrt{R_S R_{IN}}} \tag{3.31}$$

The element values for the bias network are chosen so that it forms a maximally flat high-pass filter with a cutoff frequency of $0.8f_{in}$. It is particularly important for adequate stability of the multiplier that no high-Q series resonances be present in the bias network, especially at low frequencies. Bias circuit components should be chosen with this in mind. Component values are given by [38]

$$L_B = \frac{24.4}{f_{in}} \qquad C_B = \frac{8.85 \times 10^{-3}}{f_{in}} \tag{3.32}$$

The bias network must be located at a distance of much less than $\lambda_{in}/4$ from the diode, to avoid unsuitable driving impedances for the diode at frequencies less than f_{in}.

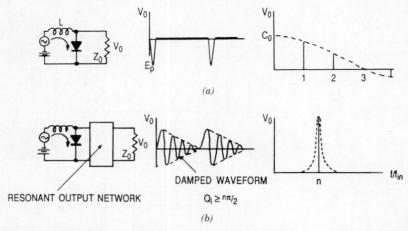

(a)

(b)

RESONANT OUTPUT NETWORK

DAMPED WAVEFORM

$Q_l \geq n\pi/2$

Figure 3.19 Form of the diode waveform and the approximate frequency spectrum of the impulse for (*a*) no output filtering and (*b*) output filtering with $Q_n = \pi n/2$. (Adapted from Ref. 38, courtesy of Hewlett-Packard.)

The equivalent frequency-domain spectrum of the pulse train is shown in Fig. 3.19a. The impulse generator can clearly be considered to behave as a "comb"-generating device in the frequency domain. The pulse width determines the variation in power between any two adjacent frequencies of the comb, with a narrower pulse resulting in a flatter amplitude line spectrum and a higher frequency for the first zero crossing. Pulses of as short as 70 ps can be produced with practical step-recovery diodes.

The pulse width is usually chosen to lie between $1/2f_{out}$ and $1/f_{out}$ for multiplier applications, thus optimizing the output spectrum of the comb generator near the desired output frequency. The output from the impulse circuit is fed to a resonant output network with loaded Q adjusted so that most of the energy in the impulse is delivered to the network during one cycle of the input signal. The loaded Q required to achieve this is approximately $\pi n/2$, where n is the multiplication order. Most of the energy of the output spectrum of the impulse generator (nearly 75%, if the diode series resistance can be neglected) has now been concentrated around f_{out}, as shown in Fig. 3.19b. The output network is most easily implemented as a quarter-wave transmission-line resonator or as a series L-C resonant network, where the required output voltage is developed across a capacitor to ground. Design information for these two types of resonators is given in Fig. 3.20.

In most applications additional filtering of the output signal will be required, in order to reduce the amplitude of the unwanted residual components of the comb spectrum to an acceptable level. Various high-Q filter designs are available for this task, including cavity filters and structures implemented in stripline and microstrip. The reader is referred to the references for further information on the design of such filters [39].

The selection of an appropriate diode for a particular multiplier application is a most important step in multiplier design. The important parameters of the diode are as follows.

Power-Handling Ability. The power-handling capability of a step-recovery diode in a multiplier circuit is determined by one of two limitations. The first is a constraint imposed by the maximum power dissipation allowable in the device. The second is a constraint imposed by the reverse breakdown voltage limit for the diode, which limits

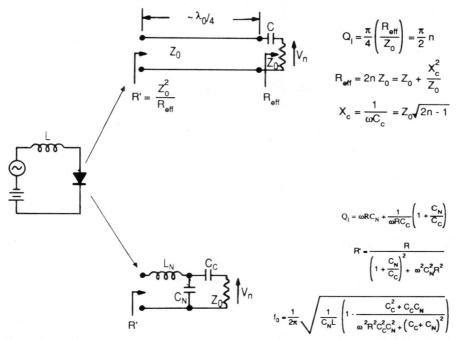

$$Q_1 = \frac{\pi}{4}\left(\frac{R_{eff}}{Z_0}\right) = \frac{\pi}{2}\,n$$

$$R_{eff} = 2n\,Z_0 = Z_0 + \frac{X_c^2}{Z_0}$$

$$X_c = \frac{1}{\omega C_c} = Z_0\sqrt{2n-1}$$

$$Q_1 = \omega R C_N + \frac{1}{\omega R C_C}\left(1 + \frac{C_N}{C_C}\right)$$

$$R' = \frac{R}{\left(1 + \frac{C_N}{C_C}\right)^2 + \omega^2 C_N^2 R^2}$$

$$f_0 = \frac{1}{2\pi}\sqrt{\frac{1}{C_N L}\left(1 - \frac{C_C^2 + C_C C_N}{\omega^2 R^2 C_C^2 C_N^2 + (C_C + C_N)^2}\right)}$$

Figure 3.20 Design data for two possible resonant output networks that would be suitable for filtering the diode output.

the maximum amplitude of the voltage impulse. Both of these limitations are strongly influenced by the conversion efficiency of the multiplier and limit the maximum output power attainable with a given design.

The impulse generator efficiency is determined by four parameters: recombination loss in the diode, finite matching circuit loss, forward-to-reverse bias transition loss, and series resistance loss. Of these, only transition loss and series resistance loss are of great significance. Recombination loss can be minimized by keeping $\omega\tau \gg 1$; circuit loss is small because Q_1 is ~ 1. In terms of just the two important effects, the conversion efficiency $\eta_{\text{CW-imp}}$ of the power in the input CW signal to the power in the impulse is given by [17, 38]

$$\eta_{\text{CW-imp}} \simeq \left(1 + 2N^2\,\frac{f_{\text{in}}}{f_{c1}}\right)\frac{tn_p}{tn_p + NV_{\text{br}}} \tag{3.33}$$

where $tn_p = t_p \times 10^{12}$, $f_{c1} = 159/R_s C_r$, $f_{\text{out}} = nf_{\text{in}}$, and $N = 1/(2f_{\text{in}}t_p)$.

The impulse to CW reconversion efficiency between the impulse power and the power in the damped ringing waveform is given by [38]

$$\eta_{\text{imp-CW}} = \left[\frac{4}{\pi}\frac{x\cos(\pi x/2)}{1 - x^2}\frac{1}{1 + (2R_S/Z_0)}\right]^2 \tag{3.34}$$

where $x = 2f_{\text{out}}t_p$, and Z_0 is the characteristic impedance of the resonant output quarter-wave transmission line.

Figure 3.21 shows a plot of these efficiencies for several typical step-recovery diodes. The power output limits related to the diode parameters may now be expressed as follows. If the diode is not to overheat, we must have [38]

$$P_{\text{out}} < \frac{\eta_{\text{tot}}}{1 - \eta_{\text{tot}}} P_{\text{diss,max}} \tag{3.35}$$

where $\eta_{\text{tot}} = 0.75 L_{\text{fil}}\eta_{\text{CW-imp}}\eta_{\text{imp-CW}}$ and $L_{\text{fil}} = $ output filter loss.

The diode will not exceed the reverse breakdown limit if [38]

$$P_{\text{out}} < 0.75 L_{\text{fil}} f_{\text{in}} C_r V_{\text{br}}^2 \eta_{\text{imp-CW}} \tag{3.36}$$

The factor 0.75 in Eqs. (3.34) and (3.35) arises from the assumption that the loaded Q factor of the resonant output circuit is $n\pi/2$, meaning that 75% of the total damped waveform power is then present in the nth harmonic of the input signal. For a given diode the restriction on output power changes from a breakdown limit to a dissipation limit with increasing frequency. The frequency at which the breakpoint occurs is given by

$$f_{\text{br}} = \frac{P_{\text{diss,max}}\eta_{\text{CW-imp}}}{C_r V_{\text{br}}(1 - \eta_{\text{tot}})} \tag{3.37}$$

Reverse-Biased Capacitance. The magnitude of the reverse-biased capacitance (C_r) determines the energy in the impulse as well as the impedance level of the output resonator. This capacitance is specified at -10 V reverse bias, since the capacitance of most step-recovery diodes is independent of voltage at this level. A rule of thumb for determining the value of Z_0 is

$$10\ \Omega < Z_0 < 20\ \Omega \quad \text{(50-}\Omega \text{ system)} \tag{3.38}$$

where $Z_0 = 1/2\pi f_{\text{out}} C_r$.

Fundamental Time Constants. Two fundamental time constants are of importance in the selection of an appropriate step-recovery diode for a given application. The first is the minority carrier lifetime (τ), which determines the loss that occurs during forward charge storage due to carrier recombination, as well as the value of the self-bias resistance, which develops the diode bias due to rectification current. The lifetime effects are minimized if $2\pi f_{\text{in}}\tau > 10$.

The second intrinsic time constant is the transition time (t_t), which determines the ability of the diode to achieve the required impulse width and sets the maximum output frequency limit as $t_t \leqslant 1/f_{\text{out}}$.

Package Parasitics. The package inductance (L_p) is in series with the drive inductance. Its magnitude relative to that of the external drive inductance determines the proportion of the energy in the total effective drive inductance that is coupled to the output resonator. Typically, if $L_p < Z_0/(2\pi f_{\text{out}})$, the effect may be neglected.

The package capacitance (C_p) appears in shunt with the diode capacitance and is undesirable in that it is not active in the impulse-generating process. The package capacitance should be small compared to the reverse-biased junction capacitance.

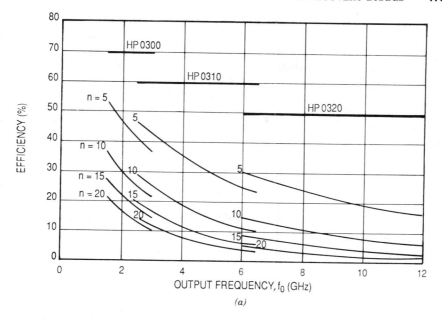

(a)

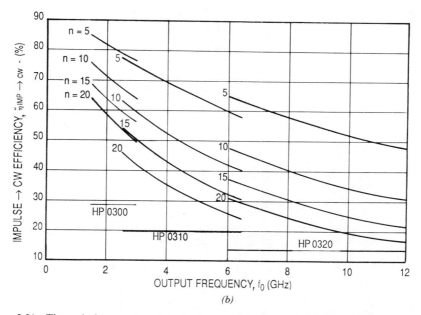

(b)

Figure 3.21 Theoretical conversion efficiencies for use in the design of multipliers constructed using Hewlett-Packard step-recovery diodes. The diode part numbers are indicated on the curves. (a) Overall efficiency from input CW to damped output waveform. (b) Efficiency of converting impulse to damped waveform. (Adapted from Ref. 38, courtesy of Hewlett-Packard.)

3.5.2 Noise in Step-Recovery Diode Multipliers

As with the varactor diode frequency multiplier, the multiplication process in a step-recovery diode results in a degradation of the ratio of the signal to FM noise at the output, relative to that of the pump input, of 6 dB for each doubling of the frequency. In addition to the effects on the input FM noise spectrum, a step-recovery diode multiplier can generate noise internally [16, 40]. In particular, the effective loaded Q of the output filter has a significant impact on the level of internally generated AM noise. Modulation of the source impedance to the filter, due to the intermittent diode conduction, can also produce FM noise in the output spectrum. There is also the possibility of phase noise being introduced by effects associated with the storage of charge when the diode is overdriven. The reader is referred to Refs. 16 and 40 for further discussion of these effects and for design techniques that can be used to reduce their impact on multiplier performance.

3.5.3 Specific Design Example—A 2.7 GHz Sextupler

The performance requirements for the practical multiplier, designed by Cooper and Wells [41], which is to be used to demonstrate the implementation of the design techniques outlined above, were

$$f_{\text{in}} = 450 \text{ MHz} \qquad f_{\text{out}} = 2700 \text{ MHz} \qquad P_{\text{out,min}} = 2\text{W}$$

Figure 3.22 Photograph of the 6 × multiplier designed by Cooper and Wells. (From Ref. 41, reproduced by permission.)

The diode chosen for this application was the HP0300 device. The characteristics of this diode are

$$C_r = 4 \text{ pF} \quad R_S = 0.12 \text{ } \Omega \quad \tau > 100 \text{ ns} \quad t_t < 600 \text{ ps} \quad L_p = 0.3 \text{ nH} \quad C_p = 0.1 \text{ pF}$$

For an output frequency of 2700 MHz, the impulse length was constrained as follows:

$$185 \text{ ps} < t_p < 370 \text{ps}$$

Hence $t_p = 270$ ps was chosen.

The theoretical values for the circuit elements of the impulse generator were determined for this value of t_p. Referring to Fig. 3.18, they are

$$L = 1.84 \text{ nH} \quad C_T = 67.7 \text{ pF} \quad R_{\text{IN}}(\text{diode}) \simeq 7 \text{ } \Omega \quad L_M = 6.6 \text{ nH} \quad C_M = 18.9 \text{ pF}$$

The resonant output network in this example is a quarter-wave slab line with a characteristic impedance of 15 Ω, design details for this type of network are given in Fig. 3.20. The loaded Q of the output resonator should be approximately 9.5 to optimize the concentration of energy in the sixth harmonic of the input signal. The equivalent Z_0 of the output line was chosen so that it was equal to $1/(2\pi f_{\text{out}} C_r)$. The effective load resistance presented to the diode is equal to $2nZ_0 = 180 \text{ } \Omega$.

The reverse breakdown voltage for the 0300 diode is 75 V and the maximum allowable power dissipation is 9 W. Assuming a filter loss of 0.2 dB gives the theoretical overall efficiency for this multiplier as 0.3, resulting in a maximum output power of 3.85 W if the maximum dissipation rating of the diode is not to be exceeded. If the voltage of the impulse is not to exceed the breakdown voltage, the maximum output power must not exceed 5.4 W. Hence the maximum output power is constrained to 3.85 W by the requirement that the dissipation in the diode not exceed 9 W.

A photograph of the multiplier is shown in Fig. 3.22. Figure 3.23 gives a schematic diagram of the unit. The optimum value of the drive inductance, L, corresponds to a few millimeters of 50-Ω line. The tuning capacitor C_T, with a value of about 70 pF, was designed to exhibit very low parasitic inductance, since it serves as a bypass for currents at the output frequency. It was constructed by wrapping polythene film around a brass block, with a 0.05-mm clearance from the top and bottom ground planes. Input matching

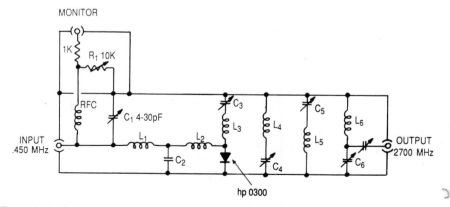

Figure 3.23 Schematic diagram of the 6 × multiplier circuit designed by Cooper and Wells. (From Ref. 41, reproduced by permission.)

TABLE 3.9 Performance Data for the 2.7 GHz Step Recovery Diode Sextupler Designed by Cooper and Wells.

Pump Power (W)	Output Power (W)	Efficiency (%)
1.8	0.65	37
3.4	1.40	41
5.0	2.00	40

Source: Ref. 41, reproduced by permission.

was achieved using a conventional *LC* network, with L_M implemented as a length of 50-Ω transmission line.

The diode was self-biased through an 8-kΩ resistor when operating at 2 W output. It was found to be important to eliminate unnecessary shunt capacitance in the bias circuit to avoid parametric oscillations in the multiplier. The output circuit of the multiplier is composed of the nominal quarter-wave resonator, which is coupled to the output through a three-section interdigital filter. The filter attenuates the adjacent harmonics at 2250 and 3150 MHz by 50 dB relative to the 2700-MHz output.

Performance data for the multiplier are summarized in Table 3.9. The multiplier was designed for use as a local oscillator for a radiometer receiver, and therefore the level of

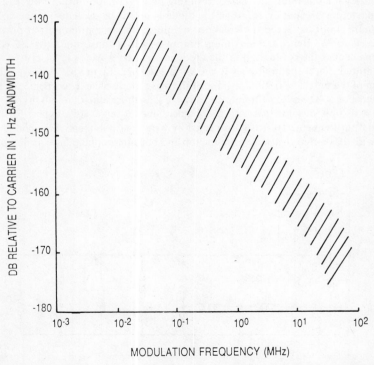

Figure 3.24 Approximate form of the output noise spectrum of the example multiplier. (From Ref. 41, reproduced by permission.)

close-in sideband noise was an important consideration. Figure 3.24 shows the approximate form of the noise spectrum of this multiplier.

3.6 PARAMETRIC AMPLIFIERS AND VARACTOR FREQUENCY CONVERTERS

Pumped varactor diodes found early widespread acceptance in low-noise amplifiers. The first parametric amplifiers (or paramps as they are popularly known) were produced at a time when the alternative for sensitive microwave reception was a silicon point-contact mixer followed by an intermediate-frequency amplifier of moderate performance by today's standards [42]. Radio astronomy, with its requirement for the utmost sensitivity, and in keeping with its experimental status at the time, provided the first significant application for parametric amplifiers [43]. The radio astronomy microwave receiver of the early 1960s consisted of a mixer followed by a vacuum-tube amplifier. The addition of a paramp ahead of the mixer gave an immediate improvement in sensitivity (a factor of 3 or 4).

Paramps gained acceptance because the improvement in sensitivity they offered outweighed the complications they introduced (the requirement for a low-loss ferrite circulator and a high-power, high-frequency pump oscillator—coupled with a reputation for being touchy to operate). As experience was gained, and with their adoption by the satellite industry [44], paramp operation became a relatively routine procedure. In applications where a greater sensitivity requirement justified the additional complication, the paramp could be cooled to take advantage of the direct dependence of noise on the varactor's physical temperature [45]. The earliest cooled paramps had the varactor and its associated microwave circuitry immersed in a bath of liquid nitrogen while the input circulator was outside the dewar at room temperature [46]. This arrangement rapidly gave way to systems in which the circulator and amplifier were cooled to temperatures as low as 15 K in closed-cycle refrigerators employing helium gas as the refrigerant.

Paramps have recently been displaced by gallium arsenide MESFET amplifiers which yield comparable noise performance with less complication. Figure 3.25 shows noise temperature versus frequency for FET amplifiers and paramps, both cooled and at room temperature. Characteristics of both devices (advantages and disadvantages) are listed for comparison in Table 3.10.

Varactor frequency converters have not found widespread application outside radio astronomy. The most notable example of their use in this field was in the VLA (an array of 27 antennas located at Socorro in New Mexico). Each antenna was equipped with dual-channel up-converters, converting signals between 1.35 and 1.75 GHz to 4.5 to 5.0 GHz for amplification in a three-stage parametric amplifier [47]. Such a system requires separate pump sources for the up-converter and parametric amplifiers. FET amplifiers have recently been retrofitted to this instrument.

3.6.1 Frequency Converters

Varactor frequency converters are characterized by the frequency relationships between the input and output and the pump supply. Four basic types are distinguished: the upper-sideband up-converter (USUC), the lower-sideband up-converter (LSUC), the upper-sideband down-converter (USDC), and the lower-sideband down-converter (LSDC). In their simplest form only three frequencies are active: the pump, input, and output. Additional frequencies (idlers) can be allowed to exist. In some cases this can lead to improved performance, such as in a four-frequency USUC [48], which can have gain greater than

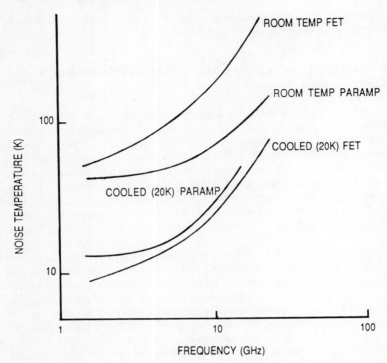

Figure 3.25 Comparative noise performance for FET and parametric amplifiers as a function of frequency.

the upper limit for a three-frequency circuit (the ratio of output to input frequency). The two lower-sideband, three-frequency converters present negative resistances at the input and output frequencies, and while this allows the possibility of achieving significant gain, care has to be taken to ensure stability. The upper-sideband converters, on the other hand, are both stable and have a conversion loss or gain that is less than the ratio of output to input frequency (equal to it for a lossless varactor). The performance of the USDC rapidly deteriorates as the separation between input and output frequency increases.

When varactor up-converters first came to prominence, they offered a means of obtaining low-noise performance at frequencies in the hundreds of megahertz region, where their high gain and low noise could overcome the noise contributed by a relatively

TABLE 3.10 Comparative Characteristics of FET and Parametric Amplifiers: Advantages and Disadvantages

Paramps	FET Amps	Comments
Low noise temperature	Low noise temperature	Comparable
Lower noise cooled	Lower noise cooled	Comparable
One circulator per stage	One isolator per amp	FET amp often needs an
High-frequency pump	Dc supply	isolator for stability
Bulky	Compact	
Expensive	Comparatively cheap	
	More stable	

poor microwave receiver [20]. For example, an up-converter with 15 dB gain and 50 K noise would give rise to an overall noise temperature of about 80 K (noise figure of 1 dB) when followed by a mixer receiver with a 6-dB noise figure. When this technology was in its infancy, this represented very good performance in the VHF and UHF region of the spectrum. Such performance was possible with the USUC when the frequency conversion range was sufficiently high (e.g., input at less than 100 MHz and output at 10 GHz). For input frequencies between 100 and 1000 MHz, a LSUC would be required to achieve sufficient gain [20]. The development of low-noise bipolar transistors and field-effect transistors provided a simpler means of achieving high sensitivity in this spectral range.

Another way of using an up-converter was to follow it with a low-noise parametric amplifier. In this case a modest gain was all that was necessary to achieve good overall performance. An early example of this application consisted of an up-converter followed by a degenerate paramp, which gave an overall system noise temperature of 160 K [43]. A more recent example is the previously mentioned case of the receivers in the VLA [47]. Each antenna in the array was fitted with receivers covering the range 4.5 to 5.0 GHz. These consisted of cooled parametric amplifiers and each could be preceded by a switch-selectable varactor USUC converting from 1.35 to 1.75 GHz. The gain of these up-converters was only 2 to 2.5 dB, but the overall system noise temperature was only 49 K. GaAsFET amplifiers have taken over from such complicated systems, providing comparable noise performance and greater operational simplicity and stability.

The discussion above has been concerned with the use of varactor converters in low-noise applications. Large-signal converters are also of interest. For information on these and details of the design of all types of varactor converters, the reader is referred to Refs. 2, 20, and 49.

3.6.2 Parametric Amplifiers

In a parametric amplifier the output frequency is the same as the input. Normally, three frequencies are present, the pump f_p, the signal f_s and an idler $f_1 = f_p \pm f_s$, but as is the case with the converters, additional idlers can be incorporated. The three-frequency configuration is analogous to the lower-sideband converters, with output being taken at the signal frequency instead of at the difference frequency, which in this case plays the role of an idler. As a consequence of current flowing at the idler frequency, a negative resistance is presented to the external circuit at the signal frequency. This negative resistance, when correctly terminated by the signal circuit, gives rise to a reflection coefficient greater than unity and is the source of the device's amplification.

Degenerate Parametric Amplifiers. There are two basic classes of paramp, degenerate and nondegenerate. In the degenerate paramp the signal frequency (f_s) is close to the idler ($f_p - f_s$), and in the limiting case, when the signal and idler bands overlap, the pump frequency is twice the signal band center frequency. The nondegenerate amplifier has the signal and idler bands separated. The latter type of amplifier is the most common, with degenerate amplifiers mainly finding application in some of the early broadband radiometers and more recently in a number of millimeter wave paramps. The degenerate amplifier's output contains noise contributions from both the signal and idler bands, which in the case when they coincide are of equal magnitude. This is analogous to the double-sideband, variable-resistance mixer in which the signal and image bands contribute equally. In radiometer applications, where the signal is broadband noise, this type of amplifier has good sensitivity, since the signal is received equally in both the signal and idler bands.

An effective input noise temperature for degenerate paramps is [20]

$$T_{\text{eff}} = \frac{T_d}{(2m_1 f_{cd}/\omega_p) - 1} \tag{3.39}$$

where T_d is the varactor temperature, and $m_1 f_{cd}$, the dynamic figure of merit, is equal to $2\pi f_0 Q_d$ [see Eqs. (3.14) and (3.15)] (m_1 is a modulation index which indicates the degree of pumping and is equal to the ratio of the first Fourier coefficient of the pumped elastance to the maximum range of elastance available from the varactor).

$$m_1 = \frac{|S_1|}{S_{\text{max}} - S_{\text{min}}} \tag{3.40}$$

This effective noise temperature can be used to compare degenerate paramps for the same type of use. However, for a more general comparison of degenerate and nondegenerate amplifiers, account must be taken of the nature of the signal; whether it is contained in a band that does not include $f_p/2$ (single-sideband operation), or includes $f_p/2$ (double-sideband operation), or whether it is noiselike, as in radio astronomy. In the double-sideband case, account must also be taken of the nature of the detector. A detailed discussion of this subject can be found in Ref. 2.

Because its signal and idler bands are coincident, the degenerate paramp comprises a comparatively simple circuit. A means must be provided for coupling in the pump power efficiently and for preventing its flow in the signal line. A short length of high-impedance transmission line resonates the varactor at the signal frequency and, at the same time, provides an idler circuit. The circuit is coupled to the input signal line by a quarter-wave impedance transformer to achieve the required condition for gain [50].

The condition for gain of a degenerate paramp is given by [2]

$$\frac{f_p}{2} = m_1 f_{cd} \tag{3.41}$$

To achieve gain the negative resistance must be transformed to a value that is close to, but less than, the resistance presented by the input circuit. As the magnitude of the transformed negative resistance approaches that of the input transmission line, the gain increases, going to infinity when the magnitudes are equal and satisfying the condition for oscillation when the net resistance is negative.

The circuit described so far is a one-port device. To make an amplifier out of it, a means of separating the amplified output from the input must be provided. A ferrite circulator is the usual way of satisfying this requirement, and this is a feature of virtually all paramps, both degenerate and nondegenerate. Early paramps used three-port circulators, but most recent designs incorporated five-port circulators to provide increased immunity from the effects of source impedance variations and to provide greater isolation between the first and second stages.

Nondegenerate Parametric Amplifiers. The most common form of parametric amplifier is the nondegenerate type, in which the signal and idler bands are clearly separated. In most recent designs the pump frequency is usually greater by a factor of 5 or more than the signal frequency. In this type of amplifier a separate circuit must be provided to support the idler.

The basic nondegenerate parametric amplifier is illustrated in Fig. 3.26. This shows schematically the varactor diode connected as a common element between signal, idler,

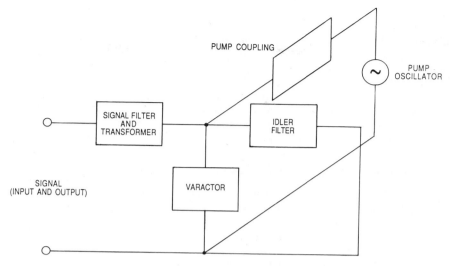

Figure 3.26   Schematic diagram of a parametric amplifier.

and pump ports. Ideally, current should not flow at any other frequencies (i.e., all other sidebands should be terminated in an open circuit). If this condition is satisfied, the only noise sources internal to the amplifier are the varactor series resistance and any resistance in the idler circuit.

Assuming this condition to be satisfied, we see that the requirement for gain in such an amplifier is

$$f_s f_i (R_s + R_i) < R_s (m_1 f_{cd})^2 \tag{3.42}$$

which, when the idler resistance $R_i = 0$, reduces to [2]

$$0 < f_s f_i < (m_1 f_{cd})^2 \tag{3.43}$$

When $R_i = 0$, the noise temperature is [2]

$$T = T_d \frac{f_s}{f_i} \frac{(m_1 f_{cd})^2 + f_i^2}{(m_1 f_{cd})^2 - f_s f_i} \tag{3.44}$$

When the paramp is pumped at an optimum pump frequency, a minimum noise temperature is obtained [2]:

$$T_{min} = T_d \frac{2 f_s}{m_1 f_{cd}} \left[\frac{f_s}{m_1 f_{cd}} + \sqrt{1 + \left(\frac{f_s}{m_1 f_{cd}} \right)^2} \right] \tag{3.45}$$

The corresponding optimum pump frequency is [2]

$$f_p = \sqrt{(m_1 f_{cd})^2 + f_s^2} \tag{3.46}$$

To obtain the best results from a given varactor it should be fully pumped, which means that the pump energy should be sufficient to swing the varactor's elastance over its maximum range. If a varactor is fully pumped, and if the pumped elastance waveform is sinusoidal, m_1 has a value of 0.25. For arbitrary elastance waveforms, m_1 has an upper limit of 0.318. In practice the degree of pumping is usually less than the maximum. This is discussed in Section 3.6.3.

In Ref. 51 six useful formulas have been collected and examined to determine their accuracy. Good agreement was obtained between calculated parameters and measurements on an actual paramp. The formulas presented are for noise temperature, gain stability, phase stability, pump power, saturation, and bandwidth.

3.6.3   Practical Considerations in Parametric Amplifier Design

The real limitations on the performance of parametric amplifiers are the result of the nature of practical varactor diodes, pump sources, ferrite circulators, and the physical realizability of circuit elements. Amplifiers have been constructed to operate at frequencies of some hundreds of megahertz to just below 100 GHz, but most have operated in the range of frequencies from 1 to 10 GHz. Table 3.11 lists examples of amplifiers operating over a wide frequency range. At the 100-GHz end of the spectrum, ferrite circulator losses, the need for very high-power, high-frequency pump sources, and the difficulty of constructing the idler circuit have made the technology unattractive compared with GaAs Schottky mixers, which are comparatively simple and capable of quite low noise performance when cooled.

Almost all parametric amplifiers have used packaged varactor diodes. Associated with the package are parasitic reactances which form part of the circuit at each of the frequencies taking part in the amplification process [60]. These parasitics play a particularly important role in the idler circuit. The reactance that the idler circuit is required to present across the varactor package depends on the relation between the idler frequency and the natural resonance of the mounted varactor. Optimum idler performance is achieved if the idler current can be confined to the diode and its package reactances. This mode of operation results in maximum idler bandwidth and minimizes idler losses. Placing a short or open circuit across the package at the idler frequency achieves this goal [61]. Referring to Fig. 3.5, when the package is short circuited, L_s, C_f, and C_j form a resonant circuit. For the type of package considered in Ref. 62, this resonance is in the vicinity of 10 GHz. In the open-circuit case, for this package, all the reactances combine to produce a resonance at around 20 GHz. These terminations can be provided by external circuit elements, such as a lumped LC circuit or a length of transmission line, but the added elements will degrade the idler performance. An alternative is to employ balanced circuits and make use of symmetry. One such arrangement consists of mounting a pair of varactors side by side with opposite polarity and to supply the pump and signal voltages in such a way that the idler currents are in antiphase [63]. This results in the idler current flowing in a loop through the two varactors, so that each effectively short circuits the other. Another approach is to mount the diodes end to end across the pump waveguide with the signal line entering through the sidewall and contacting the junction between them [61].

To achieve the optimum parametric amplifier design, microwave engineers should have at their disposal a variety of varactor chips so that varying requirements of pump frequency and idler circuit can be met. If engineers are able to design the varactors as well, they are in even better position.

The idler circuit design has an important bearing on two of the most important characteristics of a parametric amplifier, the noise performance and the bandwidth. The

TABLE 3.11 Characteristics of Some Representative Parametric Amplifiers[a,b]

Frequency (GHz)	Type	Gain (dB)	BW (MHz)	Pump (GHz)	Idler (GHz)	Noise (K)	Year	Reference	Comments
1.3	Nondegenerate	20 13 7	110 180 300	11.3	10	70 (room temp.) 29 ($T_a = 77$ K)	1964	51	Broadbanding stubs
2.7	Degenerate	15	400	5.4	2.7	60 DSB	1968	49	Broadbanding stubs, noise includes all input loss
3.95	Nondegenerate	26	500	34.7	31.2	74–85	1971	52	Two stages, broadband idler, balanced varactors, dual quarter-wave transformers
7.6	Nondegenerate	15	500	70	62.4	63	1972	53	Wafer-mounted balanced Schottky varactors
11.6	Nondegenerate	20	600	42.5	30.9	50	1974	54	Two stages, broadband idler, balanced varactors, 20 K ambient dual quarter-wave transformers, input losses included
14.95	Nondegenerate	26	500	95	80.05	75	1976	55	Two stages
24	Degenerate	14 15	>100 >85	48	24	362 (room temp.) 150 ($T_a = 20$ K)	1970	56	DSB noise temperature, including circulator loss and second stage contribution
37	Nondegenerate	18	100	101.4	64.4	404	1974	57	240 K for amplifier alone
46	Degenerate	19 22	180 200	92	46	220 (room temp.) 40 ($T_a = 20$ K)	1973	58	DSB noise temperature referred to waveguide window
94	Nondegenerate			170	76	1000 (target)	1973	59	Measured data not given

[a] This table presents a sample of amplifiers. It is not intended to be exhaustive, nor do the amplifiers listed necessarily represent the best performance at their respective frequencies.
[b] Not all authors indicate whether their noise temperature includes circulator and other losses.

equation given for noise temperature [Eq. (3.43)] assumes R_i to be zero. In fact, R_i will be nonzero, and in general at the same physical temperature as the varactor. Confining the idler to the varactor minimizes R_i and thereby its contribution to the noise.

The bandwidth that can be achieved is a function of both the signal and idler bandwidths. If the idler circuit can be made sufficiently broad, the problem of increasing the overall bandwidth reduces to the problem of designing a wideband signal circuit. One way of accomplishing this is to use a pair of broadbanding stubs forming a shunt resonant circuit, which combines with the series resonant circuit at the varactor input to provide a double-tuned response [52]. Schemes of this type can be extended to multiple-element configurations. Another approach is to use a two-stage quarter-wave transformer to achieve the condition for gain over a wider range of frequencies [64]. Bandwidth is also increased when the operating gain is reduced, and recent paramp systems have usually consisted of two or three low-gain stages, the low noise of the following stages compensating for the reduced gain of the first stage. While early amplifiers were capable of around 50 MHz bandwidth at frequencies of several gigahertz, large numbers of paramps were later made to cover the satellite band from 3.7 to 4.2 GHz.

Where wider-frequency coverage is required but instantaneous bandwidth is less important, tuned systems can be used. One way this can be achieved is by mechanically tuning the idler frequency. Alternatively, the idler could be designed to be very broad and the varactor bias varied to change the mean capacitance of the varactor junction, thereby tuning the signal circuit. This method implies that the varactor will not be fully pumped. A combination of both tuning methods could also be used.

The gain stability is an important consideration, especially in some applications, such as radio astronomy. This is greatly dependent on the stability of the pump source, for both power output and frequency. This was a particular problem with early paramps using klystron pump sources. To minimize the effect of power variations, servo loops were introduced to stabilize the pump power level incident on the varactor. This could be accomplished by sampling the pump level with a coupler and monitoring detector and controlling it by varying an attenuator in the pump line. The more popular approach was to control the attenuator to maintain a constant varactor bias. Frequency fluctuations could be taken care of in critical applications by locking the pump to a stable reference. The introduction of Gunn oscillators as pump sources resulted in a considerable improvement in stability, making stabilization schemes unnecessary.

3.6.4 Specific Design Example—A 3.25 GHz Parametric Amplifier

Figure 3.27a shows a section drawing of a nondegenerate parametric amplifier covering the frequency range 3.1 to 3.4 GHz with an instantaneous bandwidth of 40 MHz. This amplifier was designed by M. W. Sinclair of the CSIRO, Division of Radiophysics, for spectral-line radio astronomy. The varactor diode is mounted in the E-plane of a reduced-height waveguide which couples pump power from a 22 GHz reflex klystron to the varactor. A short length of high-impedance coaxial line series resonates the diode mean capacitance at the signal frequency. A three-element low-pass filter isolates the pump and idler from the input line while the pump waveguide is cut off at the idler frequency, confining the idler to the vicinity of the varactor and the idler cavity. The idler circuit consists of a tunable cavity coupled to the varactor by an iris. The position of this iris was chosen to optimize the pump coupling to the varactor. The idler frequency is determined by the combination of the package parasitic reactances, the coupling, and the tunable cavity, which consists of a micrometer-adjustable noncontacting short circuit in a cylindrical tube. The pump and idler blocking filter forms part of a quarter-wave

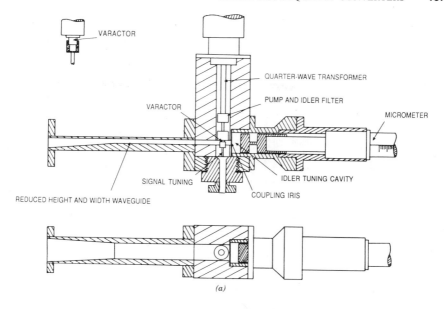

VARACTOR

QUARTER-WAVE TRANSFORMER

VARACTOR

PUMP AND IDLER FILTER

MICROMETER

SIGNAL TUNING

IDLER TUNING CAVITY

REDUCED HEIGHT AND WIDTH WAVEGUIDE

COUPLING IRIS

(a)

(b)

Figure 3.27 (a) Section diagram of a 3.25 GHz parametric amplifier; (b) photograph of the amplifier with its four-port circulator.

TABLE 3.12   Performance Data for a 3.25 GHz
Parametric Amplifier

Signal frequency	3.1–3.4 GHz
Pump frequency	22 GHz
Idler frequency	18.9–18.6 GHz
Gain	20 dB
Instantaneous bandwidth	40 MHz
Measured receiver noise[a]	118 K
Estimated amplifier noise[a]	60 K
Calculated amplifier noise	59 K

[a] Includes 10 K from second stage, 0.15 dB circulator loss, and 0.4 dB switch and coupler loss. Estimated noise is obtained by removing the effect of these losses from 118 K. The degree of agreement is fortuitous.

desired gain. No external bias is provided, the varactor being pumped until self-bias is developed. Figure 3.27b shows a photograph of the amplifier together with its four-port circulator. Its performance is summarized in Table 3.12.

REFERENCES

1. J. M. Manley and H. E. Rowe, "Some General Properties of Nonlinear Elements. Part 1. General Energy Relations," *Proc. IRE*, Vol. 44, No. 7, pp. 904–913, July 1956.

2. P. Penfield, Jr., and R. P. Rafuse, *Varactor Applications*, MIT Press, Cambridge, MA, 1962.

3. J. C. Irvin, T. P. Lee, and D. R. Decker, "Varactor Diodes," in H. A. Watson, Ed., *Microwave Semiconductor Devices and Their Circuit Applications*, McGraw-Hill, New York, 1969, pp. 149–193.

4. S. M. Sze, *Physics of Semiconductor Devices*, 2nd ed., Wiley, New York, 1981, Chap. 3.

5. M. V. Schneider, "Metal-Semiconductor Junctions as Frequency Converters," in K.J. Button, Ed., *Infrared and Millimeter Waves*, Vol. 6, *Systems and Components*, Academic Press, New York, 1982, pp. 209–275.

6. T. P. Lee, "p-n Junction Theory," in H. A. Watson, Ed., *Microwave Semiconductor Devices and Their Circuit Applications*, McGraw-Hill, New York, 1969, pp. 95–125.

7. L. E. Dickens, "Spreading Resistance as a Function of Frequency," *IEEE Trans. Microwave Theory Tech.*, Vol. MTT-15, No. 2, pp. 101–109, February 1967.

8. E. R. Carlson, M. V. Schneider, and T. F. McMaster, "Subharmonically Pumped Millimeter Wave Mixers," *IEEE Trans. Microwave Theory Tech.*, Vol. MTT-26, No. 10, pp. 706–715, October 1978.

9. K. S. Champlin and G. Eisenstein, "Cutoff Frequency of Submillimeter Schottky Barrier Diodes," *IEEE Trans. Microwave Theory Tech.*, Vol. MTT-26, No. 1, pp. 31–34, January 1978.

10. J. W. Archer, B. B. Cregger, R. J. Mattauch, and J. D. Oliver, "Harmonic Generators Have High Efficiency," *Microwaves*, Vol. 21, No. 3, pp. 84–88, March 1982.

11. B. C. DeLoach, Jr., "A New Microwave Measurement to Characterise Diodes and an 800 Gc Cutoff Frequency Varactor at Zero Volts Bias," *IEEE Trans. Microwave Theory Tech.* Vol. MTT-12, No. 1, pp. 15–20, January 1964.

12. T. P. Lee, "Evaluation of Voltage Dependent Series Resistance of Epitaxial Varactor Diodes at Microwave Frequencies," *IEEE Trans. Electron Devices*, Vol. ED-12, No. 8, pp. 457–470, August 1965.

13. N. Houlding, "Measurement of Varactor Quality," *Microwave J.*, Vol. 3, No. 1, pp. 40–45, January 1960.

14. K. Kurokawa, "On the Use of Passive Circuit Measurements for the Adjustment of Variable Capacitance Amplifiers," *Bell Syst. Tech. J.*, Vol. 41, No. 1, pp. 361–381, January 1962.

15. J. L. Moll, S. Krakauer, and R. Shen, "P-N Junction Charge-Storage Diodes," *Proc. IRE*, Vol. 50, No. 1, pp. 43–53, January 1962.

16. S. M. Krakauer, "Harmonic Generation, Rectification, and Lifetime Evaluation with the Step Recovery Diode," *Proc. IRE*, Vol. 50, No. 7, pp. 1665–1676, July 1962.

17. S. Hamilton and R. Hall, "Shunt-Mode Harmonic Generation Using Step Recovery Diodes," *Microwave J.*, Vol. 10, No. 4, pp. 69–78, April 1967.

18. J. W. Archer, "Low-Noise Receiver Technology for Near-Millimeter Wavelengths," in K. J. Button, Ed., *Infrared and Millimeter Waves*, Vol. 15, *Millimeter Components and Techniques*, Part 6, Academic Press, New York, 1986, pp. 1–86.

19. D. N. Held and A. R. Kerr, "Conversion Loss and Noise of Microwave and Millimeter Wave Mixers," *IEEE Trans. Microwave Theory Tech.* Vol. MTT-26, No. 2, pp. 49–55, February 1978.

20. M. Uenohara and J. W. Gewartowski, "Varactor Applications," in H. A. Watson, Ed., *Microwave Semiconductor Devices and Their Circuit Applications*, McGraw-Hill, New York, 1969, pp. 194–269.

21. P. H. Siegel and A. R. Kerr, "Computer Analysis of Microwave and Millimeter Wave Mixers," *IEEE Trans. Microwave Theory Tech.* Vol. MTT-28, No. 3, pp. 275–276, March 1980.

22. W. K. Gwarek, "Nonlinear Analysis of Microwave Mixers," M.S. thesis, Massachusetts Institute of Technology, Cambridge, September 1974.

23. J. W. Archer, "Low-Noise Receiver Technology for Near-Millimeter Wave Radio Astronomy," *Proc. IEEE*, Vol. 73, No. 1, pp. 109–130, January 1985.

24. O. P. Gandhi, *Microwave Engineering and Applications*, Pergamon Press, Elmsford, 1981, Chap. 13.

25. J. W. Archer, "A High Performance Frequency Doubler for 80–120 GHz," *IEEE Trans. Microwave Theory Tech.* Vol. MTT-30, No. 5, pp. 824–825, May 1982.

26. N. R. Erickson, "A High Efficiency Frequency Tripler for 230 GHz," *Dig. 12th Eur. Microwave Conf.*, Helsinki, Finland, Sept. 1982 pp. 288–292.

27. R. L. Eisenhart and P. J. Kahn, "Theoretical and Experimental Analysis of a Waveguide Mounting Structure," *IEEE Trans. Microwave Theory Tech.* Vol. MTT-19, No. 8, pp. 706–719, August 1971.

28. J. W. Archer and M. T. Faber, "High Power 80–120 GHz Doublers for a 310–345 GHz × 6 Multiplier Chain," *IEEE Trans. Microwave Theory Tech.*, Vol. MTT-33, No. 6, pp. 533–539, June 1985.

29. J. A. Calviello, "Advanced Devices and Components for Millimeter and Submillimeter Systems," *IEEE Trans. Electron Devices*, Vol. ED-26, No. 9, pp. 1273–1281, September 1979.

30. J. W. Archer, "Millimeter Wavelength Frequency Multipliers," *IEEE Trans. Microwave Theory Tech.*, Vol. MTT-29, No. 6, pp. 552–557, June 1981.

31. K. Lundien, R. J. Mattauch, J. W. Archer, and R. Malik, "Hyperabrupt Junction Varactor Diodes for Millimeter Wavelength Harmonic Generation," *IEEE Trans. Microwave Theory Tech.*, Vol. MTT-31, No. 2, pp. 235–238, February 1983.

32. T. Takada and M. Hirayama, "Hybrid Integrated Frequency Multipliers at 300 and 450 GHz," *IEEE Trans. Microwave Theory Tech.*, Vol. MTT-26, No. 10, pp. 733–737, October 1978.

33. N. R. Erickson and H. R. Fetterman, "Single Mode Waveguide Submillimeter Frequency Multiplication and Mixing," *Bull. Am. Phys. Soc.*, Vol. 27, p. 836 (abstract only), 1982.

34. J. W. Archer and M. J. Crawford, "A Synthesised 90–120 GHz Signal Source," *Microwave J.*, Vol. 28, No. 5, pp. 227–250, May 1985.

35. J. W. Archer, "An All Solid-State Receiver for 210–240 GHz," *IEEE Trans. Microwave Theory Tech.*, Vol. MTT-30, No. 8, pp. 1247–1252, August 1982.

36. J. W. Archer, "An Efficient 200–290 GHz Frequency Tripler Incorporating a Novel Stripline Structure," *IEEE Trans. Microwave Theory Tech.*, Vol. MTT-32, No. 4, pp. 416–421, April 1984.

37. J. W. Archer, "A Novel Quasi-optical Frequency Multiplier Design for Millimeter and Sub-millimeter Wavelengths," *IEEE Trans. Microwave Theory Tech.*, Vol. MTT-32, No.4, pp. 421–427, April 1984.

38. *Harmonic Generation Using Step Recovery Diodes and SRD Modules*, Hewlett-Packard Application Note 920, Hewlett-Packard, Palo Alto, CA.

39. G. L. Matthaei, L. Young, and E. M. T. Jones, *Microwave Filters, Impedance-Matching Networks, and Coupling Structures*, McGraw-Hill, New York, 1964.

40. J. C. McDade, "Measurements of Additive Phase Noise Contributed by the Step Recovery Diode in a Frequency Multiplier," *Proc. IEEE*, Vol. 54, No. 2, pp. 292–294, February 1966.

41. B. F. C. Cooper and G. A. Wells, "Six-Times Multiplier with Two Watts Output at 2700 MHz," *Proc. IREE (Aust.)*, Vol. 30, No. 10, pp. 340–341, October 1969.

42. C. T. McCoy, "Present and Future Capabilities of Microwave Crystal Receivers," *Proc. IRE*, Vol. 46, No. 1, pp. 61–66, January 1958.

43. B. J. Robinson, "Development of Parametric Amplifiers for Radio Astronomy," *Proc. IRE (Aust.)*, Vol. 24, No. 2, pp. 119–127, February 1963.

44. C. L. Cuccia, "Ultralow-Noise Parametric Amplifiers in Communication Satellite Earth Stations," in L. Young, Ed., *Advances in Microwaves*, Vol. 7, Academic Press, New York, 1971.

45. R. C. Knechtli and R. D. Weglein, "Low Noise Parametric Amplifier," *Proc. IRE*, Vol. 47, No. 4, pp. 584–585, April 1959.

46. F. F. Gardner and D. K. Milne, "A 1400 Mc/S Continuum Radiometer," *Proc. IRE (Aust.)*, Vol. 24, No. 2, pp. 127–132, February 1963.

47. S. Weinreb, M. Balister, S. Maas, and P. J. Napier, "Multiband Low-Noise Receivers for a Very Large Array," *IEEE Trans. Microwave Theory Tech.*, Vol. MTT-25, No. 4, pp. 243–248, April 1977.

48. J. A. Luksch, E. W. Matthews, and G. A. VerWys, "Design and Operation of Four-Frequency Parametric Upconverters," *IRE Trans. Microwave Theory Tech.*, Vol. MTT-9, No. 1, pp. 44–52, January 1961.

49. H. C. Okean and L. J. Steffek, "Octave Input S to Ka-Band Large Signal Up-Converter," *IEEE MTT-S Int. Microwave Symp. Dig.*, pp. 218–220, June 1974.

50. R. A. Batchelor, J. W. Brooks, and B. F. C. Cooper, "Eleven Centimeter Broadband Correlation Radiometer," *IEEE Trans. Antennas Propag.*, Vol. AP-16, No. 2, pp. 228–234, March 1968.

51. P. J. Moogk and U. Rutulis, "Six Formulas Simplify Paramp Design," *Microwaves*, Vol. 5, No. 5, pp. 36–42, May 1966.

52. J. T. DeJager, "Maximum Bandwidth Performance of a Non-degenerate Parametric Amplifier with Single-Tuned Idler Circuit," *IEEE Trans. Microwave Theory Tech.*, Vol. MTT-12, No. 7, pp. 459–467, July 1964.

53. J. C. Vokes, J. R. Dawsey, and H. A. Deadman, "Low-Noise Room-Temperature Parametric Amplifiers," *Electron. Lett.*, Vol. 7, No. 22, pp. 657–658, November 4, 1971.

54. L. E. Dickens, "A Millimeter-Wave Pumped X-Band Uncooled Parametric Amplifier," *Proc. IEEE*, Vol. 60, No. 3 pp. 328–329, March 1972.

55. J. Thirlwell, J. McPherson, and R. R. Bell, "Broadband Cryogenic Parametric Amplifier Operating at 11.6 GHz," *Electron. Lett.*, Vol. 10, No. 16, pp. 329–330, August 8, 1974.

56. H. C. Okean, J. A. DeGruyl, and E. Ng, "Ultra Low Noise Ku-Band Parametric Amplifier Assembly," *IEEE MTT-S Int. Microwave Symp. Dig.*, pp. 82–84, June 1976.

57. J. Edrich, "Parametric Amplification of Millimeter Waves Using Wafer Diodes: Results, Potentials and Limitations," *IEEE Trans. Microwave Theory Tech.*, Vol. MTT-18, No. 12, pp. 1173–1175, December 1970.

58. M. A. Balfour, A. Larson, S. Nausbaum, and J. Whelahan, "Miniaturized Nondegenerate Ka-Band Paramp for Earth to Satellite Communications," *IEEE MTT-S Int. Microwave Symp. Dig.*, pp. 225–227, June 1974.

59. J. Edrich, "20 K Cooled Parametric Amplifier for 46 GHz with Less Than 60 K Noise Temperature," *IEEE MTT-S Int. Microwave Symp. Dig.*, pp.72–74, June 1973.

60. W. J. Getsinger, "The Packaged and Mounted Diode as a Microwave Circuit," *IEEE Trans. Microwave Theory Tech.*, Vol. MTT-14, No. 2, pp. 58–69, February 1966.

61. J. Edrich, "Rauscharme Parametrische Eigenresonanzverstärker mit grosser Bandbreite" (Low-Noise Parametric Self-Resonance Amplifiers with Large Bandwidth), *Frequenz*, Vol. 20, No. 10, pp. 337–343, 1966.

62. H. C. Okean, J. R. Asmus, and L. J. Steffek, "Low-Noise 94 GHz Parametric Amplifier Development," *IEEE MTT-S Int. Microwave Symp. Dig.*, pp. 78–79, June 1973.

63. J. D. Pearson and K. S. Lunt, "A Broadband Balanced Idler Circuit for Parametric Amplifiers," *Radio Electron. Eng.* Vol. 27, No. 5, pp. 331–335, May 1964.

64. C. S. Aitchison, R. Davies, and C. D. Payne, "Bandwidth of a Balanced Micropill-Diode Parametric Amplifier," *IEEE Trans. Microwave Theory Tech.*, Vol. MTT-16, No. 1, pp. 46–47, January 1968.

4

SEMICONDUCTOR CONTROL DEVICES: PIN DIODES*

Joseph F. White

Applied Microwave
Lexington, Massachusetts

4.1 THE PIN DIODE—AN EXTENSION OF THE PN JUNCTION

4.1.1 Structure

The PIN diode should not be thought of as something physically different from the PN junction, but rather different in a sense of degree. With the abrupt junction the width of the depletion zone is inversely proportional to the resistivity of the P or N region, whichever has the lesser impurity doping concentration. As the width of the depletion zone increases, the capacitance per unit area of the junction decreases. This effect is very beneficial for a diode which is intended for use as a microwave switch because the lower the capacitance, the higher the impedance of the diode under reverse bias and the more effective the device is as an "open circuit."

The limiting case of high-resistivity material is undoped (or "intrinsic") I silicon. In practice, of course, no silicon material is without some impurities. A practical PIN diode, then, consists of an extremely high resistivity P or N zone between low resistivity (highly doped) P and N zones at its boundaries, as shown in Fig. 4.1. To distinguish unusually heavily or lightly doped material, special nomenclature has evolved. Heavily doped P and N materials are referred to as P+ and N+, respectively. To identify very lightly doped, high-resistivity P and N material, the Greek letters are used; thus high-resistivity P material is called π-type and high-resistivity N material is called v-type. Recognizing that perfectly intrinsic material is not practically obtainable, the I region of a PIN diode can consist of either v- or π-type material. The resulting diodes are indistinguishable from a microwave point of view; however, the actual junction forms at opposite ends of the intrinsic zone depending on the choice. This distinction is diagrammed for both cases in Fig. 4.1.

The first type shown in Fig. 4.1*b* shows a P+, v, N+ diode structure. If the I region is of sufficiently high resistivity, what few impurity atoms it has will be ionized and the depletion zone will extend throughout the I region and include a small penetration into

---

* This chapter is reproduced in its entirety, with slight modifications, from Joseph F. White, *Microwave Semiconductor Engineering*, Van Nostrand Reinhold, New York, 1ᵒ. 1982 where it appears as Chapter 2.

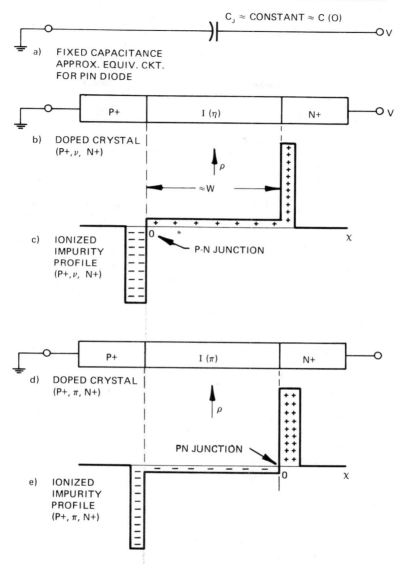

Figure 4.1   Profiles for the two PIN diode types.

both the P and N regions. Because of the heavy doping in the P+ and N+ zones the depletion zone will not extend very far into them, and the depletion zone will be essentially equal to the I-layer width, W_I. The aternate diode structure, P+, π, N+ is shown schematically in Fig. 4.1d. Here the depletion zone width is likewise approximately equal to the width of the intrinsic layer, but the junction is formed at the N+ interface rather than that of P+. Controlling the location of the junction has important consequences from the standpoint of passivating the diode chip, but no impact on performance. Most PIN diodes use ν material for the I region and the junction is formed at the P+ interface.

4.1.2 *C*(*V*) Law and Punch-Through Voltage

In the preceding section it was assumed that the I layer is of such high resistivity that, even with no applied bias, the depletion zone extends across the I layer to the P+ and N+ zones. Under such circumstances C_J is practically independent of applied voltage. At zero voltage the depletion zone has already extended through the I region; as further reverse bias is applied to the diodes, little further widening of the depletion zone proceeds because of very high impurity concentrations and correspondingly large availability of ionizable donors and acceptors in the P+ and N+ regions.

The PIN diode which actually does have so high a resistivity I layer that it is depleted at zero bias is called a *zero-punch-through* diode, because the depletion zone has "punched through" to the high-conductivity zones even before bias is applied. Such a situation, however, represents an idealization. Not all practical diodes are zero punch-through. A more *general definition of the PIN is a semiconductor diode which consists of two heavily doped P and N regions separated by a substantially higher-resistivity P or N region.*

Figure 4.2 shows schematically a practical PIN diode with ionized impurity profiles at zero bias and at punch-through. At zero bias a large portion, but not necessarily all, of the I-region impurities have been ionized and the depletion zone, $W(0)$, may be somewhat less than the I-layer width, W. As reverse-bias voltage is applied to this diode, depletion layer spreading occurs, and the capacitance, shown in Fig. 4.2*b*, decreases until the depletion layer has spread definitely to the N+ region, as shown in Fig. 4.2*c*. At this voltage the depletion layer width, $W(V_{PT})$, is approximately equal to W_I. Further spreading of the depletion layer into the low-resistivity P+ and N+ regions is, for most applications, negligible. The voltage at which the depletion zone just reaches the N+ contact is the punch-through voltage, V_{PT}.

Because in practice the resistivity levels in the P+, I, and N+ regions do not change abruptly, the resulting capacitance versus voltage characteristics have a *soft knee*. Therefore, the punch-through voltage is not directly measurable with precision. However, the practical diode usually does have two definable slopes in its $C(V)$ characteristic, when plotted using semilog paper as shown in Fig. 4.2*d*. By convention, *the voltage intersection of these two straight-line projected slopes is called the punch-through voltage.*

It is to be emphasized that this $C(V)$ characteristic is what one obtains when the measurements are made at relatively low frequencies, typically 1 MHz. At microwave frequencies, the dielectric susceptibility of silicon is much larger than the conductivity of v or π material; thus the capacitance is effectively equal to the minimum capacitance for all values of reverse bias, as is shown in the following discussion of dielectric relaxation.

4.1.3 Capacitance Measurements and Dielectric Relaxation

If the capacitance of a PIN diode that does not punch through at zero bias is measured at zero bias, a larger value of capacitance will be measured at a low frequency (such as 1 MHz) than would be measured at microwave frequencies (such as with a network analyzer measurement at 1 GHz). The reason is that silicon, in addition to being a variable conductor, also has a high dielectric constant. Therefore, its bulk differential equivalent circuit appears as a parallel combination of conductance and capacitance. The relative current division between these two equivalent-circuit parameters varies with the frequency of the applied signal, higher-frequency currents being carried mostly by the capacitive path.

To illustrate this point, consider Fig. 4.3, which shows a PIN diode below punch-through. The portions of the P+ and the I regions that are depleted represented the depletion zone, or "swept region." The remainder of the I region is "unswept" and can

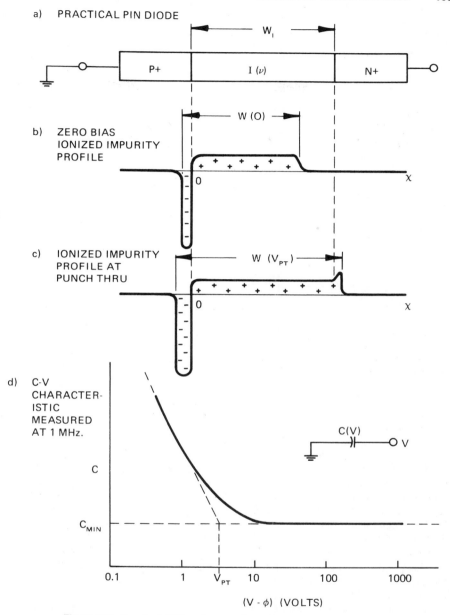

a) PRACTICAL PIN DIODE

b) ZERO BIAS
    IONIZED IMPURITY
    PROFILE

c) IONIZED IMPURITY
    PROFILE AT
    PUNCH THRU

d) C-V
    CHARACTER-
    ISTIC
    MEASURED
    AT 1 MHz.

Figure 4.2 Practical PIN and reversed punch-through characteristics.

be modeled, as shown in Fig. 4.3c, as a parallel resistance–capacitance circuit, represented by the equivalent-circuit elements, C_{US} and R_{US}.

The division of current through C_{US} and R_{US} depends on the ratio of the susceptance of C_{US} to the conductance ($1/R_{US}$). This ratio in turn depends on the dielectric constant of silicon to its bulk resistivity. The frequency at which the current division between

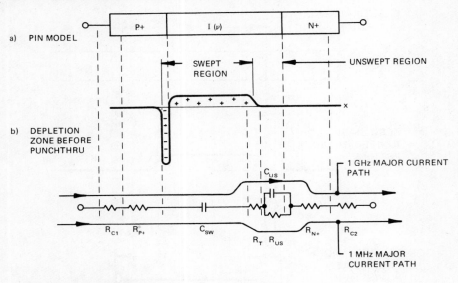

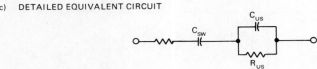

Figure 4.3 Reverse-biased PIN equivalent circuit.

these two elements is equal (i.e., when the susceptance is equal to the conductance) is defined as the *dielectric relaxation frequencies*, f_R, of the material.

When the operating frequency, f, is equal to or greater than $3f_R$, the total capacitance represented by the series combination of C_{SW} and C_{US} is approximately equal to C_J (within 10%), the parallel-plate capacitance of the totally depleted I region. This value corresponds to the minimum capacity C_{MIN} measured beyond punch-through at low frequency.

This point is a major one in the practical characterization of PIN diodes intended for microwave switching applications. It means that practical measurements of the capacitance of a PIN junction can be made at 1 MHz, and the values so attained will represent a good approximation to the actual capacitance applicable at microwave frequencies. This test only requires that sufficient bias voltage is used during the low-frequency measurement to ensure that the I region is fully depleted. A check to determine whether the I region is in fact fully depleted can be made simply by plotting the $C(V)$ characteristic for a few representative diodes from the production lot to determine at what minimum bias voltage the measured capacitance reaches what is essentially its minimum value.

The remaining required quantity to determine the applicability of the low-frequency C_{MIN} as a representation for the microwave capacitance, C_J, is an estimate of the relaxation frequency for the I region of the diodes being measured. High-purity silicon material used to make PIN diodes typically has resistivity in the range 500 to 10,000 Ω-cm prior to the diffusion and/or epitaxial growth steps used to achieve the low-resistivity P+ and N+ regions. However, after the high-temperature processing needed to realize these regions, the resistivity of the I region is always less than that of the starting crystal. Typical values for I-region resistivity are in the range 100 to 1000 Ω-cm. The dielectric relaxation frequency for the unswept portion of the I region can be written in terms of the equivalent-circuit parameters, directly from the definition, which requires that the conductance and capacitive susceptance be equal at f_R. The result is

$$f_R = \frac{1}{2\pi R_{US} C_{US}} \tag{4.1}$$

In turn, the specific values for R_{US} and C_{US} can be written in terms of the length, L, and the area, A, of this unswept region together with the bulk resistivity, ρ, and the absolute dielectric constant, $\epsilon_0 \epsilon_R$, as follows:

$$R_{US} = \frac{\rho L}{A} \tag{4.2}$$

$$C_{US} = \frac{\epsilon_0 \epsilon_R A}{L} \tag{4.3}$$

Substituting these expressions into Eq. (4.1) together with the value $\epsilon_R = 11.8$ for silicon yields Eq. (4.4), which gives the dielectric relaxation frequency directly in gigahertz when the resistivity, ρ, is known.

$$f_R = \frac{1}{2\pi \epsilon_0 \epsilon_R \rho}$$

$$f_R = \frac{153}{\rho(\Omega\text{-cm})} \qquad \text{gigahertz} \tag{4.4}$$

This expression is shown graphically in Fig. 4.4. Strictly speaking, since the final resistivity of the I layer of a practical diode depends on the actual processing steps used to fabricate the diode, one could not know beforehand what dielectric relaxation frequency would apply for a particular diode unless a method for determining the magnitude of ρ as realized in a final device were available. Usually, a PIN diode has an I-layer resistivity of at least 100 Ω-cm, which corresponds to $f_R = 1.53$ GHz. Thus for operating frequencies of 5 GHz or more, the simplified equivalent circuit in Fig. 4.3e can nearly always be applied.

An experimental method does exist for the determination of I-layer resistivity through the measurement of the punch-through voltage and knowledge of the I-layer width, which usually is known with reasonable accuracy by the diode manufacturer. To make the calculation, Eq. (4.10) is solved for V_{PT}, at which the depletion layer is equal to the I-region width, W. Recognizing that for a PIN the impurity concentration of the N+ contact, N_A, is much larger than the impurity concentration in the I region, N_D, the result becomes

$$V_{PT} = \frac{e N_D W^2}{2\epsilon_0 \epsilon_R} \tag{4.5}$$

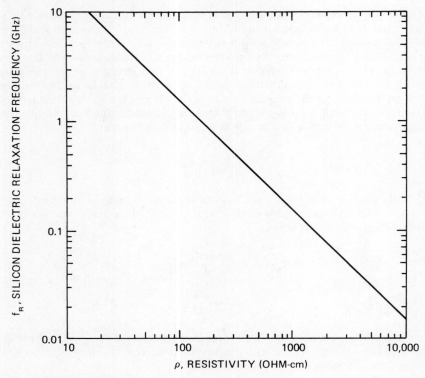

Figure 4.4   Dielectric relaxation frequency in silicon of various resistivities.

But the resistivity of the I region is related to the donor impurity density according to

$$\rho = \frac{1}{N_D e \mu_N} \tag{4.6}$$

where μ_N is the *electron mobility* (i.e., the effective drift velocity of electrons in the I region per unit applied electric field) and e is the charge of a single electron. Substituting this result in Eq. (4.5) gives

$$\rho = \frac{W^2}{2V_{PT}\epsilon_0\epsilon_R\mu_N}$$

$$\rho = \frac{(2.4 \times 10^8)W^2}{V_{PT}} \quad \Omega\text{-cm} \tag{4.7}$$

where $\epsilon_R = 11.8$ (silicon)
$\quad \mu_N = 2000$ (cm^2/V-s)*
$\quad W = $ I-region width (cm)
$\quad V_{PT} = $ punch-through voltage (V)

---

* This mobility value is representative for electrons in high-resistivity N-type silicon, as shown in Fig. 4.8. Thus this method of I region width determination is limited by the accuracy with which mobility can be estimated.

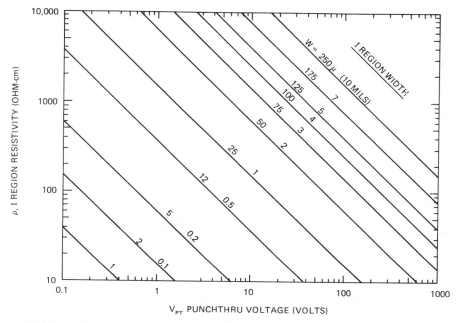

Figure 4.5 Punch-through voltage versus I-region resistivity for various I-region widths.

For example, if a particular diode having an I-region width of 0.0025 cm (1 mil) is found to have a punch-through voltage of 10 V, the resultant average resistivity is 150 Ω-cm. Equation (4.7) is shown graphically in Fig. 4.5 for various values of I-region width, W.

4.2 MICROWAVE EQUIVALENT CIRCUIT

4.2.1 Charge Control Model

Transit-Time Limit of the *I–V* Law. In Chapter I the *I–V* characteristic for a PN junction was given. The same characteristic applies for the PIN *at low frequencies,* for which the RF period is long compared with the *transit time* of an electron or hole across the I region.

The discussion to follow using a simple carrier transit-time model is only approximate. Real diodes have more complex carrier flow, which is nonuniform, subject to applied voltages (i.e., nonlinear), and so forth. The approximation is useful, as it permits estimates of frequency behavior and switching speed. The transition between low- and high-frequency behavior occurs when this transit time is equal to the RF period. To estimate the transit time, recall that the injection of carriers into the depletion zone occurs under forward bias by diffusion. That is, once forward bias is applied, it reduces the magnitude of the built-in junction potential, causing holes to diffuse from the P to the N region and electrons to diffuse in the opposite direction.

The mechanics of this diffusion charge transport are described by *diffusion constants* for holes and electrons, D_P and D_N, respectively. Diffusion, being the flow of carriers from a region of high to lower density, is described in terms of a current density proportional to the spatial gradient of charge density according to Eqs. (4.8) and (4.9).

$$\text{For holes:} \qquad \mathbf{J}_P = -eD_P(\nabla p) \qquad\qquad (4.8)$$

$$\text{For electrons:} \quad \mathbf{J}_N = -eD_N(\nabla n) \qquad\qquad (4.9)$$

where J = current density
$\qquad e$ = unit charge magnitude = $+1.6 \times 10^{-19}$ C
$\qquad D_{P,N}$ = diffusion constants for holes and electrons, respectively
$\qquad \nabla p$ = spatial gradient of hole density
$\qquad \nabla n$ = spatial gradient of electron density

To illustrate diffusion, let us estimate the approximate *transit time* for holes, the slower moving carrier, across the depletion zone of a PN junction of width W. A one-dimensional analysis is used, and Eq. (4.8) becomes

$$J_P = -eD_P \frac{dp}{dx} \qquad\qquad (4.10)$$

The minus sign is required (D is defined as a positive constant) since current flow is opposite to the direction of increasing charge density. Figure 4.6 shows a simplified model of the PIN and majority carrier profiles. The gradient dp/dx is abrupt at the P/I interface and an exact analysis would require an analytic representation of $p(x)$. However, as an approximation, we use the *average gradient of the hole density across the I region*, or

$$\frac{dp}{dx} \approx \frac{P_P}{W}$$

Equation (4.10) then becomes

$$J_P = P_P e v_P \approx -eD_P \frac{P_P}{W}$$

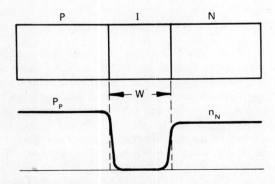

Figure 4.6   Depletion zone model used to estimate transit time frequency.

where J_P has been written explicitly using carrier velocity, v_P, and the density of carriers participating in the hole current flow. But the hole transit time, T_P, equals W/v_P; therefore,

$$\text{transit time} = T_P \approx \frac{W^2}{D_P} \tag{4.11}$$

Accordingly, we can expect that the low-frequency I–V characteristic can no longer be used at frequencies for which the RF period is comparable to T_P. If a transition frequency, f_T, is defined for the PIN diode at which

$$f_T = \frac{1}{T_P}$$

then

$$f_T = \frac{D_P}{W^2}$$

Frequently, the mobility, μ, rather than the diffusion constant, D, is evaluated for semiconductor materials. These two constants are related according to the Einstein relationship

$$D = \mu \frac{kT}{e} \qquad \text{cm}^2/\text{s} \tag{4.12}$$

where D = diffusion constant (cm^2/s)
μ = mobility (average carrier drift velocity per unit applied electric field)
k = Boltzmann's constant
T = absolute temperature (kelvin)

At 300 K (near room temperature) $kT/e = 0.026$ V; thus

$$D = 0.026\mu \text{ (at 300 K)} \tag{4.13}$$

The hole and electron mobilities vary both with impurity densities (see Fig. 4.7) and temperature (see Fig. 4.8). For the present example the hole mobility at 300 K in high-resistivity silicon is about 500 cm^2/V-s, and therefore

$$f_T = \frac{1300}{w^2} \qquad \text{megahertz} \tag{4.14}$$

where w is I region thickness in micrometers.

Thus, even for a very thin base PIN diode having only a 2.5-μm (0.1-mil) I region, $f_T = 200$ MHz. A graph showing how f_T varies with w is shown in Fig. 4.9. In practice, PIN diodes used for microwave switching have I-region widths of 25 to 250 μm (1 to 10 mils) *and accordingly, the low-frequency I–V characteristic given in Equation (I-1) is useless for evaluating microwave resistance.*

I-Region Charge and Carrier Lifetime. However, all of the concepts introduced so far to describe low-frequency behavior are easily applied to determine the microwave resistance. We shall evaluate I-region charge and use it to gauge resistance. From Fig. 4.9

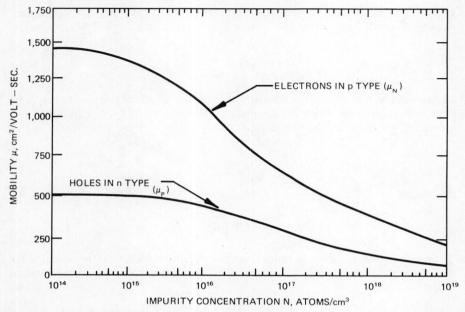

Figure 4.7 Hole and electron mobilities (at 300 K in silicon) versus impurity density. (After Ref. 1.)

it is evident that once charge, consisting of holes and electrons, has been injected into the I region under forward bias, *it cannot be removed in the brief duration of a half cycle of RF frequency* if that RF frequency is above a few hundred megahertz, even for the thinnest I-region (or base-width) diodes.

The *charge control model* for the PIN diode allows RF performance to be related to the net steady-state hole and electron charges, Q_P and Q_N respectively, in the I region. These charges are equal to the product of the (low-frequency) bias current and the respective average carrier lifetime; thus

$$Q_P = I_0 \cdot \tau_P \qquad (4.15)$$

$$Q_N = I_0 \cdot \tau_N \qquad (4.16)$$

That is, after the turn-on transient, during which the I-region charge density is established, the bias current serves as a replenishment source for holes and electrons which have recombined. Referring to Fig. 4.10, the bias current at the P/I interface consists almost entirely of holes being injected into the I region. At the I/N interface the same bias current consists mainly of electron injection into the I region.

The longer the lifetime, the less bias current is required to maintain a given charge density and, accordingly, a given microwave conductivity. Before proceeding further it is important to note that *long lifetime does not necessarily imply slow switching speed*. A properly designed driver can remove I-region charge, and thereby reverse bias the diode, in a period shorter than the lifetime. Rather, long lifetime should be considered a measure of the crystalline perfection within the diode.

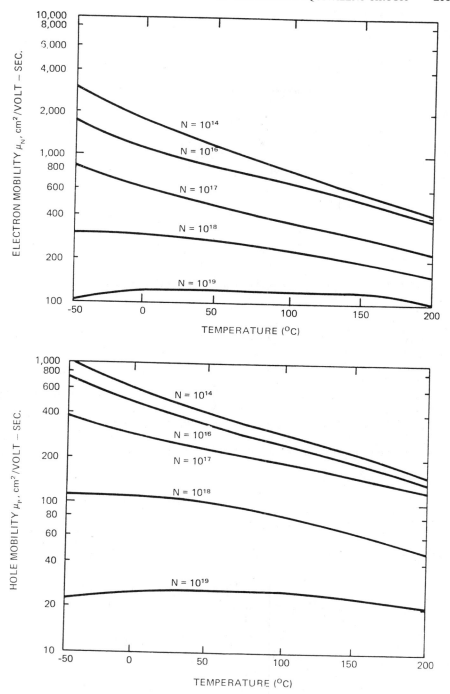

Figure 4.8   Hole and electron mobilities versus temperature for various impurity densities in section. (After Ref. 1.)

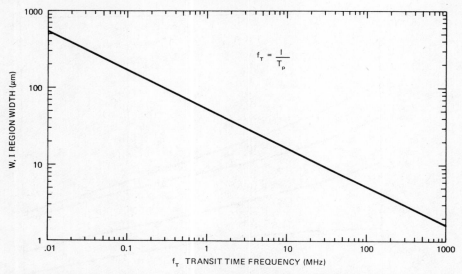

Figure 4.9   I-region width versus transit-time frequency for silicon PIN diodes at room temperature.

A pure intrinsic silicon crystal has a calculated carrier lifetime of 3.7 *seconds*. With impurity doping of 10^{15} cm^{-3}, this figure drops to 0.11 ms [1, p. 80]. In actual diodes the lifetime typically ranges from 0.1 to 10 μs, orders of magnitude less than these theoretical attainable values. To appreciate the reason for this great disparity, it is necessary to review what lifetime represents.

Lifetime is proportional to the *improbability* that an electron and hole will recombine. Imperfections in the regular array of crystal atoms create energy states within the otherwise *disallowed* bandgap of silicon. Such intermediate states provide a virtual energy "staircase" by which the recombination proceeds. In a very regular crystalline structure, energy must be given off in the transition of an electron from conduction to

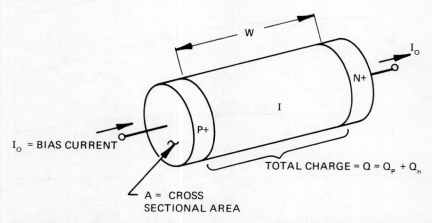

Figure 4.10   Cylindrical I-region model used to estimate I-region resistance of PIN diode.

valence bands in the form of a 1.1-eV (light-emitting) photon; the statistical probability of such an occurrence is low. But with crystalline irregularities, intermediate allowed energy states between these two bands permit a transition in a "staircase" of smaller energy transitions with corresponding low energy phonon (lattice vibrations) emissions, the overall probability of which is higher. Thus lifetime is reduced and recombination is enhanced by the presence of crystalline imperfections and/or impurities.

There are two categories of crystalline irregularities: boundary surfaces and bulk impurities. For a PIN diode the I-region boundaries consisting of the highly doped P+ and N+ represent rapid recombination surfaces for carriers which diffuse into them. Likewise, the peripheral surface boundary of the I region, although not to the same extent as the P+ and N+ regions, provides greater recombination probability than would be present for carriers were the silicon crystal of infinitely extended dimensions. Furthermore, from a bulk point of view, even the structure of an undoped silicon crystal is never ideal. There are stress lines and faults where the probability of electron–hole recombination increases. A doped crystal is all the more susceptible to such imperfections because of the temperature shocks, imperfect atomic fit of doping atoms within the silicon, and related crystal stress producing factors associated with diode manufacture.

This brief discussion of lifetime and its determining factors is qualitative. Even an approximate theoretical treatment of the effective lifetime for a real diode is impractical, although some bulk quantitive analytical treatments of semiconductor crystal lifetime have been made [3]. For the diode maker and user, resort must be made to experimental means by which *average* carrier lifetime can be measured. The conventional method for measuring PIN diode lifetime, τ, consists of injecting a known amount of charge, Q_0, into the I region and measuring the time, τ_S, required to extract it using a "constant" reverse-bias current [4, 5]. To appreciate this method, consider the equivalent circuit and charge versus time profiles shown in Fig. 4.11.

A forward-bias current, I_F, is established and permitted to flow for a period long compared to the expected lifetime, thus storing a charge, Q_0, equal to $I_F \cdot \tau$ in the diode under test. The current supplies are chosen so that $R_R \ll R_F$. Thus, when the switch, S, closes, the diode current, I_D, reverses direction and reaches a magnitude, $I_R - I_F$. The stored charge is removed by this current until it is fully depleted. If the discharge period, τ_S, is short compared to the lifetime ($\tau_S \ll \tau$), then negligible recombination occurs during the turnoff and the total stored charge is recovered. In this case, $Q_0 = I_F \cdot \tau = (I_R - I_F)\tau_S$ and the lifetime is found from

$$\tau \approx \tau_S \left(\frac{I_R}{I_F} - 1 \right) \qquad \text{where } \tau_S \ll \tau \qquad (4.17)$$

This same expression gives the approximate switching time, τ_S, of a driver that switches from forward bias, I_F, to reverse bias and has a reverse-bias transient current switching capability of $I_R - I_F$ amperes. Of course, in a practical driver circuit the forward current supply, I_F, would be switched off during reverse bias.

Practically, however, Eq. (4.17) is not always directly useable because it may be difficult to switch the diode off in a time short compared with the lifetime. Typical PIN diode lifetimes may range from 0.1 to 10 μs, requiring extremely fast switches to satisfy the requirements that τ_S, be small, say one-tenth, of the expected value of τ. To overcome this problem, a test setup is made whereby the switching time can be adjusted. In the circuit of Fig. 4.11, R_R is made variable. The ratio I_R/I_F is adjusted so that $\tau_S = \tau$; the applicable condition is determined by analysis as follows.

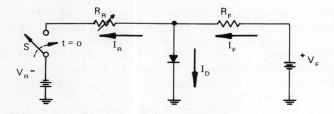

a)   SWITCHING CIRCUIT SCHEMATIC

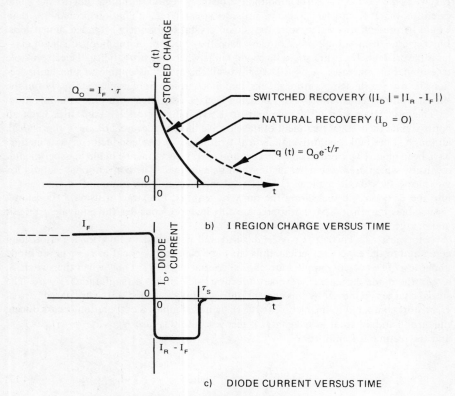

b)   I REGION CHARGE VERSUS TIME

c)   DIODE CURRENT VERSUS TIME

Figure 4.11   Lifetime measuring method.

If, at $t = 0$, I_F were turned off, the initial charge, Q_0 would decay at a rate proportional to the product of the instantaneous charge magnitude in the I region and the recombinate rate, $1/\tau$. Actually, lifetime is somewhat dependent on the bias current level; a typical variation of PIN lifetime is shown in Fig. 4.12 for a range of bias currents commonly used. Conventionally, this variation of lifetime is ignored, not because it is insignificant, but because its inclusion would not permit the simple analysis that follows. Which is not to say that the analysis is not useful. Common practice is to apply it, but one should be aware of its limitations.

With the constant lifetime assumption, suppose that the "driver" described in Fig. 4.11 provides no charge extraction current. Then, for $t > 0$, the expression describing

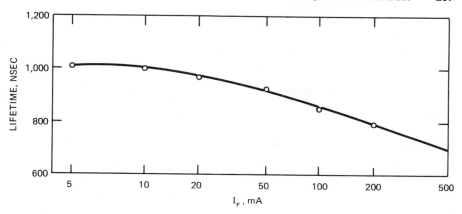

Figure 4.12 Typical variation of PIN lifetime with forward current. (After Ref. 6, p. 289.)

the instantaneous rate of charge (dq/dt) of charge in the I region, q, is

$$\frac{dq}{dt} = -\frac{q}{\tau} \quad (t > 0) \tag{4.18}$$

The solution of this differential equation is

$$q = Q_0 e^{-t/\tau} \tag{4.19}$$

This equation shows the "natural recovery" curve in Fig. 4.11b and demonstrates the definition of lifetime as τ, the time constant of charge decay. In decay $t = \tau$, q decays to $1/e$ or about 40% of its initial value. However, to make a practical measurement, it is necessary to have a measurable quantity; this requirement is most easily fulfilled by providing a reverse current during recovery. Since the reverse current also removes charge from the I region, its effect must be included in the charge defining equation. Equation (4.18) then becomes

$$\frac{dq}{dt} = \frac{-q}{\tau} + I_D \tag{4.20}$$

This expression is called the *continuity equation* for stored charge; the name reflects the fact that stored charge is neither created nor destroyed instantaneously but rather, has time continuity. This equation is general and applies for charge building with I_D positive, as well as for recovery when the diode bias current direction is reversed and I_D is negative. The solution, which can be verified by substitution into Eq. (4.20), is

$$q = Q_0 e^{-t/\tau} + I_D \tau \tag{4.21}$$

Imposing the condition that the stored charge be depleted in time $t = \tau$, as shown graphically in Fig. 4.11b, and noting that $Q_0 = I_F \tau$ and that $I_D = -(I_R - I_F)$, Eq. (4.21) gives

$$\left|\frac{I_R}{I_F}\right| = e^{-1} + 1 \approx 1.4 \tag{4.22}$$

as the test condition under which $q = 0$ at $t = \tau_S = \tau$, permitting a convenient direct measurement. In practice, I_R and I_F can be monitored by connecting an oscilloscope across a small resistance in series with the diode under test. Switch S is realized using a pulse generator with repetition rate adjusted to permit the application of forward current, I_F, for a time that is long compared to the lifetime, τ, in order that the steady-state charge, Q_0, equal to $I_F\tau$, be established before the recovery process is measured. A practical description of lifetime and switching-speed measurements is given by McDade and Schiavone [7].

Charge Control Model and Microwave Currents. We have just seen that from transit-time considerations, the PIN I-region conductivity cannot follow a microwave signal because the diffusion of charge carriers is not rapid enough to traverse the I region within the half-period of an RF cycle. Moreover, from the preceding discussion of carrier lifetime it is clear that once charge is injected into the I region it resides there for 0.1 to 10 μs— the lower limit of which is even long compared to the 0.005-μs half-period of, say, a 1-GHz signal. These two facts taken together indicate that the resistance behavior of the PIN at microwave frequencies can be described in terms of the charge present in the I region, q.

To illustrate this point, consider the diode I–V characteristic with superimposed RF excitations, as shown in Fig. 4.13. The I-V law shown is typical for a high-voltage PIN. Under a forward-bias current of 100 mA, the I region becomes sufficiently conductive that its microwave impedance drops below 1 Ω of resistance (as we describe in the next section). If a microwave current having, say, 50 A peak amplitude (500 times the bias current) is then passed through the diode, the diode is found to remain in the low-impedance condition despite the large "negative-going" half-cycle of the RF wave-form. The reason for this linear operation even under high RF current magnitudes is clear when the total charge movement produced by the RF signal is considered. As-suming a lifetime of 5 μs, typical of a high-voltage PIN, the 100-mA bias results in a stored charge of 0.5 μC. However, during the negative-going portion of a 1-GHz sinu-soid, the total charge movement is less than 0.025 μC, not even a tenth of the stored charge. This example epitomizes the charge control viewpoint that *it is the total stored charged produced by a bias which determines I-region resistance rather than the instan-taneous magnitude of an RF current.*

*Ryder** has likened the bias level on a PIN diode to "large signal" and the RF as the "small ac component," with respect to the amount of charge stored or removed from the I region. From the example above, the value of the charge control viewpoint is evident.

Under reverse bias, a relatively small voltage, about -100 V, is sufficient to hold off conduction of the diode under the application of an RF voltage whose peak voltage amplitude is as large as 1000 V. Again, the brief duration of the half-period of the RF cycle is not sufficient to cause appreciable modulation of the I region of the diode, and the diode appears as a high impedance event with this large voltage magnitude applied.

One might ask why any reverse bias is necessary at all if the diode is nearly non-conducting at zero bias. First, reverse bias fully depletes the I region and its bound-aries of charge. Thus the diode has a higher microwave Q with reverse bias. Second, the role of a reverse bias is to maintain an average field which tends to prevent the accumu-lation of significant amounts of charge in the I region. The presence of excessive charge in the space, under high RF fields, can produce impact ionization, with a "runaway"

---

* R. Ryder (Bell Telephone Laboratories, Murray Hill, New Jersey) in a talk given at the NEREM Conference in Boston, circa 1970.

DIODE STORED CHARGE

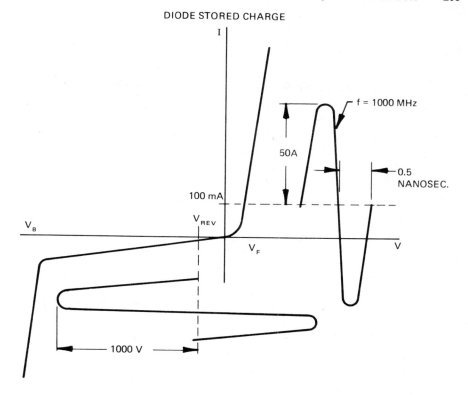

$$\text{STORED CHARGE} \quad = Q \quad = 0.1 \text{ AMPS} \times 5 \times 10^{-6} \text{ SEC.}$$
$$= 0.5 \, \mu\text{coul.}$$

$$\triangle \text{(CHARGE)} \, < \quad 50 \text{ AMPS.} \times 0.5 \times 10^{-9} \text{ SEC.}$$
$$= \quad 0.0025 \, \mu\text{coul.}$$

Figure 4.13 Example comparing charge stored by bias to charge movement due to high-level microwave signal.

current rise and resultant diode destruction. Nevertheless, under large RF excitation, impact ionization effects are often observed, resulting in a *pulse leakage current*, since it occurs only under the combined action of RF and reverse-bias excitation. It is necessary that the driver circuit have *sufficiently low impedance* to be capable of providing this pulse leakage current (usually 1 to 5 mA) in a high-power control device without causing an appreciable drop in the bias voltage supplied, if destructive diode conduction in the reverse-bias state with high RF applied voltage is to be avoided.

4.2.2 Forward-Biased I-Region Resistance

Having demonstrated the suitability of the charge control approach for determining microwave properties, let us use it to calculate the conductivity and resistance of the I region under forward bias. Conductivity, σ, is a bulk property equal to the ratio of current density, J, to applied electric field strength, E:

$$\sigma = \frac{J}{E} \tag{4.23}$$

But J is the directed average rate of flow of electric charge. In terms of I-region holes and electrons,

$$\sigma = \frac{J}{E} = e\left(\frac{v_P \cdot p}{E} + \frac{v_N \cdot n}{E}\right) \tag{4.24}$$

Also, by definition, *mobility*, μ, is the average carrier velocity per unit of applied electric field; thus

$$\sigma = e(\mu_P p + \mu_N n) \tag{4.25}$$

where $e = +1.6 \times 10^{-19}$ C = magnitude of electron's charge
$\mu_{P,N}$ = mobility of holes and electrons, respectively
p, n = respective, injected hole and electron densities in I region

The formula for the resistance of a cylindrical conductor of electrical conductivity, σ, length W along the current path, and cross-sectional area A is [8]

$$R = \frac{W}{\sigma A} \tag{4.26}$$

Using the dimensional notation of Fig. 4.10, the I-region resistance is then

$$R_I \sim \frac{W}{eA(\mu_P p + \mu_N n)} \tag{4.27}$$

Three main assumptions* have been made in this derivation of R_1:

1. The I region as a whole is electrically neutral.
2. The bias current, I_0, injects holes and electrons that *recombine with each other in the I region*; the limitations of this assumption are discussed later.
3. The carrier lifetime is sufficiently long that both the holes and electrons are uniformly distributed within the I region. Another way of stating this point is that the average hole and electron diffusion lengths, L_P and L_N, are much longer than the I-region width, W. This condition is usually valid for well-designed PIN diodes and can always be verified by using the following relation for diffusion length:

$$L = \sqrt{D_{AP}\tau} \tag{4.28}$$

where D_{AP} the ambipolar diffusion constant = $2D_P D_N/(D_P + D_N)$ and τ is the lifetime within the I region.

---

* N. H. Fletcher, "The High Current Limit for Semiconductor Junction Devices," *Proc. IRE*, Vol. 45, pp. 862–872, Jun 1957.

In silicon, D_{AP} has an effective average value for holes and electrons, the *ambipolar diffusion constant*, of 15.6 cm^2/s [9]. Thus

$$L = \begin{cases} 40\sqrt{\tau\,(\mu s)} & \mu m \\ 1.7\sqrt{\tau\,(\mu s)} & mils \end{cases} \tag{4.29}$$

For example, if the bulk lifetime is 10 μs the diffusion length is about 133 μm (5 mils).*

Under these combined assumptions, it follows that the injected hole and electron densities are equal and uniform:

$$p = n \tag{4.30}$$

and, furthermore, since they recombine with one another directly,

$$\tau_P = \tau_N \tag{4.31}$$

Then

$$R_I = \frac{W}{2eA\mu_{AP}p} \tag{4.32}$$

where $\mu_{AP} = 2\mu_P\mu_N/(\mu_P + \mu_N)$, 610 cm^2/V-s in silicon [9], is the *ambipolar mobility* (i.e., the effective average of the hole and electron mobilities). But the injected charge is directly proportional to the biase current:

$$Q_P = epAW = I_0\tau \tag{4.33}$$

Combining the last two equations gives

$$R_I = \frac{W^2}{2\mu_{AP}\tau I_0} \tag{4.34}$$

This expression is applied frequently. We note from it that R_I is theoretically independent of the I-region area, being proportional to the square of the I-region width and varying inversely with mobility, lifetime, and bias current. However, care must be taken in the application of Eq. (4.34) to practical situations. In particular, the following generalizations should be qualified:

1. *Holding all process steps the same except for varying A produces a selection of diodes with different capacitances but the same R_I for a given bias current.* This situation is true only if τ remains constant; but generally, τ decreases with a decrease in A, since I-region carriers are then nearer the periphery, where recombination can occur more rapidly.

2. R_I *decreases as* $(1/I_0)$. Again, this statement holds true only as long as τ remains constant. However, as I_0 increases, carrier density increases, and the recombination probability increases, decreasing τ. Furthermore, saturation is reached when p and n increase sufficiently that substantial injection (holes into the N+ region and electrons into the P+ region) becomes significant, in violation of the second

---

* For an analysis of the case where this assumption is not made, see Leenov's paper, Ref. 9.

assumption used to derive Eq. (4.34). Put simply, if there are high densities of electrons and holes in the I region, their chance for recombining increases, decreasing the average lifetime, τ.

3. *Above the transit-time frequency, R_I is essentially independent of frequency.* This stipulation is only approximately true for most microwave PIN applications. Skin effect causes both the contact and I-region resistances to increase somewhat with frequency.

Despite these limitations, Eq. (4.34) is very useful and is typically invoked to estimate I-region resistance at microwave frequencies. For example, consider a PIN with a 100-μm (4-mil) I region and a 5-μs lifetime operated with 100 mA bias current. Using $\mu \approx 610$ cm^2/V-s give us

$$R_I = \frac{10^{-4}}{(2)(0.1 \text{ A})(5 \times 10^{-6} \text{ s})(610 \text{ cm}^2/\text{V-s})}$$

$$= 0.16 \ \Omega \tag{4.35}$$

This result is in reasonable agreement with the measured value of 0.3 Ω for a 1.56-mm (61-mil) diameter, when one considers that the measured value includes resistive contributions of the ohmic contacts as well as those of the P+ and N+ regions. Furthermore, the lifetime at 100 mA is likely to be less than the 5-μs value which is measured at 10 mA—an additional factor contributory to a higher measured resistance than that calculated.

Using this example, let us examine the role of skin effect in the forward-biased I region. Using the parameters of the example above and solving Eq. (4.26) gives $\sigma = 3 \ (\Omega\text{-cm})^{-1}$. The skin depth, δ, in a conductor is given by [8]

$$\delta = 1/\sqrt{\pi f \mu_0 \sigma} \tag{4.36}$$

where f = operating frequency (hertz)
$\mu_0 = 4\pi \times 10^{-9}$ H/cm = free-space permeability
σ = conductivity $(\Omega\text{-cm})^{-1}$

From Eq. (4.36), the skin depth for $\sigma = 3 \ (\Omega\text{-cm})^{-1}$ at 1 GHz is 0.09 cm, about equal to the diode radius. This diode example has a junction capacitance of about 2 pF and would not usually be used at frequencies much above 1 GHz. At higher frequencies a lower capacitance, and hence reduced diameter, would be employed. Thus it can be seen that I-region* skin effect usually has but a moderate effect in PIN control devices in the frequency range 0.1 to 10 GHz.

Before leaving the subject of I-region conductivity it is interesting to note what level of carrier density, p, was injected into the I region of this sample diode to produce $R_I = 0.16 \ \Omega$. An estimate can be made using Eq. (4.32) and $\mu \approx 610$ cm^{-2}/V-s; thus

$$p = W/2eA\mu R_I = 1.7 \times 10^{16} \text{ cm}^{-3} \tag{4.37}$$

Since there is an approximately equal electron density, n, in the I region, the total free carrier density required to produce $R_I = 0.16 \ \Omega$ is 3.4×10^{16} cm^{-3}. Recalling that the

---

* Skin resistance may be more important in the P and N regions and in the leads attached to them because it affects how the currents enter the I region.

atom density is about 10^{23} cm^{-3}, this figure represents *less than one carrier per million atoms*. It is therefore easy to see why the skin depth, so significant with metallic conductors at microwave frequencies, has only a moderate effect even under "high injection" levels in the I region of the PIN diode.

4.2.3 R_R and C_J Reverse-Biased Circuit Model

Under reverse bias the I region is depleted of carriers and the PIN appears as an essentially constant capacitance to a microwave signal. The presence of dissipative losses can be taken into account by either a series or a parallel resistance element in the equivalent circuit. In a well-made PIN, the I region has sufficiently high resistivity that most of the dissipation under low RF power conditions occurs in the ohmic contacts made to the diode and in the resistances of the P+ and N+ regions. Accordingly, a fixed series resistance, R_R, used to represent these losses can be expected to offer an equivalent-circuit model that is applicable over a broader bandwidth than a parallel conductance. In any event, due to the ratio of diode capacitive reactance to practical RF circuit impedances, the dissipative losses of the PIN under reverse bias are usually much smaller than those under forward bias; thus the choice of series or parallel R–C equivalent circuit under reverse bias usually can be made according to whichever offers greater computational convenience.

Because of the high relative dielectric constant for silicon ($\epsilon_R = 11.8$), the fringing capacitance (in air) around the I region is relatively small and the capacitance calculated using the parallel-plate capacitance formula given below provides a useful estimate of junction capacitance, C_J. Thus

$$C_J \approx \frac{\epsilon_0 \epsilon_R \pi D^2}{4W} \qquad (4.38)$$

where $\epsilon_0 = 8.85 \times 10^{-14}$ F/cm = free-space permittivity
 $\epsilon_R = 11.8$ = relative dielectric constant for silicon
 D = junction diameter
 W = I-region thickness

For many design calculations—estimating thermal capacities, breakdown strength, and RF bandwidth—it is desirable to be able to interrelate the trade-offs between I-region dimensions (W and D) and junction capacitance (C_J). Figure 4.14 shows Eq. (4.38) graphically for typically available PIN I-region widths.

4.2.4 Microwave Circuit Measurements and Cutoff Frequency, f_{cs}

Equivalent Circuit Definition and f_{cs}. The microwave equivalent circuit for the unpackaged PIN diode chip to be used in this text is shown in Fig. 4.15. In most applications the PIN diode is used as a switch; therefore, the less capacitance, the better an "open circuit" it presents with reverse bias. The lower the resistances, R_F and R_R, the smaller the dissipative losses, and, under forward bias, the more the diode resembles a "short circuit." A figure of merit has been defined [10] to relate the PIN's switching effectiveness, termed *switching cutoff frequency*, f_{CS}. The utility of this definition will become apparent later in the discussion of performance limitations.

$$f_{CS} = 1/2\pi C_J \sqrt{R_F R_R} \qquad (4.39)$$

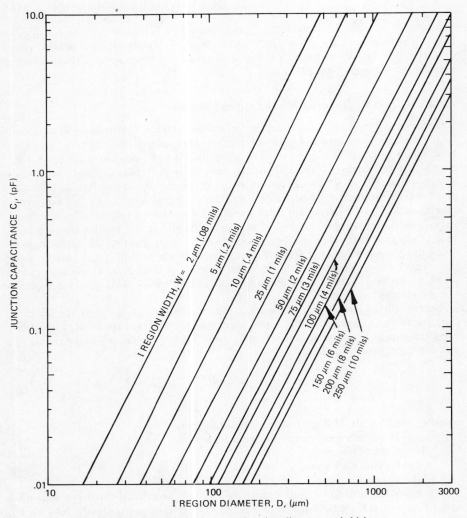

Figure 4.14 PIN diode C_J versus I-region diameter and thickness.

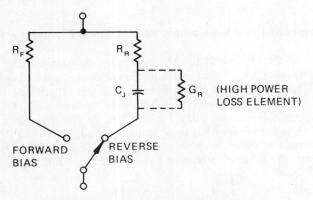

Figure 4.15 PIN diode chip equivalent circuit.

The equivalent-circuit parameters are as defined in Fig. 4.15. Because the additional loss at high power is treated here by a separate equivalent-circuit element, G_R, the *definition of f_{CS} as used here is limited to microwave power levels below the onset of nonlinear dissipation.* The effect of G_R is discussed in the next section.

In principle, the values for R_F and R_R could be evaluated and f_{CS} could be specified under high-power conditions. But it is not usually possible to obtain diodes characterized under high power as this task falls to the circuit designer; for this reason the separation of low- and high-power characterization is more consistent with actual practice.

Isolation Measurements. It was shown earlier that C_J measurements made at low frequency (≈ 1 MHz) with sufficient reverse bias to deplete the I region provide a usable indication of the microwave capacitive reactance to be expected. However, the *resistances under forward, R_F, and reverse, R_R, bias conditions must always be determined by direct microwave measurements* since they include not only the inherent I-region loss effects of the diode but of the P+ and N+ regions as well as contact resistances, none of which is predictable with desirable analytic precision. Since in most control device circuits, the greater microwave dissipation occurs under forward bias, the determination of R_F usually warrants the greater attention.

Many diode resistance measurement methods have been described [11, 12]. Ultimately, the diode loss in the actual circuit of use is what is desired. For determining R_F, in either a test circuit or the actual circuit of use, the terminals where the diode is to be connected are short circuited and the loss of the circuit without diodes (i.e., the *cold circuit loss*) is measured. The additional circuit loss with diodes installel can then be attributed to the diodes themselves, and if the FR currents through the diodes can be estimated, the equivalent-circuit parameters can be determined. Of course, by this time the insertion loss of the circuit under test is known and the value of knowledge of the diode equivalent-circuit parameters is of use only in future design applications. Nevertheless, such direct evaluation in the circuit of end use is often required especially where diodes are circuit mounted in chip or beam lead configurations, requiring permanent bonding into the circuit to make the adequate ohmic contact necessary for accurate resistance determinations.

The two most common methods used to characterize PIN diodes, outside the circuit of end use, are the *isolation* and *reflection* (or "slotted line") measurements. To make an isolation measurement, the diode is used to interrupt a transmission line. When the diode is mounted in shunt with the line (Fig. 4.16), this method provides a sensitive measurement of the forward resistance, R_F. The isolation produced by a line shunting admittance Y is

$$\text{isolation} = \frac{P_A}{P} = \frac{V_A^2/Z_0}{V_L^2/Z_0} = \left| 1 + \frac{YZ_0}{2} \right|^2$$

$$= 1 + GZ_0 + \frac{G^2 Z_0^2}{4} + \frac{B^2 Z_0^2}{4} \tag{4.40}$$

where $Y = Z^{-1} = G + jB$.

To achieve the maximum test sensitivity, any series inductance introduced when mounting the diode across a transmission line is series resonated by a tunable capacitor. For this reason the measurement is most practical in the frequency range 0.5 to 1.0 GHz. Under these conditions the net series reactance, jX, of the mounted diode is zero and Eq. (4.40) reduces to:

$$\text{Isolation} = 1 + \frac{Z_0}{R} + \frac{Z_0^2}{4R^2} \tag{4.41}$$

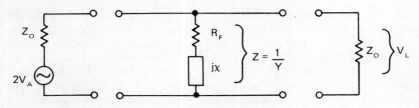

a)   FORWARD BIAS SHUNT (TO MEASURE R_F)

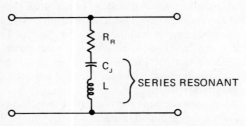

b)   REVERSE MODE (DeLOACH) METHOD (R_R, APPROX. VALUE OF C_J)

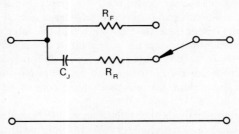

c)   SERIES DIODE (C_J, APPROX. VALUE OF R_F) ISOLATION (TWOPORT) DIODE MEASUREMENT CIRCUITS

Figure 4.16   Equivalent circuits for diode measurements.

For example, if a series-resonated diode having a forward resistance of 1 Ω shunts a 50 Ω line, the normalized conductance, GZ_0 equals $Z_0 R_F = 50$. The resulting isolation equals 676, or approximately 28 dB. The resistance-limited isolation described by Eq. (4.41) is encountered often in both diode measurement and SPST switch design. For convenient reference it is shown graphically in Fig. 4.17.

There are some fine points to be considered in performing this measurement. First, if the diode is mounted in a package, the package capacitance transforms the effective resistance of the diode. This effect is usually negligible in the frequency band 0.5 to 1.0 GHz. Second, the circuitry used must not have significant leakage paths whereby power can reach the load from the generator by alternate paths such as higher-order waveguide modes, fringing electric fields, and so forth. This condition is readily tested by measuring isolation with the diode replaced by a short circuit of dimensions similar to the diode. Apart from ensuring that the leakage through the device is within acceptable

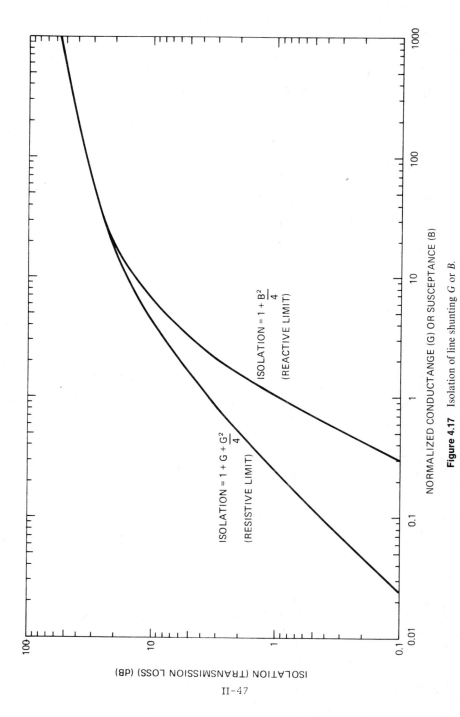

Figure 4.17 Isolation of line shunting G or B.

limits, the isolation value so obtained gives an indication of the circuit *contact resistance. It is common practice to subtract contact resistance when quoting diode resistance.*

An interesting variation of the shunt-mounted isolation measurement occurs when C_J series resonates with the mounted inductance of the diode (Fig. 4.16b). Then the diode shorts the transmission line under *reverse bias* and the isolation is a measure of R_R. From the isolation bandwidth an estimate of C_J is possible. This technique is usually used in waveguide at high frequencies, 5 to 15 GHz, to effect resonance with C_J. Switches built this way are often called *reverse mode*; the measurement technique is called the DeLoach method [11]. The reverse-mode switching circuit is important for duplexer and radar receiver protector designs where isolation in the zero-biased diode state is required, as well as in other *fail-safe* applications where it is desirable that should there be a failure of the driver to bias the diode, the high reflection state of the diode switch is obtained.

If the diode is mounted in series with the line (Fig. 4.16c) the high isolation condition gives a measurement of capacitive reactance, X_C, equal to $-(2\pi f C_J)^{-1}$, and hence C_J. For an impedance $Z = R + jX$ in series with a line of characteristic impedance, $Z_0 = 1/Y_0$, the isolation, by duality, is given by the dual of Eq. (4.41):

$$\text{isolation} = \left| 1 + \frac{ZY_0}{2} \right|^2 = 1 + RY_0 + \frac{(RY_0)^2}{4} + \frac{(XY_0)^2}{4} \tag{4.42}$$

If $|X_C| > 15R_R$, as is almost always the case, the RY_0 terms in Eq. (4.42) can be ignored with an error of less than 1%, and the reactance versus isolation can be read directly from the reactance-dominated characteristic curve shown in Fig. 4.17. This method is especially useful for measuring the circuit-mounted capacitance of low-capacitance devices such as beam lead diodes.

Series-mounted diodes require special equivalent-circuit treatment. Fig. 4.18 shows schematically the electric field contours of a capacitor representing a reverse-biased diode both within and without a series coaxial line mounting. Measured in free space, all E field lines terminate on the diode terminals directly, and a capacitance, C_0, is measured. When mounted in the coax line, however, some E field lines intercept the outer conductor. The effect is that the effective series capacitance, C, is less than C_0. An additional shunt capacitance, C_2, appears, but in most cases the effect of C_2 on the transmission line is negligible, since it serves to replace the distributed capacitance of the section of center conductor removed to install the diode. However, the fact that the mounted series capacitance, C_1, is less than the capacitance associated with the diode, C_0, means that a higher isolation is obtained in a switching circuit (generally, a benefit). Moreover, in a phase shifter circuit a different phase shift than that anticipated will be obtained if this effect is overlooked.

The accuracy of the series measurement can be related to the loss and isolation measurement accuracies. Typically, the series isolation (or loss) measurement can be made to within an accuracy of ± 0.1 dB for losses below 3 dB, and ± 0.3 to 0.5 dB for isolation values from 10 to 40 dB. Thus R_F can be determined to an accuracy of about $\pm 2\%$; X_C, to about $\pm 5\%$ of the magnitude of Z_0. For most PIN diodes operated at sufficient forward bias to saturate the I region, $R_F < 1/\Omega$; this measurement method would require impractically low Z_0 for meaningful measurements. However, for R_F measurements at low bias levels or with beam lead diodes wherein $R_F = 2$ to 10 Ω, the measurement is very practical using standard 50-Ω line. For example, a beam lead diode having $R_F = 5 \, \Omega$ produces a 10% insertion loss of about 0.5 dB. The same diode, having $C_J = 0.03$ pF, has a reactance of $-j795 \, \Omega$ at 10 GHz, yielding, in the same test fixture, isolation under reverse bias at 10 GHz of 24 dB.

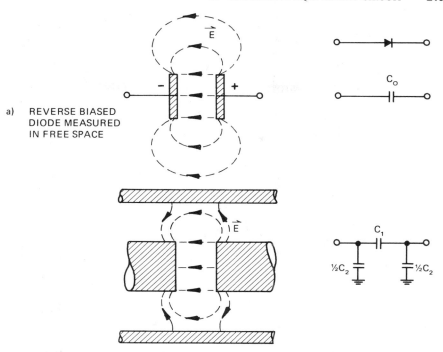

a) REVERSE BIASED
    DIODE MEASURED
    IN FREE SPACE

b) MOUNTED IN COAXIAL LINE

Figure 4.18 Charge of effective series capacitance with circuit mounting.

Reflection Measurements. Most of the principles described for isolation measurements are also applicable to reflection measurements wherein the diode is used to terminate the line (Fig. 4.19); the reflection coefficient (Γ equal to $\rho e^{j\phi}$) measurement is used to deduce diode parameters. Other things being equal, the sensitivity of this measurement method is about four times that of the matched load method described previously in the determination of R_F and R_R for a given line impedance and dissipation; therefore, it is used for most standard diode characterizations. Diode reactances under both forward and reverse bias can be determined from the reflection coefficient argument.

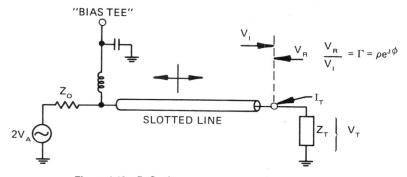

Figure 4.19 Reflection measurement equivalent circuit.

The added sensitivity arises because if the magnitude of Γ_L is either high or low compared to Z_0, the current (I_L) or voltage (V_L) at the end of the line is nearly double the value $(V_A/Z_0$ or V_A, respectively) experienced under matched load $(Z_L = Z_0)$ conditions; the relative power absorbed in the diode consequently increases fourfold. For both a load impedance $Z_L = R_L + jX_L$ and a line with Z_0 characteristic impedance, the reflection coefficient at the load position is [13, 14]

$$\Gamma = \rho e^{j\phi} = \frac{Z_L - Z_0}{Z_L + Z_0} = \frac{(R_L - Z_0) + jX_L}{(R_L + Z_0) + jX_L} \tag{4.43a}$$

$$\phi = \tan^{-1}\left(\frac{X_L}{R_L - Z_0}\right) - \tan^{-1}\left(\frac{X_L}{R_L + Z_0}\right) \tag{4.43b}$$

$$\rho = \sqrt{\frac{(R_L - Z_0)^2 + X_L^2}{(R_L + Z_0)^2 + X_L^2}} \tag{4.43c}$$

The fractional dissipation in Z_L is

$$\text{Insertion loss} = 1 - \rho^2 = \frac{4R_L Z_0}{(R_L + Z_0)^2 + X_L^2} \tag{4.44}$$

Under forward bias, using $Z_0 = 50\ \Omega$, both R_L and X_L are usually much less than Z_0. The fractional power loss is approximately equal to $4R/Z_0$. Under the same approximation, the insertion loss ratio by the series isolation measurement is approximately equal to $1 + R/Z_0$ and the *fractional loss* is approximately R/Z_0, only one-fourth that of the reflection measurement. For example, with $Z_0 = 50\ \Omega$ and $R_F = 1\ \Omega$, the measured loss is about 0.4 dB with the reflection method and 0.1 dB with the isolation method, giving (with ± 0.1 dB accuracy) the determination of R_F with ± 0.25-Ω and ± 1.0-Ω accuracies, respectively. Accordingly, when the diode is mounted in series with the line, the reflection measurement is usually employed for determining R_F.

This method is also used for determining R_R and C_J, but the calculations are less convenient than for R_F because the series reactance, X, cannot be ignored. Furthermore, the impedance transformation effects of a diode package (the package being necessary if the diode is to be conveniently mounted at the end of a slotted line) are not negligible in the reverse-biased condition. For routine diode evaluation the reflection coefficient magnitude, ρ, and phase, ϕ, are measured and the exact equations relating diode C_J and R_R are solved using a computer program. It is common practice to use a coaxial line and obtain a zero-impedance reference by short circuiting the line at the leading surface of the diode package (as shown in Fig. 4.20). Both a phase reference ($\phi = 180°$) and a loss reference ($\rho = 1.0$) are thereby established. The packaged diode impedance, Z_L, is then evaluated by solving Eq. (4.45) for Z_L:

$$Z_L = R_L + jX_L = Z_0 \frac{1 + \Gamma}{1 - \Gamma} = Z_0 \frac{1 + \rho e^{j\phi}}{1 - \rho e^{j\phi}} \tag{4.45}$$

It is illustrative of the measurement method to plot the reflection coefficients obtained under forward and reverse bias on the Smith chart. For this example, consider a PIN with $C_J = 2$ pF, $R_R = R_F = 0.3\ \Omega$ mounted in a package having $C_P = 1$ pF and $L_{INT} = 0.3$ nH. At 3 GHz the *chip* impedances are $0.3\ \Omega$ and $(0.3 - j26.5)\ \Omega$ under forward and reverse bias; they are transformed by the package to $(0.37 + j6.30)\ \Omega$ and $(0.15 - j15)\ \Omega$, respectively. When normalized to $Z_0 = 50\ \Omega$ the corresponding reflec-

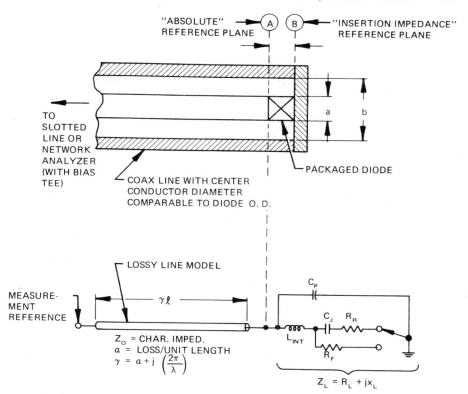

Figure 4.20   Schematic and equivalent-circuit detail for slotted-line measurement of packaged PIN diode.

tion coefficients are [from Eq. (4.44)] $\Gamma = 0.986/165.6°$ with forward bias and $\Gamma = 0.976/-145.8°$ with reverse bias. These results are shown graphically in Fig. 4.21. The proximity of the points to the Smith chart periphery ($\rho = 1.0$) underscores the need for careful measurements if R_F and R_R are to be evaluated accurately.

While the application of the principles of this method is straightforward, great care must be exercised if results with useful accuracy are to be obtained. Since a single-frequency measurement of Γ produces only two bits of data, ρ and ϕ, it is necessary either to have foreknowledge of the values of package *parasitics* (internal inductance, L_{INT}, and package capacitance, C_P) or to perform the reflection measurement at more than one frequency in order to solve for C_P, C_J, R_R, and R_F. In practice the former method is usually followed. C_P is first determined using an empty diode package; this result for C_P is used with an internally shorted package having a wire or strap lead similar to that to be employed with the diodes to be measured. In this respect, the eventual accuracy of evaluation of C_J and R_R is dependent on the reproducibility of C_P and L_{INT} and their relative reactances compared to that of C_J. Furthermore, since the point of measurement and diode reference plane are separated by a line with finite loss, the resulting *lossy line transformation* must be taken into account when R_F and/or R_R are small ($<2\%$ of Z_0), as is usually the case. This requirement necessitates a computer program to reduce the data if such measurements are to be made routinely.

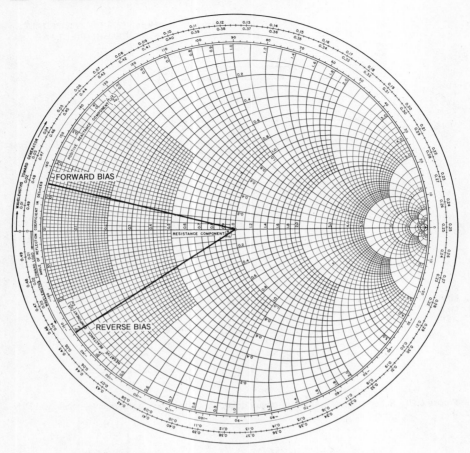

Figure 4.21 Reflection coefficients for measurement example.

Diode Inductance Measurements and Definitions. It should be noted that values measured for diode impedances depend to some extent on the test fixture—especially with inductive reactance, which can *only* be specified in terms of a return path. For example, the inductance per unit length of coaxial line having an outer conductor diameter, b, and an inner conductor diameter, a, is [8]

$$L = \frac{\mu_0}{2\pi} \ln \frac{b}{a} \tag{4.46}$$

and the characteristic impedance Z_0 is [8]

$$Z_0 = \frac{1}{2\pi} \sqrt{\frac{\mu_0}{\epsilon_0}} \ln \frac{b}{a} \tag{4.47}$$

For 50 Ω, the ratio b/a equals 2.3 for air dielectric coax. Suppose that a packaged diode having 2.5 mm (0.1 in) length and effective diameter a of 1.25 mm ((0.05 in) is first measured under forward bias in a 50-Ω line having b equal to 15 mm (0.6 in); the inductance is 0.62 nH. If, however, the same measurement is performed using a smaller-diameter 50-Ω coaxial line in which b equals 7.5 mm (0.3 in), the inductance is 0.45 nH.

Not only the absolute circuit dimensions but also the reference plane definition affects the inductance determination. For example, the determination of inductance above corresponds to reference plane A selection in Fig. 4.20. If, however, reference plane B were selected—by replacing with an equivalent length of center conductor to obtain a short-circuit measurement reference—*insertion impedance* would be obtained. Insertion impedance is Z_L less the Z_S of the short-circuit terminated length of the measurement line, l, neglecting line loss:

$$Z_S = jZ_0 \tan\left(\frac{2\pi l}{\lambda}\right) \tag{4.48}$$

where λ is the wavelength at test frequency.

If, as is usually the case, $2\partial l < \lambda$, the value of the tangent term can be replaced by its argument (within 3%). Furthermore, Z_S is an inductive reactance ($j2\pi f L_S$); for a coaxial line

$$L_S \text{ (nanohenries)} \approx 0.0033 \cdot l \text{ (mm)} \cdot Z_0 \text{ (\Omega)}$$

$$\approx 0.084 \cdot l \text{ (in.)} \cdot Z_0 \text{ (\Omega)} \tag{4.49}$$

For the example cited, $l = 2.5$ mm, $L_S = 0.41$ nH, and the diode respective *insertion inductance** values determined from measurements in the two line sizes are 0.21 and 0.04 nH, respectively.

These examples highlight the importance, especially for inductance measurements, of specifying both the measurement fixture and reference plane selection. Similar reasoning indicates that if the actual circuit of use does not duplicate these conditions—and usually it does not—calculations of performance sensitive to inductance will be inaccurate unless the new conditions are taken into account.

4.3 HIGH RF POWER LIMITS

4.3.1 Forward-Biased Limits

Under forward bias the PIN diode chip usually has an RF resistance of 1 Ω or less. Failure of the diode in this bias state will occur if the dissipative heating ($I^2 R_F$) is sufficient to cause the diode temperature to rise sufficiently to induce metallurgical changes. For silicon and its dopants, this point is not reached until a temperature of about 1000°C. However, the metal contacts at the silicon boundaries introduce failure mechanisms in the vicinity of 300 to 400°C, at which temperatures common contact metals form eutectic alloys with silicon. For example, the gold–silicon eutectic occurs at 370°C [15]. Repeated or continuous exposure of silicon to the eutectic temperature in the presence of the corresponding metal can produce conducting filaments of metal–silicon alloy, which eventually "grow" across the I region of a PIN diode, short circuiting it. This structural change of the diode crystal is the most common diode failure mechanism with heat, even with reverse breakdown–induced failures, described subsequently.

---

* Also sometimes called *excess inductance.*

Failure of a diode does not occur instantaneously when an overstress is applied unless the resulting temperature greatly exceeds 300°C, as can occur with filamentry heating produced by avalanche breakdown in the reverse-bias condition. This situation is also discussed subsequently. Except for the rapid failure induced by avalanche breakdown, thermally produced failures proceed over a time period related to the ratio of the operating temperature, T, to that which causes nearby instantaneous burnout. Rather extensive experiments carried out on computer diodes have shown that the mean time to failure can be described by the empirical relationship [16] given by

$$t_M(T) = Ae^{+Q/kT} \tag{4.50}$$

where $t_M(T)$ = mean time to failure at operating temperature T
$\quad\quad A$ = constant
$\quad\quad Q$ = "activation energy" constant
$\quad\quad k$ = Boltzmann's constant
$\quad\quad T$ = average device temperature in kelvin ($= °C + 273$)

This expression is called the Arrhenius law. It can be applied when the variation of operating life with temperature is determined by *only one* failure mechanism—for example, the formation of a particular alloy of the metallization system with the silicon.

To apply this relationship, the failure temperature, T_F, is first determined for the diode type; it depends on the semiconductor material (usually, silicon for a PIN) and the metallization system. Next, the device is operated at a lower temperature for a period until 50% of the samples under test fail, establishing a data point along the temperature–time graph. Additional data points at different temperatures are determined to allow for averaging of experimental data. This process, called *step-stress temperature testing*, is time consuming because data points corresponding to hundreds and thousands of operating hours are required if the failure curve is to be established with sufficient accuracy to permit meaningful extrapolation to long-life operation—on the order of years.

Care must be exercised that only the common *thermal* failure mode applies throughout the step stress tests. Careful analysis, usually including sectioning of failed diodes, is required to confirm the failure mode of each diode specimen used to establish the failure curve. The resulting temperature–time data plotted on semilog paper form a straight line, permitting extrapolation for longer periods. Figure 4.22 shows a typical plot for a surface-glass passivated, mesa-type, high-voltage PIN diode used in a phased array application. Notice that with a 200°C junction temperature (often cited as a safe operating limit for semiconductor devices) the anticipated mean life is 1000 h (or 0.1 year), while for 140°C the anticipated mean life is extended to 1,000,000 h (or 114 years). Accordingly, the role of operating temperature must be given careful consideration if the estimate of anticipated life is to be meaningful.

4.3.2   Reverse-Biased Limits

The reliability criteria apply for reverse-biased operation as just discussed for forward bias. The junction temperature is again the result of ambient and RF heating. Unlike the forward-biased condition, however, the fractional RF insertion loss does not remain nearly constant once the applied RF voltage has a magnitude comparable to either the reverse bias and/or the diode's reverse breakdown voltage. Under these conditions diode dissipation (is nonlinear and increases more rapidly than RF power, producing at times a runaway insertion loss). The onset of this rapidly increasing insertion loss nonlinearity

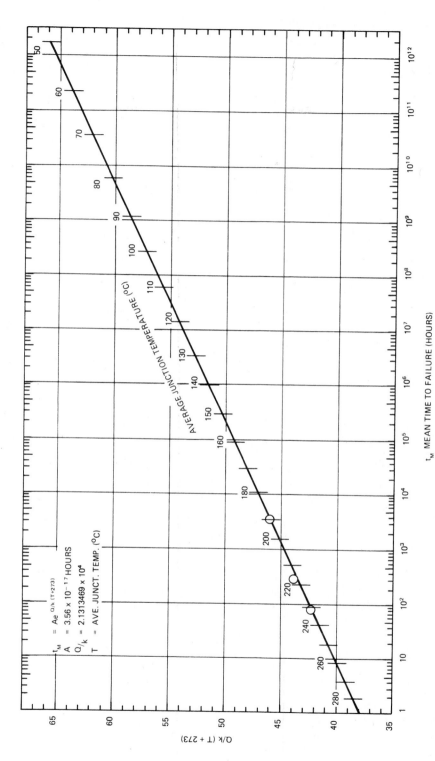

Figure 4.22 PIN life expectance versus temperature. (Courtesy of P. Ledger, Microwave Associates, Inc.)

t_M MEAN TIME TO FAILURE (HOURS)

$$t_M = A e^{Q/k(T+273)} \text{ HOURS}$$
$$A = 3.56 \times 10^{-17} \text{ HOURS}$$
$$Q/_k = 2.1313469 \times 10^4$$
$$T = \text{AVE. JUNCT. TEMP. (°C)}$$

AVERAGE JUNCTION TEMPERATURE (°C)

$Q/k (T + 273)$

225

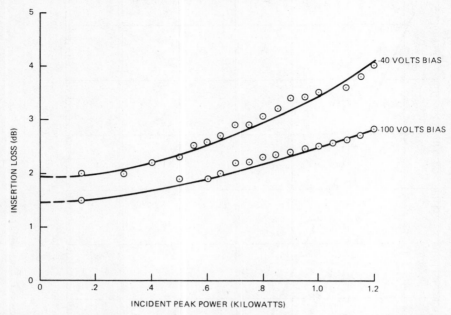

Figure 4.23 Insertion loss versus peak power for a 4-bit X-band phase shifter under reverse bias in all bits.

can be used as a practical measurement that the destructive temperature has been reached in the reverse-biased state, since diode failure usually occurs if the incident RF power level is increased much beyond this level. Figure 4.23 shows a typical insertion loss versus RF power characteristic obtained with a reverse-biased diode phase shifter. The mechanisms of reverse-biased diode failure under RF voltage stress are not sufficiently evaluated for a definitive theory to be developed, largely because of the difficulty of performing such measurements, with enough samples to have adequate statistical data. Qualitively, two conditions in which I-region charge is generated occur and which one predominates depends, as described in Fig. 4.24, upon the relative magnitudes of the peak RF voltage, V_P; the bias voltage, V_{BIAS}; and the diode breakdown voltage, V_{BD}, as described in Fig. 4.24.

The condition shown in Fig. 4.24 with voltage, V_P, is representative of typical operation near the failure limit. The RF voltage has a large excursion into the forward direction. Although, as has been shown earlier, the duration of this half-cycle is insufficient to result in conduction by the diffusion transit (injection) of P-region holes and N-region electrons across the I region, *some charge* is introduced into the I region from these boundaries, and not all of it is extracted by the combined action of the reverse bias and the negative-going half of the RF cycle before the next forward-going RF voltage excursion. However small the magnitude of the injected charge may be, it can increase multiplicatively with each succeeding RF cycle through *impact ionization*; electron–hole pair production results when mobile carriers accelerated by the high RF field strike silicon atoms in the I-region lattice with sufficient energy to promote valence band electrons to the conduction band. This cause of increased insertion loss can be identified experimentally by its bias voltage dependence. Increasing the magnitude of reverse-bias voltage sweeps such injected charge out of the I region more effectively and thereby extends to

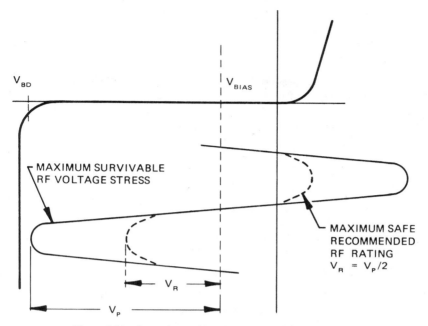

Figure 4.24 Operation at high RF voltage and reverse bias.

a higher applied RF voltage the onset of rapid insertion loss increase, which precedes what, for present purposes, is called the *injection mode* failure mechanism.

The second mechanism causing nonlinear insertion loss is the *direct impact-ionization mode*, occuring when the combined RF + bias voltage exceeds the diode bulk breakdown (i.e., $V_P + V_{BIAS} > V_{BD}$). In this case no partial injection is needed to initiate impact ionization; the requisite electron–hole pairs are obtained directly by high-electric-field ionization of I-region silicon atoms. One might think this mechanism would be eliminated by reducing the bias voltage, since this action would reduce the combined magnitude ($V_{BIAS} + V_P$); but in most practical cases, where the bias is 10 to 20% of V_{BD}, reduction of V_{BIAS} would only precipitate the injection-mode failure. An exception, of course, is when the bias is kept at half the breakdown (i.e., $V_{BIAS} = V_{BD}/2$). Then the RF waveform makes no injecting excursion into the forward direction. But for high-power switching applications, a driver circuit to accomplish this end requires prohibitively high voltage transistors (500 V or more); the overall expense of the RF control circuit with driver could be more readily reduced by using a larger number of PIN diodes operated with less RF voltage stress.

Practically, the maximum sustainable RF voltage, V_P, must be determined by measurement. Taking high-power loss data for the diode—RF frequency, pulse length, and duty cycle of intended use—is the most direct and effective technique. In no case can $V_P + V_{BIAS}$ exceed V_{BD}, where V_{BD} is the bulk breakdown voltage of the diode. The bias voltage is selected to be as large as is possible in a practical driver circuit (usually, 10 to 20% of the diode's reverse bulk breakdown voltage) and the RF power level (from which the corresponding voltage, V_P, can be calculated) is set at the point at which a statistical sample of diodes have been found to undergo rapid loss increase and/or failure. Failures due to either of the two modes described are usually evidenced by a permanent short formed by a conducting filament across the I region.

Most high-power switching applications use the PIN diodes in a transmissive circuit with a matched load. Accordingly, failure or removal of the load, transmission-line arcing, or any mechanism that affects the load match can result in a voltage reflection and possible RF voltage enhancement at the diode. Neglecting losses in the switching circuit and diodes, this reflection voltage enhancement can double the stress on the diodes. Such a condition, even if encountered only briefly, usually precipitates diode failure. Therefore, it is good practice to rate the diode at a stress level $V_R \leqslant V_P/2$ (see Fig. 4.24) in order for the device to be able to survive such a total reflection. Since power is proportional to the square of voltage, *PIN diode devices should be rated at one-fourth or less of the power level at which, with matched load, they would be expected to undergo near instantaneous failure.* Even if provision has been made to minimize the likelihood of a totally reflecting load, consideration should be given to the following factors before opting for a power safety factor of less than 4 to 1:

1. Diode failure is not an exactly reproducible event even with PINs made by the same process within the same lot. A 2-to-1 variation in burnout is typical for a given process. Thus a production run of diodes may have a considerably lower (or higher) burnout than experienced with a prototype test lot.

2. Most high-power tests are conducted at room temperature, while practical devices usually must perform at considerably higher temperature, reducing the power safety margin that is inferred from a room-temperature test.

3. High-power RF testing is often of short duration, an hour or less, due to the generally limited availability of high-power testing facilities. However, semiconductor devices are usually expected to have useful lifetimes of years. Derating, according to Fig. 4.22, is necessary to accomplish long life.

4. PIN devices operated at or below one-fourth of their burnout power are typically found to be able to survive temporary driver failures wherein the high-power RF signals are applied with the diodes at zero bias.

Generally, PIN devices to control pulsed high RF power are limited in the reverse-biased state by the maximum safe rated RF voltage stress, V_R, as is seen in the circuit discussions to follow. While a device that fails to meet circuit performance expectations may cause some user disappointment, it has been the author's observation that nothing quite equals the state of dissatisfaction resulting when solid-state control devices fail catastrophically due to overrating. No doubt it is for reasons such as this that it has been industry practice in large phased array systems to design PIN phase shifters to survive operation into a short-circuit load of any phase.

Only by carefully rating these devices can the good reliability that has come to be expected—indeed, often assumed without question—of solid-state control be sustained. Accordingly, the designer should adopt as a minimum a policy of both designing diode control devices to sustain operation into a short circuit of any phase and testing throughout production to ensure that this level, at least statistically, is maintained for the complete population of devices built.

REFERENCES

1. A. B. Phillips, *Transistor Engineering and Introduction to Integrated Semiconductor Circuits*, McGraw-Hill, New York, 1962.
2. W. Shockley, *Electrons and Holes in Semiconductors*, D. Van Nostrand, Princeton, NJ, 1953, pp. 318–324.

3. J. S. Blakemore, *Semiconductor Statistics*, Pergamon Press, Elmsford, NY, 1962. Chaps. 5–9.

4. *The Step Recovery Diode, Hewlett-Packard Application Note*, Hewlett-Packard, Inc., Palo Alto, Ca., December 1963.

5. J. L. Moll, S. Krakauer, and R. Shen, "PN Junction Charge Storage Diodes," *Proce. IRE*, Vol. 50, pp. 43–53, January 1962.

6. H. A. Watson, Ed., *Microwave Semiconductor Devices and Their Circuit Applications*, McGraw-Hill, New York, 1969.

7. J. C. McDade, and F. Schiavone, "Switching Time Performance of Microwave PIN Diodes," *Microwave J.*, pp. 65–68, December 1974.

8. S. Ramo, and J. R. Whinnery, *Fields and Waves in Modern Radio*, Wiley, New York, 1944 and 1953. Also revised by S. Ramo, J. R. Whinnery, and T. Van Duzer as *Fields and Waves in Communication Electronics*, Wiley, New York, 1965.

9. D. Leenov, "The Silicon PIN Diode as a Microwave Radar Protector at Megawatt Levels," *IEEE Trans. Electron Devices*, Vol. ED-11, No. 2, pp. 53–61, February 1964.

10. M. E. Hines, "Fundamental Limitations in RF Switching and Phase Shifting Using Semiconductor Diodes," *Proc. IEEE*, Vol. 52, pp. 697–708, June 1964.

11. B. C. Deloach, Jr., "A New Microwave Measurement Technique to Characterize Diodes and an 800 GHz Cutoff Frequency Varactor at Zero Volts Bias," *1963 IEEE MTT Symp. Dig.*, pp. 85–91.

12. W. J. Getsinger, "The Packaged and Mounted Diode as a Microwave Circuit," *IEEE Trans. Microwave Theory and Tech.* Vol. MTT-14, No. 2, pp. 58–69, February 1966. Also see by the same author, "Mounted Diode Equivalent Circuits," *IEEE Trans. Microwave Theory and Tech.* Vol. MTT-15, No. 11, pp. 650–651, November 1967.

13. Jerome L. Altman, *Microwave Circuits*, Van Nostrand Reinhold, New York, 1964.

14. R. E. Collin, *Foundations for Microwave Engineering*, McGraw-Hill, New York, 1966.

15. Max Hansen, *Constitution of Binary Alloys*, McGraw-Hill, New York, 1958.

16. D. S. Peck and C. H. Zierdt, Jr., "The Reliability of Semiconductor Devices in the Bell System," *Proc. IEEE*, Vol. 62, No. 2, pp. 185–211, February 1974.

5

SEMICONDUCTOR CONTROL DEVICES: PHASE SHIFTERS AND SWITCHES

Ajay I. Sreenivas and Ron Stockton

Ball Aerospace Systems Division
Boulder, Colorado

5.1 INTRODUCTION

Phase shifters are essential components in electronically steered phased array antennas [1–4], which are extensively used in radar and communication systems. There are four main types of digital phase shifters: loaded line, reflection, switched line, and low-pass/high-pass. The principles and equations governing the design of these types of phase shifters are presented in Section 5.2 to 5.5. In all types the desired phase shift is introduced by altering the bias state of the control devices, which are typically PIN diodes or GaAs FETs.

The emphasis of this chapter is on digital phase shifters. While switched-line and low-pass/high-pass designs are not suitable for analog operation, the loaded-line and reflection-type phase shifters can be designed with continuous phase shift by using varactor diodes as control elements. The theory of operation for analog phase shifters is presented in Section 5.6. Miscellaneous topics related to digital phase shifters are covered in Section 5.7.

5.2 LOADED-LINE PHASE SHIFTERS

A loaded-line phase shifter [5–9] consists of two identical, yet variable shunt admittances, Y_i, separated by a transmission line of characteristic impedance, Z, and electrical length, θ, as shown in Fig. 5.1a. In its most elementary form variable admittances are replaced with two-state loading, which is achieved by switching between two reactance states. For practical designs the reactances are separated by about one-quarter wavelength of transmission line. The advantages of loaded-line phase shifters are simplicity and low loss. However, bandwidth constraints usually restrict their application to phase shifts of 45° or less.

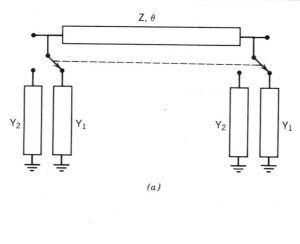

(a)

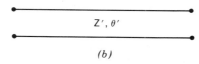

(b)

Figure 5.1   Basic loaded-line phase shifter circuit: (*a*) schematic; (*b*) equivalent circuit.

5.2.1   Design Principles

The theory of operation of a loaded-line phase shifter is based on two factors:

1. Any symmetric pair of discontinuities placed one-quarter wavelength apart on a transmission line will have mutually canceling reflections provided that the magnitude of the reflection coefficient is small compared to unity.
2. Shunt capacitive elements electrically lengthen a transmission line, while shunt inductances have the opposite effect.

5.2.2   Design Equations

Under lossless conditions ($G_i = 0$, $i = 1, 2$) the parameters of the equivalent circuit in Fig. 5.1*b* are given by

$$Z' = Z/[1 - (ZB_i)^2 + 2ZB_i \cot \theta]^{1/2} \tag{5.1}$$

$$\theta'_i = \arccos[\cos \theta - ZB_i \sin \theta] \qquad i = 1, 2 \tag{5.2}$$

For a perfect match and the desired phase shift, Φ, the design parameters Z, θ, and B_i ($i = 1, 2$) must be chosen such that

$$Z' = Z_0 \tag{5.3}$$

and

$$\Phi = \theta'_1 - \theta'_2 \tag{5.4}$$

where Z_0 is the system characteristic impedance. The general solution is given by

$$Z = Z_0 \cos(\Phi/2)/\sin \theta \qquad (5.5)$$

$$B_1 = Y_0[\cos \theta \sec(\Phi/2) + \tan(\Phi/2)] \qquad (5.6)$$

$$B_2 = Y_0[\cos \theta \sec(\Phi/2) - \tan(\Phi/2)] \qquad (5.7)$$

where

$$Y_0 = 1/Z_0 \qquad (5.8)$$

Opp and Hoffman [5] divide the load switching scheme into three classes. Class 1 is the general case where $B_1 \neq B_2 \neq 0$. Class 2, known as load/unload (L/U), is the case where B_1 or B_2 is zero, and class 3, known as complex conjugate loading (CC), is the case where $B_2 = -B_1$. While the solution in general is not unique, the latter two special cases result in the following design equations:

Class 2 (L/U);

$$Z = Z_0 \qquad (5.9a)$$

$$\theta = (\pi + \Phi)/2 \qquad (5.9b)$$

$$B_1 = 2Y_0 \tan(\Phi/2) \qquad (5.9c)$$

$$B_2 = 0 \qquad (5.9d)$$

Class 3 (CC):

$$Z = Z_0 \cos(\Phi/2) \qquad (5.10a)$$

$$\theta = \pi/2 \qquad (5.10b)$$

$$B_1 = -B_2 = Y_0 \tan(\Phi/2) \qquad (5.10c)$$

5.2.3   Theoretical Performance under Ideal Switching Conditions

Class 2 (L/U) Phase Shifters. Figure 5.2 shows the calculated performance characteristics of class 2 (L/U) loaded-line phase shifters based on Eq. (5.9a)–(5.9d). The phase error is calculated as the absolute deviation from constant phase. It must be noted that the phase error, with this design, will be significantly lower if the requirement is for true time delay rather than constant phase. Table 5.1 lists the bandwidth of lossless 45°, 22.5°, and 11.25° class 2 phase shifters with maximum allowable VSWR and phase error as parameters. The bandwidth is defined as

$$\text{percentage bandwidth} = 200.0 \times \frac{(f_2 - f_1)}{(f_1 + f_2)} \qquad (5.11)$$

where f_1 and f_2 define the frequency range over which the specified limits on phase error and VSWR are satisfied. Bandwidths higher than those given in Table 5.1 can be realized by optimizing the design parameters for conditions other than 0° phase error and 1:1 VSWR at the design frequency.

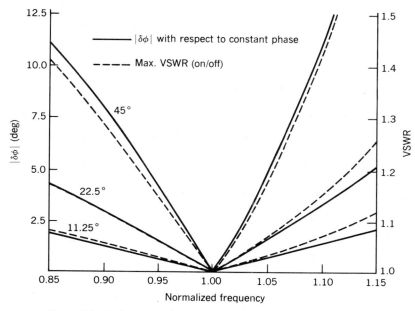

Figure 5.2 Performance characteristics of class 2 (L/U) phase shifters.

TABLE 5.1 Bandwidth of Class 2 (L/U) Loaded Line Phase Shifters under Lossless Conditions with Ideal Switch

Max VSWR	Max $|\delta\Phi|$ (deg)	Percent Bandwidth for True Time-Delay Operation[a]			Percent Bandwidth for Constant Phase Operation[a]		
		45°	22.5°	11.25°	45°	22.5°	11.25°
1.2	2	9.2	31.3	—	4.5	12.8	30.6
1.2	3	12.5	31.3	—	6.7	19.4	46.6
1.2	4	12.5	31.3	—	9.0	25.8	—
1.2	5	12.5	31.3	—	11.3	31.0	—
1.3	2	9.2	—	—	4.5	12.8	30.6
1.3	3	14.4	—	—	6.7	19.4	46.6
1.3	4	18.3	—	—	9.0	25.8	—
1.3	5	18.5	—	—	11.3	33.0	—
1.4	2	9.2	—	—	4.5	12.8	30.6
1.4	3	14.4	—	—	6.7	19.4	46.6
1.4	4	20.9	—	—	9.0	25.8	—
1.4	5	24.3	—	—	11.3	33.0	—
1.5	2	9.2	—	—	4.5	12.8	30.6
1.5	3	14.4	—	—	6.7	19.4	46.6
1.5	4	20.9	—	—	9.0	25.8	—
1.5	5	31.8	—	—	11.3	33.0	—

[a] Dashes indicate more than 45%.

TABLE 5.2 Configuration and Bandwidth Characteristics of Class 3 (Complex Conjugate Loaded) Phase Shifter Circuits

Type[b]	Stub Configuration	Stub Lengths (deg) θ_1	θ_2	Percentage Bandwidth VSWR < 1.2:1 $|\delta\Phi| < 2°$ $\Phi^a = 45°$	$\Phi^a = 22.5°$	VSWR < 1.5:1 $|\delta\Phi| < 5°$ $\Phi^a = 45°$	$\Phi^a = 22.5°$	Phase Shift Type[c]
1		$\Phi/2$	$90 - \Phi$	7.7	5.7	20.2	14.2	CP
2		$90 - \Phi/2$	Φ	9.2 / 11.4	16.3 / 18.5	19.7 / 22.1	29.8 / 32.1	CP / TTD
3		$90 + \Phi/2$	$90 - \Phi$	7.5	5.9	14.3	15.6	CP
4		$\Phi/2$	$180 - \Phi$	2.0	2.4	6.4	6.1	CP
5		$\Phi/2$	$90 - \Phi/2$	7.7	5.7	20.2	14.2	CP
6		$90 - \Phi/2$	$90 + \Phi/2$	9.2 / 11.4	16.3 / 18.5	19.7 / 22.1	29.8 / 32.1	CP / TTD
7		$\Phi/2$	$180 - \Phi/2$	2.0	2.4	6.4	6.1	CP

[a] Φ, bit size in degrees.

[b] Type 1 results in smallest total stub length, type 2 requires two ground connections on each leg, types 5, 6, and 7 require SPDT switches.

[c] CP, constant phase; TTD, true time delay.

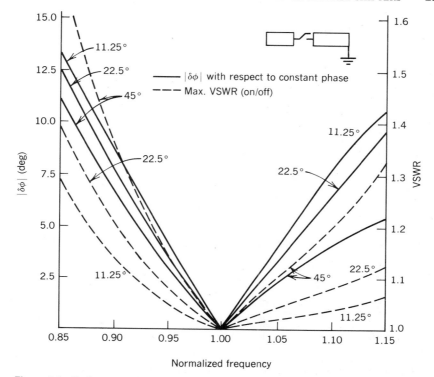

Figure 5.3 Performance characteristics of class 3 (CC) type 1 loaded-line phase shifters.

Class 3 (CC) Phase Shifters. Numerous possible configurations for class 3 (CC) operation are shown in Table 5.2 along with the equations that govern the stub lengths. Also listed are the percentage bandwidths under different constraints and modes of operation. The type 1 circuit requires minimum overall stub length. However, it is not suited for true time delay (TTD) operation. The type 2 circuit has the best overall bandwidth characteristics and is well suited for TTD operation, but it requires two ground connections on each leg. Types 3 and 4 generally result in narrow bandwidth. Types 5, 6, and 7 require single-pole double-throw (SPDT) switches and are not practical unless special situations warrant such a configuration. The bandwidths were calculated using Eq. (5.10a)–(5.10c), which neglect losses and assume ideal switching conditions. The performance characteristics of type 1 phase shifters with various bit sizes are plotted in Fig. 5.3 under lossless conditions. Figure 5.4 shows the calculated performance of two 45° type 2 phase shifters. One is optimized for maximum bandwidth with less than 5° absolute phase error and a VSWR of less than 1.5:1. The other is optimized under the constraint that absolute phase error be less than 2° and VSWR less than 1.2:1. The optimized design parameters are Z_c, θ_c, θ_1, and θ_2. The circuits resulted in bandwidths of 26.6% and 13.4%, respectively. The optimum value of θ_c was found to be 90° in both the cases. However, this may not be true in general.

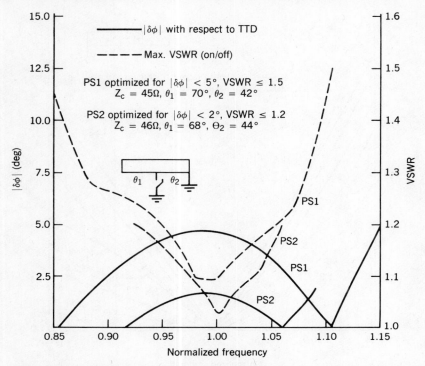

Figure 5.4   Performance characteristics of two class 3 (CC) 45° type 2 phase shifters.

5.2.4 Design Considerations Using PIN Diodes

The ability of a PIN diode to change from a low resistance under forward bias to a low-loss capacitor with reverse bias makes it attractive for phase shifting and switching applications. Figure 5.5 shows the equivalent circuit of a packaged PIN diode in the forward- and reverse-biased states.

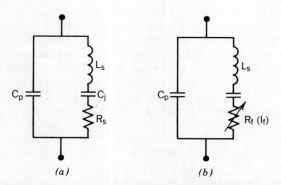

Figure 5.5   PIN diode equivalent circuit: (*a*) reverse bias; (*b*) forward bias.

When reverse biased the junction capacitance, C_j, is given by

$$C_j = \frac{\epsilon A}{W_I} \tag{5.12}$$

where W_I is the width of the I-layer, ϵ the permeability, and A the cross-sectional area. In the forward-biased state the series resistance, R_s, is the sum of all the resistances of the undepleted silicon regions and the contact resistance of both ohmic contacts. The parasitics, associated with the packaging, contribute to C_p and L_s. For well-designed diodes R_s ranges from a fraction of an ohm to 3.0 Ω, depending on junction size. Since the reactance associated with L_s is generally small, the PIN diode can be represented as a fixed-value low-loss capacitor in the reverse-biased state. This is particularly true in the low-frequency microwave region.

In the forward-biased state the resistance, R_f, is given by the sum

$$R_f = R_{sf} + R_I \tag{5.13}$$

where R_{sf} is a constant resistance, which is generally the same as R_s for abrupt junction diodes and R_I is the resistance associated with the intrinsic region, given by

$$R_I = \frac{W_I^2}{2I_f\bar{\mu}\tau_L} \tag{5.14}$$

with I_f being the forward-biased current, $\bar{\mu}$ the average carrier mobility, and τ_L the carrier lifetime in the I-region. For sufficiently large I_f, the intrinsic resistance R_I can be made negligibly small so that R_f approaches R_{sf}. However, this is accompanied by increased dc power dissipation and decreased switching speed. For low-voltage PIN diodes a bias current of 10 mA is sufficient to make $R_f \simeq R_{sf}$.

Design Neglecting Losses. When the diode series ON resistance is 2 Ω or less, the initial design is determined by neglecting the series resistance and using the simplified equivalent circuits shown in Fig. 5.6, for the diode, consisting of jX_f in the ON state and $-jX_R$ in the OFF state.

Class 2 (L/U). The main transmission-line parameters (Z and θ) and the loading stub parameters (Z_s and θ_s) completely define L/U-type phase shifter. The stub parameters are not unique and given by

$$Y_s \tan \theta_s = B_s \tag{5.15a}$$

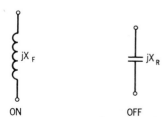

ON   OFF   **Figure 5.6** PIN diode simplified equivalent circuit.

where

$$Y_s = 1/Z_s \tag{5.15b}$$

$$B_s = \frac{b + b(1 + \delta)^{1/2}}{2(1 + \delta_2)} \tag{5.15c}$$

$$b = Y_0(1 - X_f B_R)(2 \tan \Phi/2) \tag{5.15d}$$

$$B_R = -1/X_R$$

$$\delta_1 = 4B_R(1 + \delta_2)[2 \tan(\Phi/2)]/b^2 \tag{5.15e}$$

$$\delta_2 = X_f[2 \tan(\Phi/2) + B_R] \tag{5.15f}$$

The main transmission-line parameters are given by

$$Z = \frac{B_n \tan(\Phi/2) + [1 + B_n^2 \sec^2(\Phi/2)]^{1/2}}{1 + B_n^2} \tag{5.16a}$$

$$\theta = (\pi + \Phi)/2 \tag{5.16b}$$

TABLE 5.3   Design Equations Using Nonideal but Lossless Switching Elements

Stub Configuration	Design Equation[a]
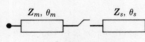 Series switch short-circuit stub	Series switch $$Z_m = \frac{X t_m (X_R - X_F)^{1/2}}{[X_R - X_F - 2X(1 + t_m^2)]^{1/2}}$$ For short-circuit stub $$-Z_s \tan \theta_s = \frac{Z_m^2 t_m + Z_m(X_R - X) + XX_R t_m}{Z_m + X t_m}$$
Series switch open-circuit stub	For open-circuit stub $$\frac{+Z_s}{\tan \theta_s} = \frac{Z_m^2 t_m + Z_m(X_R - X) + XX_R t_m}{Z_m + X t_m}$$
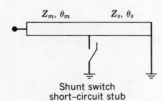 Shunt switch short-circuit stub	Shunt switch $$Z_m = \frac{[(B_R - B_F) - 2B(1 + t_m)^2]^{1/2}}{B t_m (B_R - B_F)^{1/2}}$$ For short-circuit stub $$+Z_s \tan \theta_s = \frac{Y_m + B t_m}{Y_m^2 t_m + Y_m(B_R - B) + BB_R t_m}$$
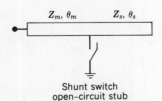 Shunt switch open-circuit stub	For open-circuit stub $$\frac{-Z_s}{\tan \theta_s} = \frac{Y_m + B t_m}{Y_m^2 t_m + Y_m(B_R - B) + BB_R t_m}$$

[a] $X = 1/B$, $t_m = \tan \theta_m$, $Y_m = 1/Z_m$, $B_F = 1/X_F$.

where

$$B_n = B_s B_R / (B_s + B_R) \tag{5.16c}$$

Class 3 (CC). The design equations defining the matching section parameters Z_m, θ_m and the stub parameters Z_s, θ_s are given in Table 5.3 for four important stub configurations. The main-line parameters Z, θ are assumed to satisfy Eqs. (5.10a)–(5.10c). Again, the design parameters are not unique.

5.2.5 Design with GaAs FET Switches

GaAs FETs are three-terminal devices, as shown in Fig. 5.7. In general, they are characterized by higher ON resistance and larger OFF capacitance compared to PIN diodes. As a result, phase shifter designs must take the finite ON resistance into account, and closed-form design equations are cumbersome. Although the physical nature of the GaAs FET device warrants a complex equivalent-circuit representation, for almost all practical design purposes, the GaAs FET may be represented by a parallel RC circuit with a separate set of R and C values in the ON and OFF states. Under lossy conditions, a perfect match in both phase states is generally not possible. The following equations serve as a starting point for the design under lossy conditions.

$$Z = Z_0 [\cos(\Phi/2)/\sin \theta]/(1 - \Delta^2)^{1/2} \tag{5.17a}$$

$$B_i = Y_0 [\cos \theta \sec(\Phi/2)(1 - \Delta^2)^{1/2} \pm \tan(\Phi/2)] \tag{5.17b}$$

$$\Delta = G_i Z_0 \cos(\Phi/2) \tag{5.17c}$$

In the design of loaded-line phase shifters, with lossy switching devices, the designer is faced with the problem of finding an impedance transforming network that transforms

Figure 5.7 GaAs FET switch with resonating loop inductor.

the two impedances Z_1 and Z_2 of the switching device to Z_1' and Z_2', respectively. The *abcd* parameters of such a lossless transforming network are given by the following equations [9]:

$$a = \alpha d \tag{5.18a}$$

$$b = \beta d \tag{5.18b}$$

$$c = \gamma d \tag{5.18c}$$

$$d = 1/(\alpha + \beta\gamma)^{1/2} \tag{5.18d}$$

where

$$\alpha = R_1' - \gamma(R_1'X_1 + X_1'R_1)/R_1 \tag{5.18e}$$

$$\beta = X_1' + \gamma(R_1R_1' - X_1X_1') - \alpha X_1 \tag{5.18f}$$

$$\gamma = (R_1'R_2 - R_2'R_1)/[R_2(R_1'X_1 + X_1'R_1) - R_1(R_2'X_2 + X_2'R_2)] \tag{5.18g}$$

$$Z_{1,2} = R_{1,2} + jX_{1,2} \tag{5.18h}$$

$$Z_{1,2}' = R_{1,2}' + jX_{1,2}' \tag{5.18i}$$

5.3 REFLECTION PHASE SHIFTER

The basic reflection-type phase shifter [10–15], also known as a hybrid coupler phase shifter, consists of variable or switched impedance elements Z_a and Z_b terminating the coupled ports of a 3-dB quadrature hybrid. The output is from the isolated port of the hybrid, as shown in Fig. 5.8a.

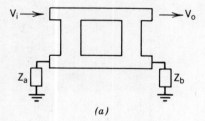

(a)

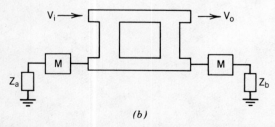

(b)

Figure 5.8 Reflection phase shifter: (*a*) basic configuration; (*b*) configuration with matching network.

The reflection design is most appropriate in applications where the desired phase shifts exceed of 45°. In a multibit digital phase shifter, the larger (90° and 180°) bits typically employ a reflection-type design.

5.3.1 Principle of Operation

In Fig. 5.8a the output V_o from the hybrid is given by

$$V_o = j(V_i/2)[\Gamma_a(Z_a) + \Gamma_b(Z_b)] \tag{5.19}$$

where V_i is the input signal and Γ_a and Γ_b are the complex reflection coefficients corresponding to the terminating impedances Z_a and Z_b, respectively. In practice $Z_a = Z_b = Z$, where Z is made up of a switching device connected to an open- or short-circuited stub. If Z_1 and Z_2 represent the impedances of the combination (of the switching device and stub) in the ON and OFF states, the following equality must be satisfied for proper phase shift operation:

$$\Gamma_1(Z_1) = \Gamma_2(Z_2)e^{j\Delta\Phi} \tag{5.20}$$

where $\Delta\Phi$ is the desired phase shift. Neglecting the losses in the hybrid, the phase shifter loss is then given by

$$\text{loss (dB)} = 20 \log |\Gamma| \tag{5.21}$$

where $|\Gamma| = |\Gamma_1| = |\Gamma_2|$.

When the switching device is ideal, the desired phase shift is realized simply by switching a pair of open-circuited stubs of electrical length $\Delta\Phi/2$ in and out of the circuit. However, in practice Eq. (5.20) is not satisfied without introducing an impedance transforming circuit as shown in Fig. 5.8b. The purpose of this network, M, is to transform Z_1 and Z_2 to Z_1' and Z_2' such that

$$\Gamma_1(Z_1') = \Gamma_2(Z_2')e^{j\Delta\Phi} \tag{5.22}$$

When Eq. (5.22) is satisfied, the reflection coefficient at the input is the same as that of the hybrid alone. The principal advantages of the reflection phase shifter are:

a. Only two switching devices per bit are required.
b. The input match is dependent only on the design of the hybrid, thereby allowing independent optimization of the phase shifting terminations for best phase shift and insertion loss characteristics over the desired bandwidth. However, in practice, the nonideal characteristics of the hybrid complicate the optimization process.
c. Unlike the loaded-line phase shifter, the incremental phase shift of a reflection phase shifter is not limited to small values.

5.3.2 Design with Ideal Switches

Table 5.4 shows common stub configurations along with the required stub lengths, for proper phase shift, under ideal switching conditions.

TABLE 5.4   Common Stub Configurations for Reflection Phase Shifter with Ideal Switching Devices[a]

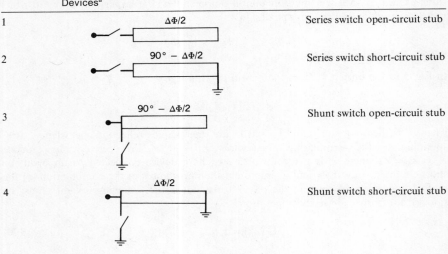

1		Series switch open-circuit stub
	$\Delta\Phi/2$	
2		Series switch short-circuit stub
	$90° - \Delta\Phi/2$	
3		Shunt switch open-circuit stub
	$90° - \Delta\Phi/2$	
4		Shunt switch short-circuit stub
	$\Delta\Phi/2$	

[a] $\Delta\Phi$ is the smallest number representing the absolute phase shift.

TABLE 5.5   Design Equations for Reflection Phase Shifters Using Nonideal Lossless Switching Devices

No.	Stub Configuration		Defining Equations
1	Z_s, θ_s Series switch short-circuit stub	$X_s = Z_s \tan \theta_s$	$X_s^2 + bX_s + c = 0$ $b = (X_F + X_R)Z_0$
2	Z_s, θ_s Series switch open-circuit stub	$X_s = Z_s \cot \theta_s$	$c = 1 + \dfrac{X_F X_R}{Z_0^2} - \dfrac{X_F - X_R}{Z_0} \cot \dfrac{\Phi}{2}$
3	Z_s, θ_s Shunt switch short-circuit stub	$X_s = Z_s \tan \theta_s$ $B_s = 1/X_s$	$B_s^2 + bX_s + c = 0$ $b = \dfrac{Z_0}{X_F} + \dfrac{Z_0}{Z_R}$
4	Z_s, θ_s Shunt switch open-circuit stub	$X_s = -Z_s \cot \theta_s$ $B_s = 1/X_s$	$c = 1 + \dfrac{Z_0^2}{X_F X_R}$ $+ \left(\dfrac{Z_0}{X_F} - \dfrac{Z_0}{X_R}\right) \cot \dfrac{\Phi}{2}$

5.3.3 Nonideal But Lossless Switch

PIN diodes with very low ON resistance fall under this class. The design formulas for the basic configurations are listed in Table 5.5,

5.3.4 Nonideal and Lossy Switching Devices

As mentioned in Section 5.3.1, a transforming network is generally required for reflection phase shifters using nonideal switching devices. By means of an elegant bilinear transformation Atwater [13] translated the problem of the reflection phase shifter design into that of an impedance-matching problem. In Ref. 13 an expression is derived for a fictitious impedance, Z_m, uniquely characterized by the phase shifter parameters Z_1, Z_2, Φ; therefore, when a matching network transforms Z_m to Z_0 (the characteristic impedance of the hybrid), it also transforms Z_1 and Z_2 to Z'_1 and Z'_2, satisfying Eq. (5.20).

A graphical method of determining Z_m has been described by Watanabe et al. [14]. The matching scheme would be impractical if the solution for Z_m does not exist or if the real part of Z_m is negative. Although any network that matches Z_m to Z_0 will result in proper phase shift operation at the design frequency, the bandwidth depends on the characteristics of the switching device, matching network, and the hybrid itself. Thus a solution satisfying Eq. (5.20) should be considered as only a starting point for further optimization using computer-aided design (CAD) techniques. Switching devices with a large ON resistance and/or a large OFF capacitance generally result in poor bandwidth characteristics. In the case of GaAs FETs, this problem may be alleviated by using an inductor to resonate the OFF capacitance. At millimeter wave frequencies, the Q-factor of a line inductor may be tailored to balance the ON and OFF insertion loss characteristics [15]. Figure 5.9 is a photograph of an X-band 5-bit monolithic digital phase shifter incorporating resonating line inductors. This phase shifter incorporates a loaded-line design for the 11.25°, 22.5°, and 45° bits and reflection designs for the 90° and 180° bits. The resonating inductors also keep the source and the drain terminals of the switching FETs at ground potential without any vias. This is a very important yield consideration in the fabrication of monolithic phase shifters.

In the loaded-line bits, the source and drain are tied together with a short section of very lossy line fabricated as part of the monolithic process. The purpose of the lossy lines is to equalize the insertion loss in the ON and OFF states. The design equations for these line inductors are presented next.

5.3.5 Design of Resonating Inductors

The design equations for wire, ribbon, and spiral inductors are given in Ref. 16. The application of such inductors is generally limited to the lower microwave region. For monolithic millimeter wave applications the resonating inductors are conveniently made in the form of high-impedance microstrip transmission lines. Figure 5.10 shows such a U-shaped line inductor along with two equivalent representations. In the first representation, the curved part of the inductor is treated as an uncoupled line of length πr, with r being the mean radius. The curvature effects are neglected in this representation. In the second representation, the curved portion is replaced by an equivalent short-circuited coupled section of length $\pi r/2$ and spacing $2r$. This leads to an overall simplified representation in terms of a single short-circuited coupled transmission line. The input impedance of such a transmission line is given by

$$Z_{in} = j2/(\cot \theta_o/Z_{oo} + \tan \theta_e/Z_{oe}) \tag{5.23}$$

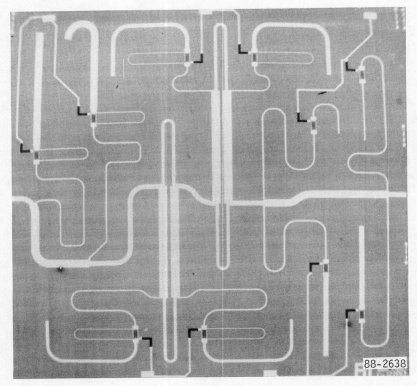

88-2638

Figure 5.9   X-band monolithic phase shifter using resonating line inductors and folded quadrature hybrids.

where Z_{oo} and Z_{oe} are the odd- and even-mode characteristic impedances and θ_o and θ_e are the corresponding electrical lengths.

The characteristics of very lossy lines may conveniently be analyzed by neglecting coupling effects. The complex characteristic impedance Z and the propagation constant γ of such lines are given by

$$Z = Z_0(1 + \delta^2)^{1/4}e^{-j\theta} \tag{5.24a}$$

$$\gamma = j\beta_0(1 + \delta^2)^{1/4}e^{-j\theta} \tag{5.24b}$$

$$\mathrm{Re}(\gamma) = \beta_0 \sin\theta / \sqrt{\cos 2\theta} \tag{5.24c}$$

$$\mathrm{Im}(\gamma) = \beta_0 \cos\theta / \sqrt{\cos 2\theta} \tag{5.24d}$$

where

$$\delta(\omega) = R/\omega L \tag{5.24e}$$

$$Z_0 = \sqrt{L/C} \tag{5.24f}$$

$$\beta_0 = \omega\sqrt{LC} \tag{5.24g}$$

$$\theta = \tfrac{1}{2}\tan^{-1}\delta \tag{5.24h}$$

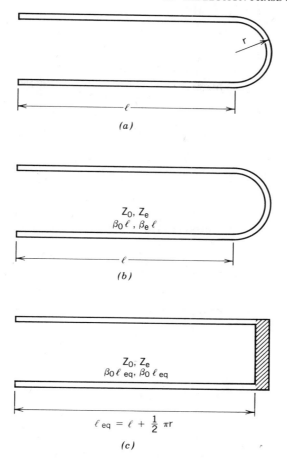

(a)

Z_0, Z_e
$\beta_0 \ell$, $\beta_e \ell$

(b)

Z_0, Z_e
$\beta_0 \ell$ eq, $\beta_0 \ell$ eq

ℓ eq $= \ell + \dfrac{1}{2} \pi r$

(c)

Figure 5.10 Resonating line inductors: (a) U-shaped inductor; (b), (c) equivalent representations.

R, L, and C are the characteristic resistance, inductance, and capacitance parameters of the transmission line.

The S-parameters of a lossy line inductor characterized by Z, γ, and l are given by

$$S = \frac{1}{D_s} \begin{bmatrix} (Z_n^2 - 1)Sh(\gamma l) & 2Z_n \\ 2Z_n & (Z_n^2 - 1)Sh(\gamma l) \end{bmatrix} \tag{5.25a}$$

where

$$D_s = 2Z_n Ch(\gamma l) + (Z_n^2 + 1)Sh(\gamma l) \tag{5.25b}$$

$$Z_n = Z/Z_0 \tag{5.25c}$$

and Z and γ are complex in general.

5.3.6 3-dB Couplers

The following four types of couplers are used in the design of reflection phase shifter:

a. *Branch-Line Coupler.* The branch-line coupler is the most common type of 3-dB quadrature coupler, used in reflection phase shifters. Its design and performance characteristics have been reported extensively in the literature [17–21].

b. *180° Rat Race Coupler.* The rat race coupler provides higher bandwidth than that of a conventional branch-line coupler [22]. An extra quarter-wave section will be required to obtain quadrature phase relationships at the output ports.

c. *Proximity Coupler* [23, 24]. The main disadvantage of the branch-line coupler is its narrow bandwidth. A broader-band coupler may be built by having the main transmission line parallel and in close proximity to a secondary line. However, to achieve 3-dB coupling, the secondary line needs to overlap on the primary line. Uniform spacing is maintained by a thin dielectric layer. With such a design a useful bandwidth of one octave can be obtained with a single-section quarter-wave coupler.

d. *Lange Coupler* [25–36]. The Lange coupler is capable of providing several octaves of bandwidth at the expense of increased loss and fabrication complexity.

5.4 SWITCHED-LINE PHASE SHIFTERS

As shown in Fig. 5.11, the switched-line phase shifter [37–39] uses two single-pole double-throw (SPDT) switches to select alternate transmission lines of length l_0 and

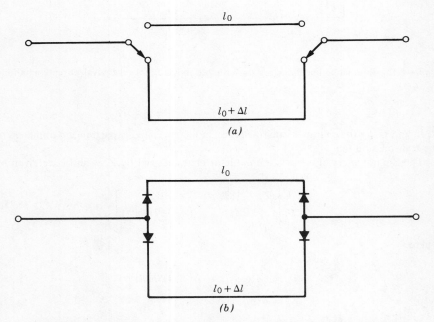

Figure 5.11 Switched line phase shifter: (*a*) schematic; (*b*) diode implementation.

$l_0 + \Delta l$, to achieve phase shifting action. The phase difference $\Delta\phi$ between the two states is given by

$$\Delta\phi = \beta(\Delta l) \qquad \text{radians} \tag{5.26}$$

where β is the propagation constant at the operating frequency and Δl is the difference in physical lengths between the two paths. The advantages of a switched-line design are:

- The design and principle of operation are very simple.
- It is the only practical means of achieving true time-delay operation, particularly where phase shifts in excess of 360° are required.
- A switched-line phase shifter can also be designed for constant phase shift using a Schiffman coupled section [40–42].

The disadvantages of the switched-line design are:

- It requires four switch devices per bit compared to only two devices for a reflection-type or loaded-line bit.
- The loss of a switched-line phase shifter is high, since the switching devices are in the transmission path.
- Because of unequal transmission paths the insertion loss is different for the two states of a switched-line bit. Unless special measures are taken, this results in PM-to-AM conversion.
- At the higher microwave and millimeter wave frequencies the imperfect character-istics of the switching devices make the switched-line design very difficult.

5.4.1 Ideal Devices

The design and operation of a switched-line phase shifter with ideal devices is very straightforward. The length, l_0, of the reference path is based on practical layout consid-erations. The differential length Δl is determined by the amount of phase shift required [Eq. (5.26)]. One important consideration is the coupling between the two parallel line segments of the phase shifting path. The coupling effects can be made negligible by choosing large enough spacing, typically three to five board thicknesses between adja-cent microstrip lines. Another practical consideration is the finite spacing between the T-junction and the switching diode. If this distance is of significant electrical length, the resultant loading effect will cause mismatch and increased insertion loss as well as phase errors.

5.4.2 Effect of Finite Impedances of Switching Devices

Practical diodes are characterized by finite ON resistance and OFF reactance. The effect of finite impedances may be studied in terms of the scattering parameters of the phase shifter circuit shown in Fig. 5.12. Z_1 and Z_2 represent the ON and OFF impedances, re-spectively, of the switching diodes. The scattering parameters of this circuit can be con-veniently derived by using the even- and odd-mode principle [43] and are given by

$$S_{21} = S_{12} = \frac{Z_{ie} - Z_{io}}{(Z_0 + Z_{ie})(Z_0 + Z_{io})} \tag{5.27a}$$

$$S_{11} = S_{22} = \frac{Z_{io}Z_{ie} - Z_o^2}{(Z_0 + Z_{ie})(Z_0 + Z_{io})} \tag{5.27b}$$

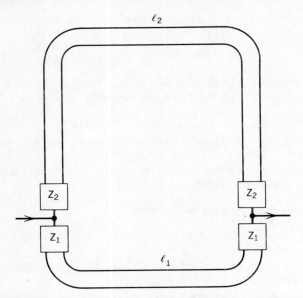

Figure 5.12 Switched line phase shifter with diodes characterized by finite ON/OFF impedances.

where

$$Z_{io} = \left[\frac{1}{Z_{1,2} + Z_0 \tanh(\gamma_1/2)} + \frac{1}{Z_{2,1} + Z_0 \tanh(\gamma_2/2)} \right]^{-1} \qquad (5.27c)$$

$$Z_{ie} = \left[\frac{1}{Z_{1,2} + Z_0 \coth(\gamma_1/2)} + \frac{1}{Z_{2,1} + Z_0 \coth(\gamma_2/2)} \right]^{-1} \qquad (5.27d)$$

Z_0 is the characteristic impedance $\gamma = \alpha + j\beta$ is the complex propagation constant of the lines. Equations (5.27a)–(5.27d) take line losses into account. In Eqs. (5.27c) and (5.27d) the first subscripts correspond to the case where the reference path is ON and the second subscripts correspond to the case where the phase shifting (delay) path is ON. Given the diode ON/OFF impedances and the bit size, Eqs. (5.27a)–(5.27d) may be used to examine the effect of the reference path length, l_1, on the performance of the phase shifter. In general, the performance of a 90° bit is sensitive to the reference path length with severe performance degradation for certain lengths. These critical lengths are dependent on the diode impedances and line-loss characteristics. The 180° bit is the least sensitive to the reference path length. Selection of the reference path length between 20 and 50° seem to provide minimum phase errors and balanced ON/OFF losses [39]. An approximate expression for insertion loss of a switched-line phase shifter is given by

$$\text{loss (dB)} = 40 \log \left[1 + \frac{1}{2} \left(\frac{R_{ON}}{Z_0} + \frac{Z_0}{R_{OFF}} \right) \right] \qquad (5.28)$$

which is valid when the ON/OFF impedances are predominantly real and $R_{ON}/R_{OFF} \ll 1$. The line losses must be added to the loss given by Eq. (5.28).

5.4.3 Design Considerations at Millimeter Wave Frequencies

At millimeter wave frequencies the parasitics associated with PIN diodes and the non-ideal characteristics of GaAs FETS prove to be deleterious, and special design techniques must be used to realize acceptable performance. Some of these techniques particularly suitable for monolithic devices are described below.

Resonating Inductors. Resonating inductors can be used to obtain better isolation performance from the switching device. Using CAD techniques the resonating inductors can be accurately modeled and laid out with great precision. In a GaAs FET switch the inductor also serves the dual purpose of keeping the source and drain at same potential. The success of the design depends on the accuracy with which the OFF parameters of the device are known and the accuracy of the inductor model. It is advantageous to choose an inductor geometry that is amenable to accurate modeling. One disadvantage of using resonating inductors is the increased area requirement for the phase shifter.

Matching Networks. Incorporation of stub matching networks into a switched-line phase shifter entails the following steps: A single-pole double-throw (SPDT) switch is first designed by treating it as a two-port device. Resonating inductors in parallel with the switching FETs are designed for maximum insertion loss through the OFF FET at the desired frequency. The S-parameters of the input and ON ports of the SPDT switch are calculated with the OFF port terminated in a load. The S-parameters are quite insensitive to the load impedance at the OFF port. After the S-parameters have been calculated for the input and ON ports, a set of equations is used to find the source and load impedances that produce optimum power transfer through the two-port. These equations are usually applied to power amplifier designs [44, 45], but they are appropriate for any two-port network. Combinations of transformers and stub tuners are used to convert the desired source and load impedances to those required for optimum power transfer. The tuning circuit designed for the ON port of the SPDT is also placed at the OFF port, since this will have little effect on switch performance. After the SPDT switch has been designed, a pair is combined with appropriate lengths of transmission line to form the phase shifter bits.

In general optimum performance results by selecting a device with as low as possible ON resistance and improving the isolation by means of a resonating inductor. Matching networks result in only second order improvement and also increase area requirements.

Multiple Devices in Series Shunt Configuration. One method of increasing OFF isolation is to cascade two diodes or FETs. However, this doubles the series ON resistance, as well, with an attendant increase in insertion loss. Conversely, if two devices are operated in parallel, the ON resistance decreases while the isolation degrades due to increased OFF capacitance. One way of avoiding this problem is to use two devices: one with low ON resistance and the other with high OFF resistance, in a series-shunt configuration as shown in Fig. 5.13.

With this approach the shunt diodes are designed for high isolation with less regard to ON resistance and the series diodes are designed for very low ON resistance. To turn on the reference path, diodes designated D_{sA} and D_{pA} are forward biased and D_{sB} and D_{pB} are reverse biased. Since the shunt mounted diodes are located a quarter wavelength away from the T-junctions, the forward biasing of these diodes provides an open circuit at the junction for path B. The reverse-biased diodes, D_{pB}, and the forward-biased diodes

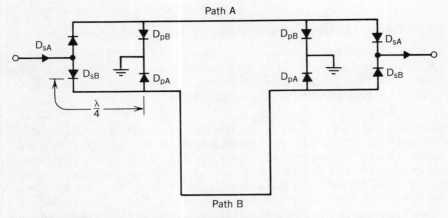

Figure 5.13 Switched line phase shifter with diodes in series shunt configuration.

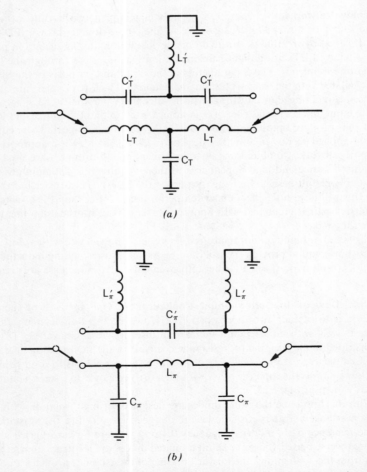

Figure 5.14 Low-pass/high-pass phase shifter schematic: (*a*) T-network implementation; (*b*) Π-network implementation.

D_{sB} result in low loss due to their inherent high isolation and low ON resistance characteristics, respectively.

To change the bit state the bias conditions are reversed. Although diodes are shown in Fig. 5.13, the approach lends itself to GaAs FET implementation.

5.5 HIGH-PASS/LOW-PASS PHASE SHIFTER

A high-pass filter comprised of series capacitors and shunt inductors acts as a phase-advancing network. Conversely, a low-pass filter with series inductors and shunt capacitors serves as a phase delay network. These networks are the basis of the high-pass low-pass phase shifter [46–49]. Phase shifting is accomplished by alternately switching in a high/low-pass filter section into the transmission path as shown in Fig. 5.14. Thus a straightforward implementation of a high-pass/low-pass phase shifter involves the replacement of the reference path and delay path in a switched-line phase shifter with high-pass and low-pass filters, respectively. At low frequencies lumped elements may be used to implement the filters. However, at millimeter wave frequencies the semiconductor devices themselves affect both the switching and filtering functions. The broadband characteristics associated with this type of phase shifter result from the complementary nature of the frequency response of the low-pass and high-pass filters. Lumped-element implementation results in compact size at low frequencies. However, at high microwave and millimeter wave frequencies these phase shifters are known more for their broad bandwidths rather than for their size.

5.5.1 Design Equations for the Lossless Case

The transmission coefficient S_{21} of the low-pass filter in Fig. 5.14a is given by

$$S_{21} = \frac{2}{2(1 - B_n X_n) + j(B_n + 2X_n - B_n X_n^2)} \tag{5.29a}$$

where X_n and B_n are the normalized reactance and susceptance given by

$$X_n = \omega L_T / Z_0 \tag{5.29b}$$

$$B_n = \omega C_T Z_0 \tag{5.29c}$$

with ω being the radian frequency.

Under lossless conditions a perfect match results when the magnitude of S_{21} is unity. For this condition

$$B_n = 2X_n / (1 + X_n^2) \tag{5.30}$$

or at a given frequency

$$C_T = 2L_T / (Z_0^2 + \omega^2 L_T^2) \tag{5.31}$$

When Eq. (5.30) is satisfied the transmission phase is given by

$$\arg(S_{21}) = \tan^{-1} \frac{2X_n}{X_n^2 - 1} \tag{5.32}$$

If the values of the corresponding high-pass elements are chosen such that

$$L'_T = \frac{1}{\omega^2 C_T} \tag{5.33a}$$

$$C'_T = \frac{1}{\omega^2 L_T} \tag{5.33b}$$

then the transmission phase of the high-pass filter will be same as that given by expression (5.32) except with a change in sign. The resulting phase shift from low-pass to high-pass is then

$$\Delta\Phi = 2 \tan^{-1}\left(\frac{2X_n}{X_n^2 - 1}\right) \tag{5.34}$$

The elements of the equivalent Π-network implementation shown in Fig. 5.14b are given by

$$C_\Pi = L_T/Z_0^2 \tag{5.35a}$$

$$L_\Pi = C_T Z_0^2 \tag{5.35b}$$

$$C'_\Pi = \frac{1}{\omega^2 L_\Pi} \tag{5.35c}$$

$$L'_\Pi = \frac{1}{\omega^2 C_\Pi} \tag{5.35d}$$

From the design point of view the element values L_T and C_T are given by

$$L_T = (Z_0/\omega)\tan(\Delta\Phi/4) \tag{5.36a}$$

$$C_T = (1/\omega Z_0)\sin(\Delta\Phi/2) \tag{5.36b}$$

Table 5.6 is a summary of the design equations.

TABLE 5.6   Low-Pass/High-Pass Phase Shifter Design Equations for Lossless Elements

	Low-Pass Filter Elements	High-Pass Filter Elements
T-implementation	$L_T = \left(\dfrac{Z_0}{\omega}\right)\tan\left(\dfrac{\Delta\Phi}{4}\right)$	$L'_T = \dfrac{Z_0}{\omega\sin(\Delta\Phi/2)}$
	$C_T = \left(\dfrac{1}{\omega Z_0}\right)\sin\left(\dfrac{\Delta\Phi}{2}\right)$	$C'_T = \dfrac{1}{\omega Z_0}\cot(\Delta\Phi/4)$
Π-implementation	$L_\Pi = \left(\dfrac{Z_0}{\omega}\right)\sin\left(\dfrac{\Delta\Phi}{2}\right)$	$L'_\Pi = \dfrac{Z_0}{\omega}\cot\left(\dfrac{\Delta\Phi}{4}\right)$
	$C_\Pi = \left(\dfrac{1}{\omega Z_0}\right)\tan\left(\dfrac{\Delta\Phi}{4}\right)$	$C'_\Pi = \dfrac{1}{\omega Z_0} - \dfrac{1}{\sin(\Delta\Phi/2)}$

5.5.2   Diode Implementations

The low-pass/high-pass phase shifter implementations shown in Fig. 5.14 require four diodes and six lumped elements. However, at the higher microwave and millimeter wave frequencies, the nonideal diode characteristics can be used to design low-pass high-pass phase shifters with only three diodes per bit. There are four configurations, which are shown in Fig. 5.15. Strictly speaking, these are high/low-pass bandpass phase shifters. In one state the networks act like bandpass filters, while the other state is a high- or low-pass filter depending on the implementation. Another common feature of all these implementations is that all three diodes are either forward biased or all reverse biased. In the forward-biased condition the diodes are characterized by a small resistance R_f and in the reverse-biased condition they are characterized by a series combination of a diode OFF capacitance, C, and a small resistance, R_r. A brief description of individual implementations is followed by design equations in Table 5.7.

High-Pass Bandpass T-Network Implementation. When the diodes are reverse biased they act as capacitors, C_d, with a small resistance, R_r, in series, as illustrated in Fig. 5.15a. The filter elements L_A and C_A and the diode capacitance C_d are chosen to achieve the desired phase shift with negligible transmission loss at the center frequency. The presence of finite resistance R_r introduces a small loss with negligible impact on phase shift. The design equations presented in Table 5.7 neglect the effect of this resistance. When the diodes are forward biased the filter elements C_A and L_A resonate at the center

TABLE 5.7   Design Equations for High/Low-Pass Bandpass Phase Shifters with Three Switching Diodes

Configuration	Defining Equations for C_d, L, and C	Defining Equations for L_A and C_A
High-pass bandpass T-network (Fig. 5.15a)	$C_d = C = \dfrac{1}{\omega Z_0} \cot\left(\dfrac{\Delta\Phi}{2}\right)$ $L = \dfrac{Z_0}{\omega \sin(\Delta\Phi)}$	$\omega C_A = \dfrac{1 + \sqrt{1 + 4\omega^2 LC}}{2\omega L}$ $L_A = 1/(\omega^2 C_A)$
High-pass bandpass Π-network (Fig. 5.15b)	$C_d = C = \dfrac{1}{\omega Z_0} \dfrac{1}{\sin(\Delta\Phi)}$ $L = \dfrac{Z_0}{\omega} \cot\left(\dfrac{\Delta\Phi}{2}\right)$	$\omega C_A = \dfrac{1 + \sqrt{1 + 4\omega^2 LC}}{2\omega L}$ $L_A = 1/(\omega^2 C_A)$
Low-pass bandpass T-network (Fig. 5.15c)	$L = \dfrac{Z_0}{\omega} \tan\left(\dfrac{\Delta\Phi}{2}\right)$ $C_d = 1/(\omega^2 L)$ $C = \dfrac{1}{\omega Z_0} \sin(\Delta\Phi)$	$C_A = \dfrac{C + \sqrt{C^2 + 4CC_d}}{2}$ $L_A = \dfrac{1}{\omega^2(C_A - C)}$
Low-pass bandpass π-network (Fig. 5.15d)	$L = \dfrac{Z_0}{\omega} \sin(\Delta\Phi)$ $C_d = 1/(\omega^2 L)$ $C = \dfrac{1}{\omega Z_0} \tan\left(\dfrac{\Delta\Phi}{2}\right)$	$C_A = \dfrac{C + \sqrt{C^2 + 4CC_d}}{2}$ $L_A = \dfrac{1}{\omega^2(C_A - C)}$

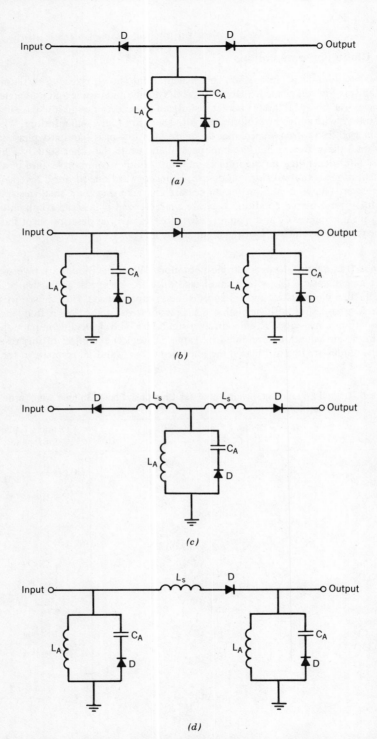

Figure 5.15 Low/high-pass bandpass phase shifter implementation with three diodes per bit: (*a*) high-pass bandpass T-network configuration; (*b*) high-pass bandpass Π-network configuration; (*c*) low-pass bandpass T-network configuration; (*d*) low-pass bandpass Π-network configuration.

frequency, producing zero insertion phase with a very small insertion loss due to R_f. This should be contrasted with low-pass/high-pass implementation, where one state produces phase delay and the other state produces phase advance.

High-Pass Bandpass Π-Network Implementation. The operation is similar to the high-pass bandpass T-network implementation. The arrangement of the elements in Fig. 5.15b and their values are different.

Low-Pass Bandpass T-Network Implementation. In the reverse-biased state of the network shown in Fig. 5.15c, the diode capacitance, C_d, resonates with the series inductor, L_s. The combination of C_A, C_d, and L_A in the shunt arm form a parallel resonant circuit, resulting in a bandpass network with zero insertion phase at the center frequency. The values of L_s, L_A, and C_A are chosen so that the forward-biased condition transforms the network into a low-pass filter with the desired phase shift at the center frequency.

Low-Pass Bandpass Π-Network Implementation. The operation is similar to the low-pass bandpass T-network implementation described above except for the arrangement of the circuit elements and their magnitudes. This design approach is illustrated in Fig. 5.15d.

5.6 ANALOG PHASE SHIFTERS

Analog phase shifters permit continuous phase variation as a function of the control voltage. This offers higher phase resolution, which has application in advanced radar functions such as searching and tracking multiple targets.

Analog phase shifters are generally built in a reflection configuration using varactor diodes whose doping profile can be controlled to achieve hyper-abrupt junctions with a larger capacitance change as a function of applied voltage. The capacitance versus voltage (C versus V) characteristic of a varactor diode is typically related to the diode's gamma (γ) as follows:

$$C = \frac{C_{jo}}{(1 - V/\Phi)^\gamma} \tag{5.37}$$

where c_{jo} is a constant and Φ is a constant potential of the diode.

Hyper-abrupt varactor diodes may be fabricated with a constant gamma over a limited voltage range with a high-capacitance variation. If a reflection phase shifter is built using two such diodes, terminating the direct and coupled ports of a quadrature coupler with characteristic impedance, Z_0, then the phase shift, Φ, is given by

$$\Phi = -\pi + \tan^{-1}(X_d) \tag{5.38}$$

where X_d is the normalized diode reactance given by

$$X_d = \frac{1}{\omega C Z_0} = \frac{(1 + V/\Phi)^\gamma}{\omega C_{jo} Z_0} \tag{5.39}$$

and ω is the radian frequency.

Diodes with gamma values greater than unity have nearly linear phase shift characteristics as a function of applied bias voltage. Thus the problem reduces to that of fabricating diodes with a large gamma that remains constant over the desired range of applied voltage. A number of hybrid as well as monolithic analog phase shifter designs have been reported in the literature [50–56].

5.7   MISCELLANEOUS CONSIDERATIONS

5.7.1   Biasing Schemes

The switching devices require dc bias for turning the device ON and OFF. The bias must be applied such that the dc path does not short circuit the RF signal or otherwise result in degraded RF performance. In practice the bias is applied through a choke inductor that blocks RF while acting as a dc short. Lumped inductors are also used at low frequencies. A short-circuited quarter-wavelength stub is convenient means of implementing the biasing choke at microwave frequencies. This could take either stripline or a microstrip form. Figure 5.16 is a loaded-line phase shifter bit demonstrating the biasing arrangement commonly used in microstrip implementations. The dc ground return is provided through a plated-through hole at the end of a quarter-wave stub *AB*. *CD* and *C'D'* are matching stubs which in combination with the open-circuited stubs *DE* and *D'E'* provide proper phase shift. *PQ* and *QR*, which are both a quarter-wave long, are used to provide proper biasing without affecting the RF performance. Since *PQ* and *QR* are a quarter-wavelength, the open circuit at *R* is reflected as an RF short circuit at *Q* and an RF open at *P*. Thus the biasing sections *PQ* and *QR* do not present any load at *P*. The location of point *P* can be anywhere along *DE* as long as *PQ* is a quarter-wave long. One limitation of this biasing scheme is its inherent narrow bandwidth. However, the bandwidth can be increased by making *QR* a radial line stub.

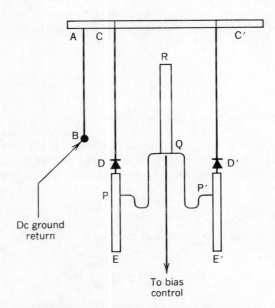

Figure 5.16   A typical Phase shifter biasing network.

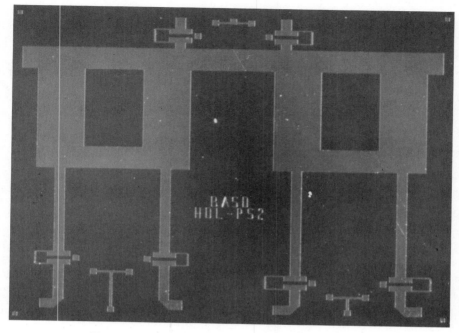

Figure 5.17 Monolithic millimeter wave 3-bit phase shifter.

Biasing of switching FETs is somewhat simpler since RF isolation can be provided by placing a large resistor between the gate terminal and the dc bias control line. This is practical because the voltage-controlled FETs draw neglegible current, on the order of a few tens of microamperes at the most. The monolithic 5-bit phase shifter shown in Fig. 5.9 makes use of such a biasing scheme. Figure 5.17 is a photograph of a 3-bit millimeter wave monolithic phase shifter ultilizing similar biasing scheme. Another feature to be noted in this photograph is that the dc return to source and drain, which are tied together by means of resonating inductors, is provided through a quarter-wave stub grounded through a via hole, shown black in the figure.

5.7.2 Power Handling

The maximum RF power that a PIN diode can handle is limited by either the diode breakdown voltage or its thermal dissipation capability, in most cases the latter. In a good design, a phase shifter can handle RF power levels many times greater than the power dissipated in the diode. The maximum power, P_d, that a PIN diode can dissipate is given by

$$P_d = (T_{jm} - T_a)/(\Theta_{jc} + \Theta_{ca}) \tag{5.40}$$

where T_{jm} = maximum operating diode junction temperature
T_a = ambient temperature
Θ_{jc}, Θ_{ca} = thermal resistances from the junction to case and case to ambient, respectively

The diode package design and diode mounting determine Θ_{jc} and Θ_{ca}, respectively.

The ratio of incident power to dissipated power is a function of the phase shifter design and the diode ON/OFF impedance characteristics. A simplified expression for this ratio [57] applicable to a double-pole double-throw switch is given by

$$\frac{P_d}{P_i} = \frac{Z_0}{R_f}\left(1 + \frac{R_f}{2Z_0}\right)^2 \tag{5.41}$$

where P_i = average incident power
P_d = average dissipated power
R_f = ON resistance of the diode
Z_0 = characteristic impedance of the line

The peak power, P_{max}, that a shunt diode can control is given by

$$P_{max} = \left(\frac{V_{BR} - V_{BIAS}}{8Z_0}\right)^2 \tag{5.42}$$

where V_{BR} is the breakdown voltage of the diode and V_{BIAS} is the magnitude of the reverse bias applied to the diode. An applied bias of zero volts results in maximum controlled power. However, this may not be practical in fast switching applications where reverse bias is required to reduce the switching time. Further, a zero reverse bias may result in excessive current through the diode during positive RF voltage swings.

The expressions given in Eqs. (5.41) and (5.42) are only approximate. It is best to simulate and analyze the complete phase shifter circuit, or at least one complete bit, to determine the power dissipation in, and voltage swings across, the diodes under worst-case conditions.

Loaded-line and reflection-type phase shifters can generally control much higher power compared to switched-line phase shifters. Power-handling considerations for reflection-type phase shifters have been treated by Burns and Stark [58]. The S-parameters of loaded-line phase shifters can be conveniently determined in terms of odd- and even-mode impedances, and the dissipated power in the diodes may then be obtained, by neglecting the line losses, as

$$P_d = (1 - |S_{11}|^2 - |S_{12}|^2)P_i \tag{5.43}$$

where P_i is the incident power.

5.7.3 Switching Speed

The switching speed of semiconductor-controlled phase shifters is generally much higher than those of ferrite phase shifters. GaAs FET devices can switch much faster than PIN diodes, due to the very high mobility of the carriers in a GaAs FET device. In determining the operational speed of a phase shifter, one must consider various delays—driver delay, driver rise time, and PIN diode delay—as well as the diode switching time, which is the time required to change the state of the I-region from "no stored charge" high impedance to "large stored charge" low impedance, and vice versa. A typical PIN diode switching waveform is illustrated in Fig. 5.18. It shows the various contributions [59, 60] to the delay time.

The reverse to forward switching time of a PIN diode is typically a few percent of the specified lifetime of the diode, and is much smaller than switching time from forward to reverse. When in the high-impedance state the I–V characteristics of the diode are inductive and the driver circuit must deliver a current spike, with substantial overvoltage,

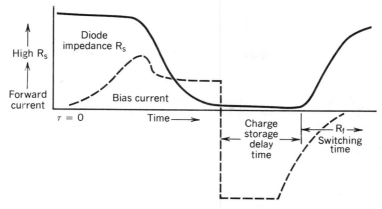

Figure 5.18 Typical PIN diode switching waveforms.

in order to reduce the switching time. This is generally accomplished by incorporating a "speed-up" capacitor in parallel with the dropping resistor at the output of the driver circuit. When switching from forward to reverse mode, the switching time is minimized by providing a reverse current on the order of 10 to 20 times the forward bias with a moderately high reverse-biased voltage. The actual "RF switching time" will be minimized by a larger negative bias and/or by a low forward bias. It should be noted, however, that a low forward bias will also mean somewhat increased insertion loss in the ON state.

It is preferable to design the bias circuit to have the same characteristic impedance as the RF line, to minimize the reflection and ringing effects. Extraneous capacitance, in the form of blocking and bypass elements, must be kept to a minimum in applications requiring maximum switching speeds. Again, a trade may exist between RF performance and the switching time. To minimize the delays associated with shunt capacitance in the control lines between the driver circuit and the diodes, it is best to build the driver circuit on the back side of the phase shifter board and make the bias connections in the form of simple feed-throughs.

5.7.4 Bandwidth Considerations

The switched line and low-pass/high-pass phase shifter are most suited for broadband applications. Reflection- and loaded-line phase shifters are inherently narrow-band. However, loaded-line bits with small phase shifts, 22° and less can be designed to have broader bandwidths.

The bandwidth of a true time-delay switched-line phase shifter is limited by the bias circuitry and by the resonances in the phase shifter. Despite these limitations, switched-line phase shifters can be designed with bandwidths in excess of one octave. However, in any broadband design some RF performance must be compromised in one or more of the following areas:

- Insertion loss
- Insertion loss flatness
- Input mismatch
- Phase versus frequency characteristics
- Circuit complexity

Switched-line phase shifters can also be designed with constant phase shift versus frequency characteristics over broadband by using a dispersive compensating scheme in the reference path. One method entails the use of two quarter-wave short-circuited shunt stubs connected to the reference path at a quarter-wavelength apart [4]. A second approach is to use a Schiffman section in the reference path [42, 61].

High-pass/low-pass-type phase shifters can be designed for broader bandwidths by a judicious choice of switching element characteristics [49]. This usually entails in a larger number of diodes (or GaAs FETs), typically 6 or more per bit, which are not necessarily identical. Broadbanding of analog phase shifters is generally done by cascading several identical broadband phase shifters, each capable of providing a variable phase shift of limited magnitude.

REFERENCES

1. J. F. White, "Diode Phase Shifters for Array Antennas," *IEEE Trans. Microwave Theory Tech.*, Vol. MTT-22, pp. 658–674, June 1974.

2. M. E. Davis, "Integrated Diode Phase Shifter Elements for an X-Band Phased Array Antenna," *IEEE Trans. Microwave Theory Tech.*, Vol. MTT-23, pp. 1080–1084, December 1975.

3. F. G. Terrio, R. J. Stockton, and W. D. Sato, "A Low Cost *p-i-n* Diode Phase Shifter for Airborne Phased-Array Antennas," *IEEE Trans. Microwave Theory Tech.*, Vol. MTT-22, pp. 688–692, June 1974.

4. R. W. Burns et al., "Low Cost Design Techniques for Semiconductor Phase Shifter," *IEEE Trans. Microwave Theory Tech.*, Vol. MTT-22, No. 6, pp. 675–688, June 1974.

5. F. L. Opp and W. F. Hoffman, "Design of Digital Loaded-Line Phase-Shift Networks for Microwave Thin-Film Applications," *IEEE J. Solid-State Circuits*, Vol. SC-3, pp. 124–130, June 1968.

6. T. Yahara, "A Note on Designing Digital Diode-Loaded Line Phase Shifter," *IEEE Trans. Microwave Theory Tech.*, Vol. MTT-20, pp. 703–704, October 1972.

7. W. A. Davis, "Design Equations and Bandwidth of Loaded Line Phase Shifters," *IEEE Trans. Microwave Theory Tech.*, Vol. MTT-22, pp. 561–563, May 1974.

8. I. J. Bahl and K. C. Gupta, "Design of Loaded-Line *p-i-n* Diode Phase Shifter Circuits," *IEEE Trans. Microwave Theory Tech.*, Vol. MTT-28, pp. 219–224, March 1980.

9. H. A. Atwater, "Circuit Design of the Loaded-Line Phase Shifter," *IEEE Trans. Microwave Theory Tech.*, Vol. MTT-33, No. 7, pp. 626–634, July 1985.

10. R. E. Fisher et al., "Digital-Reflection-Type Microwave Phase Shifters," *Microwave J.*, Vol. 12, No. 12, pp. 63–68, May 1969.

11. K. Kurokawa and W. O. Schlosser, "Quality Factor of Switching Diodes for Digital Modulation," *Proc. IEEE*, Vol. 38, pp. 180–181, January 1980.

12. P. Wahi and K. C. Gupta, "Effect of Diode Parameters on Reflection-Type Phase Shifters," *IEEE Trans. Microwave Theory Tech.*, Vol. MTT-24, pp. 619–621, September 1976.

13. Harry L. Atwater, "Reflection Coefficient Transformations for Phase-Shift Circuits," *IEEE Trans. Microwave Theory Tech.*, Vol. MTT-28, No. 6, pp. 563–568, June 1980.

14. K. Watanabe et al., "Graphical Design of *p-i-n* Diode Phase Shifters," *IEEE Trans. Microwave Theory Tech.*, Vol. 29, No. 8, pp. 829–831, August 1981.

15. A. Sreenivas, *Use of Shunt Resistors and Lossy Line Inductors to Improve the Performance of Reflection and Loaded Line Type Phase Shifters*, BASD SER. No. 3319-15, November 1986.

16. K. C. Gupta et al., *Computer Aided Design of Microwave Circuits*, Artech House, Dedham, MA, 1981, pp. 207–211.

17. B. Schiek, "Hybrid Branchline Couplers: A Useful New Class of Directional Couplers," *IEEE Trans. Microwave Theory Tech.*, Vol. 22, No. 10, pp. 865–869, October 1974.

18. T. Okashi et al., "Computer Oriented Synthesis of Optimum Circuit Pattern of 3-dB Hybrid Ring by Planar Circuit Approach," *IEEE Trans. Microwave Theory Tech.*, Vol. MTT-29, pp. 194–202, March 1981.

19. V. K. Tripathi et al., "Analysis and Design of Branchline Hybrids with Coupled Lines," *IEEE Trans. Microwave Theory Tech.*, Vol. MTT-32, No. 4, pp. 427–432, April 1984.

20. M. Kirsching and R. H. Jansen, "Accurate Wide Range Design Equations for the Frequency-dependent Characteristics of Parallel Coupled Microstrip Lines," *IEEE Trans. Microwave Theory Tech.*, Vol. MTT-32, pp. 83–90, January 1984; corrections, Vol. MTT-33, p. 288, March 1985.

21. B. Petrovic, "A New Type of Microstrip Directional Coupler," *Microwave J.*, Vol. 29, pp. 197–201, April 1986.

22. H. Howe, Jr., *Stripline Circuit Design*, Artech House, Dedham, MA 1974, Chap. 3.

23. Ibid., pp. 153–159.

24. F. C. de Ronde, "Wide-Band High Directivity in MIC Proximity Couplers by Planar Means," *IEEE MTT Symp. Dig.*, pp. 480–482, May 1980, Washington, D.C.

25. J. Lange, "Interdigitated Stripline Quadrature Hybrid," *IEEE Trans. Microwave Theory Tech.*, Vol. MTT-17, pp. 1150–1151, December 1969.

26. R. Waugh and D. La Combe, "Unfolding the Lange Coupler," *IEEE Trans. Microwave Theory Tech.*, Vol. MTT-20, pp. 777–779, November 1972.

27. W. P. Ou, "Design Equations for an Interdigitated Directional Coupler," *IEEE Trans. Microwave Theory Tech.*, Vol. MTT-23, pp. 253–255, February 1975.

28. V. Rizzoli, "Stripline Interdigitated Couplers: Analysis and Design Considerations," *Electron. Lett.*, Vol. 11, pp. 392–393, August 1975.

29. S. J. Hewitt and R. S. Pengelly, "Design Data for Interdigital Directional Couplers," *Electron. Lett.*, Vol. 12, pp. 86–87, February 1976.

30. D. D. Paolino, "Design More Accurate Interdigitated Coupler," *Microwaves*, Vol. 15, pp. 34–38, May 1976.

31. J. A. G. Malherbe, "Interdigital Directional Couplers with an Odd or Even Number of Lines and Unequal Characteristic Impedances," *Electron. Lett.*, Vol. 12, pp. 464–465, September 1976.

32. A. Presser, "Interdigitated Microstrip Coupler Design," *IEEE Trans. Microwave Theory Tech.*, Vol. MTT-26, pp. 801–805, October 1978.

33. D. Kajfez et al., "Simplified Design of Lange Coupler," *IEEE Trans. Microwave Theory Tech.*, Vol. MTT-26, pp. 806–808, October 1978.

34. V. Rizzoli and A. Lipparini, "The Design of Interdigitated Couplers for MIC Applications," *IEEE Trans. Microwave Theory Tech.*, Vol. MTT-26, pp. 7–15, January 1978.

35. R. C. Waterman, Jr., W. Fabian, R. A. Pucel, Y. Tajima, and J. L. Vorhaus, "GaAs Monolithic Lange and Wilkinson Couplers," *IEEE Trans. Electron Devices*, Vol. ED-28, pp. 212–216, February 1981.

36. R. M. Osmani, "Synthesis of Lange Couplers," *IEEE Trans. Microwave Theory Tech.*, Vol. MTT-29, No. 2, pp. 168–170, February 1981.

37. E. M. Rutz and J. E. Dye, "Frequency Translation by Phase Modulation," *IRE WESCON Conv. Rec.*, Pt. 1, pp. 201–207, 1957.

38. E. J. Wilkinson, L. I. Parad, and W. R. Connerney, "An X-Band Electronically Steerable Phased Array," *Microwave J.*, Vol. 7, pp. 43–48, February 1964.

39. R. V. Garver, "Broad-Band Diode Phase Shifters," *IEEE Trans. Microwave Theory Tech.*, Vol. MTT-20, pp. 314–323, May 1972.

40. R. C. Hansen and H. Jasik, Eds., *Antenna Engineering Handbook*, McGraw-Hill, New York, 1984, pp. 20–26 to 20–62.

41. B. M. Schiffman, "A New Class of Broadband Microwave 90-Degree Phase Shifters," *IRE Trans. Microwave Theory Tech.*, Vol. MTT-6, pp. 232–237, April 1958.

42. C. M. Gravling, Jr., and B. D. Geller, "A Broad-Band Frequency Translator with 30-dB Suppression of Spurious Sidebands," *IEEE Trans. Microwave Theory Tech.*, Vol. MTT-18, pp. 651–652, September 1970.

43. K. C. Gupta, R. Garg, and I. J. Bahl, *Microstrip Lines and Slot Lines*, Artech House, Dedham, MA, 1979, pp. 189–192.

44. William A. Suter, "Active Two-Port Program Speeds Amplifier Design," *Microwaves*, Vol. 19, pp. 79–83, March 1980.

45. *S-Parameter Techniques for Faster, More Accurate Network Design*, Hewlett-Packard Application Note 95-7, Hewlett-Packard, Palo Alto, CA.

46. P. Onno and A. Plitkins, "Lumped Constant, Hard Substrate, High Power Phase Shifters," presented at the IEEE MIC (Materials and Design) Seminar, Monmouth College, West Long Branch, NJ, June 1970.

47. P. Onno and A. Plitkins, "Miniature Multi-kilowatt PIN Diode MIC Digital Phase Shifters," *IEEE MTT-s Int. Microwave Symp. Dig.*, pp. 22–23, 1971.

48. P. Stabile et al., *EHF Phase Shifter*, RADC-TR-84-169, August 1984.

49. Y. Ayasli et al., "Wide-Band Monolithic Phase Shifter," *IEEE Trans. Electron Devices*, Vol. ED-31, No. 12, pp. 1943–1947, December 1984.

50. S. Hopfer, "Analog Phase Shifter for 8-18 GHz," *Microwave J.*, Vol. 22, No. 3, pp. 48–50. March 1979.

51. B. Ulriksson, "Continuous Varactor-Diode Phase Shifter with Optimized Frequency Response," *IEEE Trans. Microwave Theory Tech.*, Vol. MTT-27, pp. 650–654. July 1979.

52. R. K. Mains, G. I. Haddad, and D. F. Peterson, "Investigation of Broadband, Linear Phase Shifters Using Optimum Varactor Diode Doping Profiles," *IEEE Trans. Microwave Theory. Tech.*, Vol. MTT-29, pp. 1158–1164, November 1981.

53. R. V. Garver, "360 Degree Varactor Linear Phase Modulator," *IEEE Trans. Microwave Theory Tech.*, Vol. MTT-17, pp. 137–147, March 1969.

54. E. C. Niehenke, V. V. DiMarco, and A. Friedberg, "Linear Analog Hyperabrupt Varactor Diode Phase Shifters," *IEEE Microwave Theory Tech. Symp. Dig.*, pp. 657–660, 1985.

55. D. E. Dawson, A. C. Conti, S. H. Lee, G. F. Shade, and L. E. Dickens, "An Analog X-Band Phase Shifter," *IEEE Microwave and Millimeter Wave Monolithic Circuits Symp. Dig.*, pp. 6–10, 1984.

56. C. L. Chen, W. E. Courtney, L. J. Mahoney, M. J. Manfra, A. Chu, and H. A. Atwater, "A Low-Loss Ku-Band Monolithic Analog Phase Shifter," *IEEE Trans. Microwave Theory Tech.*, Vol. MTT-35, pp. 315–320, March 1987.

57. R. V. Garver, "Theory of TEM Diode Switching," *IRE Trans. Microwave Theory Tech.*, Vol. MTT-9, pp. 224–238, May 1961.

58. R. W. Burns and L. Stark, "PIN Diodes Advance High-Power Phase Shifting," *Microwaves*, Vol. 4, No. 11, pp. 38–48, November 1965.

59. *PIN Diode Basics*, Hewlett-Packard Application Note 80200, Hewlett-Packard, Palo Alto, CA.

60. J. C. McDade and F. Schiavone, "Switching Time Performance of Microwave PIN Diodes," *Microwave J.*, Vol. 17, September 1974.

61. R. B. Wilds, "Try $\lambda/8$ Stubs for Fast Fixed Phase Shift," *Microwaves*, Vol. 18, pp. 67–68, December 1979.

6

TRANSFERRED ELECTRON DEVICES

Cheng Sun

*California Polytechnic State University
San Luis Obispo, California*

6.1 INTRODUCTION

The transferred electron effect, also referred to as the Gunn effect, was discovered by J. B. Gunn [1] in 1963. Gunn observed current oscillations as the applied voltage exceeded a certain threshold voltage on the n-type GaAs and InP samples. The frequency of these oscillations could be made to lie in the microwave range by choosing the proper sample doping level and thickness. His measurements showed that a high-field domain formed near the cathode, propagated across the sample, and disappeared at the anode. As the domain was absorbed at the anode, the current increased to the threshold value and a new domain nucleated at the cathode contact. The current would drop and the process would be repeated.

The transferred electron effect was explained in 1964 by Kroemer [2] to be the negative differential mobility mechanism described by Ridley and Watkins [3] in 1961. The theory of achieving negative differential mobility was further advanced by Hilsum [4].

The theory shown by Ridley and Watkins [3], and Hilsum [4] indicated that the negative differential mobility of electrons could be obtained in GaAs by exciting high-mobility electrons from a lower conduction valley to a low-mobility, upper conduction valley.

While GaAs has dominated the development work for the transferred electron in recent years because of its maturity, materials such as InP, ZnSe, CdTe, and GaAsP have also demonstrated the same effect. By applying pressure to certain crystal samples, the energy bands of InAs, for example, can be changed so that the transferred electron effect exhibiting the negative differential mobility mechanism was observed. In fact, GaAs samples subjected to hydrostatic pressure caused a reduction in the threshold voltage, as discovered by Hutson et al. [5] and Allen et al. [6].

Transferred electron devices have been used extensively for low-noise local oscillators, low-power transmitters, and wideband tunable sources. For a single device, continuous-wave (CW) power levels of several hundred milliwatts in X-, Ku-, and Ka-band frequency ranges and about 75 mW in W-band frequencies are commercially available. Higher power levels have also been achieved with power combining techniques by combining multiple devices.

In this chapter we present an overview on transferred electron effect devices. In Section 6.2 we review the negative differential mobility mechanism. The modes of operation of transferred electron oscillators and amplifiers are described in Sections 6.3 and 6.4, respectively. In Section 6.5 we discuss the equal-area rule and in Section 6.6, present equivalent-circuit models. Microwave circuits and combining techniques are covered in Sections 6.7 and 6.8. In Section 6.9 we summarize epitaxial growth of the material and device fabrication. Finally, recent developments and future trends are given in Section 6.10.

6.2   NEGATIVE DIFFERENTIAL MOBILITY

The existence of the negative differential mobility effect associated with the transferred electron devices was explained by Ridley and Watkins [3] and Hilsum [4], and is often called the Ridley–Watkins–Hilsum mechanism. For n-type GaAs, this effect is due to its energy band structure, which allows the transfer of conduction electrons from a higher-mobility, lower-energy valley to lower-mobility, higher-energy satellite valleys along the $\langle 100 \rangle$ crystallographic directions.

A simple two-valley band structure model of n-type GaAs is shown in Fig. 6.1. The effective masses of electrons in the lower and upper valleys are $m_1^* \simeq 1.2m_e$, $m_2^* \simeq 0.068m_e$, respectively, where m_e is the mass of a free electron. The electron mobility in the lower valley μ_1 is approximately 50 times greater than the mobility in the upper valley μ_2. The energy separation between the two valleys is 0.36 eV. The steady-state conductivity of the n-type GaAs is

$$\sigma = q(\mu_1 n_1 + \mu_2 n_2) \tag{6.1}$$

where n_1 and n_2 are electron densities in the lower and upper valley bands, respectively.

The total electron density n is given by

$$n = n_1 + n_2 = \text{constant} \tag{6.2}$$

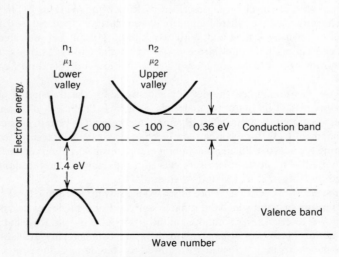

Figure 6.1   Simplified band structure of GaAs.

In Eq. (6.1), μ_1, n_1 and μ_2, n_2 are functions of the applied electric field E. Differentiation of Eqs. (6.1) and (6.2) with respect to E reduces to

$$\frac{d\sigma}{dE} = q\left(\mu_1 \frac{dn_1}{dE} + \mu_2 \frac{dn_2}{dE}\right) + q\left(n_1 \frac{d\mu_1}{dE} + n_2 \frac{d\mu_2}{dE}\right) \tag{6.3}$$

$$\frac{dn_1}{dE} = -\frac{dn_2}{dE} \tag{6.4}$$

If we assume that μ_1 and μ_2 are proportional to E^P, where p is a constant [7], Eqs. (6.3) and (6.4) can be combined to

$$\frac{d\sigma}{dE} = q(\mu_1 n_1 + \mu_2 n_2)\frac{p}{E} + q(\mu_1 - \mu_2)\frac{dn_1}{dE} \tag{6.5}$$

From Ohm's law, $J = \sigma E$, one obtains the following equation by differentiation and rearrangement:

$$\frac{1}{\sigma}\frac{dJ}{dE} = 1 + \frac{E}{\sigma}\frac{d\sigma}{dE} \tag{6.6}$$

To provide negative conductivity, dJ/dE must be negative in Eq. (6.6). This condition can be solved by combining Eqs (6.1), (6.5), and (6.6):

$$\left[\left(\frac{\mu_1 - \mu_2}{\mu_1 + f\mu_2}\right)\left(-\frac{E}{n_1}\frac{dn_1}{dE}\right) - p\right] > 1 \tag{6.7}$$

where $f = n_2/n_1$. dn_1/dE in Eq. (6.7) is negative because more electrons will be transferred from the lower to the upper valley. Equation (6.7) is satisfied if the following requirements are met:

1. $\mu_1 > \mu_2$ or electrons must start in a lower valley and transfer to an upper valley when an electric field is applied.
2. The exponent p should be negative and large. If lattice scattering is dominant, p is negative. When the impurity scattering is dominant, p becomes positive and is therefore not desirable for the transferred electron effect. A more detailed discussion on scattering mechanisms is given by Bulman et al. [8].

For a material to give rise to a negative differential mobility, its band structure must satisfy the following criteria based on the discussion above:

1. Two conduction subbands must exist. The energy difference between the lower and upper valley bands must be much greater than the thermal energy.
2. The electron mobility μ_2 for the upper valley must be lower than the electron mobility μ_1 for the lower valley. The effective mass m_2^* for the electrons in the upper valley must be higher than that for electrons in the lower valley m_1^*.
3. The energy difference between the upper and lower valleys should be much smaller than the semiconductor bandgap (1.43 eV for GaAs); otherwise, the semiconductor would break down and become highly conductive before the electrons can be transferred to the upper valleys.

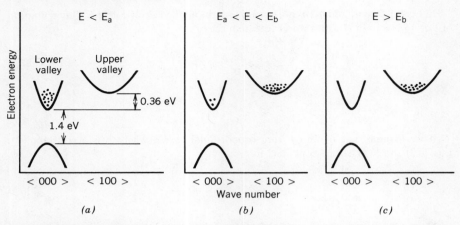

Figure 6.2   Electron distributions for a two-valley semiconductor under different electric fields.

Because of its advanced technology, n-type GaAs is the most frequently used negative differential mobility material. Other compound semiconductors, such as InP, CdTe, ZnSe, GaAsP, and InAs, also meet the requirements above.

A simple graphical representation for the electron concentrations of a two-valley semiconductor as a function of electric field is illustrated in Fig. 6.2.

$$\text{For Fig. 6.2}a: \quad n_1 \simeq n \text{ and } n_2 \simeq 0 \quad \text{for } 0 < E < E_a \tag{6.8}$$

$$\text{For Fig. 6.2}b: \quad n_1 + n_2 \simeq n \qquad \text{for } E_a < E < E_b \tag{6.9}$$

$$\text{For Fig. 6.2}c: \quad n_1 \simeq 0 \text{ and } n_2 \simeq n \quad \text{for } E > E_b \tag{6.10}$$

and the current density takes on the asymptotic values:

$$J = \begin{cases} qn\mu_1 E & \text{for } 0 < E < E_a \tag{6.11} \\ qn\mu_2 E & \text{for } E > E_b \tag{6.12} \end{cases}$$

If $\mu_1 E_a$ is greater than $\mu_2 E_b$, a region of negative differential mobility will exist between E_a and E_b, as shown in Fig. 6.3. J_T corresponds to the threshold current density and

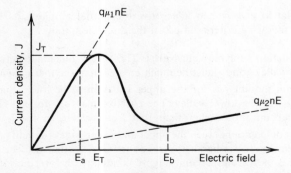

Figure 6.3   Current density versus electric field of a two-valley semiconductor.

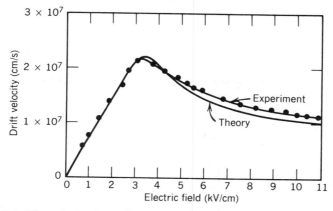

Figure 6.4 Theoretical and experimental velocity-electric field characteristic of GaAs.

E_T is the threshold field at the onset of the negative differential conductivity. E_T is 3200 V/cm for GaAs and 10,500 V/cm for InP.

Experimental measurements of the drift velocity versus electric field for GaAs were made by Ruch and Kino [9] to show the existence of the negative differential mobility, as shown in Fig. 6.4. Butcher and Fawcett [10] made a numerical calculation which showed excellent correlation with the measurements, as indicated in Fig. 6.4. McCumber and Chynoweth [11] determined that the magnitude of negative differential mobility decreases with increasing temperature and vanishes around 800 K.

6.3 MODES OF OPERATION FOR OSCILLATORS

Since the discovery of the transferred electron effects, many modes of oscillation dependent on material and circuit parameters have been unveiled. A summary of these oscillating modes is outlined in this section.

6.3.1 Gunn Mode or Dipole Mode

The Gunn mode is characterized by the formation of a domain or dipole first observed by Gunn [1]. In a low-impedance circuit, the oscillation frequency was related to v_s/L, where v_s is the domain velocity and L the device length.

From the current continuity equation and Poisson's equation,

$$\nabla \cdot \mathbf{J} = -\frac{\partial \rho}{\partial t} \tag{6.13}$$

where

$$\mathbf{J} = \sigma_0 \mathbf{E} \tag{6.14}$$

$$\nabla \cdot \mathbf{E} = \frac{\rho}{\varepsilon} \tag{6.15}$$

By combining Eqs. (6.13)–(6.15), one can solve the macroscopic space charge density ρ,

$$\rho = \rho_0 e^{-t/\tau_d} \tag{6.16}$$

where $\tau_d = \epsilon/\sigma_0 = \epsilon/qn_0\mu_n =$ dielectric relaxation time

$\epsilon =$ semiconductor dielectric permittivity

$n_0 =$ doping concentration

$\mu_n =$ electron mobility

$q =$ electron charge

$\sigma_0 =$ conductivity of the sample

For GaAs semiconductor material with $\sigma_0 = 1$ $(\Omega\text{-cm})^{-1}$ and $\epsilon = 13.1\epsilon_0$, ϵ_0 is the permittivity of air, τ_d is approximately 10^{-12} s, and macroscopic space-charge neutrality is quickly satisfied at every point in the material. Therefore, any deviations or fluctuations in carrier density will quickly disappear.

For a material with negative differential mobility, on the other hand, τ_d is negative and space-charge fluctuations build up in time rather than being neutralized. Figure 6.5

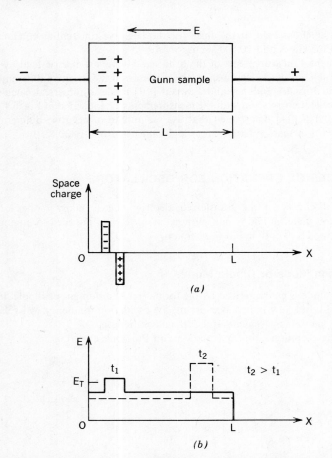

Figure 6.5 (a) Charge dipole at a local nonuniformity; (b) electric field buildup due to charge dipole in time t_1 and t_2 with the device biased beyond E_T.

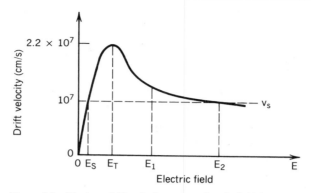

Figure 6.6 Electron drift velocity versus electric field for GaAs.

shows an example of an instability building up in the semiconductor sample. If a small nonuniformity or fluctuation in the electron density occurs at the cathode of the device, a charge dipole will form locally as shown in Fig. 6.5a. If the Gunn device is biased in the negative differential mobility region, the localized electric field will build up as shown in Fig. 6.5b. This dipole or domain will grow and drift to the anode. When the domain reaches the anode, a current pulse will appear in the external circuit. Another domain now nucleates at the cathode and the process is repeated. Therefore, the pulse frequency is inversely proportional to the drift length L.

After the initial growth of the domain, a stable condition is eventually realized. At this condition the electrons drift at a constant velocity v_d everywhere in the sample, and the domain travels down the sample without further growth.

The stable condition may be explained with the velocity versus field diagram for n-type GaAs as shown in Fig. 6.6. Assume that the sample is biased at E_1. Since the field in the dipole region is higher than E_1, the electrons will move slower than those in the neutral regions. Hence the electron density will vary, causing accumulation of electrons on the left side and depletion on the right side of the dipole layer, causing the field in the dipole layer to rise until it reaches E_2. Since the total voltage applied to the sample remains the same, an increase of the electric field in the dipole layer will result in a decrease of the electric field outside the domain. A stable condition is finally reached when the field outside the domain drops to E_s, with the same velocity v_s as that for the domain shown in Fig. 6.6.

For the Gunn mode of operation, the space-charge domain must grow to its stable condition before it reaches the ohmic contact at the anode of the device. This implies that the dielectric relaxation time τ_d must be less than the transit time through the sample length L,

$$\frac{\epsilon}{\sigma_0} = \frac{\epsilon}{q|\mu_n|n_0} < \frac{L}{v_s} \tag{6.17}$$

or

$$n_0 L > \frac{\epsilon v_s}{q|\mu_n|}$$

For n-type GaAs, $v_s \simeq 10^7$ cm/s, $\epsilon = 13.1\epsilon_0$, and $\mu_n \simeq -100$ cm^2/V-s; $n_0 L$ product must be approximately $> 10^{12}$ cm^{-2}.

The frequency of oscillation f is related to the transit time T_t of carriers through the device:

$$f = \frac{1}{T_t} = \frac{v_s}{L} \tag{6.18}$$

or

$$fL = 10^7 \text{ cm/s}$$

The Gunn mode of operation is simple to operate since only a dc bias is required. However, the efficiency is low, only a few percent. The operation of a Gunn mode is indicated in Fig. 6.7b, where E_T is the threshold electric field, E_s the sustaining electric field, and T the period of the oscillation [12].

6.3.2   Resonant Gunn Mode

For the Gunn or dipole mode discussed in Section 6.3.1, the applied electric field is above the threshold value and is constant because the RF circuit used has a low impedance. For the resonant Gunn mode, the device is operated in a resonant circuit whose resonant frequency is selected to be approximately equal to the transit-time frequency. If the field never drops below the sustaining field E_s, a domain will be formed and can propagate to the anode during each cycle, as shown in Fig. 6.7c. For this mode, $n_0L \simeq 10^{12} \text{ cm}^{-2}$ and $fL = 10^7 \text{ cm/s}$. The efficiency is still low, typically less than 10%.

6.3.3   Inhibited Mode or Delayed Mode

When the resonant circuit impedance is increased so that the electric field drops below the threshold value but remains above the sustaining electric field, a new domain may be delayed until the field rises above the threshold E_T as indicated in Fig. 6.7d. For this mode, $10^6 < fL < 10^7 \text{ cm/s}$, and $T > T_t$. Since the time between output current pulses has been increased, the efficiency is improved over the Gunn and resonant Gunn modes. The oscillating frequency can also be slightly altered by the resonant circuit. Theoretical calculations have shown that the efficiency of the delayed-domain in mode can reach 27% [13].

6.3.4   Quenched Domain Mode

If the bias field drops below the sustaining field E_s for the negative part of the cycle, the domain may be quenched before it reaches the anode, as shown in Fig. 6.7e. In this quenched domain mode, another domain is nucleated when the bias field goes above E_T and the process repeats again. Consequently, the frequency of operation is raised since the transit time of electrons has been effectively shortened. The oscillation frequency of the resonant circuit can be several times the transit-time frequency. A theoretical efficiency of 13% for the quenched domain mode has been calculated [13].

6.3.5   LSA Mode

The limited space-charge accumulation (LSA) mode [14] operates in a sample where only a small accumulation layer is formed before it is quenched by the RF voltage as shown in Fig. 6.7f. For the LSA mode, the frequency is so high that domains do not

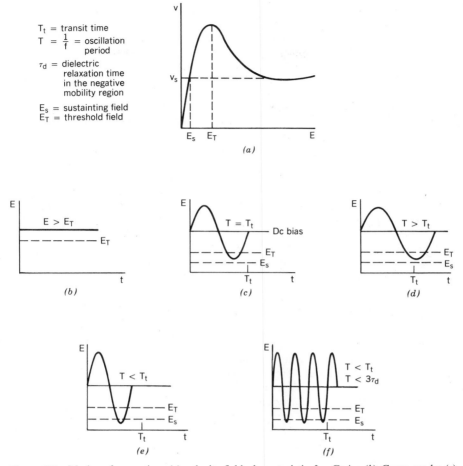

Figure 6.7 Modes of operation: (*a*) velocity-field characteristic for GaAs; (*b*) Gunn mode; (*c*) resonant Gunn Mode; (*d*) inhibited or delayed mode; (*e*) quenched domain mode; (*f*) LSA mode.

have enough time to form, and the entire length of the sample is subjected to an electric field greater than the threshold field E_T. Therefore, a large part of the sample length is maintained in the negative conductance state during most of the RF cycle.

The requirements for the LSA mode are that the period of the oscillating frequency be no more than a few times the dielectric relaxation time τ_d in the negative conductance region, but must be larger than the τ_{d_+} for the low-field positive conductance region to dissipate any accumulation of electrons while the signal is below the threshold field E_T. For $\mu_n = -100 \text{ cm}^2/\text{V-s}$, $\tau_d = \epsilon/q n_0 |\mu_n|$, $T = \tau_d$, $n_0/f = 7 \times 10^4 \text{ s cm}^{-3}$ is calculated. For $\mu_n = 8000 \text{ cm}^2/\text{V-s}$ in the positive conductance region, $\tau_{d_+} = \epsilon/q n_0 \mu_n$,

$T = \tau_{d^+}$, and $n_0/f = 10^3$ is calculated. Selecting the low-frequency limit such that $T \simeq 3\tau_d (n_0/f = 2 \times 10^5)$ and the high-frequency limit so that $T = 20\tau_{d^+} (n_0/f = 2 \times 10^4)$ will provide the requirement for the LSA mode,

$$2 \times 10^4 < \frac{n_0}{f} < 2 \times 10^5 \text{ cm}^{-3} \text{ s}$$

These limits have been experimentally verified by Copeland [15]. Maximum efficiencies in the range of 18 to 23% [14, 15] have been calculated for this LSA mode.

A distinct advantage for the operation of the LSA mode over other modes is its high efficiency and high output power capability. Since the entire LSA sample is biased in the negative conductance region rather than a narrow region, as in the domain modes, a larger percentage of the sample is useful for generating microwave power, resulting in an improvement of efficiency. The operating voltage for the LSA mode can be applied considerably higher before the impact ionization occurs because the peak field in the LSA mode is less than that in the domain modes. Consequently, the input power to the device and the output power are higher.

The major problem with the LSA mode is its sensitivity to circuit tuning and doping fluctuations [16]. A dipole domain might form to create a high dielectric field to damage the device.

6.3.6 Hybrid Mode

The hybrid mode [17, 18] operates between the LSA and domain modes. This mode is characterized by a space-charge growth that occurs during each cycle of oscillation but does not reach the stable domain condition of the domain modes. During the time in which the applied voltage is above threshold for the hybrid mode, the presence of the domain causes the current in the external circuit to be less than that of the LSA mode but greater than that of domain modes. As a result, the efficiency of the hybrid mode is also a compromise between them. The hybrid mode operation is easier to achieve than the LSA mode because of its less stringent n_0/f conditions [17] and doping homo-

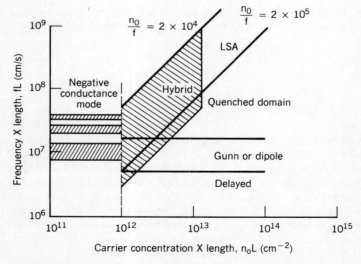

Figure 6.8 Mode chart for transferred electron devices. (From Ref. 15.)

geneity requirements. Furthermore, the hybrid mode is less sensitive to the circuit conditions.

To summarize all the modes for transferred electron oscilllators, Copeland [15] has shown a mode chart as depicted in Fig. 6.8. The ordinate and abscissa of the chart are the fL and n_0L products of the device, respectively. The locus of a constant n_0/f ratio is a straight line at 45° with the axes.

6.4 MODES OF OPERATION FOR AMPLIFIERS

Transferred electron devices can be used as negative resistance reflection amplifiers in two basic types. One type is to use an oscillator with the presence of a traveling domain as a negative resistance element [19]. This arrangement has not been very useful since spurious signals due to the domain oscillation are not desirable for many system applications. The second type is to stabilize a transferred electron device so that oscillations are suppressed, and the negative resistance is achieved by biasing the device in the negative differential mobility region. This type of amplifier is of more interest because of its simplicity.

Four methods of stabilizing the amplifiers have been suggested:

1. *Subcritical n_0L.* Section 6.3.1 has discussed that the stability condition for GaAs is given by

$$n_0L < 10^{12} \text{ cm}^{-2} \tag{6.19}$$

   Devices that satisfy inequality in Eq. (6.19) are called subcritical, while devices with $n_0L > 10^{12}$ cm^{-2} are called supercritical.
2. *Circuit Stabilization.* McCumber and Chynoweth [11] have shown that all transferred electron devices are unconditionally stable under constant current conditions. Under constant voltage conditions, devices can be made stable with a series positive resistor [20].
3. *Diffusion Stabilization.* Jeppesen and Jeppsson [21] realized the importance of diffusion effects in stabilizing the transferred electron devices. Diffusion tends to balance the space-charge growth; thus the device can be stabilized.
4. *Injection-Limiting Cathode.* Thim [22] reported that a uniform electric field from cathode to anode inside a transferred electron device is the optimum field configuration. A stable and optimum operation of devices can be achieved by introducing a doping notch near the cathode contact. The width and depth of the notch can be made to provide a uniform electric field distribution [23]. Other methods of injection limiting, such as using a *p-n* junction at the cathode or a low-height Schottky barrier, have also been suggested [24, 25].

Two types of stable amplification modes are available for Gunn devices: the subcritically doped mode and the supercritically doped mode. For the subcritically doped mode, the n_0L product of the GaAs device is less than about 10^{12} cm^{-2}, and the dipole domain for the device will not be formed, as indicated in the mode chart of Fig. 6.8. However, a negative resistance still exists in the device at microwave frequencies, and the device can be used as an amplifier in the vicinity of the transit-time frequency and its harmonics without oscillation [26]. If the device is incorporated in a circuit with adequate positive feedback, the circuit can be used as an oscillator. Hakki [27] has demonstrated that a Gunn device can amplify at microwave frequencies or it can be used simultaneously as an amplifier and local oscillator. Because of the low value of

$n_0 L$, the output power and efficiency are quite low compared with the supercritically doped mode of amplification.

For the supercritically doped mode of amplification, the value of $n_0 L$ of the device must be greater than 10^{12} cm^{-2}. With this $n_0 L$ product the dipole domain is formed. The device oscillates at the transit-time frequency, whereas it can be amplified at some other frequencies. Walsh et al. [28] have demonstrated this mode of amplification. Broadband amplification performances were achieved for 4.5 to 8.0 GHz, 8 to 12 GHz, and 12 to 15 GHz [29], respectively. Linear gains of 6 to 12 dB and saturated power up to 500 mW have been obtained. All the performances were realized by providing a broadband circuit for the amplification frequencies and a suitable stabilizing circuit to prevent the Gunn oscillation frequencies from entering the amplifying frequency band.

Although devices operated in the supercritically doped amplifier mode provided much better power and efficiency performance than the devices operated in the subcritically doped mode, typical efficiencies still range from 2 to 3%. These are significantly lower than the efficiencies that can be achieved by FET amplifiers now available for the same frequency ranges. The main applications of Gunn amplifiers appear to lie in the high-millimeter-wave frequency bands, where FET devices are not mature or available.

6.5 EQUAL-AREA RULE

As discussed in previous sections, the stable domain within the device travels at a constant velocity. On the left and right sides of the domain, the electric field is below threshold. Inside the domain, the field rises quickly through the accumulation layer to a peak value in excess of E_T and decreases through the depletion layer. Somewhere inside the domain the field will be maximum at $E = E_d$, and at this point the carrier concentration will equal N_d, the ionized donor concentration. An exact analysis of the domain physics requires a nonlinear solution of Poisson's equation and current continuity equation [30, 31],

$$\frac{\partial E}{\partial x} = \frac{q}{\epsilon}(n - N_d) \tag{6.20}$$

where n is the intantaneous carrier concentration and N_d is the ionized donor concentration.

$$J = qnv(E) - qD\frac{dn}{dx} + \epsilon\frac{\partial E}{\partial t} \tag{6.21}$$

where J is total current density from a Gunn device and is the sum of the drift, diffusion, and displacement current densities, $v(E)$ is the field-dependent carrier velocity, and D is the diffusion constant.

While the current versus voltage behavior of the domain physics for the Gunn device has been solved by computer analysis, Butcher [32, 33] has analyzed this problem as the equal-area rules to be described below.

In addition to the Poisson equation and the current continuity equation given in Eqs. (6.20) and (6.21), respectively, the boundary conditions satisfied by a stable domain propagation are

$$E = E_0 = \text{constant at } X = \pm\infty \tag{6.22}$$

As a result,

$$n = N_d \qquad \frac{\partial n}{\partial x} = 0 \qquad \frac{\partial E}{\partial t} = 0 \quad \text{at } X = \pm \infty \tag{6.23}$$

and Eq. (6.21) reduces to

$$J = q N_d v_0 \tag{6.24}$$

where

$$v_0 = v(E_0) \tag{6.25}$$

Inside, the domain must have the following form of solution:

$$E = E(x - v_d t) \tag{6.26}$$

$$n = n(x - v_d t) \tag{6.27}$$

where v_d is the drifting velocity of the domain. From Eq. (6.26) one obtains

$$\frac{\partial E}{\partial t} = -v_d \frac{\partial E}{\partial x} \tag{6.28}$$

Also, from Eq. (6.27) we have

$$\frac{\partial n}{\partial x} = \frac{\partial n}{\partial E} \frac{\partial E}{\partial x} \tag{6.29}$$

Since the total current density in the device must be constant, this implies that Eq. (6.21) can be written as

$$q N_d v_0 = q n v(E) - q D \frac{\partial n}{\partial x} + \epsilon \frac{\partial E}{\partial t} \tag{6.30}$$

Substituting Eqs. (6.28) and (6.29) into (6.30) yields

$$q(N_d v_0 - n v(E)) = -\left(q D \frac{dn}{dE} - \epsilon v_d \right) \frac{\partial E}{\partial x} \tag{6.31}$$

Substituting $\partial E / \partial x$ from Eq. (6.20) and dividing Eq. (6.31) by $n N_d$ and integrating both sides from infinity to any point in the domain, we have the following result:

$$\int_{N_d}^{n} \left(\frac{1}{N_d} - \frac{1}{n} \right) dn = \frac{\epsilon}{q N_d D} \int_{E_0}^{E} \left\{ [v(E) - v_d] - \frac{N_d}{n} (v_0 - v_d) \right\} dE \tag{6.32}$$

When the diffusion constant is assumed to be independent of electric field, Eq. (6.32) can be expressed as

$$\frac{n}{N_d} - 1 - \ln \frac{n}{N_d} = \frac{\epsilon}{q N_d D} \int_{E_0}^{E} \left\{ [v(E) - v_d] - \frac{N_d}{n} (v_0 - v_d) \right\} dE \tag{6.33}$$

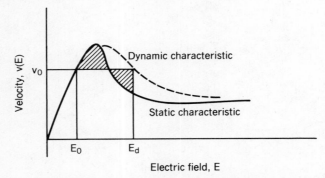

Figure 6.9 Equal-areas rule for transferred electron devices.

In Eq. (6.33), if $E = E_d$, then $n = N_d$ and the left side of the equation must vanish. As seen in Fig. 6.9, the integration of Eq. (6.33) can be made from E_0 to E_d through the accumulation layer, or through the depletion layer, where $n < N_d$. Both results must zero according to Eq. (6.33). Since the first term in the integral is independent of n, one must have

$$v_0 = v_d \tag{6.34}$$

and

$$\int_{E_0}^{E_d} [v(E) - v_d] \, dE = 0 \tag{6.35}$$

From Eq. (6.35), one can obtain the equal-area rule that the shaded areas in the two regions of Fig. 6.9 must be equal. Two important conclusions can be derived from the equal-area rule:

1. If the diffusion constant D of the electrons is not a function of electric field, the domain travels at the same velocity v_0 as the electrons in the uniform region outside the domain.

2. A dynamic characteristic curve, as shown by the dashed line in Fig. 6.9, relates v_0 to the peak field E_d. If the variation of the diffusion constant with electric field is included, most behaviors of the domain remain the same, except that the domain travels at a velocity which is slightly different from the velocity of the electrons drifting outside the domain [34].

The total voltage V applied to the sample can be written

$$V = V_x + E_0 L \tag{6.36}$$

where V_x is the domain potential defined as the voltage across the domain, and L is the length of the sample.

$$V_x = \int_0^L [E(x) - E_0] \, dx \tag{6.37}$$

Assuming a triangular-shaped field domain with the peak field E_d and the uniform field outside domain E_0, the width W of the domain calculated from Poisson's equation given by Eq. (6.20) is

$$W = \frac{\epsilon}{qN_d}(E_d - E_0) \tag{6.38}$$

The domain voltage V_x is approximately the area under the triangular shape:

$$V_x \simeq \frac{1}{2} W(E_d - E_0) = \frac{\epsilon}{2qN_d}(E_d - E_0)^2 \tag{6.39}$$

Consider a GaAs Gunn diode with a dipole domain operating with the following parameters:

$$L = 10 \ \mu m$$

$$\mu_n = 6000 \ \text{cm}^2/\text{V-s}$$

$$N_d = 2 \times 10^{15} \ \text{cm}^{-3}$$

$$\text{area} = 3 \times 10^{-4} \ \text{cm}^2$$

$$\text{instantaneous current } I = 1.152 \ \text{A}$$

The domain velocity v_0 can be calculated from

$$v_0 = \frac{I}{qAN_d} = 1.2 \times 10^7 \ \text{cm/s}$$

One can find E_0, corresponding to v_0 from a given velocity versus electric field characteristics for GaAs, $E_0 = 2 \times 10^3$ V/cm and $E_d = 10.5 \times 10^3$ V/cm from the equal-area rule. The domain voltage V_x calculated from Eq. (6.39) is

$$V_x = \frac{\epsilon}{2qN_d}(E_d - E_0)^2 = 0.13 \ \text{V}$$

The total voltage V applied to the device from Eq. (6.36) is found to be

$$V = E_0 L + V_x = (2 \times 10^3)(10 \times 10^{-4}) + 0.13 = 2.13 \ \text{V}$$

and the frequency of oscillation for the Gunn mode (transit time mode) is

$$f = \frac{v_0}{L} = \frac{1.2 \times 10^7}{10 \times 10^{-4}} = 12 \ \text{GHz}$$

6.6 EQUIVALENT-CIRCUIT MODELS

For a circuit designer, the equivalent-circuit models for Gunn devices are useful to serve as a starting point for the circuit design. Equivalent circuit models may be grouped for devices operating in domain modes or LSA modes.

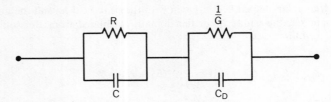

Figure 6.10 Equivalent circuit for Gunn devices by Hobson. (From Ref. 35; previously published by the Institution of Electrical Engineers in *Electronic Letters*, Vol. 3, June 1966, reproduced by permission.)

6.6.1 Domain Modes

Although many researchers working in this field have analyzed the domain modes of transferred electron devices with detailed computer calculations, the results are difficult to use because the device parameters employed in the computer simulations are not always applicable. Instead, simpler equivalent circuits with lumped elements are devised to represent the operation of the device. Hobson [35] suggested that the diode can be expressed as a series combination of a low-field and high-field equivalent circuit. As shown in Fig. 6.10. C_D and $1/G$ are the active components in the circuit. C_D is the domain capacitance and G is the negative conductance responsible for the power-generating part of the device. The components C and R are related to the dielectric relaxation time and low-field resistance of the sample, respectively. For a device with 1-Ω-cm GaAs with $N_d = 10^{15}$ cm^{-3}, an area of 100 μm square and 10 μm long, $C_D \simeq 0.30$ pF, $1/G \simeq -500 \, \Omega$, $R \simeq 10 \, \Omega$, and $C \simeq 0.10$ pF. Carroll and Giblin [36] used the analog model for the Gunn device operation. By incorporating ideal switches and transistors, the model can calculate voltage and current waveforms associated for different loading requirements. The operating modes of the Gunn device can be identified by comparing the waveforms calculated from the analog model with those occurring in the actual sample.

Robrock [37] devised a lumped-circuit model to describe the phenomena of domain nucleation, modulation, and quenching in GaAs samples of uniform doping. As shown in Fig. 6.11, a voltage-dependent current source I_D is used to represent the propagating high-field domain, which is characterized by its excess voltage V_X. Dynamic effects are included in the differential domain capacitance C_D. The remaining part of the bulk sample is represented by a parallel circuit formed by the ohmic resistance R and the low-field capacitance C.

Since $E_0 L$ is the voltage drop across the low-field ohmic resistance R of the bulk material of the device,

$$E_0 L = RI \tag{6.40}$$

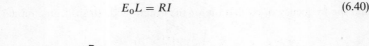
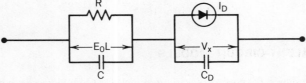

Figure 6.11 Equivalent circuit for Gunn devices by Robrock. (From Ref. 37; copyright 1970 IEEE, reproduced by permission.)

Under quasi-static conditions, the current I flowing through the bulk device of donor density N_d and area A is given by

$$I = qN_dAv(E_0) \tag{6.41}$$

where $v(E_0)$, the velocity of carriers outside the domain, is available from velocity versus electrical field characteristics of the device. Thus R can be solved from Eqs. (6.40) and (6.41).

$$R = \frac{EL}{qN_dAv(E_0)} \tag{6.42}$$

The low-field capacitance C can be expressed as

$$C = \frac{\epsilon A}{L} \tag{6.43}$$

To generate the quasi-static current versus voltage characteristics that will be used to determine the current source I_D in the equivalent circuit, the following steps are involved:

1. A curve of $v(E_0)$ versus E_0L is first produced using the velocity electric field characteristic for the sample used by a scaling factor L to provide E_0L.
2. The appropriate V versus E_0 curve will then be made by mapping the E_0 values to the corresponding velocity $v(E_0)$. This curve is also computer generated by Copeland [38] as shown in Fig. 6.12.
3. The total voltage V is obtained by Eq. (6.36) producing a plot of $v(E_0)$ versus V. The I_D–V characteristic is then obtained by using Eq. (6.41).

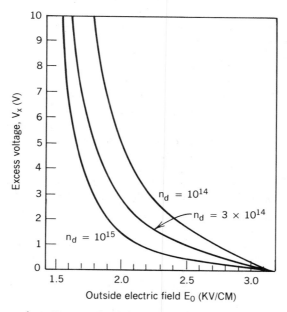

Figure 6.12 Excess voltage V_x versus electric field E_0 outside the domain by Copeland. (From Ref. 38; copyright 1966 IEEE, reproduced by permission.)

The differential capacitance C_D of a high-field domain shunting the current source in the equivalent circuit is defined as

$$C_D = \frac{dQ_D}{dV_x} \tag{6.44}$$

Again we assume that the effects of diffusion are neglected. The resulting electric field distribution associated with the dipole domain provides a triangular shape, with width W and domain voltage V_x given in Eqs. (6.38) and (6.39), respectively.

Combining Eqs. (6.38) and (6.39) yields

$$V_x = \frac{qN_d}{2\epsilon} W^2 \tag{6.45}$$

The charge Q_D contained within the depletion region of the domain is

$$Q_D = qN_d WA \tag{6.46}$$

C_D can then be obtained by using Eqs. (6.44) and (6.45)

$$C_D = \frac{\epsilon A}{W} \tag{6.47}$$

The equivalent circuits provided by Hobson and Robrock were compared with fair agreement [39]. A domain width of 2 μm used in Hobson's simulation corresponds to a V_x value of 2.76 V according to Eq. (6.45), where $\epsilon_r = 13.1$. With a doping of $N_d = 10^{15}$ cm^{-3} and $V_x = 2.76$ V, and E_0 value of 1.8 kV/cm is obtained from Copeland's curve shown in Fig. 6.12. For this value of E_0, $v(E_0)$ is 1.4×10^7 cm/s from Eq. (6.41). R is calculated to be 10.2 Ω using Eq. (6.42), C is 0.077 pF by Eq. (6.43), $I = 176$ mA by Eq. (6.41), and $C_D = 0.448$ pF by Eq. (6.47). These values agree approximately with Hobson's model, $C_D = 0.30$ pF, $R = 10 \Omega$, and $C \simeq 0.10$ pF.

6.6.2 LSA Mode

For the LSA modes, the key features that must be included in the equivalent circuit are the lack of domain formation and the long, uniform high-field region in the sample. These are two features that allow a high RF voltage to be developed, thereby producing a high power–impedance product. The equivalent circuit used to represent the LSA mode is given in Fig. 6.13 [(14,) (36) (40)], which is a negative resistance $(-R)$ in parallel with the low-field capacitance C. The negative resistance $(-R)$ calculated by Copeland [15] is dependent on RF electric field E_1 applied to the sample, the low-field mobility μ_0,

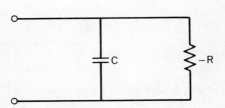

Figure 6.13   Equivalent circuit for LSA mode.

RF power output P_{RF} from the sample, and low-field positive resistance R_0,

$$\frac{R}{R_0} = \frac{qE_1^2\mu_0}{2P_{RF}} \tag{6.48}$$

For a fixed dc bias field of 10^4 V/cm, and for $E_1 = 2 \times 10^3$ to 9×10^3 V/cm, $R/R_0 = -12$ to -20 [15].

In designing an LSA oscillator, circuits must be designed to induce the device operating in the LSA mode and to prevent domain formation. A delay-line loading circuit [41] and a fast rise-time bias pulse [40] for pulsed LSA oscillators have proven successful.

6.7 MICROWAVE CIRCUITS

Three types of microwave circuits are commonly used to accomodate the transferred electron oscillators and amplifiers: waveguide, coaxial, and microstrip.

6.7.1 Waveguide Circuits

At high frequencies, from X-band to millimeter wave ranges, a common method of coupling the Gunn devices into a waveguide cavity is to mount the packaged diode under a post as shown in Fig. 6.14. The circuit provides high stability, low FM noise, and good mechanical tuning range by adjusting a movable short in the circuit. The

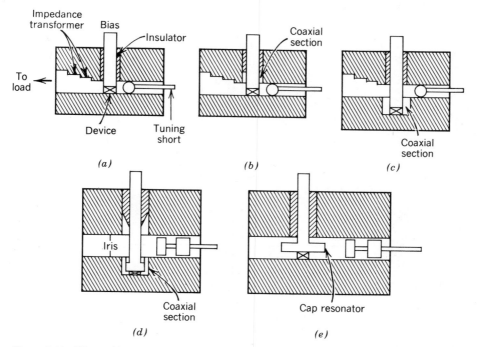

Figure 6.14 Waveguide circuits for transferred electron devices. (From Ref. 42, reproduced by permission.)

additional impedance matching for transferred electron devices may be achieved by using the reduced height waveguide to lower the impedance, by incorporating coaxial spacers around devices, and by providing matching impedance transformers at the output as shown in Fig. 6.14a, b, and c.

Another waveguide circuit suggested by Harkless [43] is shown in Fig. 6.14d, where a lossy absorber is inserted to eliminate all undesirable oscillating modes. Figure 6.14e is a full-height waveguide circuit with diode mounted under a radial cap on the center post. The frequency of the oscillator is controlled mainly by the dimensions of the cap. The circuit has very low losses and is frequently used in millimeter wave applications.

Although waveguide circuits for transferred electron devices have many advantages, they do have disadvantages over coaxial and microstrip circuits. They are bulky, costly to machine, and likely to experience mode jumping and spectral purity problems.

The diode mounting structure shown in Fig. 6.14a to d has been used extensively for solid-state devices such Gunn and IMPATT devices for high frequencies at X-band and beyond. Many theoretical analyses are available to predict the impedance presented by the circuit to the packaged diode terminals. Eisenhart et al. [44] suggested an equivalence between a coaxial entry and a gap in the post. Based on this equivalence and the equivalent-circuit model developed by Eisenhart and Khan [45, 46], the theoretical anal-

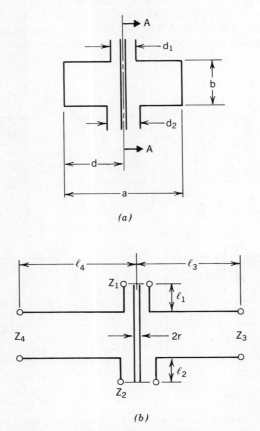

(a)

(b)

Figure 6.15 (a) Coaxial-waveguide diode mounting structure; (b) side view on AA plane. (From Ref. 50; copyright 1980 IEEE, reproduced by permission.)

ysis for the coaxial-waveguide junctions shown in Fig. 6.14a to d can be obtained. A dyadic Green's function is provided for a rectangular waveguide using the Lorentz reciprocity theorem, which determines the electric field and current density at two locations in the waveguide. This approach permits the calculation of the impedance seen by the device placed on a post located anywhere in the waveguide. To facilitate mathematical computations, Eisenhart and Khan [45] assumed that the post may be replaced by an infinitesimally thin strip of width W. They found that a post with diameter d can be replaced by a strip with $W = 1.8d$. Using this circuit approach along with the previously described model for a Gunn device, Eisenhart and Khan [47] have developed a Gunn oscillator where jumps in the oscillation frequency were predicted according to their theory.

Another approach to analyze the cross-coupled coaxial-waveguide diode mounting structure is given by Lewin [48, 49]. Based on his analysis, an equivalent circuit was developed for this mounting structure. Chang and Ebert [50] modified the equivalent circuit and verified experimentally with reasonably good agreement. The equivalent circuit was also used for the individual IMPATT module design for a W-band power combiner.

A general cross-coupled coaxial-waveguide mounting structure is shown in Fig. 6.15 and its equivalent circuit in Fig. 6.16. The coaxial line in the upper and lower sections has different diameters d_1 and d_2. $Z_1, Z_2, Z_3,$ and Z_4 are the load impedances at each port, respectively. Z_0 is the characteristic impedance of the waveguide, and Z_{01} and Z_{02} are those of the coaxial lines. Z_{0p} is an inductive component due to the post in waveguide

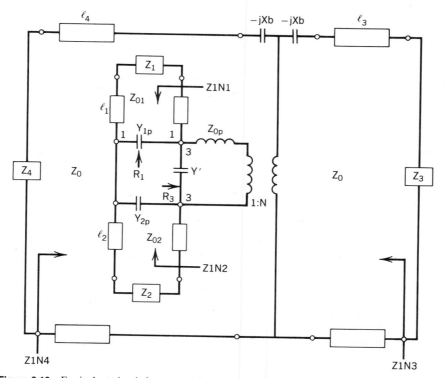

Figure 6.16 Equivalent circuit for a coaxial-waveguide diode mounting structure. (From Ref. 50; copyright 1980 IEEE, reproduced by permission.)

excited by TE_{n0} modes. Y', Y_{1p}, and Y_{2p} are the admittances due to the effects of wave-guide–coaxial junctions. The circuit elements given in the equivalent diagram shown in Fig. 6.16 are repeated here for clarity.

$$Y' = j \sum_{m=1}^{\infty} \frac{\cos m\pi}{X_m} e^{-m\pi r/b}$$

$$Y_{1p} = j \sum_{m=1}^{\infty} \frac{1}{X_m} e^{-m\pi r/b} - j \sum_{m=1}^{\infty} \frac{\cos m\pi}{X_m} e^{-m\pi r/b} = Y_{2p}$$

$$Z_{0p} = j \frac{Z_0}{4} \left[k_0^2 - \left(\frac{\pi}{a}\right)^2 \right]^{1/2} \sum_{n=1}^{\infty} \frac{\cos(n\pi r/a) - \cos[n\pi(2d \pm r)/a]}{(n^2\pi^2/a^2 - k_0^2)^{1/2}}$$

$$X_m = \frac{b^2}{4} \frac{\eta}{k_0 ab} \left(\frac{m^2\pi^2}{b^2} - k_0^2\right) \left(\frac{a}{\pi}\right) \left([K_0(r\Gamma_m) - K_0(2d\Gamma_m)] \right)$$

$$X_b = Z_0 \frac{a}{\lambda g} \left(\frac{2\pi r}{a}\right)^2 \sin^2 \frac{\pi d}{a}$$

$$N = \sqrt{\csc \frac{\pi d}{a} \csc \frac{\pi(d \pm r)}{a}}$$

where

$$\Gamma_m = \left(\frac{m^2\pi^2}{b^2} - k_0^2\right)^{1/2}$$

$$Z_0 = 2\eta \frac{b}{a} \frac{\lambda g}{\lambda}$$

and

$$\lambda_g = \frac{2\pi}{\sqrt{k_0^2 - (\pi/a)^2}}$$

$$\eta = 120\pi$$

$$k_0 = 2\pi/\lambda$$

ZIN1 is the input impedance locking into the circuit at the coaxial end with the other three ports terminated by Z_2, Z_3, and Z_4. ZIN2, ZIN3, and ZIN4 are defined the same way, respectively. R_1 and R_3 are the real parts of the impedance locking into ports 1–1 and 3–3, as shown in Fig. 6.16. If a Gunn device is placed at Z_2 to produce microwave power, the power generated will be split into two parts. One goes to the waveguide load and the other reaches the absorber located at Z_1, as suggested by Harkless, to suppress undesirable oscillation modes. The ratio R of the power, which provides the coupling between the coaxial line section and waveguide cavity, is then equal to R_3/R_1. The objective is to maximize R by using optimum circuit parameters, while the circuit is stabilized to prevent all spurious oscillations at the same time.

The accuracy of this equivalent circuit model has been checked with X-band experimental results made by Eisenhart and Khan [45]. Figures 6.17 and 6.18 show the comparison between them. The discrepancy between the experimental results and theoretical analysis for the equivalent circuit is due primarily to allowing only the TE_{10} mode for propagation in the rectangular waveguide. By including three modes [51] to propagate

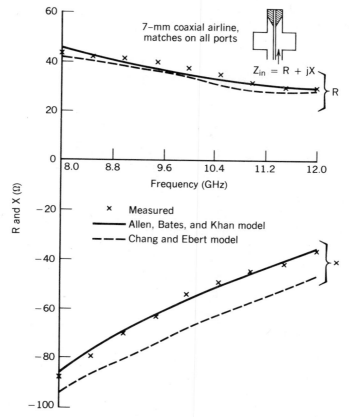

Figure 6.17 Comparison of calculated and measured input impedance for a coaxial-waveguide diode mounting structure.

in the X-band waveguide, the discrepancy was much improved, as also indicated in Figs. 6.17 and 6.18. Bialkowski [52] also modeled the coaxial–waveguide mounting structure using a radial–medal wave analysis and showed very good agreement between theory and measurements. The equivalent circuit for the waveguide-mounted oscillator may be used to predict the tuning characteristics of the oscillator.

Kurokawa [53] suggested that the active device can be represented by

$$Z_D = -R_D + jX_D \tag{6.49}$$

where R_D and X_D are the real and imaginary parts of the device impedance, which is a function of amplitude of oscillation, frequency, and bias voltage.

The total circuit impedance seen by the device is

$$Z_L = R_L + jX_L \tag{6.50}$$

The criteria for free-running oscillation are

$$|-R_D| \geqslant R_L \tag{6.51}$$

$$X_D + X_L = 0 \tag{6.52}$$

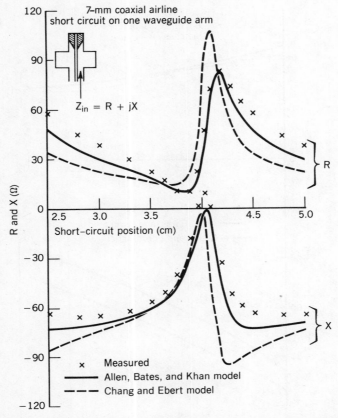

Figure 6.18 Comparison of calculated and measured input impedance at 10 GHz.

If the circuit impedance Z_L is known at various frequencies, theoretical tuning curves can be obtained along with an equivalent circuit for the device impedance. This approach proves to be adequate in explaining the general behavior of the tuning characteristics of the Gunn oscillator, such as mode jumping and frequency saturation.

6.7.2 Microstrip Circuit

Microstrip circuits for transferred electron devices are usually designed in a TEM mode and have more copper, dielectric, and radiation losses than the waveguide and coaxial-line counterparts. As a result, these losses result in low-Q (quality) factors. Therefore, microstrip circuits are not suitable for stable and low-noise oscillators. On the other hand, microstrip circuits are quite useful for broadband applications because of their low-Q property. In addition, the microstrip circuits have the advantages of small size, light weight, and good mechancial stability over other circuits.

6.7.3 Coaxial-Line Circuits

Coaxial cavities made from coaxial-line circuits are well suited to match the circuit impedance with the device impedance. The dimensions of the coaxial cavity must be carefully designed so that the waveguide modes will not be excited in the operating frequency ranges for the transferred electron devices. The length of the coaxial line required to match the device impedance is usually inductive. The dc bias needed by the Gunn device is provided by a low-pass filter in the bias line. The output is coupled either with a loop or a capacitive probe and can be adjusted for the optimum condition.

Although coaxial-line circuits also operate at a TEM mode, they have several advantages over microstrip circuits. Q-factors are generally higher because of the lower copper, dielectric, and radiation losses associated with the coaxial cavities.

The electrical characteristics of the circuits are relatively straightforward because the propagating TEM mode is well defined compared with the waveguide and microstrip circuits. A disadvantage over microstrip is that physically bulky cavities must be constructed around the Gunn device.

In addition to the foregoing three main types of microwave circuitry, many other circuits have been suggested. Suspended stripline circuits are especially useful for high-frequency transferred electron devices since they have lower losses and most of the advantages of microstrip circuits. Lumped-element circuits were tried on microstrip substrates to provide wider tuning ranges to eliminate tuning limitations incurred by distributed components. Coupled TEM line circuitry [54] is also used to facilitate mechanical mounting of packaged Gunn devices and the circuit also provides good impedance matching for the device.

6.7.4 Electronically Tuned Transferred Electron Oscillators

Electronic tuning of transferred electron oscillators can be accomplished by using a magnetically tuned ferrite sphere or a voltage-tuned varactor diode. The ferrite sphere can be fabricated from yttrium iron garnet (YIG) or barium ferrite (BaFe).

Magnetic Tuning. Figure 6.19 shows a typical YIG-tuned transferred electron oscillator circuit [55], which illustrates the arrangement for coupling the diode to the sphere. Broadband tuning of this CW oscillator was achieved by magnetically coupling the RF current of the oscillating diode to the YIG sphere so that the frequency is controlled by the ferromagnetic resonance of the sphere. As shown in Fig. 6.20, good frequency linearity for the YIG-tuned oscillator was obtained. Linearity and hysteresis of the oscillator are determined by the magnetic components and driving circuits which usually limit the sweep rate to 1 MHz μs^{-1}. As a result, YIG-tuned oscillators are not adequate for system applications requiring high modulation frequencies and rapid frequency

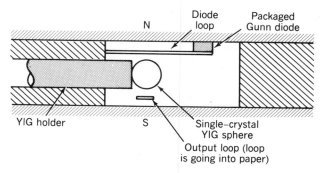

Figure 6.19 YIG-tuned Gunn oscillator. (From Ref. 55; copyright 1969 IEEE, reproduced by permission.)

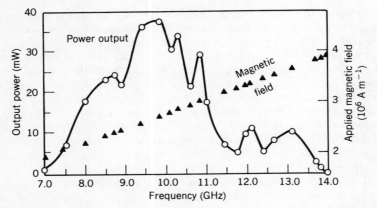

Figure 6.20   Performance of a YIG-tuned Gunn oscillator. (From Ref. 55; copyright 1969 IEEE, reproduced by permission.)

changes. The resonant frequency is temperature sensitive. However, if the YIG sphere is oriented with a correct crystallographic direction parallel to the magnetic field, its resonance is independent of temperature [55]. With proper compensation of Gunn devices, a temperature stability of ± 0.1 MHz degree^{-1} may be achieved.

Magnetically tuned barium ferrite (BaFe) spheres have also been used for wideband tuning of transferred electron oscillators [56] in millimeter wave frequencies. Compared with YIG, BaFe has a higher built-in anisotropy field and lower quality factor Q. Consequently, a BaFe tuned Gunn oscillator requires lower biasing current for magnetic coils to tune the sphere and provides slightly lower RF power output.

Varactor Tuning. Varactor tuning provides a faster response than YIG-tuned Gunn oscillators. Tuning rates up to 1 GHz μs^{-1} have been obtained routinely. However, its disadvantages is the nonlinear tuning characteristics associated with the varactor capac-

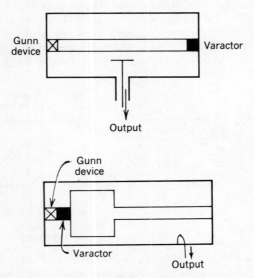

Figure 6.21   Varactor-tuned Gunn oscillator coaxial cavity circuits.

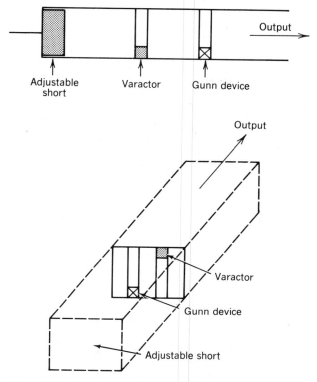

Figure 6.22   Varactor-tuned Gunn oscillator waveguide circuits.

itance dependence on the bias voltage. Linearizing circuits must be incorporated for system applications where tuning linearity and tuning rates are both of importance.

Cawsey [57] analyzed the tuning range of a single-tuned circuit and showed that tuning range, power output, and oscillator stability are related. A change in any one of these parameters affects the others in a given oscillator circuit. He also showed that the tuning range is inversely proportional to the circuit Q-factor. To reduce the susceptance slope or to improve tuning bandwidth, a parallel-tuned oscillator can be introduced with a series-tuned compensating circuit and a series-tuned oscillator can be introduced with a parallel-tuned compensating circuit. Aithchison and Celsthorpe [58] used this technique to increase the tuning range of a varactor-tuned Gunn oscillator.

Varactor-tuned Gunn oscillators have been fabricated in coaxial line [59], waveguide [60], microstrip [61], and lumped [62] circuits. Figure 6.21 shows some typical varactor-tuned coaxial circuits. A tuning range as high as 30% has been realized. Compared with coaxial circuits, waveguides circuits offer a potentially higher Q-factor but with lower tuning range. Some varactor-tuned oscillator circuits in waveguide are shown in Fig. 6.22.

6.8 POWER COMBINING

The output power from a single transferred electron device is limited by fundamental impedance and thermal problems. It is necessary to combine several diodes to achieve high power levels to meet certain system requirements. Power combining techniques for microwave and millimeter wave frequencies have been reviewed by Russell [63] and Chang and Sun [64], respectively. Although combining many IMPATT devices provides higher-power and higher-efficiency transmitters, Gunn device power combiners produce lower-noise signal sources for many system applications. Some power combining circuits for transferred electron devices are summarized below.

6.8.1 Resonant Cavity Combiners

A resonant cavity combiner with a 12-IMPATT diode combiner operating at X-band was first suggested and demonstrated by Kurokawa and Magalhaes [65]. The circuit consisted of a rectangular-waveguide cavity with cross-coupled coaxial waveguide diode mounting modules in the waveguide walls. Later, Harp and Stover [66] modified the combiner configuration by replacing the rectangular resonant waveguide cavity with a cylindrical resonant cavity for increased packaging density to accommodate a large number of diodes in a small volume.

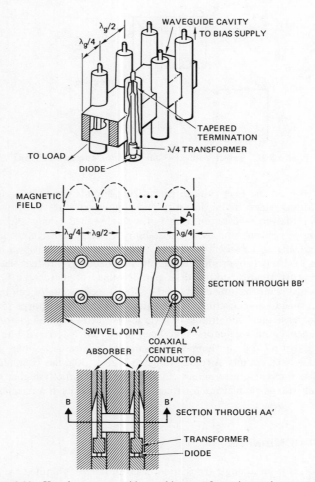

Figure 6.23 Kurokawa waveguide combiner configuration and cross sections.

The rectangular waveguide resonator described by Kurokawa and Magalhaes [65] is shown in Fig. 6.23. Each diode is mounted at one end of a coaxial line that is coupled to the magnetic field at the sidewall of a waveguide cavity. The other end of the coaxial line is terminated by a tapered absorber, which is used to prevent spurious oscillations. To be coupled properly to the waveguide cavity, the coaxial circuits must be located at the magnetic field maxima of the cavity and the diode pairs must be spaced $\lambda_g/2$ apart along the waveguide.

The key building block is the cross-coupled coaxial-waveguide diode module mounting structure. The design procedures of Chang and Ebert [50], Allen, et al. [51], and Bialkowski [52] were outlined in Section 6.7. With eight Gunn-diode modules combined in a rectangular cavity, an output power of 1 W has been achieved at 45 GHz with a combining efficiency of 91% [67]. At 33 GHz, over 500 mW has been obtained over 3-GHz bandwidth by combining four diodes with individual output powers of 120 to 150 mW [68]. The diodes were mounted in the corners of a rectangular cavity operating in the TE_{101} mode. The cavity can be tuned by a ceramic rod that enters the cavity along one of the sidewalls.

6.8.2 Parallel-Device Combiners

Direct parallel combining of Gunn devices [69] can be used to obtain a higher-output power level. Figure 6.24 shows the circuit for the parallel-device combiner. Impedance matching to the diodes is achieved by using a reduced-height waveguide, a two-stage

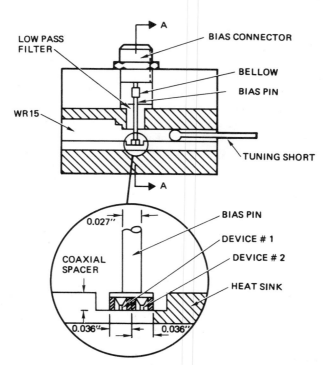

Figure 6.24 Parallel-device Gunn combiner circuit. (From Ref. 69; copyright 1979 IEEE, reproduced by permission.)

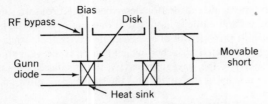

Figure 6.25 Cap resonator Gunn combiner circuit. (From Ref. 70; copyright 1979 IEEE, reproduced by permission.)

quarter-wave transformer at the output circuit, a tuning short, and a coaxial spacer. When a single device was used in the circuit, the maximum power obtained was 20.6 dBm. An output power of 23.75 dBm was achieved with two devices in parallel, which is approximately a 3-dB improvement over that of a single device.

6.8.3   Cap Resonator Circuit Combiners

Cap resonators are useful for combining Gunn diodes. Figure 6.25 shows the combining configuration in which two resonant-cap structures are mounted in a common waveguide with a Gunn diode mounted under each cap [70]. Approximately 80-mW power was obtained at 73 GHz using two 50-mW Gunn diodes. This technique has been extended to 90 GHz by combining four InP diodes with 260-mW output power and 93% combining efficiency [71].

6.8.4   Push-Pull Combiners

Push-pull combining techniques for Gunn devices have been reported at both low and high frequencies [54, 72]. Because the impedance of the devices operated in the push-pull operation is doubled, matching of the devices is easier, resulting in a power combining with good performance. Figure 6.26 shows a push-pull combiner consisting of a rectangular cavity with two Gunn devices mounted near the walls [72]. At the load-coupling point, the magnetic fields add in-phase. The cavity was designed for 42 GHz. A power output of 260 mW with greater than 90% combining efficiency was achieved.

In addition to the power combining techniques described above, other methods of combining Gunn devices in dielectric waveguide [73] and in optical cavities [74] have been demonstrated successfully.

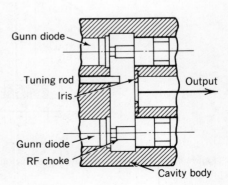

Figure 6.26 Push-pull Gunn combiner circuit. (From Ref. 72; copyright 1972 IEEE, reproduced by permission.)

6.9 EPITAXIAL MATERIAL GROWTH AND DEVICE FABRICATION

Typical epitaxial growth and device fabrication techniques for GaAs Gunn devices are reviewed in this section. Vapor-phase epitaxy (VPE) and liquid-phase epitaxy (LPE), the most established methods to grow GaAs materials are readily available in the industry. The molecular beam epitaxy (MBE) technique has also been used successfully to fabricate these devices. Not included in the discussion is the metal–organic chemical vapor deposition (MOCVD) method, which has been used to fabricate InP Gunn devices with good performance [75].

All epitaxial techniques require a bulk single crystal grown by Czochralski or Bridgeman methods to define the single-crystal growth. The bulk material has a large compensated impurity density with a low electron mobility and negative temperature coefficient of resistivity. The latter property would cause thermal runaway of devices in operation. Therefore, bulk material is usually unsuitable for the reliable operation of a transferred electron device. A detailed discussion on the growth of bulk GaAs is given by Howes and Morgan [76] and Bulman et al. [8].

6.9.1 Vapor-Phase Epitaxial Growth

Two common methods for vapor-phase epitaxial (VPE) growth of GaAs are the arsenic trichloride [77] and arsine [78] systems. The components of the compounds in both systems are introduced into the vapor-phase form that will produce a single crystal on the substrate surface.

Arsenic Trichloride Systems. Figure 6.27 shows a VPE apparatus. The system consists of two separate zones operated at different temperatures. Hydrogen and arsenic trichloride ($AsCl_3$) are introduced into the system with the following reaction:

$$4AsCl_3 + 6H_2 \longrightarrow As_4 + 12HCl$$

Arsenic vapor reacts with the Ga source placed in the source boat at a temperature near 800°C until saturation is achieved with a crust of GaAs formed over the entire surface. The HCl also reacts with Ga to produce GaCl:

$$2\,Ga + 2HCl \longrightarrow 2GaCl + H_2$$

GaCl and As_4 vapor combine in zone 2 at 750°C, and epitaxial crystal growth occurs on the GaAs substrate:

$$6GaCl + As_4 \longrightarrow 4GaAs + 2GaCl_3$$

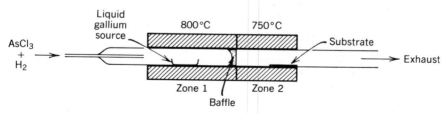

Figure 6.27 Schematic diagram of a furnace for vapor-phase epitaxial growth with arsenic trichloride system. (From J. R. Knight et al., "The Preparation of High Purity GaAs by Vapor Phase Epitaxial Growth," *Solid State Electronics*, Vol. 8; copyright 1965 Pergamon Journals, Ltd., reproduced by permission.)

Varying the amount of arsenic trichloride flow makes it possible to control the ambient net donor density of the epitaxial material. A decrease in growth temperature and an increase in arsenic trichloride cause a decrease in the electron concentration [79]. The n-type doping required for transferred electron devices may be obtained with sulfur or selenium, which may be added to the system as a controlled flow of hydrogen sulfide or hydrogen selenide. Another possibility is to introduced dopants in the molten Ga source. Materials with doping densities in the 10^{14} to 10^{15} cm^{-3} range and room-temperature mobilities as high as 8000 to 9000 cm^2/V-s have been achieved using this technique.

Arsine System. A similar system using the vapor-phase method to produce epitaxial material successfully is the arsine process shown in Fig. 6.28. The GaCl is generated by passing HCl over the Ga source:

$$2Ga + 2HCl \xrightarrow{>800°C} 2GaCl + H_2$$

As$_4$ is produced from the pyrolysis of arsine (AsH$_3$):

$$4AsH_3 \longrightarrow As_4 + 6H_2$$

Finally, As$_4$ reacts with GaCl at about 750°C to form the epitaxial layer:

$$6GaCl + As_4 \longrightarrow 4GaAs + 2GaCl_3$$

The source saturation step is not required in the arsine system. The critical temperature control of the Ga source is also unnecessary since the reaction of HCl with Ga is almost complete at temperatures exceeding 800°C. Similar carrier concentrations and mobilities have been obtained for the arsine and arsenic trichloride systems.

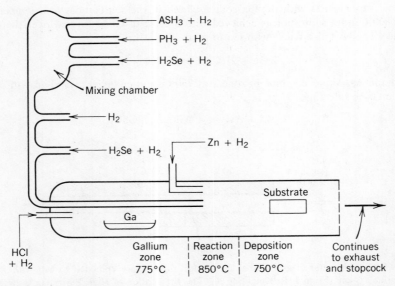

Figure 6.28 Schematic diagram of a furnace for vapor-phase epitaxial growth with arsine system. (From Ref. 78; reproduced by permission of the publisher, The Electrochemical Society, Inc.)

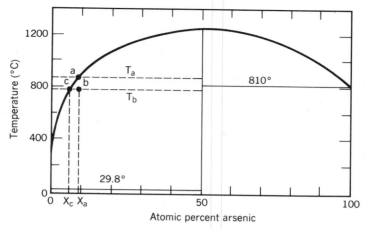

Figure 6.29   Schematic of GaAs phase diagram. (From M. J. Howes and D. V. Morgan, *Gallium Arsenide Materials, Devices, and Circuits*; copyright 1985, reproduced by permission of John Wiley & Sons, Ltd.)

6.9.2   Liquid-Phase Epitaxial Growth

Liquid-phase epitaxial growth is achieved on a GaAs substrate by recrystallization of a suitable solution at the liquid–solid interface. A typical solution composition is 10 atom% As in 90 atom% molten Ga. Liquid-phase epitaxy of GaAs depends on the solubility of As in Ga-rich solutions, which decreases with decreasing temperature, as described by the Ga-As phase diagram in Fig. 6.29. At point a in the figure, the solution is saturated with As at temperature T_a. As the temperature decreases to T_b, the state of the system will provide the precipitation of GaAs until the new saturation condition at point c is reached. This precipitate may be arranged to deposit on a GaAs substrate as an epitaxial layer.

Liquid-phase epitaxy growth of GaAs was first reported by Nelson [80]. Figure 6.30 shows the apparatus of a liquid epitaxial reactor in which the substrate and a melt are placed at opposite ends of a graphite boat in a tilt system. The growth of an n-GaAs epitaxial layer involves the following steps:

1. The melt consisting of tin and GaAs is heated to about 650°C to saturate the tin with GaAs.

2. The system is tilted so that the melt will be in contact with the substrate. After the system is cooled, GaAs intially dissolves from the substrate surface until a solution equilibrium is obtained. With further cooling, epitaxial growth on the GaAs occurs.

3. The system is tilted to its original position to end growth.

Two common growth techniques are used for the liquid-phase epitaxy: (1) the transient method and (2) the steady-state method. The transient method keeps the temperature of the liquid-solid-vapor system uniform. The temperature is then uniformly decreased so that recrystallization of the solute occurs. The steady-state method keeps the system in steady state with a fixed temperature difference between the solution and substrate so that solute crystallizes on the cooler substrate. Both vertical and horizontal systems have been employed [8]. Epitaxial layers grown below about 800°C without

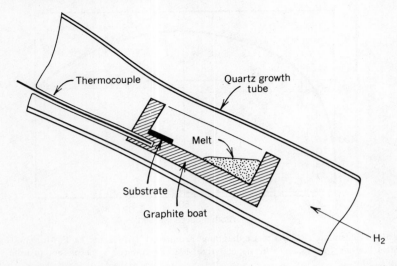

Figure 6.30   Schematic diagram of a tipping liquid-phase epitaxy system. (From Ref. 80.)

intentional doping are usually *n*-type and in the range 10^{14} cm^{-3}. Room-temperature mobilities between 7500 and 9300 cm^2/V-s are routinely obtained. To produce the desired electron density for transferred electron devices, it is usually necessary to introduce suitable levels of impurity elements into the molten solution. Sn, Se, or Te are commonly used dopants. Good doping uniformity may be achieved over lengths of 100 μm, making liquid-phase epitaxy attractive for low-frequency or LSA devices.

6.9.3   Molecular Beam Epitaxy

Transferred electron devices have recently been fabricated by molecular beam epitaxy (MBE) [81] with good results. Molecular beam epitaxy is basically a sophisticated extension of the vacuum evaporation technique that grows elemental, compound, and alloy semiconductor films by impinging directed thermal-energy atomic or molecular beams on a crystalline surface under ultrahigh-vacuum conditions.

A simplified system of MBE is shown in Fig. 6.31 for GaAs. Separate effusion ovens are used for Ga, As, and the dopants. All the effusion ovens are enclosed in an ultrahigh vacuum chamber with pressure around 10^{-10} torr. The temperature of each oven is adjusted to provide the desired evaporation rate. The substrate holder rotates continuously to achieve uniform epitaxial layers.

Haydl et al. [82] have made Gunn devices for 50 to 110 GHz using MBE. The material was grown at a temperature of about 580°C, which is low compared with the 700 to 800°C for LPE and VPE techniques. The low-temperature operation is advantageous since the outdiffusion from the substrate is minimized. Millimeter wave Gunn diodes require an active layer thickness of 1 to 3 μm with a doping concentration between 0.5 and 2×10^{16} cm^{-3}. Tin was used as the dopant. With these diodes, 20 mW at 90 GHz, 10 mW at 100 GHz, and 4 mW at 110 GHz have been achieved. These results are comparable with those obtained from devices using GaAs grown by the conventional LPE and VPE methods [83].

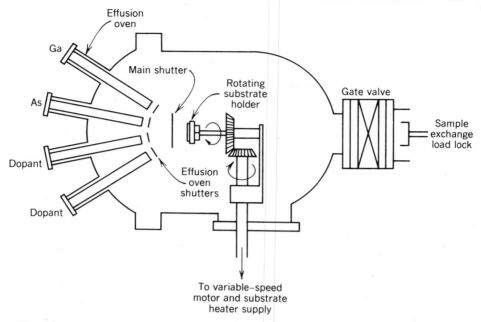

Figure 6.31 Schematic diagram of a molecular beam epitaxial system. (From A. Y. Cho and J. R. Arthur, "Molecular Beam Epitaxy," *Prog. Solid-State Chem.*, Vol. 10; copyright 1975 Pergamon Journals, Ltd., reproduced by permission.)

6.9.4 Buffer Layers

For ideal transferred electron devices, the epitaxial process should form a smooth graded n-n^+ interface when grown on an n^+ substrate. In practice, however, high-resistance layers often exist at the n-n^+ interface. To prevent this problem, an n^+ buffer epitaxial layer of a few micronmeters is grown between the substrate and the active n-region to produce a reasonable yield for the device. Also, a contact buffer layer is frequently grown on top of the n-layer to ensure a reliable operation of the Gunn device.

6.9.5 Metal Contact and Diode Fabrication

After epitaxial growth, metal contacts are applied to both the epitaxial contact layer and the back of the highly doped substrate. Electrical contact to the device usually requires low-resistance ohmic contacts, even though it was suggested that a barrier contact at the cathode for some devices may provide superior performance [84].

The ohmic contacts can be formed by a photolithographic pattern of their configuration and evaporation of suitable metals, followed by an alloying process. Germanium-gold alloy is commonly used with a small amount of Ni. Alloying is performed in a short heating cycle in the furnace at about 450°C. Contact is finally made to the alloyed metals with a thermocompression bonding of gold wire or ribbon. The wafers are then diced by wire saw or cleaving. Alternatively, the wafers may be etched into mesas before the devices are separated for mounting.

TABLE 6.1 Semiconductor Characteristics for GaAs and InP Gunn Devices

Properties	Material	
	InP	GaAs
Low field moblity (cm^2/V-s at 500K)	3000	5000
Energy gap (ev at 300K)	1.34	1.43
Thermal conductivity (W/cm-°C at 300K)	0.68	0.54
Breakdown field (kV/cm at $N_D = 10^6$)	500	400
Peak-to-valley ratio (300K)	3.5	2.2
Threshold field (kV/cm)	10.0	3.2
Effective transit velocity (cm/s)	1.2×10^7	0.7×10^7
Temperature dependence of electron velocity	$-0.1\%/°C$	$-0.5\%/°C$
Inertial energy time constant (ps)	0.75	1.5
Energy relaxation time (ps)	0.2	0.4

6.10 CONCLUSIONS AND FUTURE TRENDS

Significant accomplishments have been made in the past two decades on transferred electron devices in the areas of material growth, device fabrication, circuit development, and system applications. With increasing demands on future microwave and millimeter wave systems in radar and communication systems, new developments in transferred electron devices will undoubtedly continue. More emphasis is likely to be placed in the millimeter wave frequencies, since field-effect transistors can fulfill many applications better in microwave frequencies. It is anticipated that future trends can be summarized as follows:

Indium Phosphide (InP) Transferred Electron Devices Development. The steady improvement of growing high-purity InP material required for Gunn devices has led to some significant advances in the performance of these devices. A detailed discussion on InP Gunn devices compared with GaAs has been described elsewhere [85]. Table 6.1 compares the important semiconductor characteristics of InP and GaAs. The key characteristics for its high efficiency, high power, and high-frequency response are the high peak-to-peak valley ratio and fast energy-transfer-time constant. Other important factors affecting output power for InP are its high threshold field, higher electrical breakdown field, and higher thermal conductivity. Table 6.2 summarizes the present InP Gunn oscillator and stable reflection amplifier performance for different frequencies

TABLE 6.2 Performance of Present InP Gunn Oscillators and Amplifiers

Oscillators	Frequency (GHz)	Power (mW)	Efficiency (%)
	35	500	15
	56	385	11
	94	125	6
Amplifiers	Frequency (GHz)	$P_{saturated}$ (mW)	Power Added Efficiency (%)
	35	1000	15
	56	250	11
	94	50	6

[86]. These efficiency and output power results are at least a factor of 2 better than the equivalent GaAs Gunn devices. The devices used in Table 6.2 are made from VPE materials and are operated in the fundamental frequency mode. A higher-frequency response can be achieved if the harmonic mode of the device is used. With further developments in device design and material growth, such as MOCVD, better power and efficiency performance will be expected in the future.

Monolithic Gunn Device Development. Many system applications require small, lightweight, low-cost local oscillators for receivers and low-power transmitters for FM/CW missile seekers. Monolithic Gunn oscillators [87–89] fabricated on the semi-insulating substrate, such as GaAs, are promising candidates to meet these requirements. Figure 6.32 shows two millimeter wave monolithic Gunn oscillator circuits. The epitaxial layers

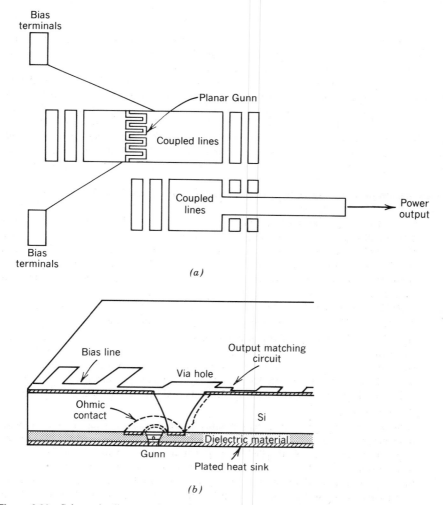

Figure 6.32 Schematic diagrams for monolithic Gunn oscillators: (*a*) 35-GHz Gunn oscillator (from Ref. 87; copyright 1987 IEEE, reproduced by permission); (*b*) 41-, 44-, and 68-GHz Gunn oscillators (from Ref. 89; copyright 1987 IEEE, reproduced by permission.)

TABLE 6.3 Summary of Monolithic Gunn Oscillators

Frequency (GHz)	Power Output (dBm)	Efficiency (%)	Comments	Reference
36	2.3	0.42	Planar Gunn coupled microstrip lines	87
35	1.9	0.5	Planar Gunn coplanar waveguide resonators	88
44.9	2	2	Flip-chip Gunn microstrip lines	89
68.4	0	1	Flip-chip Gunn microstrip lines	89

of active Gunn devices in Fig. 6.32 are grown on GaAs semi-insulating substrate by the conventional VPE method. The matching and bias circuits are also fabricated monolithically on the same chip as the active devices.

Table 6.3 provides a summary of the monolithic Gunn oscillator development. All the monolithic Gunn developments have been concentrated in the GaAs thus far. InP monolithic Gunn work is expected to receive acceptance since it has the possibility of extending frequency capabilities over that of equivalent GaAs Gunn oscillators. InP is also an attractive material in that InP transferred devices can be integrated with InP field effect transistors (FETs) for future system applications. As a first step toward monolithic realization, some planar InP Gunn devices have already been fabricated [90, 91].

Investigation of New Materials, New Device Structures, and New Mechanisms

1. *New Materials.* Transferred electron effect oscillations have been found in the ternary alloy $Ga_{0.47}In_{0.53}Aa$ [92, 93]. When this alloy is grown on InP substrates, the mobility and peak velocity are greater than GaAs and its threshold field is smaller; thus improved performance can be expected. Friscourt et al. [94] made a theoretical calculation by solving Boltzmann's transport equation and concluded that n^+-n-n^+ GaInAs devices exhibit higher output power and efficiency in the 30-GHz region than GaAs and InP Gunn devices, but have inferior performance beyond 40 GHz because of their high-energy relaxation time.

2. *New Device Structure.* Effects of heterojunction cathode contacts on transferred electron devices have been theoretically investigated by Friscourt et al. [95] He showed that Gunn devices with the reverse-biased or a limiting contact, formed by n^+ GaInAsP or GaInAs on InP active layer can exhibit efficiency levels twice as high as those predicted for n^+-n-n^+ fundamental InP devices. Experimental proof of these findings is likely to follow.

3. *New Mechanisms.* The mechanism of conventional Gunn devices involves transfer of electrons at high electric field to low-mobility conduction band valleys in momentum space. Hess et al. [96] proposed a negative resistance device using a mechanism where the electrons transfer from a high-mobility state to a low-mobility state in real space. This mechanism is based on hot-electron emission thermionic emission from high-mobility GaAs into low-mobility Al_xGa_{1-x} As in an alternating GaAs and Al_xGa_{1-x}As layers structure. Theoretical and experimental research in the area of new mechanism investigation are expected to continue.

REFERENCES

1. J. B. Gunn, "Microwave Oscillations of Current in III–V Semiconductors," *Solid State Commun.*, Vol. 1, p. 88, September 1963.

2. H. Kroemer, "Theory of Gunn Effect," *Proc. IEEE*, Vol. 52, p. 1736, December 1964.

3. B. K. Ridley and T. B. Watkins, "The Possibility of Negative Resistance Effects in Semiconductor," *Proc. Phys. Soc. London*, Vol. 78, pp. 294–304, August 1961.

4. C. Hilsum "Transferred Electron Amplifiers and Oscillators," *Proc. IRE*, Vol. 50, pp. 185–189, February 1962.

5. A. R. Hutson, A. Jayaraman, A. G. Cheynoweth, A. S. Coriell, and W. L. Feldman, "Mechanism of the Gunn Effect from a Pressure Experiment," *Phy. Rev. Lett.*, Vol. 14, p. 639, 1965.

6. J. W. Allen, M. Shyam, Y. S. Chen, and G. L. Pearson, "Microwave Oscillations in $GaAs_xP_{1-x}$ Alloys," *Appl. Phys. Lett.*, Vol. 7, p. 78, August 1965.

7. F. Soohoo, *Microwave Electronics*, Addison-Wesley, Reading, MA, 1971.

8. P. J. Bulman, G. S. Hobson, and B. C. Taylor, *Transferred Electron Devices*, Academic Press, London, 1972.

9. J. R. Ruch and G. S. Kino, "Measurement of the Velocity-Field Characteristic of Gallium Arsenide," *Appl. Phys. Lett.*, Vol. 10, pp. 40–42, January 15, 1967.

10. P. N. Butcher and W. Fawcett, "Calculation of the Velocity-Field Characteristic of Gallium Arsenide," *Phys. Lett.*, Vol. 21, pp. 489–490, June 15, 1966.

11. D. E. McCumber and A. G. Cheynoweth, "Theory of Negative-Conductance Amplification and of Gun Instabilities in Two-Valley Semiconductors," *IEEE Trans. Electron Devices*, Vol. ED-13, pp. 4–21, January 1966.

12. B. G. Streetman, *Solid State Electronic Devices*, 2nd ed., Prentice-Hall, Englewood Cliffs, NJ, 1980.

13. H. W. Thim, "Computer Study of Bulk GaAs Devices with Random One-Dimensional Doping Fluctuations," *J. Appl. Phys.*, Vol. 39, p. 3897, 1968.

14. J. A. Copeland, "A New Mode of Operation for Bulk Negative Resistance Oscillators," *Proc. IEEE*, Vol. 54, pp. 1479–1480, October 1966.

15. J. A Copeland, "LSA Oscillator Diode Theory," *J. Appl. Phys.*, Vol. 38, pp. 3096–3101, July 1967.

16. J. A. Copeland, "Doping Uniformity and Geometry of LSA Oscillator Diodes," *IEEE Trans. Electron Devices*, Vol. ED-14, p. 497, 1967.

17. H. C. Huang and L. A. Mackenzie, "A Gunn Diode Operated in the Hybrid Mode," *Proc. IEEE*, Vol. 57, p. 261, 1969.

18. G. S. Hobson, *The Gunn Effect*, Clarendon Press, Oxford, 1974.

19. H. W. Thim, "Linear Microwave Amplification with Gunn Oscillators," *IEEE Trans. Electron Devices*, Vol. ED-14, pp. 517–522, September 1967.

20. F. Sterzer, "Stabilization of Supercritical Transferred Electron Amplifiers," *Proc. IEEE*, Vol. 57, pp. 1781–1783, October 1969.

21. P. Jeppesen and B. I. Jeppsson, "The Influence of Diffusion on the Stability of the Supercritical Transferred Electron Amplifier," *Proc. IEEE*, Vol. 60, p. 452, 1972.

22. H. W. Thim, "Noise Reduction in Bulk Negative Resistance Amplifiers," *Electron. Lett.*, Vol. 7, pp. 106–108, 1971.

23. J. Magarshak, A. Rabier, and R. Spitalnik, "Optimum Design of Transferred Electron Amplifier Devices in GaAs," *IEEE Trans. Electron Devices*, Vol. ED-21, pp. 652–654, October 1974.

24. M. M. Atalla and J. L. Moll, "Emitter Controlled Negative Resistance in GaAs," *Solid State Electron.*, Vol. 12, pp. 119–129, 1969.

25. T. Hariu, S. Ono, and Y. Shibata, "Wideband Performance of the Injection Limited Gunn Diode," *Electron. Lett.*, Vol. 6, pp. 666–667, 1970.

26. H. W. Thim and M. R. Barber, "Microwave Amplification in GaAs Bulk Semiconductor," *IEEE Trans. Electron Devices*, Vol. ED-13, pp. 110–114, January 1966.

27. B. W. Hakki, "Amplification in Two-Valley Semiconductors, *J. Appl. Phys.*, Vol. 38, pp. 808 –818, February 1967.

28. T. E. Walsh, B. S. Perlman, and R. E. Enstrom, "Stabilized Supercritical Transferred Electron Amplifiers," *IEEE J. Solid-State Circuits*, Vol. SC-5, pp. 374–376, December 1969.

29. B. S. Perlman, C. L. Updahyayula, and R. E. Marx, "Wideband Reflection-Type Transferred Electron Amplifiers," *IEEE Trans. Microwave Theory Tech.*, Vol. MTT-18, pp. 911–921, November 1970.

30. M. R. Lakshminarayana and L. D. Partian, "Numerical Simulation and Measurement of Gunn Device Dynamic Microwave Characteristics," *IEEE Trans. Electron Devices*, Vol. ED-27, pp. 546–552, March 1980.

31. J. A. Copeland, "Stable Space-Charge Layers in Two-Valley Semiconductors," *J. Appl. Phys.*, Vol. 37, pp. 3602–3609, August 1966.

32. P. N. Butcher, "Theory of Stable Domain Propagation in the Gunn Effect," *Phys. Lett.*, Vol. 19, pp. 546–547, December 1965.

33. P. N. Butcher, W. Fawcett, and C. Hilsum, "A Simple Analysis of Stable Domain Propagation in the Gunn Effect," *Br. J. Appl. Phys.*, Vol. 7, pp. 841–850, 1966.

34. I. B. Bott, and W. Fawcett, "The Gunn Effect in Gallium Arsenide," in L. Young, Ed., *Advances in Microwaves*, Academic Press, New York, 1968, p. 251.

35. G. S. Hobson, "Small-Signal Admittance of a Gunn Effect Device," *Electron. Lett.*, Vol. 2, pp. 207–208, June 1966.

36. J. E. Carroll and R. A. Giblin, "A Low Frequency Analog for a Gunn Effect Oscillator," *IEEE Trans. Electron Devices*, Vol. ED-14, pp. 640–656, October 1967.

37. R. B. Robrock II, "A Lumped Model for Characterizing Single and Multiple Domain Propagation in Bulk GaAs," *IEEE Trans. Electron Devices*, Vol. ED-17, pp. 93–102, February 1970.

38. J. A. Copeland, "Electrostatic Domains in Two-Valley Semiconductors," *IEEE Trans. Electron Devices*, Vol. ED-13, pp. 189–192, January 1966.

39. W. A. Davis, *Microwave Semiconductor Design*, Van Nostrand Reinhold, New York, 1984.

40. S. Y. Narayan and F. Sterzer, "Transferred Electron Amplifiers and Oscillators," *IEEE Trans. Microwave Theory Tech.*, Vol. MTT-18, pp. 773–783, November 1970.

41. J. A. Copeland and R. R. Spiwak, "LSA Operation of Bulk GaAs Diodes," *Int. Solid-State Circuit Conf. Dig. Tech. Pap.*, pp. 26–27, 1967.

42. K. J. Button, *Infrared and Millimeter Waves*, Vol. 1, Academic Press, New York, 1979.

43. E. T. Harkless, U.S. Patent 3,534,293, October 13, 1970.

44. R. L. Eisenhart et al., "A Useful Equivalence for a Coaxial-Waveguide Junction," *IEEE Trans. Microwave Theory Tech.*, Vol. MTT-26, pp. 172–174, March 1978.

45. R. L. Eisenhart and P. J. Khan, "Theoretical and Experimental Analysis of a Waveguide Mounting Structure," *IEEE Trans. Microwave Theory Tech.*, Vol. MTT-19, pp. 706–719, August 1971.

46. R. L. Eisenhart, "Discussion of a 2-Gap Waveguide Mount," *IEEE Trans. Microwave Theory Tech.*, Vol. MTT-24, pp. 987–990, December 1976.

47. R. L. Eisenhart and P. J. Khan, "Some Tunning Characteristics and Oscillation Conditions of a Waveguide-Mounted Transferred Electron Oscillator," *IEEE Trans. Electron Devices*, Vol. ED-19, pp. 1050–1055, September 1972.

48. L. Lewin, "A Contribution to the Theory of Probes in Waveguides," *Proc. Inst. Elec. Eng.*, Monogr. 259R, pp. 109–116 October 1957.

49. L. Lewin, *Theory of Waveguides*, Wiley, New York, 1975, Chap. 5.

50. K. Chang and R. L. Ebert, "W-Band Power Combiner Design," *IEEE Trans. Microwave Theory Tech.*, Vol. MTT-28, pp. 295–305, April 1980.

51. P. J. Allen, B. D. Bates, and P. J. Khan, "Analysis and Use of Harkless Diode Mount for IMPATT Oscillators," *IEEE MTT-S Int. Microwave Symp. Dig.*, pp. 138–141, 1981.

52. M. E. Bialkowski, "Modeling of a Coaxial-Waveguide Power-Combining Structure," *IEEE Trans. Microwave Theory Tech.*, Vol. MTT-34, pp. 937–942, September 1986.

53. K. Kurokawa, "Some Basic Characteristics of Broadband Negative Resistance Oscillator Circuits," *Bell Syst. Tech. J.*, pp. 1937–1955, July 1969.

54. H. J. Kuno, J. F. Reynolds, and B. E. Berson, "Push-Pull Operation of Transferred Electron Oscillators," *Electron. Lett.*, Vol. 5, pp. 178–179, 1969.

55. M. Omori, "Octave Tunning of a CW Gunn Diode Using a YIG Sphere," *Proc. IEEE*, Vol. 57, p. 97, January 1969.

56. Y. S. Lau and D. Nicholson, "Barium Ferrite Tuned Indium Phosphide Gunn Millimeter Wave Oscillators," *IEEE MTT-S Int. Microwave Symp. Dig.*, pp. 183–186, 1986.

57. D. Cawsey, "Wide Range Tuning of Solid-State Microwave Oscillators," *IEEE J. Solid-State Circuits*, Vol. SC-5, pp. 82–84, April 1970.

58. C. S. Aitchison and R. V. Celsthorpe, "A Circuit Technique for Broadbanding the Electronic Tunning Range of Gunn Oscillators," *IEEE J. Solid-State Circuits*, Vol. SC-12, p. 21, February 1977.

59. C. D. Corbey et al., "Wideband Varactor Tuned Coaxial Oscillators," *IEEE Trans. Microwave Theory Tech.*, Vol. MTT-24, p. 31, January 1976.

60. B. J. Downing and F. A. Myers, "Broadband Varactor Tuned X-Band Gunn Oscillator," *Electron. Lett.*, Vol. 7, p. 407, July 1971.

61. G. E. Brehm and S. Mai, "Varactor-tuned Integrated Gunn Oscillators," presented at the *International Solid-State Circuits Conference*, Philadelphia, 1968.

62. L. D. Cohen and E. Sard, "Recent Advances in the Modelling and Performance of Millimeter Wave InP and GaAs VCOs and Oscillators," *IEEE MTT-5 Int. Microwave Symp. Dig.*, pp. 429–432, 1987.

63. K. J. Russell, "Microwave Power Combining Techniques," *IEEE Trans. Microwave Theory Tech.*, Vol. MTT-27, pp. 472–478, May 1979.

64. K. Chang and C. Sun, "Millimeter-Wave Power Combining Techniques," *IEEE Trans. Microwave Theory Tech.*, Vol. MTT-31, pp. 92–153, February 1983.

65. K. Kurokawa and F. M. Magalhaes, "An X-band 10-Watt Multiple-IMPATT Oscillator," *Proc. IEEE*, pp. 102–103, January 1971.

66. R. S. Harp and H. L. Stover, "Power Combining of X-band IMPATT Circuit Modules," *IEEE-ISSCC Dig. Tech. Pap.*, Vol. 16, pp. 118–119, February 1973.

67. Y. Ma and C. Sun, "1-W Millimeter-Wave Gunn Diode Combiner," *IEEE Trans. Microwave Theory Tech.*, Vol. MTT-28, pp. 1460–1463, December 1980.

68. K. R. Varian, "Power Combining in a Single Multiple-Diode Cavity," *IEEE MTT-S Int. Microwave Symp. Dig.*, pp. 344–345, June 1978.

69. C. Sun, E. Benko, and J. W. Tully, "A Tunable High Power V-Band Gunn Oscillator," *IEEE Trans. Microwave Theory Tech.*, Vol. MTT-27, pp. 512–514, May 1979.

70. A. K. Talwar, "A Dual-Diode 73 GHz Gunn Oscillator," *IEEE Trans. Microwave Theory Tech.*, Vol. MTT-27, pp. 510–512, May 1979.

71. J. J. Sowers, J. D. Crowley, and F. B. Fank, "CW InP Gunn Diode Power Combining at 90 GHz," *IEEE MTT-S Int. Microwave Symp. Dig.*, pp. 503–505, June 1982.

72. T. G. Ruttan, "42 GHz Push-pull Gunn Oscillator," *IEEE Proc.*, Vol. 60, pp. 1441–1442, November 1972.

73. J. J. Potoczniak, H. Jacobs, C. L. Casio, and G. Novick, "Power Combiners with Gunn Diode Oscillator," *IEEE Trans. Microwave Theory Tech.*, Vol. MTT-30, pp. 724–728, May 1982.

74. L. Wandinger and V. Nalbandian, "Millimeter-Wave Power Combiner Using Quasi-optical Techniques," *IEEE Trans. Microwave Theory Tech.*, Vol. MTT-31, p. 189, February 1983.

75. M. A. di Forte-Poisson et al., "High-Power High-Efficiency LP-MOCVD Gunn Diodes for 94 GHz," *Electron. Lett.*, Vol. 20, pp. 1061–1062, December 1984.

76. M. J. Howes and D. V. Morgan, *Gallium Arsenide Materials, Devices, and Circuits*, Wiley, New York, 1985.

77. J. R. Knight et al., "The Preparation of High Purity GaAs by Vapor Phase Epitaxial Growth," *Solid State Electron.*, Vol. 8, pp. 178–180, 1965.

78. J. J. Tietjen and J. A. Amick, "The Preparation and Properties of Vapor-Deposited Epitaxial $GaAs_{1-x}P_x$ Using Arsine and Phosphine," *J. Electrochem. Soc.*, Vol. 113, pp. 724–772, July 1966.

79. J. V. DiLorenzo, "Vapor Growth of Epitaxial GaAs: A summary of Parameters Which Influence Purity and Morphology of Epitaxial Layers," *J. Cryst. Growth*, Vol. 17, pp. 184–206, 1972.

80. H. Nelson, "Epitaxial Growth from the Liquid State and Its Application to the Fabrication of Tunnel and Laser Diodes," *RCA Rev.*, Vol. 24, pp. 603–615, December 1963.

81. A. Y. Cho and J. R. Arthur, "Molecular Beam Epitaxy," *Prog. Solid-State Chem.*, Vol. 10, p. 157, 1975.

82. W. H. Haydl, R. S. Smith, and R. Bosch, "50-110 GHz Bunn Diodes Using Molecular Beam Epitaxy," *IEEE Electron Device Lett.*, Vol. EDL-1, No 10, pp. 224–226, October 1980.

83. T. G. Ruttan, "Gunn-Diode Oscillator at 95 GHz," *Electron. Lett.*, Vol. 11, pp. 293–294, July 1975.

84. S. P. Yu et al., "Transit-Time Negative Conductance in GaAs Bulk-Effect Diodes," *IEEE Trans. Electron Devices*, Vol. ED-18, pp. 88–93, 1971.

85. F. B. Frank et al., "High Efficiency InP Millimeter-Wave Oscillators and Amplifiers," *Eur. Microwaves Conf. Proc.*, September 1984, pp. 575–580.

86. F. B. Fank, Private communication.

87. A. Chu et al., "Low Cost Millimeter Wave Monolithic Receivers," *IEEE Microwave Millim. Wave Monolithic Circuits Symp.*, 1987, pp. 63–67.

88. N. Wang, S. W. Schwartz, and T. Hierl, "Monolithically Integrated Gunn Oscillator at 35 GHz," *Electron. Lett.*, Vol. 20, p. 603, July 1984.

89. J. C. Chen, C. K. Pao, and D. W. Wong, "Millimeter-Wave Monolithic Gunn Oscillators," *IEEE Microwave Millim.-Wave Monolithic Circuits Symp.*, 1987, pp. 11–13.

90. T. Weng et al., "Two and Three Terminal Planar InP TEDs," *IEEE Electron Device Lett.*, Vol. EDL-1, pp. 69–71, May 1980.

91. S. C. Binari, P. E. Thompson, and H. L. Grubin, "Self-Aligned Notched Planar InP Transferred-Electron Oscillators," *IEEE Electron Device Lett.*, Vol. EDL-6, pp. 22–24, January 1985.

92. W. Kowalsky and A. Schlachetzki, "InGaAs Gunn Oscillator," *Electron. Lett.*, Vol. 20, pp. 502–503, June 1984.

93. T. Takeda, N. Shikagawa, and A. Sasaki, "Transferred-Electron Oscillation in InGaAs," *Solid-State Electron.*, Vol. 23, pp. 1003–1005, September 1980.

94. M. R. Friscourt, P. A. Rolland, and R. Faquembergue, " Theoretical Investigation of n^+-n-n^+ $Ga_{0.47}In_{0.53}As$ TEOs up to Millimeter-Wave Range," *IEEE Electron Device Lett.*, Vol. EDL-5, pp. 434–436, 1984.

95. M. R. Friscourt, P. A. Rolland, and M. Pernisek, "Heterojunction Cathode Contact Transferred-Electron Oscillators," *IEEE Electron Device Lett.*, Vol. EDL-6, pp. 497–499, October 1985.

96. K. Hess et al., "Negative Differential Resistance through Real-Space Electron Transfer," *Appl. Phys. Lett.*, Vol. 35, No. 6, pp. 469–471, September 1979.

7

IMPATT AND RELATED TRANSIT-TIME DEVICES

Kai Chang

Texas A&M University
College Station, Texas

H. John Kuno

Hughes Aircraft Company
Torrance, California

7.1 INTRODUCTION

The term "IMPATT" stands for "*imp*act-ionization *a*valanche *t*ransit *t*ime." The IMPATT device is one of the most powerful solid-state microwave sources. The IMPATT diode uses impact-ionization and transit-time properties to produce the negative resistance required for oscillation and amplification of microwave signals.

The concept of the IMPATT diode was first proposed by Read in 1958 for a relatively complex diode structure n^+pip^+, which is now commonly called the Read diode [1]. It was not until 1965 that the experimental observation of IMPATT oscillation from a *p-n* junction diode was reported by Johnston et al. [2]. Since then a large amount of effort has been directed toward a theoretical understanding of the physical mechanism and the development of practical microwave and millimeter wave IMPATT devices, oscillators, and amplifiers. IMPATT devices have been fabricated with Si, GaAs, and InP semiconductor material. The Si-IMPATT devices have been operated at frequencies up to 394 GHz [3].

Although three-terminal devices are now more commonly used at microwaves, the IMPATT diode is still very useful as a power source at millimeter wave frequencies. At present, it generates the highest power output among all solid-state devices at millimeter wave frequencies. At lower microwave frequencies, FETs and bipolar transistors, which are three-terminal devices, have replaced IMPATTs for many applications. For this reason, we focus our discussion on millimeter wave applications.

In this chapter we present an overview of the IMPATT device and its applications. The device physics of the IMPATT diode is described in Section 7.2. Included are both small-signal and large-signal models. In Section 7.3 we describe the typical device design and fabrication. In Sections 7.4 and 7.5 we discuss the applications of the IMPATT diode in oscillator and amplifier circuits. The use of power combining techniques to combine several IMPATT diodes in circuits is given in Section 7.6. In Section 7.7 we present possible applications of the IMPATT diode as a nonlinear reactor or a frequency multiplier. Noise properties and techniques to improve noise characteristics are discussed in Section 7.8, followed by a description in Section 7.9 of related transit-time devices, such as TRAPATT, MITATT, TUNNETT, and BARITT diodes. Finally, Section 7.10 is devoted to new developments in InP IMPATT diodes, traveling-wave IMPATT devices, heterojunction IMPATT diodes, monolithic IMPATT devices, active arrays, and other topics.

7.2 DEVICE PHYSICS AND MODELING

Microwave oscillation and amplification are due to the frequency-dependent negative resistance arising from the phase delay between current and voltage waveforms due to avalanche breakdown and transit-time effects. Since Read first proposed a negative resistance diode consisting of an n^+pip^+ (or p^+nin^+) structure, many other configurations have been developed. For silicon diodes, the single-drift diode (p^+nn^+) and the double-drift diode (p^+pnn^+) are the two most popular structures. For GaAs diodes, single-drift Read diodes ($p^+n^+nn^+$ or $p^+n^-n^+nn^+$), double-drift Read diodes ($p^+pp^+n^+nn^+$ or $p^+pp^+pnn^+nn^+$), and hybrid Read diodes ($p^+pn^+nn^+$ or $p^+pnn^+nn^+$) have been fabricated. In this section the physics of avalanche breakdown in a p-n junction will first be discussed. A Read diode will be used for an analytical description of the operation theory of an IMPATT diode. For complicated structures, computer programs are generally used to generate small-signal and large-signal device parameters.

7.2.1 Avalanche Breakdown in *p-n* Junction

Avalanche breakdown normally imposes an upper limit on the reverse voltage applied to most *p-n* junction diodes. In an IMPATT device, the same avalanche breakdown, which is caused by avalanche multiplication (or impact ionization), can be used effectively to generate microwave power.

A *p-n* junction is shown in Fig. 7.1 together with its I–V characteristics. As the reverse-biased voltage is near V_B, the electric field around the *p-n* junction reaches a very high value, and avalanche breakdown occurs. The ionization rate α for holes and electrons can be represented by

$$\alpha = Ae^{-(b/E)^m} \tag{7.1}$$

where m, b, and A are constants depending on the material, and E is the electric field. In general, the ionization rates for holes (α_p) and electrons (α_n) are not equal (i.e., $\alpha_n \neq \alpha_p$). For a *p-n* junction, the multiplication factor M_p of holes can be written as [4]

$$1/M_p = 1 - \int_0^w \alpha_p \exp\left[-\int_0^x (\alpha_p - \alpha_n)\, dx'\right] dx \tag{7.2}$$

where w is the width of depletion region. A similar result applies to the multiplication factor M_n of electrons

$$1/M_n = 1 - \int_0^w \alpha_n \exp\left[-\int_x^w (\alpha_n - \alpha_p)\, dx'\right] dx \tag{7.3}$$

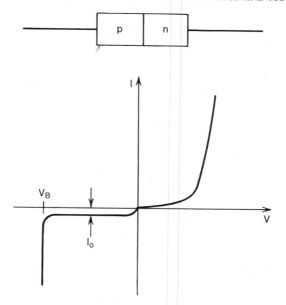

Figure 7.1 *p-n* junction and its *I–V* curve.

The avalanche breakdown occurs as M_p or M_n approaches infinity. The breakdown condition is thus given by

$$\int_0^w \alpha_p \exp\left[-\int_0^x (\alpha_p - \alpha_n)\, dx' \right] dx = 1 \tag{7.4}$$

or

$$\int_0^w \alpha_n \exp\left[-\int_x^w (\alpha_n - \alpha_p)\, dx' \right] dx = 1 \tag{7.5}$$

Solving Eq. (7.4) and (7.5) will give avalanche breakdown voltages. For a semiconductor with equal ionization rates for electrons and holes, Eq. (7.4) and (7.5) reduce to

$$\int_0^w \alpha\, dx = 1 \tag{7.6}$$

7.2.2 Physical Explanation of Operation Theory [5]

The operation theory can best be understood by considering a Read diode as shown in Fig. 7.2. The electric field distribution inside the device and the ionization rates are shown in the same figure. The device can be divided into two regions: avalanche region and drift region. The avalanche region is a small high-field region inside the *n* layer near the *p-n* junction, where impact ionization occurs and electrons and holes are generated. The drift region is the low-field region where carriers drift at saturated velocity. Due to the time delay in avalanche and drift, there is a phase shift between the voltage and

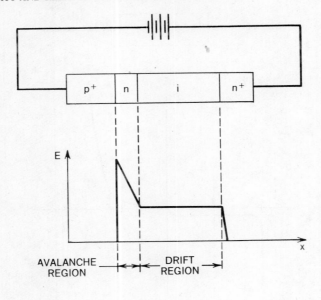

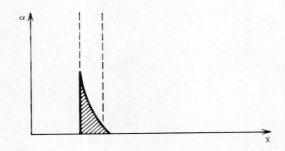

Figure 7.2   Read diode (p^+nin^+) and its E-field distribution.

current. The total phase shift is equal to

$$\theta = \omega(\tau_A + \tau_d) \tag{7.7}$$

where τ_A is the time delay attributed to avalanche multiplication and τ_d is due to finite drift time. For negative resistance to occur, θ needs to be greater than 90°. Consider the ionization rate as

$$\alpha = A \exp\left[-\left(\frac{b}{E_{dc} + E_{rf}}\right)^m\right] \tag{7.8}$$

Here the total field is equal to $E_{dc} + E_{rf}$ and let $E_{dc} \simeq$ breakdown field. Then the addition of an ac field will cause avalanche breakdown due to the exponential increase in electron–hole pairs. This additional ac field could be introduced due to a noise signal in the oscillation startup. Thus the current generated by the avalanche process will have its maximum when the RF field goes through zero. Therefore, as shown in Fig. 7.3, the

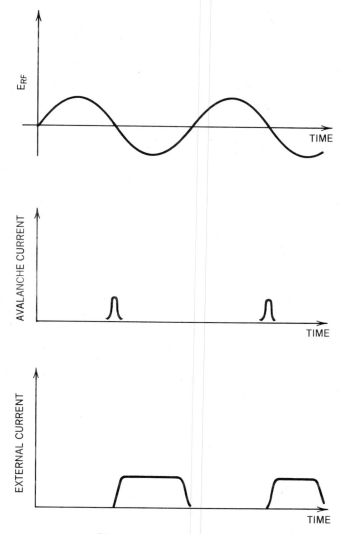

Figure 7.3 Current waveforms.

avalanche process contributes a 90° inductive lag to the current generated in the avalanche region. This current is then injected into the drift region of the diode. The current induced in the external circuit by this change is shown in the bottom of Fig. 7.3. It is obvious that the external current is more than 90° out of phase with respect to the RF voltage. A negative resistance is therefore created.

7.2.3 Generalized Model

Consider a *p-n* junction under reverse avalanche breakdown conditions as shown in Fig. 7.4 for a single-drift or double-drift diode [6]. Assuming a one-dimensional analysis and neglecting the diffusion effects, a set of equations governing the dynamics of electrons

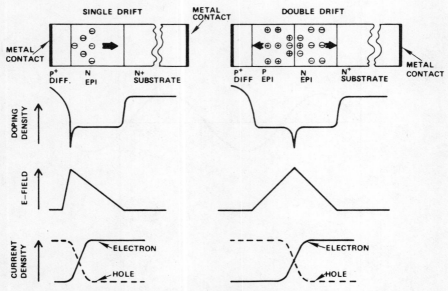

Figure 7.4   IMPATT diode model. (From Ref. 6, reproduced by permission.)

and holes are given by [7] Poisson's equation:

$$\frac{\partial E}{\partial x} = \frac{q}{\epsilon}(N_D - N_A + p - n) \tag{7.9}$$

and by the continuity equations for electrons and holes:

$$\frac{\partial n}{\partial t} = \frac{1}{q}\left(\frac{\partial J_n}{\partial x} + \alpha_n J_n + \alpha_p J_p\right) \tag{7.10}$$

$$\frac{\partial p}{\partial t} = \frac{1}{q}\left(\frac{-\partial J_p}{\partial x} + \alpha_n J_n + \alpha_p J_p\right) \tag{7.11}$$

where N_D, N_A = doping concentration in the n and p materials
p, n = hole and electron carrier densities, respectively
ϵ = dielectric constant
J_p, J_n = hole and electron current densities, respectively
t = time
x = distance from the junction
q = electron charge

The total current density is

$$J = qnv_n + qpv_p + \epsilon\frac{\partial E}{\partial t} \tag{7.12}$$

where v_p and v_n are drift velocities of holes and electrons, respectively.

7.2.4 Small-Signal Analysis

For small-signal analysis, it is assumed that the ac signal is small compared with the dc component. Small-signal analysis provides useful information on the impedance and frequency responses of the device. The electric field and current density J can be written as a dc component superimposed by an ac signal, that is,

$$E = E_0 + E_1 e^{j\omega t} \tag{7.13}$$

$$J = J_0 + J_1 e^{j\omega t} \tag{7.14}$$

$$\Delta E = E_1 e^{j\omega t} \tag{7.15}$$

$$\alpha = \alpha_0 + \frac{\partial \alpha}{\partial E} \Delta E = \alpha_0 + \alpha' E_1 e^{j\omega t} \tag{7.16}$$

$$n = n_0 + n_1 e^{j\omega t} \tag{7.17}$$

$$p = p_0 + p_1 e^{j\omega t} \tag{7.18}$$

J_1 and E_1 represent ac components which are smaller than dc components (E_0, J_0). The higher-order terms for ac components are neglected. ω is the angular frequency of operation and $\alpha' = \partial \alpha / \partial E$.

Substituting Eqs. (7.13)–(7.18) into Eqs. (7.9)–(7.11), we can separate the dc and ac terms. The ac components are obtained by

$$-\frac{\partial E_1}{\partial x} = j\left(\frac{\omega}{v_p}\right) E_1 + \frac{1}{\epsilon}\left[\left(\frac{1}{v_n}\right) + \left(\frac{1}{v_p}\right)\right] J_{n1} - \frac{J_1}{\epsilon v_p} \tag{7.19}$$

$$-\frac{\partial J_{n1}}{\partial x} = (\alpha'_n J_{n0} + \alpha'_p J_{p0} - j\omega\epsilon\alpha_{p0})E_1 + \left(\alpha_{n0} - \alpha_{p0} - j\frac{\omega}{v_n}\right) J_{n1} + \alpha_{p0} J_1 \tag{7.20}$$

where the derivatives with respect to t are replaced by $j\omega$ and $J_n = J_{n0} + J_{n1}$, $J_1 = J_{n1} + J_{p1}$.

The dc terms are

$$\frac{\partial E_0}{\partial x} = -\frac{1}{\epsilon}\left(\frac{1}{v_n} + \frac{1}{v_p}\right) J_{n0} + \frac{q}{\epsilon}(N_D - N_A) + \frac{J_0}{\epsilon v_p} \tag{7.21}$$

$$-\frac{\partial J_{n0}}{\partial x} = (\alpha_{n0} - \alpha_{p0})J_{n0} + \alpha_{p0} J_0 \tag{7.22}$$

These equations can be solved numerically using a finite-difference method subject to boundary conditions. The ac impedance Z_1 of the devices can be obtained by

$$Z_1 = \frac{1}{Y_1} = \frac{V_1}{J_1 A} \tag{7.23}$$

where A is the cross-sectional area of the device. The ac voltage V_1 across the depletion region is given by

$$V_1 = \int E_1 \, dx \tag{7.24}$$

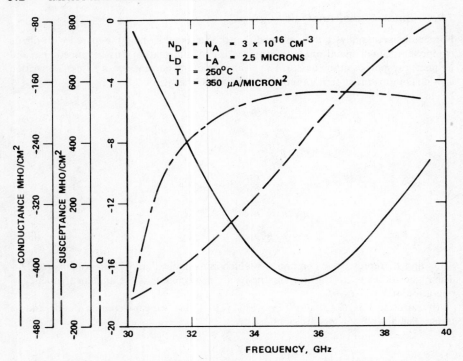

N_D = N_A = 3×10^{16} CM^{-3}
L_D = L_A = 2.5 MICRONS
T = 250°C
J = 350 μA/MICRON^2

Figure 7.5 Calculated small-signal RF characteristics for 35-GHz pulsed IMPATTs.

From the ac solution on J_1 and E_1, the ac impedance of the device as a function of frequency can be found from Eqs. (7.23) and (7.24). Note that N_A and N_D depend on the doping profile, which is a function of x. The parameters α_n, α_p, v_p, and v_n vary with temperature.

The numerical analysis is generally carried out with the aid of computers. For a specified current density, junction temperature, and doping profile, the computer program calculates and plots the dc electric field as a function of distance. It then uses this dc solution to calculate the device small-signal conductance and susceptance per unit area and the device Q factor as a function of frequency for a specified frequency range [8]. Figure 7.5 shows a typical computer output for a 35-GHz pulsed double-drift silicon IMPATT diode with $N_A = N_D = 3 \times 10^6$ cm^{-3}, $J = 350 \mu A/\mu m^2$, and a junction temperature of 250°C. The lengths for the n and p regions are 2.5 μm. It can be seen that the conductance of the diode is negative over a broad bandwidth. The susceptance part goes through a resonance below which it is inductive and above which it is capacitive. The device will be designed to achieve a maximum Q at the desired frequency.

Although the RF power limit, efficiency, and many other phenomena of an IMPATT device are determined by a large-signal analysis, the small-signal values provide a convenient basis for impedance information under various conditions. It has been used extensively for designing devices and circuits.

7.2.5 Simplified Small-Signal Model

To illustrate the physical significance of the avalanche multiplication and the transit-time effects in an IMPATT diode, a simplified small-signal model based on the Read diode structure was developed [1, 4, 9].

In this model, the device is divided into three regions: (1) the avalanche region, in which the thickness is very thin, the ionization rate is uniform, and transit-time delay is negligible; (2) the drift region, in which no carriers are generated and all carriers entering from the avalanche region move at saturated velocities; and (3) the inactive region, which introduces additional parasitic resistance. Furthermore, we assume that

$$\alpha_n = \alpha_p = \alpha \tag{7.25}$$

$$v_n = v_p = v_s \tag{7.26}$$

where v_s is the saturated velocity.

Avalanche Region. In the avalanche region, add Eqs. (7.10) and (7.11) and integrate from $x = 0$ to $x = x_A$ (where x_A is the width of the avalanche region, as shown in Fig. 7.6); then we can derive the equation

$$\frac{dJ}{dt} = \frac{2J}{\tau_A}(\alpha x_A - 1) \tag{7.27}$$

where the boundary conditions at $x = 0$ and $x = x_A$ are used and the transit time is defined as $\tau_A = x_A/v_s$. Substituting Eqs. (7.13)–(7.16) into (7.27), we have

$$J_1 = \frac{2\alpha' x_A J_0 E_1}{j\omega\tau_A} \tag{7.28}$$

where J_1 is the avalanche current, which is in phase quadrature with electric field E_1. The total current is

$$J_T = \frac{2\alpha' x_A J_0 E_1}{j\omega\tau_A} + j\omega\epsilon E_1 \tag{7.29}$$

where the second term on the right-hand side accounts for the displacement current. Therefore, it can be seen that two components of the total current are in the avalanche region. For a given field, the avalanche current J_1 is reactive and varies inversely with ω as in an inductor. The other component, $j\omega\epsilon E_1$, is also reactive and varies proportionally with ω as a capacitor. The equivalent circuit for the avalanche region can be

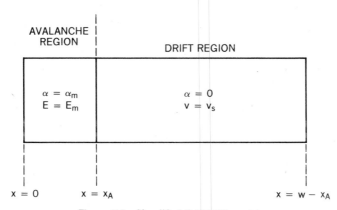

Figure 7.6 Simplified IMPATT model.

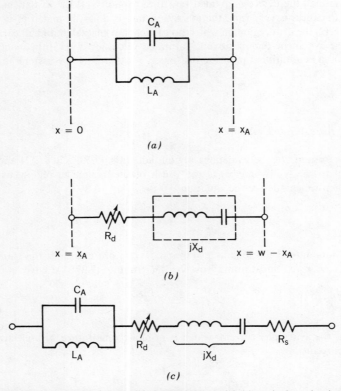

Figure 7.7   Equivalent circuit of a simplified IMPATT model: (*a*) equivalent circuit of avalanche region; (*b*) equivalent circuit of drift region; (*c*) total equivalent circuit.

represented by an inductance L_A in parallel with a capacitance C_A, as shown in Fig. 7.7a, where L_A and C_A are given by

$$L_A = \frac{\tau_A}{2J_0 \alpha' A} \tag{7.30}$$

$$C_A = \frac{\epsilon A}{x_A} \tag{7.31}$$

A resonant frequency is given by

$$f_r = \frac{1}{2\pi} \frac{1}{\sqrt{L_A C_A}} = \frac{1}{2\pi} \sqrt{\frac{2\alpha' v_s J_0}{\epsilon}} \tag{7.32}$$

and the impedance of the avalanche region is

$$Z_A = L_A \| C_A = \frac{1}{j\omega C_A} \left[\frac{1}{1 - (f_r/f)^2} \right] \tag{7.33}$$

Drift Region. The current injected into the drift region is equal to

$$J(x) = J_T(x) = j\omega\epsilon E_1(x) + J_1 \exp\left(-j\omega \frac{x}{v_s}\right) \tag{7.34}$$

Define the current ratio

$$r = \frac{J_1}{J} = \frac{1}{1 - (f/f_r)^2} \tag{7.35}$$

Then we have

$$E_1(x) = J \frac{1 - re^{-j\omega(x/v_s)}}{j\omega\epsilon} \tag{7.36}$$

The ac voltage across the drift region is

$$V_d = \int_{x_A}^{w-x_A} E_1(x)\,dx = \frac{(w - x_A)J}{j\omega\epsilon}\left[1 - \frac{1}{1 - (f/f_r)^2}\left(\frac{1 - e^{-j\theta_d}}{j\theta_d}\right)\right] \tag{7.37}$$

where $\theta_d = \omega(w - x_A)/v_s = \omega\tau_d$.

Therefore, the impedance of the drift region is

$$Z_d = \frac{V_d}{JA} = R_d + jX_d = \frac{1}{\omega C_d}\left[\frac{1}{1 - (f/f_r)^2}\left(\frac{1 - \cos\theta_d}{\theta_d}\right)\right]$$
$$+ \frac{j}{\omega C_d}\left[-1 + \frac{1}{1 - (f/f_r)^2}\left(\frac{\sin\theta_d}{\theta_d}\right)\right] \tag{7.38}$$

where $C_d = A\epsilon/(w - x_A)$. The equivalent circuit is shown in Fig. 7.7(b). From Eq. (7.38) it is seen that $R_d < 0$ for $f > f_r$ except for the nulls at $\theta_d = 2n\pi$, where $R_d = 0$. The imaginary part, jX_d, will be an inductance or a capacitance, depending on the value inside the brackets.

Total Impedance. The total impedance of the device is the sum of those of the three regions:

$$Z = Z_A + Z_d + R_s \tag{7.39}$$

The equivalent circuit is shown in Fig. 7.7c. R_s represents the series resistance due to the semiconductor.

7.2.6 Large-Signal Analysis

In order to understand various nonlinear effects, transient responses, power output, efficiency, and frequency tuning of the oscillator and amplifier, a large-signal analysis needs to be used. The detailed large-signal performance can be obtained by solving Eqs. (7.9)–(7.11) with appropriate boundary conditions. Large-signal analyses and nonlinear effects have been studied by many researchers. A few notables among them are given in Ref. 10–19.

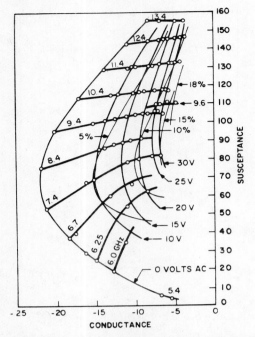

Figure 7.8   Large-signal calculation of diode admittance as a function of frequency and ac voltage amplitude. Current density: 200A/cm². (From Ref. 12; copyright 1969 IEEE, reproduced by permission.)

As an example, the device admittance and resultant efficiency calculated for various ac voltages are given in Fig. 7.8. The device used for this simulation is a p^+nvn^+ diode.

7.3   DEVICE DESIGN AND FABRICATION

IMPATT devices can be made in many different configurations of doping profiles. The two mostly common semiconductor materials used are Si and GaAs, although InP and Ge can also be used. This section first describes various doping profiles, followed by discussions on the device design, fabrication, and packaging considerations.

7.3.1   Doping Profiles for Si Diodes

The two most commonly used doping profiles for silicon IMPATTs are the single-drift (SDR) and double-drift (DDR) profiles. The single-drift, uniform doping profile is the simplest structure. Its doping and electric field profiles are shown in Fig. 7.9. Although the n^+pp^+ IMPATT was also fabricated and showed better efficiency [20], the p^+nn^+ type is easier to fabricate since the n^+ substrate is generally used. The avalanche region voltage is approximately one half of the total applied voltage.

The double-drift diode was first proposed by Scharfetter et al. [21] and also by Seidel and Scharfetter [22]. The doping and electric field profiles are shown in Fig. 7.10. A double-drift diode consists of a p-type IMPATT and n-type IMPATT connected in

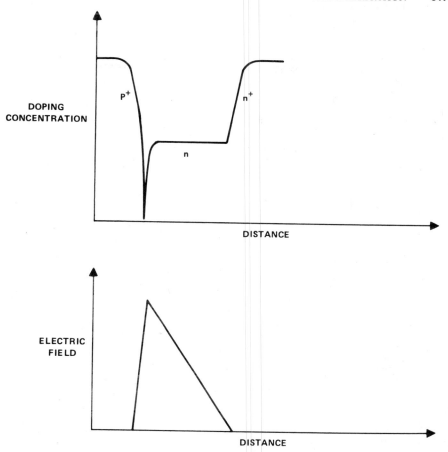

Figure 7.9 Doping and electric-field profiles for a single-drift flat structure.

series. In the double-drift diode, however, the hole and electron drift regions share an avalanche zone. The resultant improvement in the Q-factor enables the diode to operate at a higher efficiency. In addition, since a double-drift diode is equivalent to two single diodes connected in series, a double-drift diode with twice as large a junction area as a single diode maintains the same impedance level as that of the single diode. This results in four times as much output power as the single diode.

7.3.2 Doping Profiles for GaAs Diodes

The SDR and DDR structures used for Si diodes can also be used for GaAs. However, other profiles, such as Read-type diodes, are more popular for GaAs since a higher efficiency can be achieved. The Read-type structure is more difficult to fabricate in silicon [23].

Single-Drift Read Profile. The Read diode is a better device than the flat-profiled IMPATT because of its confined avalanche zone and the well-defined drift region, which gives rise to a better phase relationship between RF voltage and current. A single-drift

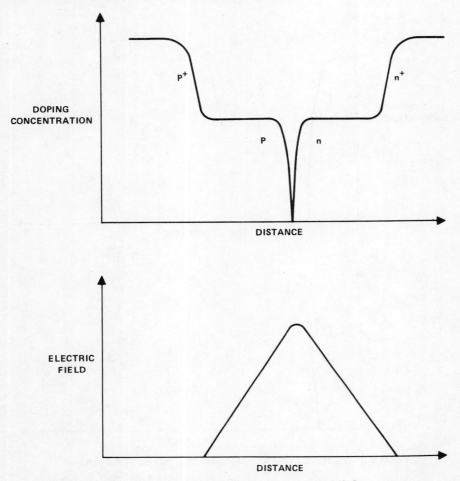

Figure 7.10   Doping and electric-field profiles for a double-drift flat structure.

Read structure can be further subdivided into low–high and low–high–low profile categories according to their slightly different doping profiles. A n-type single drift Read low–high structure with its associated electric field is shown in Fig. 7.11. The n-type diode is preferred to the p-type since the electron mobility is much higher in GaAs material, resulting in lower resistance arising from the depletion layer modulation. Figure 7.12 shows the single-drift Read low–high–low structure.

Double-Drift Read Profile.   The double-drift Read structure can be viewed as one p-type SDR diode in series with an n-type SDR diode as in the p^+pnn^+ double-drift diode. Consequently, higher output power can be achieved since the diode can be operated with a larger area while maintaining the same impedance level as a SDR diode. Figure 7.13 shows the low–high double-drift Read profile and its associated electric field profile. The low–high–low structure is shown in Fig. 7.14. Theoretically, the improvement in power and efficiency of DDR Read over SDR diodes are approximately 4 and 1.5 times,

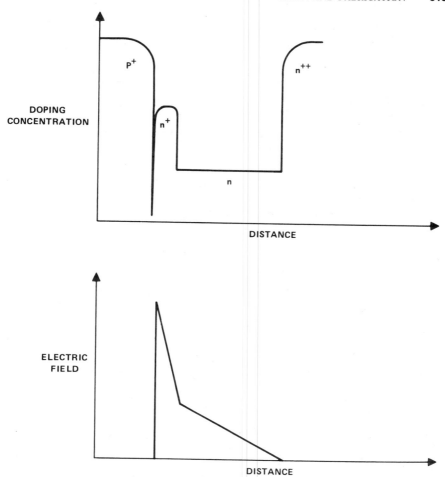

Figure 7.11 Doping and electric-field profiles for a single-drift Read low–high structure.

respectively. But the thermal and electric limitations reduce these advantages. The hole mobility in GaAs is about 20 times smaller than the electron mobility. The additional series resistance introduced in the p region of the DDR Read structure reduces the power and efficiency. The increase in thermal resistance due to the additional p-layer of GaAs material also places a limitation on the available efficiency of a DDR Read diode.

Hybrid Read Profile. The hybrid Read structure is a combination of single-drift flat structure and single-drift Read structure. The device resembles a p-type single-drift flat structure in series with an n-type single-drift Read diode. As shown in Figs. 7.15 and 7.16, two types of hybrid profiles can be realized, one is a flat–high–low structure, the other is a flat–low–high–low structure. The RF performances of these two structures are similar. Since the electron mobility is much higher than the hole mobility in GaAs, the single-drift Read side is always an n-type and the p region in the flat side is designed to be punchthrough at operating current to reduce the series resistance.

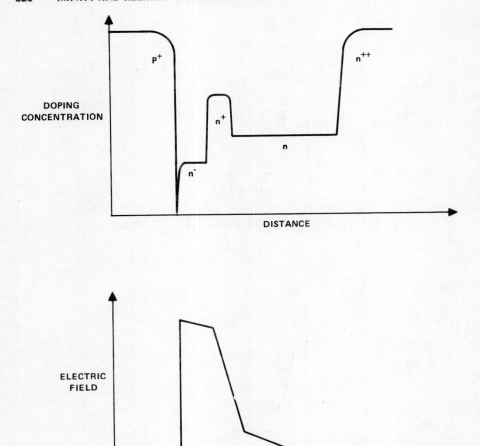

Figure 7.12 Doping and electric-field profiles for a single-drift Read low–high–low structure.

Similar to the double-drift Read diode, a larger-area diode can be used to increase the power output. The same thermal limitation applied to the double-drift Read structure also places an upper limit on the efficiency of the hybrid Read diode. Consequently, the hybrid Read diode should have the same efficiency but slightly higher power compared to the single-drift Read diode. Compared with the double-drift Read diode, this structure is easier to fabricate and the power output capability is similar.

Schottky-Barrier Diode. For a GaAs device, a Schottky-barrier junction formed by a metal–semiconductor junction can be used to replace the p^+n junction. IMPATT diodes made with a Schottky barrier have been reported with good output power and efficiency [24, 25].

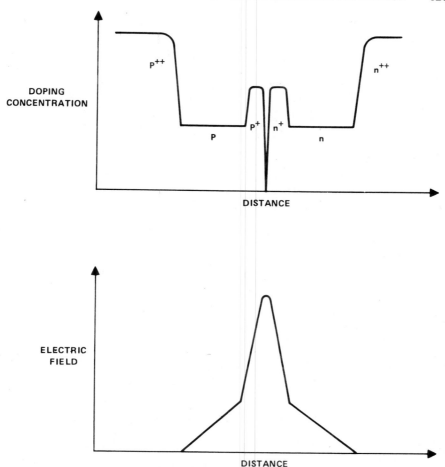

Figure 7.13 Doping and electric-field profiles for a double-drift Read low–high structure.

7.3.3 Device Design Example for a Silicon 94-GHz IMPATT Double-Drift Pulse Diode [8, 26]

The device design is based on an iterative process shown in Fig. 7.17. The theoretical value for diode parameters such as doping densities and epi-layer thickness is first obtained via a small-signal computer calculation. IMPATT diode wafers are then fabricated using these diode parameters. After diode fabrication, the diode profile is characterized by $C–V$ measurement or SIMS analysis. Finally, RF testing of diodes will yield information and correlate performance with device parameters.

For pulsed diode design, the primary consideration is the impedance–frequency characteristic of the diode as a function of current density. Because IMPATT operation is strongly dependent on the bias current density, the frequency for peak negative conductance is a function of the operating current. As the current density increases, the optimum

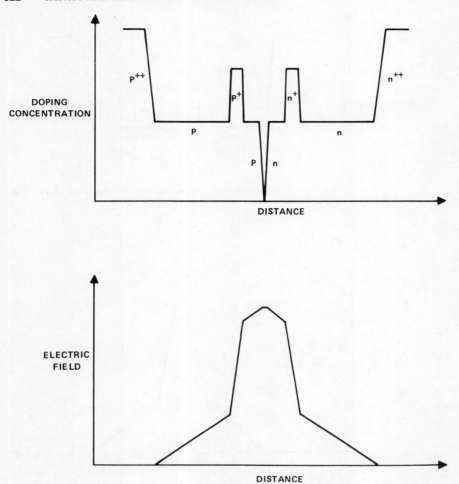

Figure 7.14 Doping and electric-field profiles for a double-drift Read low–high–low structure.

frequency and the diode output power also increase. For CW diodes the maximum current density is limited thermally, but for pulsed diodes this limit is extended many times, depending on the pulse width and duty factor. For an extremely narrow pulse, low-duty operation, the diode is no longer thermally limited. The current density can be extended further until space-charge effects cause power saturation and efficiency reduction. In other words, the ultimate diode output power is limited electronically. The theoretical design of silicon IMPATT diodes for a given frequency is normally carried out with a small-signal computer analysis. Strictly speaking, optimum diode design requires knowledge of the large-signal characteristics of the devices, and its performance is strongly dependent on the circuit parameters as well. Since an exact analysis and accurate prediction of the circuit response is difficult at millimeter wave frequencies, the diode design is based on the small-signal analysis, modifying the design subsequently according to the experimental results. For a specified current density, junction temperature, and junction doping profile, the small-signal computer program calculates and plots the dc electric

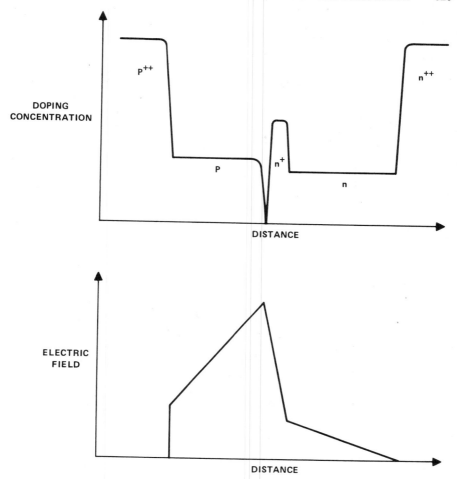

Figure 7.15   Doping and electric-field profiles for a hybrid Read flat–high–low structure.

field as a function of distance. The computer program then uses this dc solution to calculate the small-signal RF conductance and susceptance per unit area, and the device Q, as a function of frequency for a specified frequency range. The designer uses this program by running it iteratively for different values of input parameters, such as doping density, until a condition is reached for which the plot of device Q versus frequency displays its maximum near the desired frequency of operation. The parameters of the device that produce this are then taken as the design values. Note that the optimum frequencies cover a relatively wide range (i.e., the device Q varies slowly with frequency around its maximum point).

To design a pulsed diode properly, it is necessary to predetermine the operating current density. This value of current density that is used as an input parameter in the computer program must be determined independently, such as by theoretical means, which would generally include a thermal analysis, or by an estimate derived from the operating conditions of existing IMPATT diode. After several iterations in calculation, diode junction area measurement of an actual IMPATT diode, and consideration of the

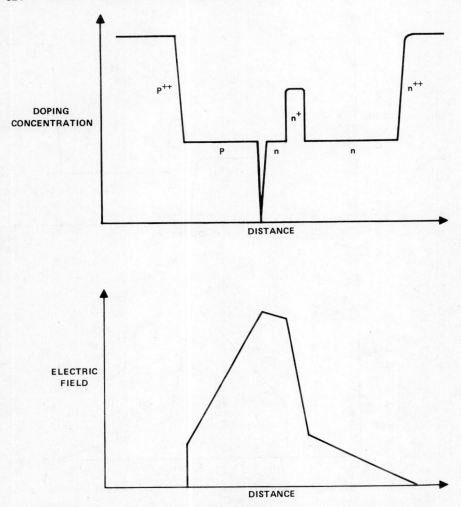

Figure 7.16 Doping and electric-field profiles for a hybrid Read flat–low–high–low structure.

transient thermal effects with a 15-W power goal, an optimum current density of 1.0×10^5 A/cm^2 was chosen for the design example. With a given operating current density and a maximum junction temperature of 250°C, the diode doping profile and epitaxial layer thickness can be calculated using small-signal analysis. A double-drift region structure is shown in Fig. 7.18. Figure 7.19 presents the small-signal device admittance as a function of frequency for $J = 1.0 \times 10^5$ A/cm^2 and $T_j = 250$°C. The doping concentration was selected to be 1.3×10^{17} atoms/cm^3. The device admittance was obtained from a computer simulation based on the small-signal analysis. The ionization rates of Grant [28] and the drift velocities given by Canali et al. [29] were used in the calculation. It is seen from Figure 7.19 that the device Q reaches its maximum near 94 GHz. The total epitaxial layer thickness was determined to be 0.90 μm. The breakdown voltage was computed to be 15 V. Table 7.1 is a summary of the 94-GHz pulsed double-drift diode design. The measured diameter of a 4.5-pF diode is about 100 μm.

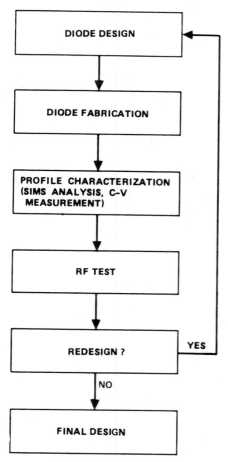

Figure 7.17   Diode design optimization procedure. (From Ref. 27; copyright 1981 IEEE, reproduced by permission.)

7.3.4   Device Fabrication Example for a Silicon Double-Drift IMPATT Diode [8, 26]

Although different techniques can be used to fabricate an IMPATT diode, a typical fabrication procedure is given here as an example. Considering the fabrication of a p^+nn^+ diode, the procedure can be divided into several tasks: epitaxial growth and evaluation, low-temperature diffusion, substrate thinning, and metallization.

Epitaxial Growth and Evaluation.   The growth and evaluation of high-quality epitaxial silicon films on low-resistivity substrates is fundamental to the fabrication of high-performance millimeter wave IMPATT diodes. Starting with an n^+ substrate, the growth of silicon epitaxial material for IMPATT diodes is accomplished with the pyrolysis of silane through the following simplified reaction:

$$SiH_4 \longrightarrow Si + 2H_2$$

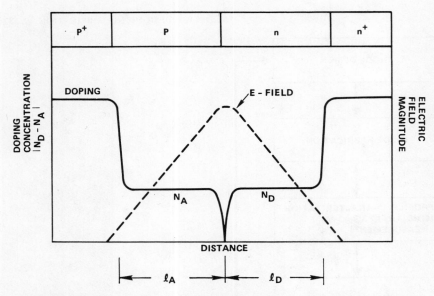

Figure 7.18 Doping profile and electric-field configuration for double-drift diodes.

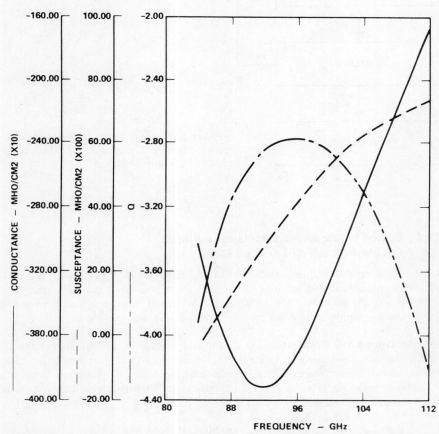

Figure 7.19 Computer output of the small-signal admittance for a 94-GHz pulsed double-drift IMPATT diode. (From Ref. 8; copyright 1980 IEEE, reproduced by permission.)

TABLE 7.1 94-GHz Pulsed Double-Drift Diode Design Parameters

Doping concentration ($N_A = N_D$)	1.3×10^{17} cm^{-3}
Epi thickness ($l_A = l_D$)	0.45 μm
Breakdown voltage	15.0 V
Junction diameter	100–110 μm
Current density	1.0×10^5 A/cm^2
Bias voltage	20.0 V
Bias current	10–12 A
Junction capacitance	4.0–5.0 pF
Peak power outputs	10–15 W
Heat sink	Copper for pulse width $\leq$ 100 ns

Reaction is complete at temperatures above 800°C. The growth temperature used is 1000°C. This temperature is optimum for obtaining good thickness control and reproducibility while limiting outdiffusion of dopant from the substrate. For a very thin layer, the use of molecular beam epitaxy (MBE) could be advantageous. Devices grown by MBE have been reported for both Si and GaAs IMPATTs [30, 31].

A key step in high-frequency IMPATT fabrication is evaluation of the epitaxial layer deposited; the epi layer must not only have the proper thickness and resistivity necessary to obtain the desired frequency characteristics, but the impurity concentration as a function of distance through the layer must be uniform, with no concentration irregularities or variations in excess of $\pm 10\%$. To monitor the quality of the epi material, a representative impurity-doping profile is commonly obtained for all epi runs utilizing capacitance versus voltage measurements made on selectively located *p-n* junction regions.

Capacitance per unit area at zero bias is a function of the background impurity concentration; as the reverse voltage is increased, the junction depletion layer width increases while the capacitance decreases. The relationship between the effective doping and variation of capacitance with voltage is given by

$$N_{\mathrm{eff}} = \frac{1}{q\epsilon A^2}\left(\frac{C^3}{dC/dV}\right) \tag{7.40}$$

where q is the electron charge, ϵ the dielectric constant, and A the diode area. The doping concentration at various levels below the epi surface can be obtained from the capacitance at various depletion layer widths. It can be seen that in this case, the spread of the depletion layer under bias into the heavily doped region is negligible, so that a one-sided abrupt junction approximation can be made. The profile measurement gives the doping concentration and epi thickness in the lightly doped epitaxial layer.

For double-drift diodes, the magnitudes of N_A and N_D are approximately the same, so that depletion occurs equally on both sides of the junction. The resulting profile plot gives information on the total depletion width and an effective doping density N_{eff}, given by

$$\frac{1}{N_{\mathrm{eff}}} = \frac{1}{N_A} + \frac{1}{N_D} \tag{7.41}$$

There is no effective direct way to separate N_A and N_D from the profile data obtained by $C\text{–}V$ measurement.

Secondary ion mass spectroscopy (SIMS) analysis can be used to supplement the $C\text{–}V$ measurement method. In SIMS analysis, a section of the semiconductor wafer is

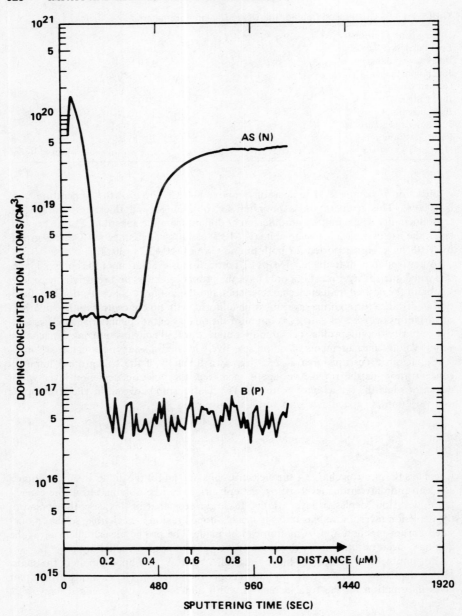

Figure 7.20   Doping profile of a 200 to 250-GHz IMPATT single-drift diode from a SIMS analysis. (From Ref. 27; copyright 1981 IEEE, reproduced by permission.)

mounted in a sample holder that is placed in a SIMS sample vacuum chamber. An incident ion beam of cesium impinges on the semiconductor target in a 20 mil × 20 mil square raster which bombards a square hole slowly and at a constant rate into the sample wafer. The secondary ions that are driven off from the central region of this square crater are analyzed by a computer-controlled mass spectrometry system to determine the doping concentration. This equipment can produce relative concentration profiles of the dopant elements, as well as of almost any other desired elements, such as possible contaminants within the sample. Absolute concentration doping profiles can readily be obtained by calibrating the system with a sample containing the desired element of known concentration. The absolute accuracy of the result depends on the accuracy to which the concentration of the elements are known in the calibration sample. By controlling the key parameters of the instruments a nearly constant etching rate of the surface can be achieved. A final measurement of total depth of the etched away surface with a Dektak surface profile measuring system then gives a very accurate depth calibration of the doping profile with a resolution of 0.01 μm.

An example of the SIMS method of junction profiling is shown in Fig. 7.20 for a single-drift diode. In the production of this wafer an arsenic-doped n-type silicon substrate was used, the n epitaxial layer was grown, and the p-n junction was formed by means of boron diffusion. The complete, unambiguous doping profile, including the junction depth, epi thickness, doping concentration, and uniformity, is shown.

Low-Temperature Diffusion. After the growth of multiple epitaxial layers, a shallow p^+ diffusion was followed to form a highly doped region for ohmic contact. To ensure that the p dopant (boron) diffusion does not degrade the epitaxial p-type doping profile, the diffusion temperature must be kept low (about 1000°C). At this temperature the boron diffusion is relatively slow, as is the outdiffusion of arsenic from the substrate. Because of the difference in diffusion coefficients, diffusion from the surface from a high-concentration boron source is typically more than five times faster than outdiffusion from the substrate.

Diffusion at lower temperatures has a similar behavior. The 1000°C temperature was chosen primarily because of ease of depth control. For higher temperatures control is more difficult because the total time for a shallow diffusion is less (on the order of 1 to 2 min). The time required for temperature stabilization is considerably longer than this, resulting in unwanted outdiffusion from the substrate. The p^+ layer can be obtained by ion implantation as well as diffusion.

Substrate Thinning. At high frequencies the diode series resistances can be significant. The series resistance is caused by the unswept epitaxial material and skin effect. The unswept epitaxial layer can be reduced to a minimum by accurate diode design. The skin-depth resistance must be taken into consideration in high-frequency diode design. The skin depth can be written

$$\delta_s = \left(\frac{2}{\omega \mu \sigma} \right)^{1/2} \tag{7.42}$$

where ω is the angular frequency, μ the permeability, and σ the conductivity of the material. It can be seen that the skin depth is decreased as the frequency increases. Non-uniform current distributions due to skin depth are expected in high-frequency diode because of the high current densities.

The current nonuniformity degrades device efficiency through parasitic series resistance, and it is therefore an important parameter in device design. Figure 7.21

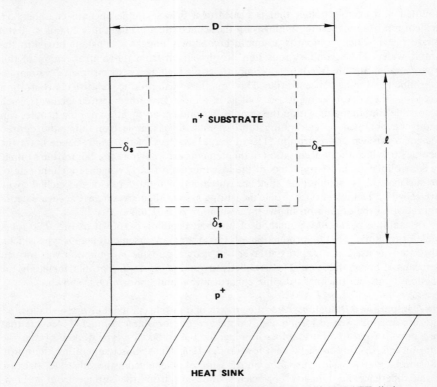

Figure 7.21   Cross-sectional view of a cylindrical single-drift IMPATT diode.

illustrates the skin-depth regions located inside the substrate of an IMPATT diode. The substrate series resistance due to the hollow thin-walled cylinder of length l is given by

$$R_s = \frac{l}{\pi \sigma \delta_s (D - \delta_s)} \tag{7.43}$$

This resistance can be minimized by reducing length l. One thinning technique is described in the following. The wafers are first mounted on a sapphire disk with black wax. Protective wax is also applied to the outer edge of the wafer to retain a thick silicon rim for handling purposes after thinning. Thinning is accomplished chemically using a barrel-etch process. The etchant used is a 3:5:3 mixture of $HF/HNO_3/HAC$ with a small content of Br^+ ions. This etchant is used for critical substrate thinning because it yields reproducible etch rates and a good surface morphology. Final thicknesses of 10 to 15 μm are achieved.

Metallization.   From the accelerated life test results, it was found that a system employing evaporated chromium (700 Å), platinum (2000 Å), gold (1500 Å), and plated gold (6000 Å) on the front p^+ side of the wafer has given consistently reliable results. The metallization on the back or substrate side of the wafer consists of evaporated chromium (700 Å), a platinum barrier layer (typically 2000 Å), and an evaporated gold layer (3000 Å). This is followed by 1 to 3 μm of plated gold on both sides to facilitate thermocompression bonding.

Photolithography is used to divide the wafer into individual pill diodes. The metallized substrate side is photomasked, forming photoresist dots. The Au and Cr are etched, leaving metal dots approximately 200 μm is diameter. Mesas are etched in the silicon with nonpreferential silicon etch using the Au and Cr as an etch mask. At the completion of this step the individual pill diodes are ready for packaging.

7.3.5 Device Packaging

Depending on the operating frequencies, various packaging techniques are used to achieve the maximum performance. A good package should have the following features:

1. Low RF loss from the packaging material
2. Low electrical parasitics so that the package self-resonance is well above the operating frequency
3. Low thermal impedance between the diode chip and the remainder of the millimeter wave circuit
4. Mechanically rugged
5. Hermetically sealable
6. Reproducible

At low frequencies, a ceramic pill type of package will be sufficient. At high frequencies (above 60 GHz), a quartz-ring pill type of package is used to reduce the package capacitance. Figure 7.22 is an exploded view of the package showing the method of assembly for a quartz ring package [26]. The dimensions for a 94-GHz diode are shown in Fig. 7.23 [8]. The package parasitics are modeled by an ideal inductance L_p, representing the gold ribbon, and a capacitance C_p, representing the package capacitance, which

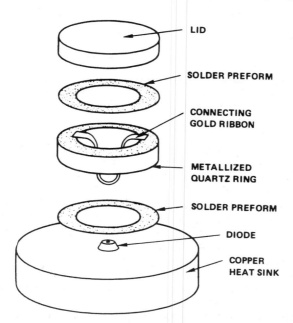

LID

SOLDER PREFORM

CONNECTING GOLD RIBBON

METALLIZED QUARTZ RING

SOLDER PREFORM

DIODE

COPPER HEAT SINK

Figure 7.22   Assembly of quartz ring package with copper disk heat sink.

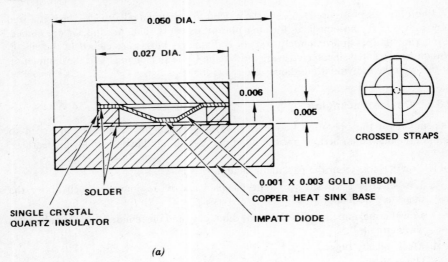

(a)

(b)

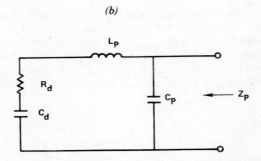

Figure 7.23 Quartz ring package and its equivalent circuit for a 94-GHz IMPATT diode. (a) Package configuration; (b) Equivalent circuit of the diode and package. (All dimensions are in inches.) (From Ref. 8; copyright 1980 IEEE, reproduced by permission).

is largely due to the quartz ring. The device impedance R_d and C_d will be transformed to an external impedance Z_p. A typical value for the package capacitance is 0.1 pF and the package inductance is 0.03, 0.04, and 0.075 nH for a crossed-strap, full-strap, and half-strap configuration, respectively.

For frequencies above 100 GHz, an open package is used to further reduce the package parasitics to the required level. A double-quartz-standoff package and direct-contacting method have been used for 140- and 217-GHz operation [32–34]. Both structures have inherent mechanical limitations. Gold balls can easily be knocked off, causing irreparable damage to the diode. To overcome this problem, a bridged, double-quartz-standoff package was developed, as shown in Fig. 7.24. In this package, a single 2-mil-wide gold ribbon contacts the diode and two quartz standoffs on opposite sides. The standoffs are also bridged by another gold ribbon, which serves as a contacting pad for the bias pin. Compared with the direct-contact method, this package is more reliable and mechanically rigid. The package has been used successfully at frequencies from 100 to 255 GHz.

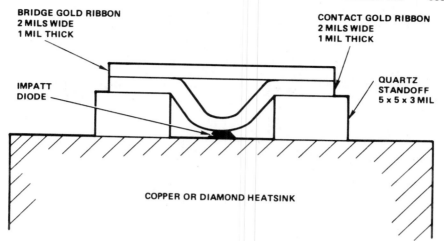

Figure 7.24 Bridged, double-quartz-standoff package. (From Ref. 27; copyright 1981 IEEE, reproduced by permission).

The diode and its package can be bonded to a copper or diamond heat sink, depending on applications. The diamond heat sink is used only for CW operation, to reduce the thermal resistance.

7.4 OSCILLATORS

One of the major applications of IMPATT diodes is for oscillators. In this section, circuit and device considerations for CW and pulse IMPATT oscillators are covered. The performance parameters of these oscillators for both silicon and GaAs diodes are given.

7.4.1 Oscillator/Amplifier Circuits

Many oscillator/amplifier circuits have been developed in waveguides for IMPATT diodes. Examples of these circuits are depicted in Fig. 7.25. They may be grouped into three basic types: reduced height, cap resonator, and cross-coupled coaxial waveguide cavity. In each circuit, a sliding short tuning element is used to achieve optimum performance at a specified frequency. The same type of circuit can be used for CW or pulsed oscillator/amplifier, but the dimensions will be optimized for each specific application.

Theoretical analyses for these circuits are available. For the reduced-height or full-height post mounting circuit, analyses have been developed by Eisenhart and Khan [35] and Fong et al. [36]. The cross-coupled coaxial-waveguide circuit can be designed using a circuit model developed by Chang and Ebert [8] or by Williamson [37]. The cap resonator circuit has recently been solved by Bates [38].

Microstrip circuits have also been developed for IMPATT diodes, although the output power is generally less than that from a waveguide circuit. Various microstrip oscillator circuits have been reported. The circuits generally consist of an impedance-matching network and a resonant circuit for stabilizing the oscillation. The resonant circuit can be a metal resonant cap [39, 40], a resonant line section [41–44], or a dielectric resonator [45]. Various circuit configurations are given in Fig. 7.26.

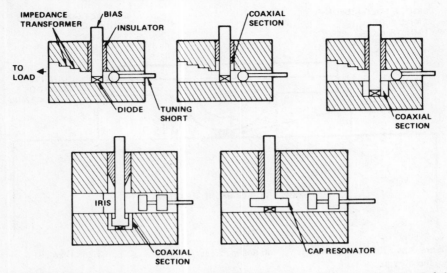

Figure 7.25   Examples of microwave/millimeter wave IMPATT oscillator circuits. (From Ref. 6, reproduced by permission).

7.4.2   General Theory

The two-terminal active devices can generally be represented in terms of impedance or admittance networks at the frequency range of operation. The device impedance or admittance should have a negative real part in the frequency range of interest. A detailed oscillator theory is described by Kurokawa [46].

The impedance of the device at the fundamental frequency can be expressed as

$$Z_D = -R_D + jX_D, \qquad R_D > 0 \tag{7.44}$$

Assuming only a fundamental component of RF voltage present in addition to the dc voltage, one can write

$$i = I_0 + I_{rf} \sin \omega t \tag{7.45}$$

where I_{rf} is the magnitude of the fundamental component of the RF current.

The device impedance is a function of frequency (f), dc current (I_0), RF current (I_{rf}), and temperature (T). Thus

$$Z_D = Z_D(f, I_0, I_{rf}, T) \tag{7.46}$$

A similar expression can be written for the device admittance by replacing the current with voltage.

A general oscillator circuit is shown in Fig. 7.27, where Z_D is the device impedance and Z_c is the circuit impedance looking at the device driving point. The transformer network includes the diode package and embedding circuits. Z_c can be expressed as

$$Z_c(f) = R_c(f) + jX_c(f) \tag{7.47}$$

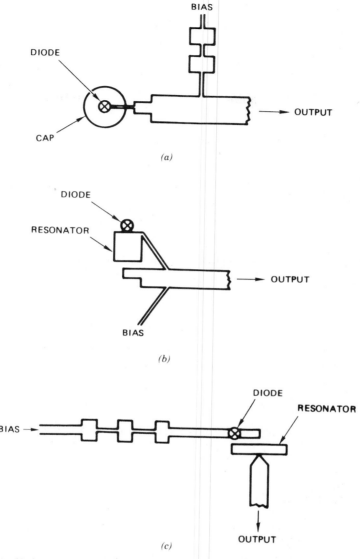

Figure 7.26 Various microstrip IMPATT oscillator circuit configurations: (*a*) resonant-cap circuit; (*b*), (*c*) resonant-line circuits.

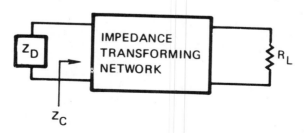

Figure 7.27 General oscillator circuit.

TABLE 7.2   Design Parameters for CW Silicon Double-Drift IMPATT

Frequency (GHz)	$N_D = N_A$ (cm^{-3})	Active Region Width (μm)	Diode Breakdown Voltage (V)
35	2×10^{16}	1.4	40
60	0.9×10^{17}	1.0	22
94	1.1×10^{17}	0.60	16
140	2.5×10^{17}	0.4	10
220	5.5×10^{17}	0.25	8.5

The oscillation occurs in the following conditions:

$$\text{Im}(Z_D) = -\text{Im}(Z_c) \tag{7.48}$$

$$|\text{Re}(Z_D)| > \text{Re}(Z_c) \tag{7.49}$$

Im and Re mean imaginary and real parts, respectively. Thus, at the oscillating frequency f_0,

$$R_c(f_0) < R_D(f_0, I_0, I_{rf}, T) \tag{7.50}$$

$$X_c(f_0) = -X_D(f_0, I_0, I_{rf}, T) \tag{7.51}$$

Equation (7.51) can be used to calculate the oscillating frequency. The condition of Eq. (7.50) controls the output power.

7.4.3   CW and Pulsed Diodes

CW and pulsed diodes operate at different current levels. A CW diode is operated at a lower current level than that of a pulsed diode. Since the impedance–frequency characteristic of the diode is a strong function of current density, the doping level for the CW diode will be different from the pulsed diode to compensate the current effects. As the current density increases, the optimum frequency increases and so does the diode output power [47]. For CW diodes the maximum current density is limited thermally, but for pulsed diodes this limit is extended many times, depending on the pulse width and duty factor. For extremely narrow pulse widths and low-duty operation, the diode is no longer thermally limited. The current density can be extended further until space-charge effects cause power saturation and efficiency reduction. Tables 7.2 and 7.3 show the doping profiles for silicon CW and pulsed double-drift diodes.

TABLE 7.3   Design Parameters for Pulsed Silicon Double-Drift IMPATT

Frequency (GHz)	$N_D = N_A$ (cm^{-3})	Active Region Width (μm)	Diode Breakdown Voltage (V)
35	3×10^{16}	1.2	37
60	1.1×10^{17}	0.8	20
94	1.3×10^{17}	0.45	15
140	3×10^{17}	0.2	8
220	7×10^{17}	0.1	6

7.4.4 Thermal Considerations for Silicon CW Diodes [6]

For reliable operation, IMPATT diode junction temperatures should be kept as low as possible. On the basis of a life test, it has been estimated that 250°C is safe for multiple-year operations, as shown in Fig. 7.28. To keep the junction below 250°C and at the same time achieve the maximum output power, the heat that dissipated in the diode must be removed efficiently. This heat removal can best be accomplished by placing a heat sink as near as possible to the junction where the heat is generated. To accomplish this goal in practice the diode is thermal-compression bonded to a copper or silver heat sink with the p^+ side down, as shown in Fig. 7.29. Similarly, a copper or silver heat sink may also be plated on the p^+ side of the diode. Since type IIA diamond provides high thermal conductivity, metallized type IIA diamond is also used as a heat sink for millimeter wave diodes. In addition, the p^+ layer should be made as thin as possible. Since the thermal resistance due to the p^+ layer is inversely proportional to the area, this requirement is accentuated particularly at millimeter wave frequencies, where the diode junctions necessarily become small.

Consider the thermal model of IMPATT diode as shown in Fig. 7.29. The model contains all the important features of actual devices, and excellent agreement has been found between calculated values of thermal resistance based on the models and experimentally determined values. The p^+ region of the cylindrical p^+nn^+ diode is bonded to

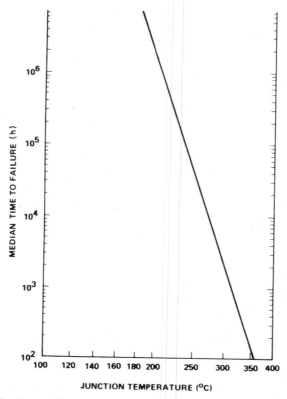

Figure 7.28 Median time to failure versus junction temperature characteristics for millimeter wave silicon IMPATT diodes derived from accelerated life test.

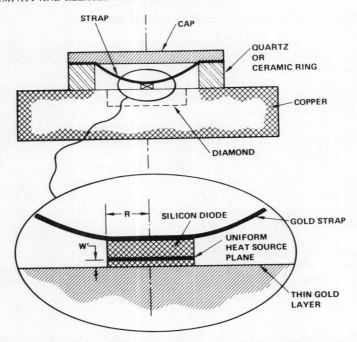

Figure 7.29 Typical package configuration and thermal model for a millimeter wave IMPATT diode.

the copper heat sink through an interface of gold. (Thin buffer metallization layers such as Cr and Pt are neglected in the analysis, as their contribution to the thermal path is small.) The mechanism of heat flow is conduction from the active region of the diode through the heat sink to ambient. Detailed calculations indicate that the effect of corrections introduced to take into account heat flow up through the substrate and gold contact ribbon is small. The temperature for the configuration is highest along the diode centerline. Since the diode radius is much larger than the length of the heat path within the diode (typically by a factor of 30), a one-dimensional mathematical treatment for the layered diode structure may be employed to calculate the steady-state distribution of this maximum temperature. The power dissipation rate and heat flux are considered to be uniform in the radial direction. The heat sink is taken to be a semi-infinite half-space. The heat dissipation profile follows the electric field profile in the active region since the current is constant. Thus the power dissipation is maximum at the p^+-n interface (in a double-drift diode) and decreases linearly to zero toward the n-n^+ interface.

For the millimeter wave IMPATT diode, it can be shown that although the power dissipation is distributed throughout the device active region, the effective heat source plane may be placed near the p-n junction without introducing a significant amount of error in the analysis. This means that the problem can be treated exactly as if all heat is generated at this plane. The thermal conductivity of the silicon k_{Si} depends on the temperature. However, detailed calculations show that the temperature variation across the active region is small, so that only a small error is introduced by using a constant value for k_{Si}. With this assumption, a solution of the one-dimensional linear heat flow equation for silicon leads simply to

$$R_{Si} = W'/k_{Si}A \qquad (7.52)$$

where W' is the distance from the effective heat source plane to the heat sink and A is the device area. For a p^+pnn^+ structure, we have $W' = W_{p^+} + W_p$.

The solution for the thermal spreading resistance for a two-layer heat sink has been obtained for the case where the thermal conductivities of the material in each layer are essentially independent of temperature. Such is the case for a gold bonding layer on a copper heat sink. The result is

$$R_{HS} = \frac{1}{\pi R k_1} \int_0^\infty \frac{1 + pe^{-2UH}}{1 - pe^{-2UH}} J_1(U) \frac{dU}{U} \qquad (7.53)$$

where $H = t/R$, $p = (k_1 - k_2)/(k_1 + k_2)$, and where k_1 and k_2 are, respectively, the thermal conductivities of the gold layer and copper heat sink. The thermal conductivity of diamonds has an inverse temperature dependence. However, an estimate of thermal resistance can be achieved by assuming a constant value of 9 W/cm K for the thermal conductivity of diamond.

The thermal analysis for the model above involves a straightforward solution of the one-dimensional heat conduction equation. The maximum diode temperature T_{max} occurs at the n-n^+ interface. The diode thermal resistance R_{TH} is defined as

$$R_{TH} = (T_{max} - T_A)/P_{diss} \qquad (7.54)$$

where T_A is the ambient temperature and P_{diss} is the total power dissipated. The thermal resistance may be approximately represented as a function of diode junction diameter d as

$$R_{TH} = (4/\pi)(1/k_{Cu}d)\{1 + (t/d)[(k_{Cu}/k_{Au}) - 1]\} + (4/\pi)(W'/k_{Si}\, d^2) \qquad (7.55)$$

where the k's are thermal conductivities of copper (or diamond), gold, and silicon; d is the diode junction diameter; and t is the thickness of the gold layer between silicon and copper. The first term of Eq. (7.55) is the contribution to R_{TH} from the copper heatsink with a small correction factor

$$(t/d)[(k_{Cu}/k_{Au}) - 1]$$

arising from the presence of the gold layer.

The thermal resistance of millimeter wave IMPATT diodes, calculated as a function of the junction diameter by using typical parameters as shown in Table 7.4, is plotted

TABLE 7.4 Parameters Used in Thermal Resistance Calculation

Quantity	Symbol	Value
Diode diameter	d	Variable
Thickness of p^+ region	w_p^+	0.3 μm
Thickness of n-region	w_n	0.6 μm
Thickness of Au layer	t	1.0 μm
Power dissipated	P	5 W
Ambient temperature	T_A	300-K
Thermal conductivity of copper	k_{Cu}	3.94 W/cm-K
Thermal conductivity of gold	k_{Au}	2.97 W/cm-K
Thermal conductivity of silicon	k_{Si}	0.8 W/cm-K
Thermal conductivity of diamond	k_{dia}	9.0 W/cm-K

Source: Ref. 6, reproduced by permission.

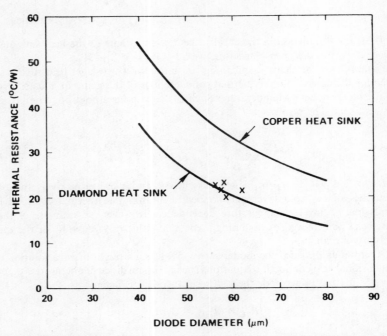

Figure 7.30 Thermal resistance versus diode diameter. (x, experimental values for diamond heat sink.)

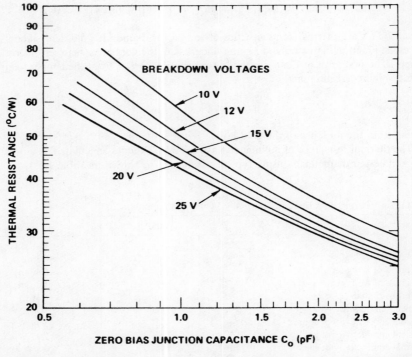

Figure 7.31 Thermal resistance versus zero-bias junction capacitance for millimeter wave IMPATT diodes (silicon diode on copper heat sink).

in Fig. 7.30. Note that the type IIA diamond heatsink provides a thermal path twice as effective as the copper heatsink. In practice, however, it is easier to measure the junction capacitance, which is a function of the diode geometry. In Fig. 7.31 the calculated thermal resistance of the millimeter wave IMPATT diodes is plotted as a function of zero-bias junction capacitance. The curve is useful for evaluating power dissipation capability for a given diode design.

7.4.5 Transient Thermal Considerations for Pulsed Diodes [6, 47]

The pulsed diode is normally operated at very low duty cycle and the diode is not thermally limited. To operate the diode at both high peak power and low junction temperature, the accurate transient thermal resistance of a diode must be determined. An equation and analysis of this transient thermal resistance have been given in Ref. 6 and 47. Figure 7.32 presents the results for a typical 94-GHz pulsed IMPATT diode as a function of pulse width for several diode junction diameters.

A unique property of the pulsed IMPATT oscillator is the frequency chirp during the bias pulse. This effect is a direct consequence of the IMPATT junction temperature variation during pulse, which results in diode impedance change. For a flat current pulse, the diode junction is at a low temperature at the beginning of the pulse and gradually heats up with its thermal time constant. This temperature variation depends on current density, junction area, and thermal resistance. As the diode heats up during the pulse, the device impedance changes accordingly. In a fixed-tuned circuit, the oscillation frequency, therefore, varies according to the junction temperature change. The faster the

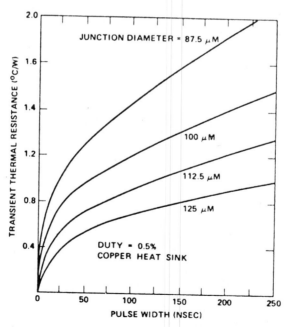

Figure 7.32 Transient thermal resistance of a typical W-band pulsed IMPATT diode plotted as a function of pulse width for several diode junction diameters. (From Ref. 47; copyright 1979 IEEE, reproduced by permission).

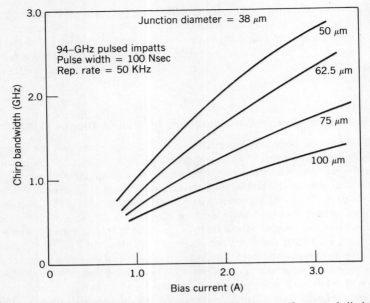

Figure 7.33 Typical chirp bandwidth as a function of bias current for several diode junction diameters. Pulse width = 100 ns; repetition rate = 50 kHz. (From Ref. 47; copyright 1979 IEEE, reproduced by permission).

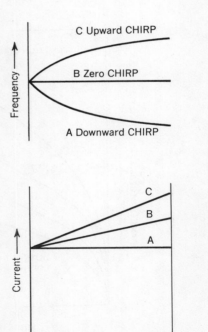

Figure 7.34 Frequency chirp characteristics of a double-drift IMPATT diode responding to a current ramp. (From Ref. 47; copyright 1979 IEEE, reproduced by permission).

diode heats up (a large transient thermal resistance), the greater the rate of chirp. A larger temperature excursion also leads to a greater amount of chirp. A typical measured chirp characteristic of 94-GHz pulsed double-drift IMPATT diode is plotted in Fig. 7.33 as a function of bias current for four diode junction diameters. It is seen that for a given bias current pulse, the chirp frequency (Δf) decreases with increasing junction diameter. This is due to the increase in the diode thermal time constant with the increasing junction diameter. For a fixed diode diameter, the chirp frequency increases with increasing bias current.

Since the diode impedance is also dependent on bias current, the amount of frequency chirp and rate of chirp can be controlled by the current pulse waveform. The frequency variation caused by the thermal effect can, therefore, be compensated by changing the operating current density. The method is illustrated in Fig. 7.34. For a flat current pulse, the oscillator frequency decreases (i.e., downward chirps). By providing an upward ramp on the current pulse, the amount of chirp can be decreased. A continuous increase in ramp slope will reach a point where the thermal and current effects cancel each other and little chirp is present. Further increase in the ramp slope beyond this point will cause frequency to chirp upward. Thus by controlling the current waveform, the frequency chirp characteristics can be controlled.

7.4.6 CW Oscillator Performance

CW IMPATT oscillators have been built from 7 to 400 GHz for silicon diodes and up to 130 GHz for GaAs diodes. The results are summarized here.

CW Silicon IMPATT Oscillators. At 10 GHz, a silicon IMPATT oscillator can produce 5 to 10 W of output power [6]. At millimeter wave, power levels of 2 to 3 W at 35 GHz [6], 1 to 2 W at 60 GHz [23], 500 mW to 1 W at 94 GHz [48], 100 mW at 140 GHz [27], 60 mW at 170 GHz [27], 78 mW at 185 GHz [3], 25 mW at 217 GHz [27], 7.5 mW at 285 GHz [3], and 200 μW at 361 GHz [3] have been achieved. The power variation as a function of frequency follows a relationship Pf = constant for frequencies below 100 GHz and Pf^2 for frequencies above 100 GHz. This indicates that the power is determined by thermal limitation for frequencies below 100 GHz and by circuit impedance limitation for frequencies above 100 GHz. The steep power falloff at high frequencies is due mainly to the increased adverse effects of diode, package, and mounting parasitics.

Since an IMPATT diode has a negative resistance over a very wide frequency band, the oscillator can be tuned over a wide frequency range using mechanical tuning or bias current tuning techniques. Figure 7.35 shows the mechanical tuning characteristics of a 60-GHz IMPATT oscillator [6] as an example. It can be seen that the oscillator can be tuned over 10 GHz. Figure 7.36 shows the bias tuning characteristics of a 140-GHz oscillator.

The efficiency of a silicon IMPATT diode oscillator is normally below 10% for a junction temperature of 250°C. Some examples are: 10% at 40 GHz, 6% at 60 GHz, and 5.8% at 94 GHz [48].

CW GaAs IMPATT Oscillators. For frequencies below 50 GHz, GaAs IMPATT diodes offer higher efficiency and power output compared with silicon diodes. At 10 GHz, output power of 20 to 30 W has been reported [49]. At higher frequencies, power levels

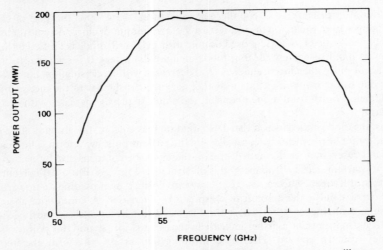

Figure 7.35 Mechanical tuning characteristics of a millimeter wave oscillator.

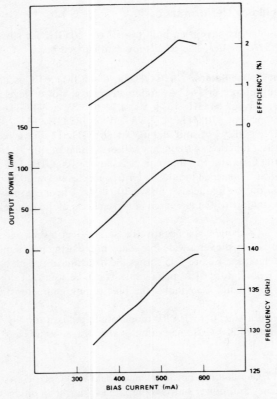

Figure 7.36 Output power, frequency, and efficiency for a CW diode. (From Ref. 27; copyright 1981 IEEE, reproduced by permission).

reported are 5 W at 20 GHz with 15 to 19% efficiency [50], 2.8 W at 44 GHz with 18% efficiency [51], and 5 mW at 130 GHz with 0.5% efficiency [25].

7.4.7 Pulsed Oscillator Performance

In many system applications, high-peak-power pulsed oscillators are required. IMPATT devices can be operated as pulsed power sources to achieve a high peak power output over a relatively short pulse width (about 100 ns) with a low duty cycle. Figure 7.37 shows a pulsed IMPATT oscillator with its associated pulse modulator.

Pulsed Silicon IMPATT Oscillators. In operating pulsed oscillators, the maximum pulse width as well as the pulse duty factor is one of the most important parameters that determine the achievable peak output power. Since IMPATT devices have small thermal time constants, as discussed in the preceding section, the junction temperature rises rapidly within a pulse. To achieve high peak power output from IMPATT oscillators, the pulse width should be kept below 100 ns and the pulse duty factor below 1%. For longer pulse widths or a higher pulse duty factor, the peak power output will decrease due to the reduced input power to keep the peak junction temperature below the upper limit consistent with the reliability.

Another important property associated with the pulsed operation of an IMPATT oscillator is the frequency chirping effect. As the junction temperature increases according to the transient thermal impedance change with a pulse cycle, the diode impedance (or admittance) changes. Typically, frequency chirping greater than 1 GHz can be obtained with pulsed IMPATT oscillators. Noting that the frequency of oscillation is also dependent on the bias current, we can control the amount of frequency chirping to meet specific system requirements by shaping the bias pulse current waveform, as shown in Fig. 7.34.

Figure 7.37 Pulsed IMPATT oscillator and modulator.

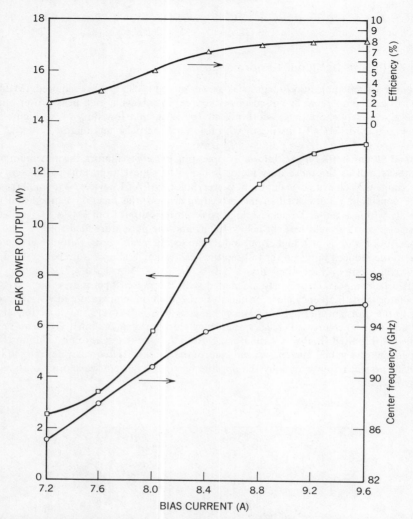

Figure 7.38 Output power, frequency, and efficiency versus bias current. (From Ref. 52; copyright 1979 IEEE, reproduced by permission).

The peak power output from a silicon IMPATT oscillator is 50 W at 10 GHz [6], 30 W at 35 GHz, 13 W at 94 GHz [52], 3 W at 140 GHz [27], 1.3 W at 170 GHz [27], and 0.7 W at 217 GHz [27]. A pulse width of 100 ns and a 25- to 50-kHz pulse repetition rate were generally used. Figure 7.38 shows the output power, frequency, and efficiency versus bias current. An oscilloscope picture of input bias current and output video pulse is given in Fig. 7.39. The efficiency is generally below 10%.

Pulsed GaAs IMPATT Oscillators. Pulsed GaAs IMPATT diodes provide a higher efficiency for frequencies below 50 GHz. One example is 16 W peak power at efficiencies up to 15% with a 5% duty cycle at 40 GHz [51].

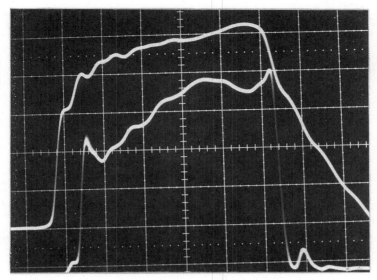

Figure 7.39 Oscilloscope picture of input bias current (top trace) and output video pulse (bottom trace). Horizontal: 20 ns/division; vertical: 2 A/division (current), 20 mV/division (video). (From Ref. 52; copyright 1979 IEEE, reproduced by permission).

7.5 POWER AMPLIFIERS

IMPATT devices have been used effectively as microwave/millimeter wave power amplifiers. For power amplification both stabilized amplifier and injection-locked amplifiers have been developed. While injection-locked amplifiers (or oscillators) are suited for high gain (>20 dB) and narrow bandwidth (<1 GHz) operation, stabilized amplifiers are for low gain (about 10 dB/stage) and broader bandwidth (>1 GHz) applications. The maximum power achievable from an amplifier is approximately the same as that obtainable from the same device operated as an oscillator.

7.5.1 Reflection-Type Stable Amplifiers [53, 54]

Two-terminal devices with negative resistance can be employed as reflection-type stable amplifiers. A circulator is used to separate the input and output ports. Figure 7.40 shows

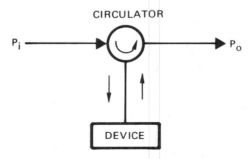

Figure 7.40 General reflection-type amplifier.

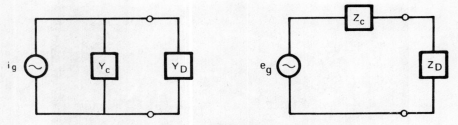

Figure 7.41 Equivalent circuit for the reflection amplifier.

a general reflection-type amplifier. The equivalent circuit for such a system can be represented in terms of a Norton or Thévenin equivalent circuit as shown in Fig. 7.41. Z_c and Y_c are the equivalent-circuit impedance and admittance, respectively, seen at the device terminal and thus include device package and transforming circuits. The gain for this type of amplifier is the reflection coefficient at the plane of the device terminal. The reflection coefficient is given by

$$\Gamma = \frac{Z_D - Z_c}{Z_D + Z_c} = \frac{Y_c - Y_D}{Y_c + Y_D} \tag{7.56}$$

The power gain is given by

$$\text{gain} = \frac{P_o}{P_i} = |\Gamma|^2 = \left| \frac{Y_c - Y_D^*}{Y_c + Y_D} \right|^2 \tag{7.57}$$

The power generation efficiency is defined by

$$\eta = (P_o - P_i)/(P_{dc} + P_i) \tag{7.58}$$

Y_D and Y_c can be expressed as

$$Y_D = -G_D + jB_D \tag{7.59}$$

and

$$Y_c = G_c + jB_c \tag{7.60}$$

Substituting into Eq. (7.57) and assuming that

$$B_c + B_D = 0 \tag{7.61}$$

Eq. (7.57) becomes

$$\text{gain} = (G_c + G_D)^2/(G_c - G_D)^2 \tag{7.62}$$

It is therefore obvious that any amount of gain can be obtained by proper choice of G_c. Note that when $G_c = G_D$, the gain becomes infinite and the device would oscillate. Therefore, for stable amplification, $G_c > G_D$. Any amount of small-signal gain can be obtained by choosing G_c properly.

7.5.2 Injection-Locked Amplifiers/Oscillators

If an external signal at frequency f_i and of power P_i is injected into a free-running oscillator whose frequency is f_0 and whose power output is P_o, then if f_i comes close to f_0, the free-running oscillator will be injection-locked by this external input signal, and all the output power will appear at f_i. The locking range (Δf) depends on the external Q of the oscillator and the power gain of the system as given by Adler [55]:

$$2\Delta f/f_0 = (2/Q_e)(P_o/P_i)^{-1/2} \tag{7.63}$$

where Δf is the one-sided locking bandwidth and Q_e is the external Q of the oscillator.

This phenomenon can be used to reduce the noise of an oscillator by locking it to an external low-noise source and can also be used as an amplifier, since within the locking range Δf, a small input signal at f_i appears as a large output signal at f_i. Of course, when power at f_i is removed, the power will continue to appear at f_0. The oscillator then works as an injection-locked amplifier. The gain of the system is given by

$$\text{gain} = P_0/P_i = (1/Q_e^2)(f_0/\Delta f)^2 \tag{7.64}$$

The gain of this type of amplifier is generally high. Locking gains of 10 to 30 dB have been reported. Equation (7.63) can be used for external-Q measurement for an oscillator since all other quantities in the equation can be measured.

Injection-locking phenomena have been studied extensively by Kurokawa [56]. The injection locking can be used to synchronize one or more oscillators to a lower power master or reference oscillator and also reduce part of the FM noise [57]. Subharmonical injection locking was also possible using a low-frequency injection signal [58].

7.5.3 Amplifier Performance

Both injection-locked and stable amplifiers have been demonstrated for CW and pulsed applications. A few examples are described in this section.

A 50-GHz silicon IMPATT diode amplifier was reported in 1968 by Lee et al. [59]. Both injection-locked and stable amplifiers were reported. The injection-locked amplifier has 20 dB of locking gain over a 500-MHz bandwidth and stable amplifier with 13 dB of gain and a 3-dB bandwidth of 1 GHz. The circuit uses a cap resonator circuit. Scherer reported a three-stage X-band injection-locked IMPATT amplifier with a total of 36 dB gain and a power output of 0.2 W over a 200-MHz bandwidth [60]. Kuno studied the nonlinear effects and large signal effects of a stable or injection-locked amplifier [17, 18]. The effects of bandwidth on the transient response of the IMPATT amplifiers as applied to phase-modulated signals and amplitude-modulated signal were investigated in detail. Figure 7.42 shows the calculated effects of the input signal level on the bandpass and phase characteristics of a stable IMPATT amplifier tuned to a small-signal gain of 10 dB. It can be seen that as the input signal level increases, the gain decreases and the bandwidth increases. A similar calculation for an injection-locked amplifier is given in Fig. 7.43. Peterson [61] reported a measurement and characterization technique that allows the design of IMPATT amplifiers operating at maximum efficiency. A 60-GHz stable amplifier with 6.9 dB of gain and a 1.9-GHz bandwidth was reported by Weller et al. [62]. The bandpass characteristics of this amplifier is shown in Fig. 7.44. A two-stage 86-GHz high-power IMPATT stable amplifier was reported with 10 dB of gain and 18 dBm of output power [63]. Recently, most high-power IMPATT amplifiers were developed with a power combiner as the output stage. Details of these amplifiers are presented in the next section, on power combiners.

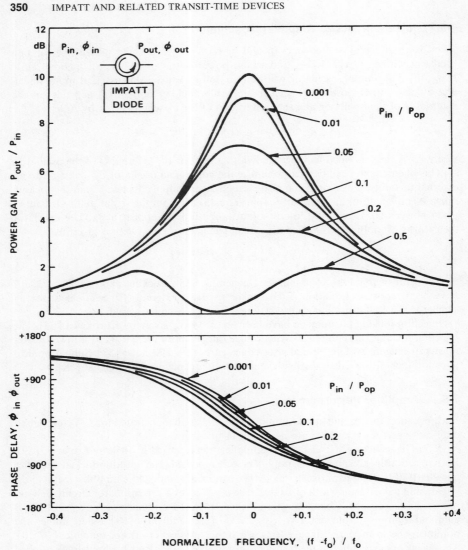

Figure 7.42 Large-signal effects on gain–bandwidth and phase-shift characteristics of stabilized IMPATT amplifier. (From Ref. 17; copyright 1973 IEEE, reproduced by permission.)

Since a circulator is always needed for an IMPATT amplifier construction, the effects of a nonideal circulator on the amplifier performance are important information. The effects have been studied by Bates and Khan [64].

Microstrip IMPATT amplifiers have also been developed using a microstrip circulator [43, 65]. Figure 7.45 shows a W-band all-microstrip IMPATT amplifier [43]. At the top is the microstrip circulator, with the input and output ports coupled to the transitions. At the bottom is the microstrip IMPATT oscillator. Using the measurement setup shown in Fig. 7.46, the injection-locking gain as a function of power gain can be measured (Fig. 7.47). The external Q can be calculated using Eq. (7.63). For this particular

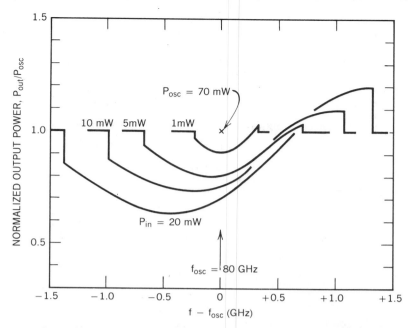

Figure 7.43 Large-signal effects on locking characteristics of an injection-locked IMPATT oscillator. (From Ref. 18; copyright 1973 IEEE, reproduced by permission.)

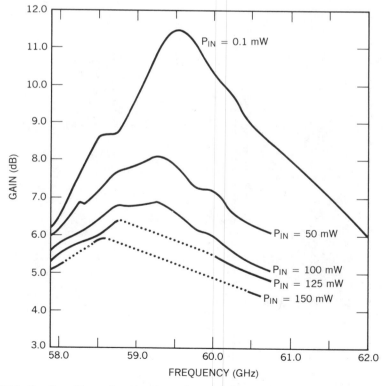

Figure 7.44 Small- and large-signal bandpass characteristics for circulator-coupled amplifier. The dc bias requirements are 475 mA current at 27.95 V. The measured diode thermal resistance is 18.6°C/W. (From Ref. 62; copyright 1978 IEEE, reproduced by permission.)

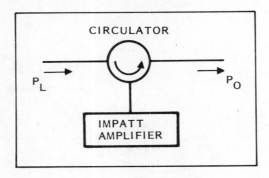

Figure 7.45   W-band microstrip IMPATT amplifier. (From Ref. 43; copyright 1985 IEEE, reproduced by permission.)

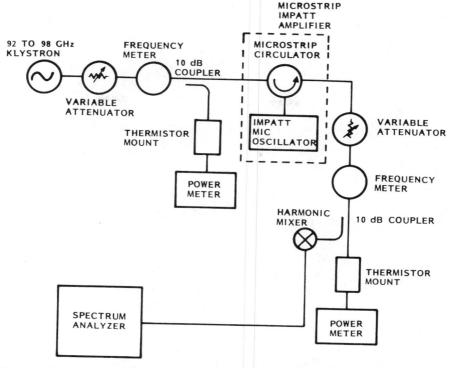

Figure 7.46 Injection-locking measurement system. (From Ref. 43; copyright 1985 IEEE, reproduced by permission.)

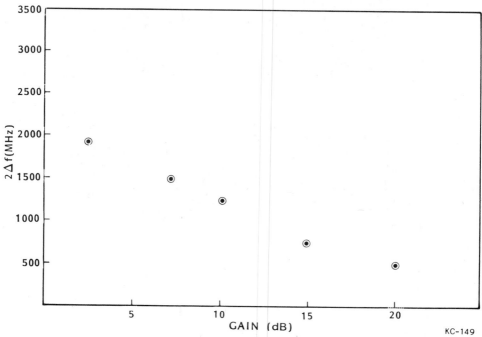

Figure 7.47 Injection-locking bandwidth ($2\Delta f$) as a function of power gain (P_0/P_L). (From Ref. 43; copyright 1985 IEEE, reproduced by permission.)

KC-149

353

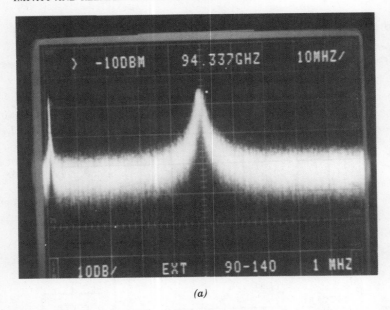

(a)

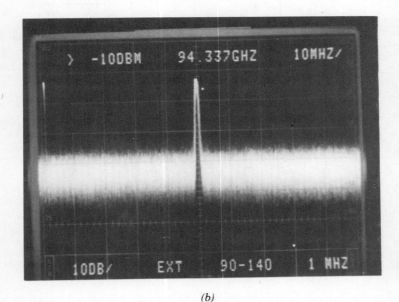

(b)

Figure 7.48 Spectra of locked and unlocked IMPATT amplifier: (*a*) free-running spectrum; (*b*) injection-locked spectrum. (From Ref. 43; copyright 1985 IEEE, reproduced by permission.)

circuit, the Q_e is approximately 37. Figure 7.48a shows the free-running spectrum of the IMPATT oscillator. After the application of the locking signal, the spectrum exhibits the low-noise characteristic shown in Fig. 7.48b.

7.6 POWER COMBINERS

Although the IMPATT diode is the most powerful millimeter wave solid-state device, the output power from a single diode is limited by fundamental thermal and impedance problems. To meet many system requirements, it is necessary to combine several diodes to achieve high-power levels. Many power combining techniques have been developed in the microwave and millimeter wave frequency range during the past 20 years. These techniques have been reviewed by Russell [66] and by Chang and Sun [67]. The methods of power combining fall mainly into four categories, as shown in Fig. 7.49 [67]: chip-level combiners, circuit-level combiners, spatial combiners, and combinations of these three. The circuit-level combiners can be further divided into resonant and nonresonant combiners. Resonant combiners include rectangular and cylindrical-waveguide resonant-cavity combining techniques. The nonresonant combiners include hybrid-coupled, conical waveguide, radial-line, and Wilkinson-type combiners. For IMPATT diodes, the three most commonly used combining techniques are rectangular resonator, cylindrical resonator, and hybrid-coupled combiners. IMPATT combiners have been built from 10 to 220 GHz.

7.6.1 Resonant Cavity Combiners

A resonant-cavity combiner was first proposed and demonstrated by Kurokawa and Magalhaes in 1971 [68] with a 12-diode power combiner that operated at X-band. The circuit consisted of a rectangular-waveguide cavity with diodes mounted in cross-coupled coaxial waveguide diode mounting modules in the waveguide walls. Kurokawa [69] also developed the oscillator circuit theory, which indicated why his circuit configuration gave stable oscillation, free from the multiple-diode moding problem. Later, Harp and Stover [70] modified the combiner configuration by replacing the rectangular resonant waveguide cavity with a cylindrical resonant cavity for increased packaging density to accommodate a large number of diodes in a small volume. This technique has been used to construct power combiners for various applications. The combiner can be used as an oscillator, injection-locking amplifier, or stable amplifier.

The resonant cavity combiner has the following advantages:

a. Combining efficiency is generally high because the power outputs of the devices combine directly without any path loss.

b. The scheme is capable of combining a number of diodes up to 300 GHz.

c. It has a compact size and can be used as a building block for multiple-level combining.

d. Built-in isolation exists between diodes to avoid mutual impedance variations by coupling to the cavity mode.

The disadvantages of resonant combiners are:

a. Bandwidth is limited to less than a few percent, although some techniques have been proposed to reduce the circuit Q and thus slightly increase the bandwidth [71, 72].

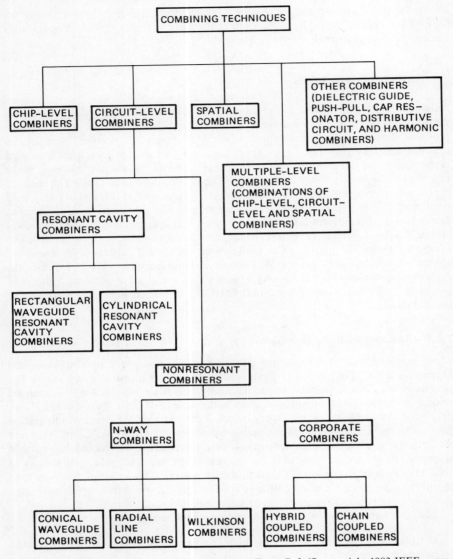

Figure 7.49 Different power combining techniques. (From Ref. 67; copyright 1983 IEEE, reproduced by permission.)

b. Number of diodes to be combined in a cavity is limited by moding problems since the number of modes increases with the cavity dimensions.

c. Electrical or mechanical tuning is difficult.

The rectangular and cylindrical waveguide resonant cavity combiners are shown in Fig. 7.50.

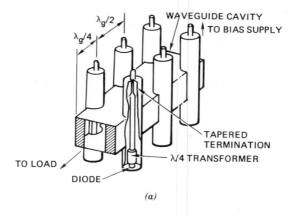

(a)

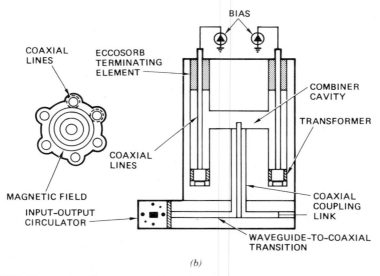

(b)

Figure 7.50 Resonant cavity power combiners: (*a*) Kurokawa waveguide combiner configuration and cross sections; (*b*) cylindrical resonant cavity combiner. (From Ref. 67; copyright 1983 IEEE, reproduced by permission.)

Resonant Rectangular Cavity Combiners. For a rectangular-waveguide resonant cavity combiner, each diode is mounted at one end of a coaxial line terminated by a tapered absorb, which serves to stabilize the oscillation. To couple properly to the waveguide cavity, the coaxial circuits must be located at the magnetic field maxima of the cavity; therefore, the diode pairs must be spaced one-half wavelength ($\lambda g/2$) apart along the waveguide (Fig. 7.50). The cavity is formed by the iris and a sliding short. Using this circuit, 10.5-W CW power at 9.1 GHz was achieved with 6.2% efficiency by combining 12 IMPATT diodes. To increase the diode capacity, two or more diodes can be positioned on either side of the peak magnetic field [73, 74].

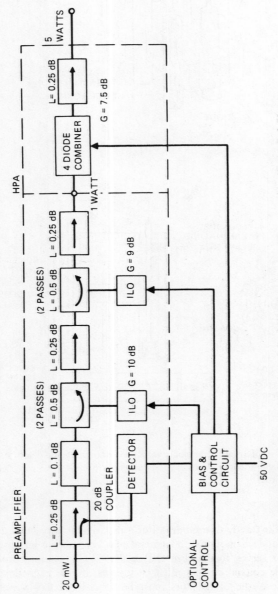

Figure 7.51 Schematic diagram of the 5-W solid-state amplifier assembly. (From Ref. 78; copyright 1985 IEEE, reproduced by permission.)

Since the inception of Kurokawa's combiner, many researchers have attempted to improve and apply the circuit in microwave and millimeter wave frequency ranges. For CW applications, a three-stage amplifier with an output stage using a 12-diode combiner was reported with 39 dB of gain and 16 W of output power at 20 GHz [75]. A two-stage amplifier using a two-diode combiner as the output stage was reported with 11 dB of gain and 3 W of output power at 44 GHz [76]. At 41 GHz, a 10-W output power was achieved with a two-stage amplifier with a 12-diode resonant rectangular cavity combiner as the output stage [77]. The power gain is 30 dB and efficiency is 10% over a 100-MHz bandwidth. The 43.5-to 45.5 GHz amplifier shown in Fig. 7.51 was developed with 5 W of output power and 24 dB of gain. The output stage is a four-diode combiner [78]. At 60 GHz, a two-diode combiner with 1.4 W of output power and a four diode combiner with 2.1 W of output power have been reported [79].

For IMPATT diodes in pulsed operation, a W-band two-diode combiner has been developed to generate 20.5 W at 92.4 GHz with 82% combining efficiency [8]. The diodes were operated with a 100-ns pulse width and 0.5% duty cycle. Each diode generated approximately 12 W in an optimized single-diode oscillator circuit. Later, the design was extended to a four-diode combiner to achieve 40 W of peak output power with 80% combining efficiency by combining four 10- to 13-W diodes [8]. The combining circuit is shown in Fig. 7.52.

At 140 GHz, a two-diode combiner was developed using a slightly oversized waveguide cavity. A peak output power of 3 W has been achieved at 145 GHz by combining two diodes with a 1- and 2.5-W power output, respectively [80]. The two-diode combiner was later modified and optimized to generate 5.2 W at 142.2 GHz from two 3-W diodes [27]. A four-diode combiner was also developed to produce 9.2 W of peak output power with a 100-ns pulse width and a 25-kHz pulse repetition rate. The combining efficiencies are about 80 to 90%. The same design was scaled up to 217 GHz in the development of a two-diode combiner with 1.05 W of output power [27].

Cylindrical Resonant Cavity Combiners. The cylindrical resonant-cavity combiner was first proposed by Harp and Stover [70]. The combiner shown in Fig. 7.50 consists of a number of identical coaxial modules on the periphery of a cylindrical resonator. The output power of each diode is combined by the cross-coupling of coaxial module and cylindrical resonator. The combined power is coupled to the load through a coaxial probe in the resonator and a coax-to-waveguide transition. This combining scheme has been established and is very successful in the microwave frequency range for its small size and symmetrical geometry (a few examples can be found in [71–72, 81–90]). However, at higher frequencies the cylindrical resonator is less desirable because of the moding problem and the requirement of an input/output coaxial probe.

A circuit design for bandwidth increase can be found in Ref. 84. Many cylindrical resonant cavity combiners have been developed at microwave frequencies. For example, a 16-diode combiner was built with 135 W peak and 45 W average output power at 10 GHz [86]. A 60-W CW six-diode combiner was developed at 5 GHz [82]. The highest-frequency cylindrical combiner built consists of a four-diode and an eight-diode combiner developed at 37 GHz with 3.6 and 5 W of CW output power, respectively [90]. One-watt silicon double-drift diodes were used in these combiners, which form the last two stages of a five-stage communication amplifier (Fig. 7.53).

7.6.2 Hybrid-Coupled Combiners

Unlike resonant combiners, the hybrid-coupled combiner has a wide bandwidth capability (larger than 5%); thus the combiner design can be achieved independent of the

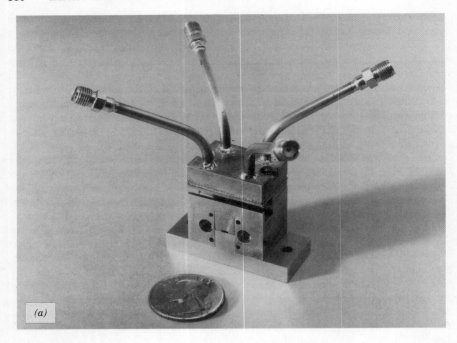

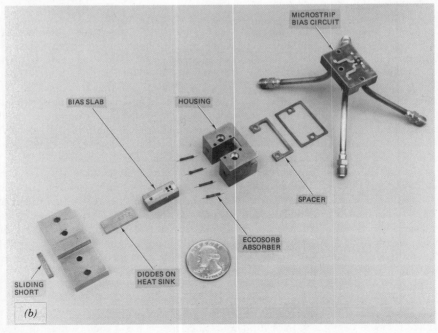

Figure 7.52 W-band, 40-W, four-diode combiner: (*a*) assembled; (*b*) disassembled. (From Ref. 8; copyright 1980 IEEE, reproduced by permission.)

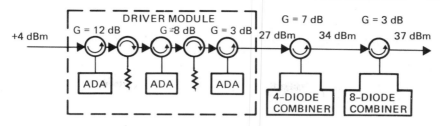

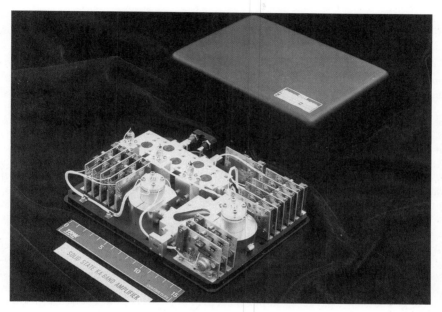

Figure 7.53 A 37-GHz, 5-W, five-stage amplifier using an eight-diode cylindrical resonant cavity combiner as output power stage. (From Ref. 67; copyright 1983 IEEE, reproduced by permission.)

hybrid characteristics. The hybrid coupler also provides isolation between sources, so that the device interaction and instability problems associated with multidevice operation are minimized. The design approach therefore reduces to that of the hybrid circuit and the diode module. The diode module can be a single-diode circuit or a combiner circuit.

The schematic diagram of a 3-dB hybrid-coupled combiner is shown in Fig. 7.54. The combiner is usually operated in the injection-locked mode for phase alignment. When input power is applied to port 1, the power is evenly coupled for ports 2 and 3, and port 4 is isolated from port 1. If ports 2 and 3 are terminated by a pair of matched amplifiers, a signal applied at port 1 is amplified and reflected from ports 2 and 3. The reflected waves are added at port 4 but canceled at port 1, due to the phase relationships between the two reflected waves. Thus the power applied to port 1 is amplified and accumulated at port 4.

An analysis by Nevarez and Herokowiz [91] provides the design guideline for this type of combiner. It was found that the amplitude balance and proper phase relationship must be achieved among individual sources at the same frequency. Therefore, in

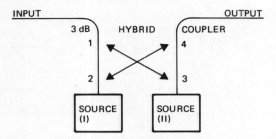

Figure 7.54 Hybrid-coupled combiner schematic diagram. (From Ref. 67; copyright 1983 IEEE, reproduced by permission.)

combiner development the individual cavity or module configuration is required to obtain proper tracking in amplitude, phase, and frequency. As the number of devices increases, the difficulty in achieving the required relationship among sources increases.

The hybrid coupler insertion loss also poses an upper limit on the number of amplifiers that can be combined; unfortunately, this loss increases with frequency. Therefore, the hybrid approach is not attractive for combining a large number of devices, especially at higher frequencies.

Use of the hybrid-coupled combining scheme has been demonstrated in the millimeter wave frequency at V- and W-band. At 60 GHz, a four-diode hybrid-coupled combiner was first reported by Kuno and English [92]. A two-stage amplifier was developed using a two-diode combiner at the first stage and a four-diode combiner at the second stage. A CW output power of 1 W has been achieved, with a small-signal gain of 22 dB and a bandwidth of 6 GHz in the 60-GHz range. The hardware of this amplifier is shown in Fig. 7.55. The amplifier was later improved to generate 2.5 W of CW output power at about 61 GHz by using four higher power double-drift IMPATT diodes [93]. At W-band, a combination of hybrid-coupled and Kurokawa combining schemes was first used by Yen and Chang to combine four two-diode combiners to generate 63 W of peak output power using eight 10- to 13-W diodes [94]. A three-stage injection-locked amplifier was developed using this eight diode combiner as the output stage (Fig. 7.56). A hybrid-coupled combiner/amplifier using a branch-line coupler in a microstrip medium has been established at microwave frequencies [95].

7.6.3 Other Combining Schemes

In addition to the commonly used combining techniques described above, several other combining schemes have been proposed and demonstrated for IMPATT diodes. A few examples of these methods are discussed here.

Josenhans first proposed combining IMPATT diodes electrically in series and thermally in parallel on a diamond heat sink [96]. An output power of 4.5 W at 13 GHz with an efficiency of 6.4% was achieved. The fundamental limitations of chip-level combining are the circuit impedance-matching and device interactions. The number of diodes is limited at high frequency, due to the small dimension and thermal interactions.

Perhaps the most successful chip-level combining was developed by Rucker et al. [97–99]. The combining geometries are shown in Fig. 7.57a. Quartz capacitors were placed in parallel with each diode chip to avoid the instability problems associated with multichip interactions. Most results reported are at X-band. An article summarizing the analytical and experimental results can be found in [99]. Recent experiments have extended this technique to 40 GHz [100].

POWER COMBINER OUTPUT STAGE   DRIVER STAGE

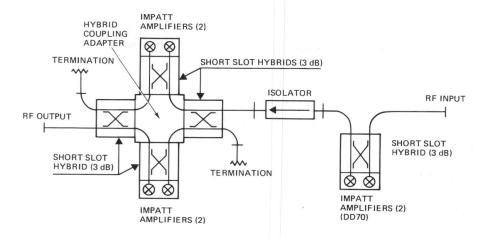

Figure 7.55 V-band two-stage IMPATT amplifier/combiner block diagram and hardware. (From Ref. 67; copyright 1983 IEEE, reproduced by permission.)

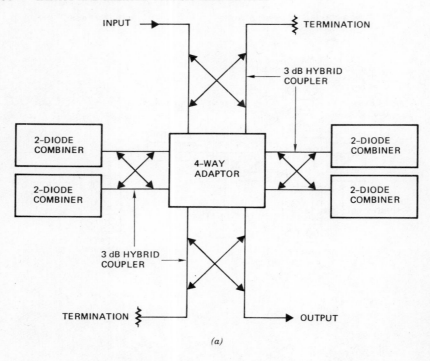

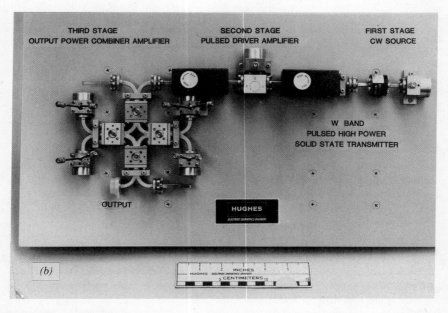

Figure 7.56 (*a*) Hybrid coupling for four two-diode combiners; (*b*) W-band 63-W peak output power three-stage injection-locked transmitter. (From Ref. 67; copyright 1983 IEEE, reproduced by permission.)

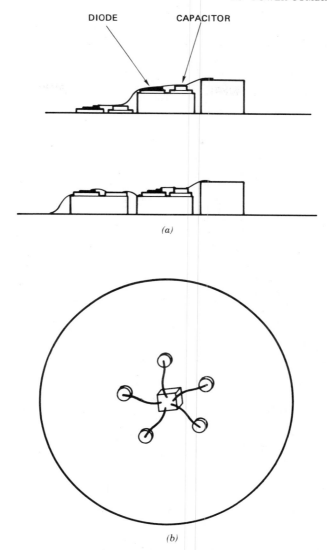

DIODE CAPACITOR

(a)

(b)

Figure 7.57 Chip-level combiner: (*a*) chip-level power combining geometries; (*b*) power combining through diode array. (From Ref. 67; copyright 1983 IEEE, reproduced by permission.)

A similar technique using the parallel diode array method was demonstrated by Swan et al. [101]. In this technique, diodes arranged in a small area are considered as a single diode from the RF point of view (Fig. 7.57*b*). Consequently, a single tuning circuit is sufficient for operation. However, as frequency increases, the lateral dimensions of the diode array are no longer small compared to the wavelength, and each diode does not share the same electromagnetic environment. Impedance matching between the array and circuit thus becomes more difficult due to the low impedance of the multidiode array.

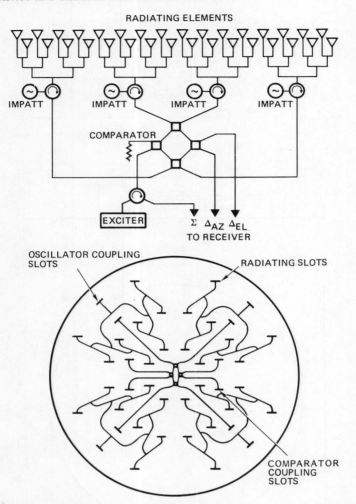

Figure 7.58 35-GHz spatial combiner block diagram and antenna array layout. (From Ref. 67; copyright 1983 IEEE, reproduced by permission.)

Spatial or quasi-optical combiners are promising for applications at high millimeter wave frequency since the size restriction and moding problems are less severe. Examples of these combiners are a 35-GHz active array developed by Durkin et al. (Fig. 7.58) [102] and a 10-GHz space power combiner by Dinger et al. [103]. The technique proposed by Hummer and Chang [104] for a spatial combiner with Gunn diodes mounted directly on microstrip patch antennas can also be used for IMPATT diodes. Quasi-optical combining techniques proposed by Mink [105], Wandiger and Nalbandian [106], and Young and Stephan [107] are also promising for IMPATT diodes.

Other combining techniques that have been used for IMPATT diodes are conical waveguide power combiners, radial-line power combiners, Rucker's combiners, push-pull combiners, and harmonic power combiners [67].

7.7   IMPATT DIODE FOR FREQUENCY MULTIPLICATION AND CONVERSION

The nonlinear properties of IMPATT diodes can be used for frequency multiplication and conversion [108]. From the small-signal analysis described in Section 7.2, it can be seen that the avalanche zone behaves as a nonlinear inductor. The "Manley Rowe" power relations show that the avalanche zone can operate as a parametric amplifier, a harmonic generator, a frequency multiplier, or a frequency up-converter or down-converter [109]. The nonlinear behavior of an IMPATT diode is due primarily to the nonlinearity of the avalanche zone. The influence of the transit zone is quite different, and its dimensions must be optimized [108].

For better nonlinearity and high-order harmonic operation, a punch-through diode at breakdown is preferred for its short transit time [108]. To eliminate series resistance, it is also desirable to have the p-n junction bounded by heavily doped p^+ and n^+ regions. Experimental output power results for an $\times 11$ multiplier versus input power are shown in Fig. 7.59. The diode used is a silicon p^+nn^+ diode with breakdown voltage $V_B = 20$ V and drift region width $W = 0.6$ μm. The output frequency is 38.8 GHz. It can be seen that a conversion efficiency of 30% was achieved for a 20-dBm input power. The use of IMPATT diodes for up-converters was also reported [108].

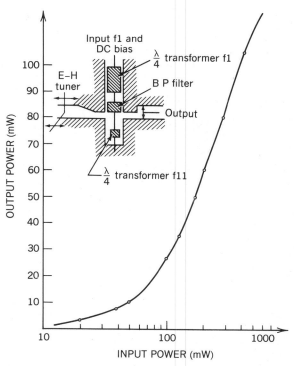

Figure 7.59   Experimental output power variations of $\times 11$ multiplier as a function of drive level. (From Ref. 108; copyright 1976 IEEE, reproduced by permission.)

7.8   NOISE CHARACTERISTICS AND SPURIOUS OSCILLATIONS

Noise characteristics of oscillators are important properties for system applications. The AM noise of the IMPATT diode is about 10 dB higher than in the Gunn oscillator. This makes the Gunn diode more suitable than the IMPATT diode for local oscillator application. The IMPATT oscillator also suffers from parametric and bias oscillation, which eventually leads to diode burnout if care is not taken to prevent it.

7.8.1   AM and FM Noise Characteristics [6]

Figure 7.60 shows the measured AM noise characteristics of typical millimeter wave oscillators. For comparison, those of Gunn oscillators and klystrons are also shown. It is interesting to note that the AM noise characteristics of IMPATT oscillators near the carrier are similar to those of Gunn oscillators and klystrons. At higher modulation frequencies (higher than several hundred megahertz), however, the IMPATT oscillator noise is higher than that of a Gunn oscillator or a klystron. For this reason IMPATT oscillators are in general difficult to use as local oscillators for mixers in receiver applications.

Shown in Fig. 7.61 is the calculated effect of local oscillator (LO) on receiver noise figures for various values of an LO noise suppression factor. For a typical IMPATT local oscillator with carrier-to-noise ratio of 150 dB/Hz, LO noise suppression of 30 dB is required to keep the adverse effect of the LO on the receiver noise figure negligible. The required LO noise suppression can be accomplished either by a bandpass filter for the LO frequency or by a balanced mixer configuration. With proper LO noise suppression techniques, IMPATT oscillators can be used effectively as local oscillators for low-noise millimeter wave mixers.

Shown in Fig. 7.62 are measured FM noise characteristics of a millimeter wave IMPATT oscillator. FM noise characteristics on the oscillator depend strongly on the

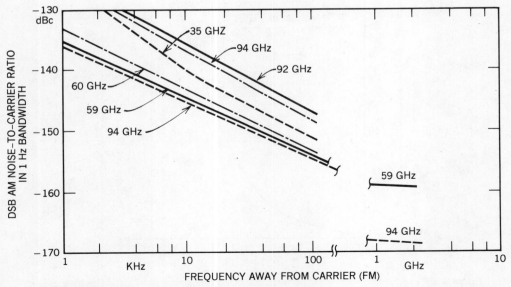

Figure 7.60   AM noise characteristics of millimeter wave oscillators. Solid line, IMPATT; dashed lines, Gunn; dashed–dotted line, Klystron. (From Ref. 6, reproduced by permission.)

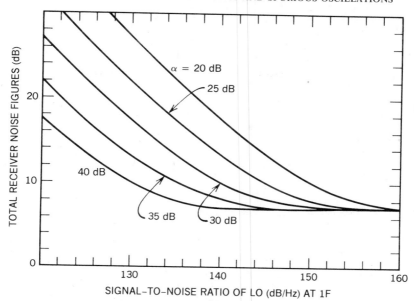

Figure 7.61 Effects of local oscillator noise on mixer noise figure. $NF = L_c(N_{IF} + N_R - 1) + P_{LO}(S/N)_{LO}/KT_0\alpha)$, $L_c(N_{IF} + N_R - 1) = 7$ dB, and $P_{LO} = 5$ dBm. (From Ref. 6, reproduced by permission.)

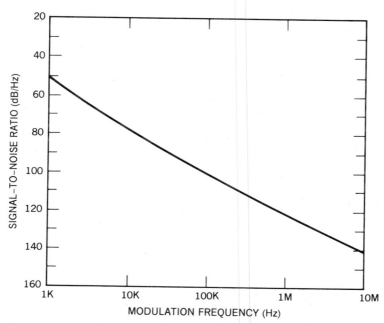

Figure 7.62 Measured FM noise characteristics of millimeter wave IMPATT oscillator. (From Ref. 6, reproduced by permission.)

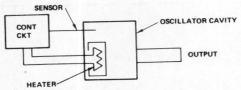

Figure 7.63 Cavity temperature-controlled frequency stabilization. (From Ref. 6; reproduced by permission.)

circuit Q. Values of the circuit Q for typical millimeter wave IMPATT oscillators range between 20 to 100. These values are based on the injection-locking gain–bandwidth characteristics measurements.

7.8.2 Frequency and Phase Stabilization Techniques [6]

A typical IMPATT oscillator has a frequency stability of $-0.5 \times 10^{-4}\,°C^{-1}$. This means that the frequency drift rate is approximately -2 MHz/°C at 40 GHz and -5 MHz/°C at 100 GHz, for example. For applications where temperature variations cause excessive frequency drifts, a number of techniques have been developed for controlling the frequency. The simplest method is to control the oscillator cavity temperature by means of a small heater and a control circuit (see Fig. 7.63). Since solid-state millimeter wave oscillator cavities have small masses, it is relatively easy to control their temperature in this way. The temperature variation can be kept to less than 1°C.

The frequency drift of an IMPATT oscillator can be reduced significantly by means of a high-Q cavity. Shown in Fig. 7.64 is a schematic diagram for the high-Q-cavity frequency stabilization technique [110].

Another approach to frequency stability is to use a frequency discriminator such as an Invar cavity filter with an AFC loop (see Fig. 7.65). This method does not require heater power but does require a more complex control circuitry.

In addition to long-term frequency stability, many systems require phase stabilities of crystal quality. For such applications two basic approaches have recently been developed for millimeter wave sources. One is a phase-locked-loop approach, and the other is an injection-locking approach using a multiplier chain. In the phase-locked oscillator (Fig. 7.66) the sample power of the millimeter wave frequency is converted down to an IF by a harmonic mixer, the phase is compared with the phase of the reference crystal oscillator, and the phase error is then corrected by means of a feedback loop to the millimeter wave oscillator. This technique has been applied successfully to millimeter wave IMPATT oscillators up to 217 GHz [111]. Shown in Fig. 7.67 is a comparison

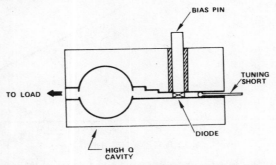

Figure 7.64 High-Q cavity for frequency stabilization. (From Ref. 6; reproduced by permission.)

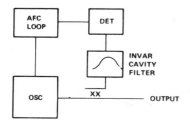

Figure 7.65 AFC loop-controlled frequency stabilization. (From Ref. 6; reproduced by permission.)

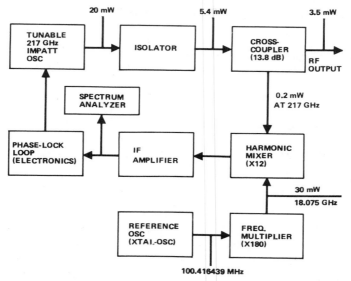

Figure 7.66 Experimental setup for a 217-GHz phase-locked source. (From Ref. 11; copyright 1969 SPIE, reproduced by permission.)

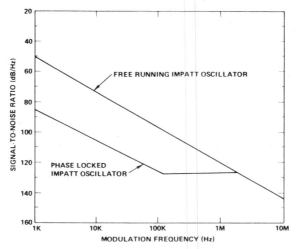

Figure 7.67 Improvement of FM noise characteristic in a 60-GHz phase-locked IMPATT oscillator.

371

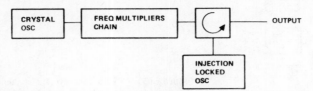

Figure 7.68 Injection locking with frequency multiplier chain for phase stabilization.

of phase noise of free-running and phase-locking millimeter wave IMPATT oscillators. A significant reduction in phase noise can be seen within the locking band. The locking bandwidth is limited by the locking loop bandwidth, which is typically 1 to 10 MHz.

Similar improvement in phase stability can be achieved by means of an injection-locked oscillator with a frequency multiplier chain, as shown in Fig. 7.68. Since trade-off between locking gain and bandwidth can be made, a broader locking bandwidth can be achieved in an injection-locked than in a phase-locked oscillator. However, an injection-locked oscillator using a multiplier chain is considerably more complex than a phase-locked oscillator at millimeter-wave frequencies.

7.8.3 Spurious Oscillations

Nonlinear effects within the IMPATT diode allow it to support spurious oscillations at any frequency. As diodes are operated for even greater power capacity, the control of spurious oscillations becomes more difficult.

The regions in which a 10-GHz IMPATT diode can support oscillations are illustrated in Fig. 7.69. Such a diode can be placed in a circuit to provide either amplification or free-running oscillation in the fundamental-frequency region around 10 GHz. The presence of this large-amplitude fundamental signal allows the possibility of spurious oscillations at frequencies in the other bands shown. From almost dc to approximately 1 GHz, bias-circuit oscillations can occur. Above the bias-oscillation range, but below the fundamental, parametric oscillations could be present, and above the fundamental, harmonics may be present.

If any spurious signals are present, effects ranging from performance degradation to actual diode burnout may occur. Nonharmonic spurious oscillations (i.e., oscillations below the fundamental frequency) have been shown to arise from two sources. Brackett [112] has shown that low-frequency instabilities arise from an interaction between bias circuit and a low-frequency negative resistance induced by the rectification properties of the nonlinear avalanche process in an IMPATT diode. This low-frequency negative resistance may act as an amplifier of bias-circuit noise, resulting in a modulation of the oscillator amplitude and frequency and producing the up-conversion of large amounts of noise to microwave frequency. Under certain conditions, the bias circuit may break into sustained oscillations, producing a combined AM and FM spectrum. The bias-

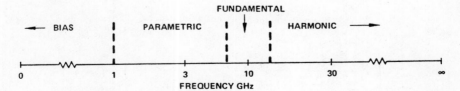

Figure 7.69 Different oscillations for a 10-GHz IMPATT oscillator.

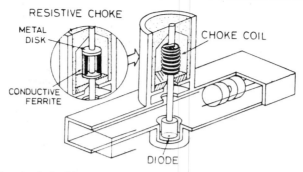

Figure 7.70   Bias circuit for bias-oscillation prevention. (From Ref. 113; copyright 1976 IEEE, reproduced by permission.)

circuit oscillations have a strong link to the low-circuit tuning-induced burnouts of the diode. To prevent these bias oscillations, an *R-L* circuit in the output lead of the power supply can be used. One of the schemes is shown in Fig. 7.70.

At frequencies above the bias-oscillation band, the nonlinear inductance of the IMPATT avalanche region could cause it to act as a parametric amplifier [114, 115]. Suppression of these parametric instabilities requires that the circuit impedance be kept low in the parametric oscillation band, to short out any idler voltage that may be present. A graphical technique for analyzing circuit stability is given by Schroeder [115].

7.9   RELATED TRANSIT-TIME DEVICES (BARITT, TRAPATT, TUNNETT, AND MITTAT)

Several transit-time devices related to IMPATT diode are discussed in this section: the BARITT, TRAPATT, MITTAT, and TUNNETT devices. These devices have several similar properties but differ in basic mechanism. They all exhibit a negative resistance or conductance in certain frequency ranges and thus can be used as oscillators or amplifiers. TUNNETT and BARITT devices can also be used as mixers, detectors, or self-mixers.

7.9.1   BARITT Diodes

A BARITT diode is a *Bar*rier *in*jection *transit-time* diode. BARITT operation was first reported by Coleman and Sze in 1971 using a metal–semiconductor–metal reach-through diode [116]. Since then both theoretical and experimental data regarding small-signal as well as large-signal characteristics of the device have been reported by various authors [117–119]. Because of the lack of avalanche delay time, the BARITT diode is expected to operate at lower power and efficiency than the IMPATT diode. On the other hand, the noise associated with carrier injection across the barrier is smaller than the avalanche noise in an IMPATT diode. This lower AM noise makes the BARITT suitable for low-power, low-noise applications such as in local oscillators [4].

The device structure and electric field profile of a BARITT diode are shown in Fig. 7.71. It is essentially a back-to-back pair of diodes biased into punch-through condition. N_D is the doping concentration in the *n*-layer, W the width of the *n*-layer, v_s the saturated velocity, E_s the field at which the velocity saturates, and E_{max} the maximum field in the devices.

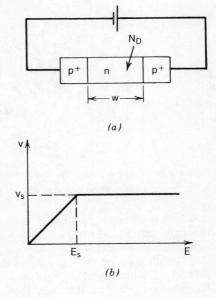

(a)

(b)

Figure 7.71 (*a*) Device structure, (*b*) velocity electric-field characteristic, and (*c*) electric-field versus distance in a BARITT device. (From Ref. 120; copyright 1976 IEEE, reproduced by permission.)

(c)

Using Poisson's equation, the punch-through voltage is given by [120].

$$V_{\mathrm{PT}} = \tfrac{1}{2}\frac{qN_D}{\epsilon}\,W^2 \tag{7.65}$$

Since the pulse of injected current occurs at the maximum voltage point, the optimum transit angle is approximately $3\pi/2$. Therefore,

$$\omega\tau = \frac{3\pi}{2} \tag{7.66}$$

where τ is the transit time through the diode. The optimum frequency is given by

$$f_{\mathrm{opt}} \approx \frac{3v_s}{4W} \tag{7.67}$$

For its low-noise characteristics, BARITT diodes have been used for self-mixing Doppler systems [120]. The device is superior to the Gunn device when prime power requirements are important. BARITT diodes were also developed for use as detectors and mixers at 35 and 95 GHz [121]. Compared with Schottky detectors, the sensitivity is comparable and the burnout performance and dynamic range are superior. As a mixer, the conversion loss is low and the noise figure is within 3 dB of that of a Schottky mixer. For its high pulse burnout performance, the BARITT diode should be useful for many radar and receiver applications.

For oscillator or amplifier applications, BARITT diodes have lower power output and efficiency than IMPATT if the carrier is injected at $\varphi = \pi/2$. However, higher efficiencies can be obtained if the carrier injection can be further delayed (i.e., $\pi/2 < \varphi \le \pi$) by using multilayered structures [122].

7.9.2 TRAPATT Diodes

TRAPATT stands for "*T*rapped *p*lasma *a*valanche *t*riggered *t*ransit." The TRAPATT mode of operation was first discovered by Prager, Chang, and Weisbrod in 1967 from silicon avalanche diodes [123]. The efficiency of the TRAPATT diode is much higher than that of the IMPATT diode, and the operating frequency is substantially lower than the transit-time frequency. Subsequent theoretical studies established that periodic avalanching of the diode begins at the high-field side and sweeps rapidly across the diode, leaving it substantially filled by a highly conducting plasma of holes and electrons whose space charge depresses the voltage to very low values [4]. Since the plasma cannot rapidly escape, this mode is called the trapped plasma mode (TRAPATT mode).

Extensive studies on the operating theory and applications have been reported [124 –129, 4]. It was found that the TRAPATT operation is quite complicated and care must be taken to control both device and circuit properties. In addition, the TRAPATT diode generally has higher noise than IMPATT and the operating frequency is practically limited to below the millimeter wave frequencies. Due to these limitations, TRAPATT diodes are rarely used today.

Oscillators and amplifiers using TRAPATT diodes have been developed up to 10 GHz. The highest pulse power of 1.2 kW, has been achieved at 1.1 GHz, with an efficiency of 25% [130]. Highest efficiency was obtained at 0.6 GHz with 75% efficiency [131]. A 8.25-GHz amplifier with peak power up to 38 W and bandwidths of 10% was also reported [132]. Since TRAPATT diode operates at low frequencies, coaxial circuit and microstrip circuits are normally used. A typical coaxial circuit is shown in Fig. 7.72.

7.9.3 TUNNETT and MITATT Diodes

The TUNNETT (*tunnel transit time*) diode, which operates as a fundamental oscillator, has shown some promise [133, 134] for low-noise local oscillator applications. TUN-NETT diode oscillators are lower in power than are IMPATT diode oscillators. However, the tunneling process in the TUNNETT diode generates much lower noise than the avalanche process in the IMPATT diode. The MITATT (*mixed tunneling and avalanche transit time*) diode provides a balance between the TUNNETT and IMPATT diodes. Medium-power output and relatively low noise (compared with IMPATT) performance can be expected from a MITATT diode oscillator. All three types of diodes have similar structures, so that the processing technology developed for IMPATT diodes can be applied directly to the fabrication of all three types of diodes.

TUNNETT and MITATT diodes show promise for local oscillator applications above the upper frequency limit of the Gunn diode, which is 140 GHz. TUNNETT diode operation has been demonstrated experimentally [133–136]. Reverse breakdown in a diode may occur due to an avalanche mechanism, a tunneling mechanism, or a

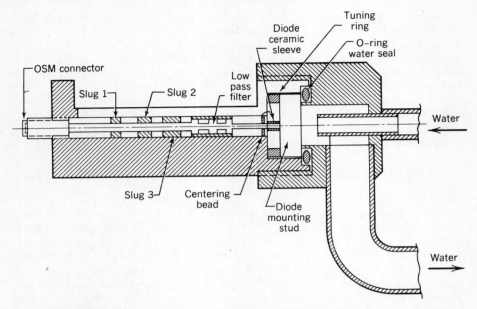

Figure 7.72 TRAPATT amplifier using coaxial circuit. (From Ref. 132; copyright 1974 IEEE, reproduced by permission.)

mixed avalanche–tunneling mechanism. The breakdown is avalanche dominated for a low electric field (< 500 kV/cm) and tunneling dominated for a high electric field (> 1000 kV/cm) [137]. For electric fields between these two limits, both the avalanche and the tunneling mechanisms are present. To achieve efficient TUNNETT or MITATT diode operation, it is necessary to design a doping profile that results in a high electric field to generate the tunneling process. The quiet tunneling process and the small time-constant features of the TUNNETT and MITATT diodes make them very useful for low-noise local oscillator applications beyond 100 GHz.

To distinguish TUNNETT diodes from IMPATT diodes, typical idealized voltage and current waveforms for transit-time devices are shown in Fig. 7.73. V_T is the terminal voltage, consisting of a dc voltage V_{dc} and an RF voltage, V_{rf}. I_{inj} and I_{ind} are the injected and induced currents, respectively. θ_w is the phase angle of the injected current pulse, θ_M is the phase angle at the center of θ_w, and θ_D is the drift region transit angle. Elta and Haddad [138] have calculated the efficiencies for three modes of operation. The device efficiency is

$$\eta = \frac{P_{rf}}{P_{dc}} = \left(\frac{V_{rf}}{V_{dc}}\right)\left[\frac{\sin(\theta_w 2)}{\theta_w 2}\right]\left[\frac{\cos\theta_M - \cos(\theta_M + \theta_D)}{\theta_D}\right] \tag{7.68}$$

For IMPATT mode operation, $\theta_M \simeq \pi$, $\theta_w = 0$ (corresponding to a δ function), the maximum efficiency η_I occurs at $\theta_D = 0.74\,\pi$ and is given by

$$\eta_I = \frac{2.27}{\pi}\left(\frac{V_{rf}}{V_{dc}}\right) \tag{7.69a}$$

For TUNNETT mode operation, $\theta_M = \pi/2$ (because the tunneling injected current is approximately in phase with the RF voltage) and $\theta_w = 0$, the optimum efficiency η_T,

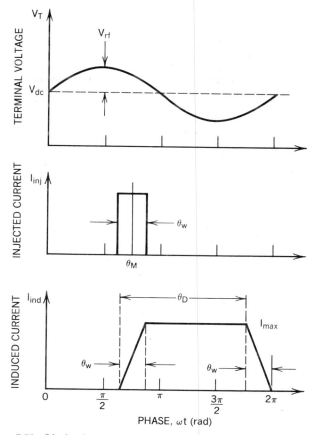

Figure 7.73   Ideal voltage and current waveforms for a transit-time device.

occurs at $\theta_D = 3\pi/2$ and is given by

$$\eta_T = \frac{2}{3\pi}\left(\frac{V_{rf}}{V_{dc}}\right) \tag{7.69b}$$

It is apparent from Eq. (7.68) and (7.69) that the TUNNETT mode will result in a lower efficiency than the IMPATT mode.

For MITATT mode operation, which corresponds to $\pi/2 < \theta_M < \pi$, the efficiency will lie between the IMPATT and TUNNETT modes and so would the noise performance. Elta and Haddad [138] have noted that for MITATT mode operation, the device efficiency will not be greatly degraded by moderate tunnel currents, but the noise performance can be greatly improved over IMPATT diodes.

The effects of tunneling in an IMPATT oscillator were first studied by Kwok and Haddad [139] based on the Gilden–Hines procedures [9]. The results from their study were not consistent with experimental results conducted by Lukaszek et al. [140]. Later, Elta and Haddad [137] modified the analysis by implementing the dead-space distances where no avalanche ionizations can occur. An equivalent circuit was also developed for a transit-time device under mixed tunneling and avalanche condition. However, their analysis is valid only for a simple diode structure; and the assumption that both electrons and holes have the same saturated velocity and impact ionization rate is not valid for silicon diodes. For accurate design of the TUNNETT and MITATT diodes, more general small-signal analysis, including tunneling and dead space, based on Misawa's [7] numerical method is needed.

In the avalanche zone of an IMPATT diode there is always a region where the field is low such that no ionization can occur. An electron (or hole) incident from one side of the avalanche region must travel a minimum distance to gain enough energy to initiate an ionizing collision. The nonionizing dead space is typically on the order of several hundred angstroms. This width is negligibly small compared to the total avalanche zone in a low-frequency IMPATT diode. However, for high-frequency diodes, the dead space becomes a significant portion of the total avalanche zone. This results in a reduction of the total ionization region and thus degrades diode efficiency. The dead space also causes a reduction in diode negative conductance [141]. If the avalanche region is smaller than the dead space, ionization will cease and the tunneling injection is the only mechanism for microwave generation.

A simplified TUNNETT and MITATT diode design is given here based on the ideal field distribution shown in Fig. 7.74. If the width of the high-field region W_T is made thin (about 500 Å) to avoid any avalanche process, the diode will be operating in the TUNNETT mode. On the other hand, if the width of W_T is made between 500 and 1000 Å, both the tunneling and avalanche processes would occur, and the diode will operate in the MITATT mode. As shown in Fig. 7.74, the electrons or holes tunnel (and avalanche, if any) through the high-field barrier and drift through a low-field region, which mainly determines the operating frequency.

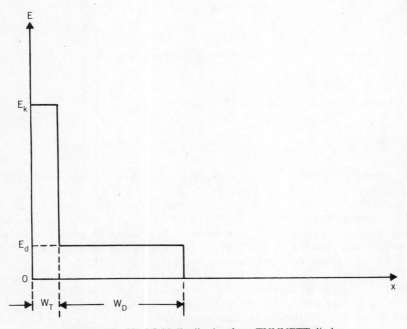

Figure 7.74   Ideal field distribution for a TUNNETT diode.

For optimum TUNNETT and MITATT operation. θ_D satisfies the following relation:

$$\theta_M + \theta_D \simeq 2\pi \tag{7.70}$$

For the TUNNETT mode $\theta_M = \pi/2$, thus θ_D is $3\pi/2$. For the MITATT mode, $\pi/2 < \theta_M < \pi$; therefore, $\pi < \theta_D < 3\pi/2$ for the same oscillating frequency.

TUNNETT and MITATT diodes have been made successfully at millimeter wave frequencies. Nishizawa et al. [133] reported a 200-GHz TUNNETT. Elta et al. [135] reported a 150-GHz MITATT with 3 mW of output power and 0.5% efficiency. The breakdown voltage of the device decreases as the temperature increases, which is a special characteristic of tunnel breakdown.

7.10 OTHER DEVICES

In the past two decades, many papers have been published on IMPATT diode and transit-time devices. To cover all these developments and results in a page-limited chapter is a difficult job. In this section, several topics or new developments not covered previously are discussed.

7.10.1 InP IMPATT Diodes

In addition to GaAs material, InP material has long been a popular and competing candidate for transferred electron devices such as Gunn diodes. Compared with GaAs, InP has many advantages, such as higher peak-to-valley ratio, higher drift velocity, lower diffusivity, reduced scattering time due to high threshold field, and narrower avalanche zone. Table 7.5 shows a comparison of silicon, GaAs, and InP material.

InP can be expected to develop as a material for high-power IMPATT oscillators with good spectral purity, high efficiency, and low noise. The efficiency of InP IMPATT is expected to be higher than that of GaAs and silicon for frequencies below 90 GHz. Above 90 GHz, silicon may be more competitive.

TABLE 7.5 Comparison of Semiconductor Material Properties

Property (Electrons)	Material		
	Silicon	GaAs	InP
Low field mobility (cm^2/V-s) (450 K)	400	5000	3000
Peak (threshold) velocity (cm/s) (450 K)	—	$\sim 1.2 \times 10^7$	$\sim 1.9 \times 10^7$
Peak-to-valley ratio (450 K)	—	2.2–2.4	3.0–3.1
Saturated drift velocity (cm/s) (450 K)	8.5×10^6	5×10^6	6×10^6
Upper cutoff Frequency for TE effect (GHz)	—	~ 100 (exp)	> 140 (thy)
Thermal conductivity (W/cm-°C) (300 K)	1.45	0.44	0.68
Energy gap (eV) (300 K)	1.11	1.43	1.34

Source: Ref. 142, reproduced by permission.

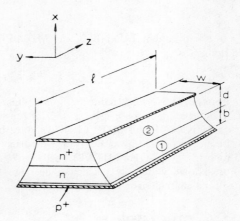

Figure 7.75 Distributed IMPATT structure. Region 1 is the active device, region 2 is the substrate, and the shaded areas are the metal contacts. (From Ref. 144, reproduced by permission from *International Journal of Electronics*.)

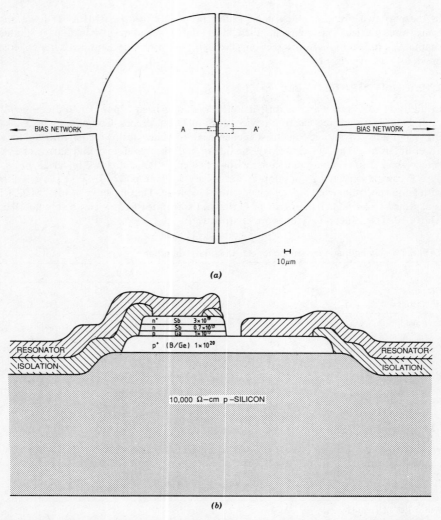

Figure 7.76 Monolithic IMPATT diode: (*a*) Circuit structure (*b*) device configuration. (From Ref. 150.)

Little work has been devoted to InP IMPATT development. Fank et al. [142] reported a p^+nn^+ IMPATT with a CW power of 1.6 W and 11.1% efficiency at 9.78 GHz. With a 10% duty cycle and 500-ns pulse width, the best performance obtained was 6.1 W of peak output power with 13.7% efficiency at 10.8 GHz. The operating voltage is generally higher for InP IMPATT diodes, due to the higher peak-to-valley ratio.

7.10.2 Traveling-Wave IMPATT Devices

A traveling-wave IMPATT device was first proposed by Midford and Bowers [143] in 1968. An analysis was provided by Hambleton and Robson in 1973 [144]. The structure is shown in Fig. 7.75, where region 1 is the active device region and region 2 is the substrate region. Analyses of similar structures have also been carried out by various researchers, including Franz and Beyer [145, 146], Fukuoka and Ioth [147], and Mains and Haddad [148]. The analyses are based on solving Maxwell's equation, distributed small-signal model, and transmission-line formulation. Experimental results were reported by Bayraktaroglu and Shih [149] in 1983 for distributed GaAs IMPATT oscillators with output power levels 1.5 W at 22 GHz, 0.5 W at 50 GHz, and 7 mW at 89 GHz.

7.10.3 Monolithic IMPATT Devices

Monolithic implementation of solid-state devices provides the potential for small size, light weight, low-cost production, improved reliability and reproducibility, and easy assembly. The monolithic realization of IMPATT diodes is lagging behind that of FETs and mixer diodes. Recently, Luy et al. reported a monolithically integrated coplanar 75-GHz silicon IMPATT oscillator [150]. Fig. 7.76 shows the circuit and device structures. The active layers were grown by silicon molecular beam epitaxy. A disk resonator was used to control the frequency, and oscillation of 76 GHz with a CW power of 1 mW was detected. Fig. 7.77 shows a SEM (scanning electron microscopy) picture

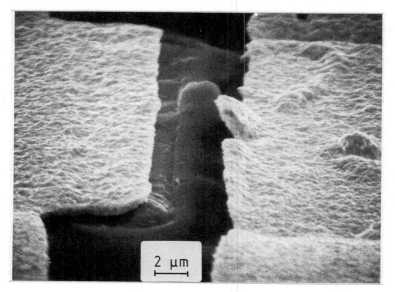

2 μm

Figure 7.77 SEM picture of the monolithic IMPATT diode. (From Ref. 150.)

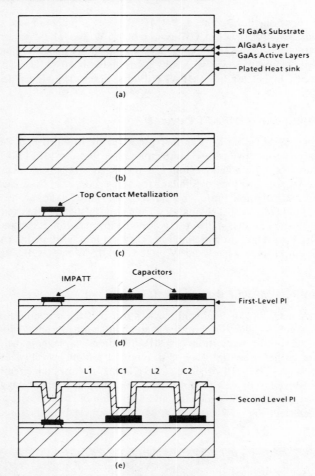

Figure 7.78 Processing sequence of monolithic IMPATT oscillator. (From Ref. 151; copyright 1985 IEEE, reproduced by permission.)

of the diode. GaAs monolithically compatible IMPATT oscillators have also been developed [151]. The processing steps are shown in Fig. 7.78. The best performance with output power of 1.25 W at 32.5 GHz was achieved with 27% efficiency.

7.10.4 Heterojunction IMPATT and MITATT Devices

Heterojunction structures have been used for FETs, HEMTs, and many optoelectronics devices. The use of a heterojunction for IMPATT or MITATT is relatively new [152]. Results from a large-signal analysis show that significant improvements in efficency can be achieved using heterojunction structures. Both single heterojunctions (GaAlAs–GaAs) and double heterojunctions (GaAlAs–GaAs–GaAlAs) have been studied [152]. Fig. 7.79 shows GaAlAs–GaAs–GaAlAs double-heterojunction transit-time device. Devices and oscillators using heterojunction structures were fabricated and

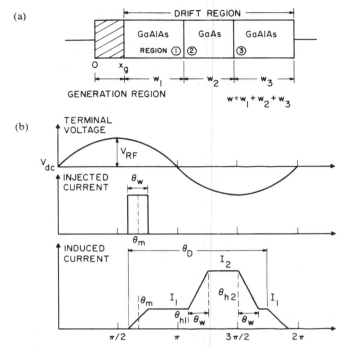

Figure 7.79 (*a*) GaAlAs–GaAs–GaAlAs double-heterojunction two-terminal transit-time device; (*b*) terminal voltage, injected current, and induced current for the device. (From Ref. 152; copyright 1987 IEEE, reproduced by permission.)

tested [152]. Preliminary results show a typical power of 45 mW at 72 GHz for a 1% duty cycle and 1-μs pulse width.

7.10.5 Active Antenna or Array Elements Using IMPATT Diodes

An active antenna element can formed by integrating the solid-state devices directly on the antenna. Many elements can be combined to build an active array or a spatial power combiner.

Recent development in microwave/millimeter wave integrated circuit has made it possible to integrate the active solid-state devices with planar antennas. Perkins [153] has mounted an IMPATT diodes on a circular microstrip patch antenna. Camilleri and Bayraktaroglu have built IMPATT diodes and resonator/antenna circuits monolithically [154]. Similar circuits have also been developed using Gunn diodes [155, 156].

7.11 SUMMARY AND FUTURE TRENDS

After over 20 years of research and development, IMPATT diode technology has gradually reached its maturity. Currently, IMPATT diodes can be produced routinely, operating from 10 to 300 GHz. IMPATT diodes are still the most powerful solid-state power sources at millimeter wave frequencies. At frequencies above 140 GHz, IMPATT or transit-time devices are the only feasible solid-state power sources operating at the fundamental frequency. With increasing future demands on millimeter wave systems in radar and communication applications, IMPATT and transit-time devices will play an important role in the successful development of these systems.

It is anticipated that continuing efforts will be directed to IMPATT and transit-time devices. Future research and development trends can be summarized as follows:

a. The development of high-efficiency IMPATT diodes will be continued. At lower millimeter wave frequencies, GaAs material is a better choice for high-efficiency diodes.

b. There is no doubt that the use of IMPATT devices will be extended into submillimeter wave frequencies. The combination of conventional microwave techniques and quasi-optical approach will be used for submillimeter wave IMPATT source development.

c. Quasi-optical and spatial power combining using integrated circuit antenna array technology will emerge as a feasible approach to combine the output power of many IMPATT devices.

d. Experimental and theoretical work on IMPATT devices and circuits will be continued to improve efficiency, reliability, and reproducibility. Good reproducibility and reliability are critical to many system applications.

e. It is anticipated that efforts will be continued on the research of new transit-time devices. Successful development of traveling-wave IMPATT devices and monolithic IMPATT devices will have a far-reaching impact on future applications.

REFERENCES

1. W. T. Read, "A Proposed High-Frequency Negative Resistance Diode," *Bell Syst. Tech. J.*, Vol. 37, pp. 401–446, March 1958.

2. R. L. Johnston, B. C. DeLoach, Jr., and B. G. Cohen, "A Silicon Diode Microwave Oscillator," *Bell Syst. Tech. J.*, Vol. 44, pp. 369–372. February 1965.

3. M. Ino, T. Ishibashi, and M. Ohmori, "CW Oscillation with p^+pn^+ Silicon IMPATT Diodes in 200 GHz and 300 GHz Bands," *Electron. Lett.*, Vol. 12, No. 6, pp. 148–149, March 18, 1976.

4. S. M. Sze, *Physics of Semiconductor Devices*, 2nd ed., Wiley, New York, 1981.

5. G. I. Haddad, P. T. Greiling, and W. E. Schroeder, "Basic Principles and Properties of Avalanche Transit-Time Devices," *IEEE Trans. Microwave Theory Tech.*, Vol. MTT-18, pp. 752–772, November 1970.

6. H. J. Kuno, "IMPATT Devices for Generation of Millimeter-Waves," in K. Button, Ed., *Infrared and Millimeter-Waves*, Vol. 1, *Sources of Radiation*, Academic Press, New York, 1979, Chap. 2.

7. T. Misawa, "Negative Resistance in *p-n* Junctions under Avalanche Breakdown Conditions. Parts I and II," *IEEE Trans. Electron Devices*, Vol. ED-13, pp. 137–151, January 1966.

8. K. Chang and R. L. Ebert, "W-Band Power Combiner Design," *IEEE Trans. Microwave Theory Tech.*, Vol. MTT-28, pp. 295–305, April 1980

9. M. Gilden and M. E. Hines, "Electronic Tuning Effects in the Read Microwave Avalanche Diode," *IEEE Trans. Electron Devices*, Vol. ED-13, pp. 169–175, January 1966.

10. W. J. Evans and G. I. Haddad, "A Large-Signal Analysis of IMPATT Diodes," *IEEE Trans. Electron Devices*, Vol. ED-15, pp. 708–717, October 1968.

11. I. L. Blue, "Approximate Large-Signal Analysis of IMPATT Oscillators," *Bell Syst. Tech. J.*, Vol. 48, pp. 383–396, February 1969.

12. D. L. Scharfetter and H. K. Gummel, "Large-Signal Analysis of a Silicon Read Diode Oscillator," *IEEE Trans. Electron Devices*, Vol. ED-16, pp. 64–77, January 1969.

13. K. Mouthaan, "Nonlinear Analysis of the Avalanche Transit-Time Oscillator," *IEEE Trans. Electron Devices*, Vol. ED-16, pp. 935–945, November 1969.

14. P. T. Greiling and G. I. Haddad, "Large-Signal Equivalent Circuits of Avalanche Transit-Time Devices," *IEEE Trans. Microwave Theory Tech.*, Vol. MTT-18, pp. 842–853, November 1970.

15. D. R. Decker, C. N. Dunn, and R. L. Frank, "Large-Signal Silicon and Germanium Avalanche-Diode Characteristics," *IEEE Trans. Microwave Theory Tech.*, Vol. MTT-18, pp. 872–876, November 1970.

16. E. F. Scherer, "Large-Signal Operation of Avalanche-Diode Amplifiers," *IEEE Trans. Microwave Theory Tech.*, Vol. MTT-18, pp. 922–932, November 1970.

17. H. J. Kuno, "Analysis of Nonlinear Characteristics and Transient Response of IMPATT Amplifiers," *IEEE Trans. Microwave Theory Tech.*, Vol. MTT-21, pp. 694–702, November 1973.

18. H. J. Kuno and D. L. English, "Nonlinear and Large-Signal Characteristics of Millimeter-Wave IMPATT Amplifiers," *IEEE Trans. Microwave Theory Tech.*, Vol. MTT-21, pp. 703–706, November 1973.

19. N. B. Kramer, "Characterization and Modeling of IMPATT Oscillators," *IEEE Trans. Electron Devices*, Vol. ED-15, pp. 838–846, November 1968.

20. R. S. Ying, "X-Band n^+pp^+ IMPATT Diodes," *Electron. Lett.*, Vol. 8, pp. 297–298, June 15, 1972.

21. D. L. Scharfetter, W. J. Evans, and R. L. Johnston, "Double-Drift-Region (p^+pnn^+) Avalanche Diode Oscillators," *Proc. IEEE*, Vol. 58, pp. 1131–1133, July 1970.

22. T. E. Seidel and D. L. Scharfetter, "High-Power Millimeter-Wave IMPATT Oscillators with Both Hole and Electron Drift Spaces Made by Ion Implantation," *Proc. IEEE*, Vol. 58, pp. 1135–1136, July 1970.

23. Y. E. Ma, E. M. Nakaji, and W. F. Thrower, "V-Band Double-Drift Read Silicon IMPATTs," *IEEE MTT-S Int. Microwave Symp. Dig.*, pp. 167–168, 1984.

24. J. R. Grierson et al., "High-Power 11 GHz GaAs Hi-Lo IMPATT Diodes with Titanium Schottky Barriers," *Electron. Lett.*, Vol. 15, pp. 13–14, January 4, 1979.

25. K. Chang, J. K. Kung, P. G. Asher, G. M. Hayashibara, and R. S. Ying, "GaAs Read-Type IMPATT Diode for 130 GHz CW Operation," *Electron. Lett.*, Vol. 17, pp. 471–473 June 25, 1981.

26. K. Chang, H. Yen, and E. M. Nakaji, *94 GHz High Power Solid State Transmitter*, Final report to BMDSC, DASG 60-78-C-0148, 1981.

27. K. Chang, W. F. Thrower, and G. M. Hayashibara, "Millimeter-Wave Silicon IMPATT Sources and Combiners for the 110-260 GHz Range," *IEEE Trans. Microwave Theory Tech.*, Vol. MTT-29, pp. 1278–1284, December 1981.

28. W. N. Grant, "Electron and Hole Ionization Rates in Epitaxial Silicon at High Electric Fields," *Solid State Electron.*, Vol. 16, pp. 1189–1203, 1973.

29. C. Canali, G. Majni, R. Minder, and G. Ottaviani, "Electron and Hole Drift Velocity Measurements on Silicon and Their Empirical Relation to Electric Field and Temperature," *IEEE Trans. Electron Devices*, Vol. ED-22, pp. 1045–1046, November 1975.

30. J. Freyer, E. Kasper, and H. Barth, "Pulsed V-Band M. B. E. Si IMPATT Diodes," *Electron. Lett.*, Vol. 16, pp. 865–866, November 6, 1980.

31. A. Y. Cho, C. N. Dunn, R. L. Kuuas, and W. E. Schroeder, "GaAs IMPATT Diodes Prepared by Molecular Beam Epitaxy," *Appl. Phys. Lett.*, Vol. 25, pp. 224–226, August 15, 1974.

32. R. S. Ying, D. L. English, K. P. Weller, E. M. Nakaji, D. H. Lee, and R. L. Bernick, "Millimeter-Wave Pulsed IMPATT Diode Oscillators," *IEEE J. Solid-State Circuits*, Vol. SC-11, pp. 279–285, April 1976.

33. K. P. Weller, R. S. Ying, and D. H. Lee, "Millimeter-Wave IMPATT Sources for the 130–170 GHz Range," *IEEE Trans. Microwave Theory Tech.*, Vol. MTT-24, pp. 738–743, November 1976.

34. C. Chao, R. L. Bernick, E. M. Nakaji, R. S. Ying, K. P. Weller, and D. H. Lee, "Y-Band (170–260 GHz) Tunable CW IMPATT Diode Oscillators," *IEEE Trans. Microwave Theory Tech.*, Vol. MTT-25, pp. 985–991, December 1977.

35. R. L. Eisenhart and P. J. Khan, "Theoretical and Experimental Analysis of a Waveguide Mounting Structure," *IEEE Trans. Microwave Theory Tech.*, Vol. MTT-19, pp. 706–719, August 1971.

36. T. T. Fong, K. P. Weller, and D. L. English, "Circuit Characterization of V-Band IMPATT Oscillators and Amplifiers," *IEEE Trans. Microwave Theory Tech.*, Vol. MTT-24, pp. 752–758, November 1976.

37. A. G. Williamson, "Analysis and Modeling of Two-Gap Coaxial Line Rectangular Waveguide Junctions," *IEEE Trans. Microwave Theory Tech.*, Vol. MTT-31, pp. 295–302, March 1983.

38. B. D. Bates, "Analysis of Multiple-Step Radial-Resonator Waveguide Diode Mounts with Application to IMPATT Oscillator Circuits," *IEEE MTT-S Int. Microwave Symp. Dig.*, pp. 669–672, 1987.

39. G. B. Morgan, "Microstrip IMPATT-Diode Oscillator for 100 GHz," *Electron. Lett.*, Vol. 17, pp. 570–571, August 6, 1981.

40. P. Yen, D. English, C. Ito, and K. Chang, "Millimeter-Wave IMPATT Microstrip Oscillators," *IEEE MTT-S Int. Microwave Symp. Dig.*, pp. 139–140, 1983.

41. B. S. Glance, "Low-Q Microstrip IMPATT Oscillator at 30 GHz," *Proc. IEEE*, Vol. 60, pp. 1105–1106, September 1972.

42. B. S. Glance and M. V. Schneider, "Millimeter-Wave Microstrip Oscillators," *IEEE Trans. Microwave Theory Tech.*, Vol. MTT-22, pp. 1281–1283, December 1974.

43. K. Chang, D. M. English, R. S. Tahim, A. J. Grote, T. Pham, C. Sun, G. M. Hayashibara, P. Yen, and W. Piotrowski, "W-Band (75–110 GHz) Microstrip Components," *IEEE Trans. Microwave Theory Tech.*, Vol. MTT-33, pp. 1375–1382, December 1985.

44. M. Dydyk, "EHF Planar Module for Spatial Combining," *Microwave J.* Vol. 26, pp. 157–176, May 1983.

45. G. B. Morgan, "Stabilization of a W-Band Microstrip Oscillator by a Dielectric Resonator," *Electron. Lett.*, Vol. 18, pp. 556–558, June 24, 1982.

46. K. Kurokawa, "Some Basic Characteristics of Broadband Negative Resistance Oscillator Circuits," *Bell Syst. Tech. J.*, Vol. 48, pp. 1937–1955, July 1969.

47. T. T. Fong and H. J. Kuno, "Millimeter-Wave Pulsed IMPATT Sources." *IEEE Trans. Microwave Theory Tech.*, Vol. MTT-27, pp. 492–499, May 1979.

48. T. A. Midford and R. L. Bernick, "Millimeter-Wave CW IMPATT Diodes and Oscillators," *IEEE Trans. Microwave Theory Tech.*, Vol. MTT-27, pp. 483–492, May 1979.

49. R. Mallavarpu and G. MacMaster, "500 GHz, 100 W X-Band Solid State Amplifier," *IEEE MTT-S Int. Microwave Symp. Dig.*, pp. 387–390, 1985.

50. M. J. Delaney, M. H. Jones, and C. Sun, "20 GHz GaAs IMPATT Diode Development for Solid State Transmitter," *IEEE MTT-S Int. Microwave Symp. Dig.*, pp. 525–527, 1985.

51. M. G. Adlerstein and S. L. G. Chu, "GaAs IMPATT Diodes Pulsed at 40 GHz," *IEEE MTT-S Int. Microwave Symp. Dig.*, pp. 481–485, 1984.

52. K. Chang, C. Sun, D. L. English, and E. M. Nakaji, "High Power 94 GHz Pulsed IMPATT Oscillators," *IEEE MTT-S Int. Microwave Symp. Dig.*, pp. 71–72, 1979.

53. K. Chang. "Millimeter-Wave Planar Integrated Circuits and Subsystems," in K. Button, Ed., *Infrared and Millimeter-Wave*, Vol. 14, Academic Press, New York, 1985.

54. K. Chang, "Millimeter-Wave Microstrip Solid-State Sources and Amplifiers," *Int. J. Infrared Millim. Wave*, Vol. 7, pp. 729–747, May 1986.

55. R. Adler, "A Study of Locking Phenomena in Oscillators," *Proc. IRE.*, Vol. 34, pp. 351–357, June 1946.

56. K. Kurokawa, "Injection Locking of Microwave Solid-State Oscillators," *Proc. IEEE*, Vol. 61, pp. 1386–1409, October 1973.

57. M. E. Hines, J. Collinet, and J. Ondria, "FM Noise Suppression of an Injection Phased-Locked Oscillator," *IEEE Trans. Microwave Theory Tech.*, Vol. MTT-16, pp. 738–742, September 1968.

58. C. H. Chien and G. C. Dalman, "Subharmonic Injected Phaselock IMPATT-Oscillator Experiments," *Electron. Lett.*, Vol. 6, pp. 240–241, April 16, 1970.

59. T. P. Lee, R. D. Standley, and T. Misawa, "A 50 GHz Silicon IMPATT Diode Oscillator and Amplifier," *IEEE Trans. Electron Devices*, Vol. ED-15, pp. 741–747, October 1968.

60. E. F. Scherer, "A Multistage High-Power Avalanche Amplifier at X-Band," *IEEE J. Solid-State Circuits*, Vol. SC-4, pp. 396–399, December 1969.

61. D. F. Peterson, "A Device Characterization and Circuit Design Procedure for Realizing High Power Millimeter-Wave IMPATT Diode Amplifiers," *IEEE Trans. Microwave Theory Tech.*, Vol. MTT-21, pp. 681–689, November 1973.

62. K. P. Weller, D. L. English, and E. M. Nakaji, "High Power V-Band Double Drift IMPATT Amplifier," *IEEE MTT-S Int. Microwave Symp. Dig.*, pp. 369–371, 1978.

63. M. Ando et al., "86 GHz High Power IMPATT Negative Resistance Amplifier," *IEEE MTT-S Int. Microwave Symp. Dig.*, pp. 312–314, 1978.

64. B. D. Bates and P. J. Khan, "Influence of Non-ideal Circulator Effects on Negative-Resistance Amplifier Design," *IEEE MTT-S Int. Microwave Symp. Dig.*, pp. 174–176, 1980.

65. V. Sokolov, M. R. Namordi, and F. H. Doerbeck, "A 4 W 56 dB Gain Microstrip Amplifier at 15 GHz Utilizing GaAs FET's and IMPATT Diodes," *IEEE Trans. Microwave Theory Tech.*, Vol. MTT-27, pp. 1058–1065, December 1979.

66. K. J. Russell, "Microwave Power Combining Techniques," *IEEE Trans Microwave Theory Tech.*, Vol. MTT-27, pp. 472–478, May 1979.

67. K. Chang and C. Sun, "Millimeter-Wave Power Combining Techniques," *IEEE Trans. Microwave Theory Tech.*, Vol. MTT-31, pp. 91–107, February 1983.

68. K. Kurokawa and F. M. Magalhaes, "An X-Band 10-Watt Multiple-IMPATT Oscillator," *Proc. IEEE*, Vol. 59, pp. 102–103, January 1971.

69. K. Kurokawa, "The Single-Cavity Multiple Device Oscillator," *IEEE Trans Microwave Theory Tech.*, Vol. MTT-19, pp. 793–801, October 1971.

70. R. S. Harp and H. L. Stover, "Power Combining of X-Band IMPATT Circuit Modules," *IEEE-ISSCC Dig. Tech. Pap.*, Vol. 16, pp. 118–119, February 1973.

71. R. Aston, "Technique for Increasing the Bandwidth of a TM_{010}-Mode Power Combiner," *IEEE Trans. Microwave Theory Tech.*, Vol. MTT-27, pp. 479–482, May 1979.

72. R. S. Harp and K. J. Russell, "Improvements in Bandwidth and Frequency Capability of Microwave Power Combinational Techniques," *IEEE-ISSCC Dig. Tech. Pap.*, Vol. 17, pp. 94–95, February 1974.

73. S. E. Hamilton, "32-diode Waveguide Power Combiner," *IEEE MTT-S Int. Microwave Symp. Dig.*, pp. 183–185, May 1980.

74. S. E. Hamilton and B. M. Fish, "Multidiode Waveguide Power Combiners," *IEEE MTT-S Int. Microwave Symp. Dig.*, pp. 132–134, June 1982.

75. Y. C. Ngan, J. Chan, and C. Sun, "20 GHz High Power IMPATT Transmitter," *IEEE MTT-S Int. Microwave Symp. Dig.*, pp. 487–488, 1983.

76. G. Jerinic, J. Fines, and M. Schindler, "3 W, Q-Band Solid-State Amplifier," *IEEE MTT-S Int. Microwave Symp. Dig.*, pp. 481–483, 1983.

77. D. W. Mooney and F. J. Bayuk, "Injection Locking Performance of a 41 GHz 10 W Power Combining Amplifier," *IEEE Trans. Microwave Theory and Tech.*, Vol. MTT-31, pp. 171–177, February 1983.

78. G. H. Nesbit and W. H. Leighton, "EHF Solid-State Amplifier," *IEEE MTT-S Int. Microwave Symp. Dig.*, pp. 391–394, 1985.

79. Y. Ma and C. Sun, "Millimeter-Wave Power Combiner at V-Band," *Proc. 7th Cornell Electr. Eng. Conf.*, August 1979, pp. 299–308.

80. Y. C. Ngan, "Two-Diode Power Combining Near 140 GHz," *Electron. Lett.*, Vol. 15, No. 13, pp. 376–377, June 21, 1979.

81. K. J. Russell and R. S. Harp, "Power Combiner Operation with Pulsed IMPATTs." *IEEE-ISSCC Dig. Tech. Pap.*, Vol. 18, February 1975.

82. R. M. Wallace, M. G. Adlerstein, and S. R. Steele, "A 60-W CW Solid State Oscillator at C-Band," *IEEE Trans. Microwave Theory Tech.*, Vol. MTT-24, pp. 483–485, July 1976.

83. K. Russell and R. S. Harp, "A Multistage High-Power Solid-State X-Band Amplifier," *IEEE-ISSCC Dig. Tech. Papers*, Vol. 21, pp. 166–167, February 1978.

84. J. Obregon, J. P. Balabaud, P. Guillon, and Y. Garault, "New Radial Power Combiner Using Tubular Dielectric Cavities," *Electron. Lett.*, Vol. 18, No. 18, pp. 771–772, September 2 1982.

85. M. Dydyk, "Efficient Power Combining," *IEEE Trans. Microwave Theory Tech.*, Vol. MTT-28, pp. 755–762, July 1980.

86. S. E. Hamilton, R. S. Robertson, F. A. Wilhelmi, and M. E. Dick, "X-Band Pulsed Solid-State Transmitter," *IEEE MTT-S Int. Microwave Symp. Dig.*, pp. 162–164, May 1980.

87. R. J. Pankow and R. G. Mastroianni, "A High Power X-Band Diode Amplifier," *IEEE MTT-S Int. Microwave Symp. Dig.*, pp. 151–161, May 1980.

88. C. A. Drubin, A. L. Hieber, G. Jerinic, and A. S. Marinilli, "A 1 KW Peak, 300 W_{avg} IMPATT Diode Injection Locked Oscillator," *IEEE MTT-S Int. Microwave Symp. Dig.*, pp. 126–128, June 1982.

89. R. Laton, S. Simoes, and L. Wagner, "A Dual TM_{020} Cavity for IMPATT Diode Power Combining," *IEEE MTT-S Int. Microwave Symp. Dig.*, pp. 129–131, June 1982.

90. F. J. Bayuk and J. Raue, "K-Band Solid-State Power Amplifier," *IEEE MTT-S Int. Microwave Symp. Dig.*, pp. 21–31, May 1977.

91. J. R. Nevarez and G. J. Herokowiz, "Output Power and Loss Analysis of 2^n Injection-Locked Oscillators Combined through an Ideal and Symmetric Hybrid Combiner," *IEEE Trans. Microwave Theory Tech.*, Vol. MTT-17, pp. 2–10, January 1969.

92. H. J. Kuno and D. L. English, "Millimeter-Wave IMPATT Power Amplifier/Combiner," *IEEE Trans. Microwave Theory Tech.*, Vol. MTT-24, pp. 758–767, November 1976.

93. Y. Ma, C. Sun, and E. M. Nakaji, "V-Band High Power IMPATT Amplifier," *IEEE MTT-S Int. Microwave Symp. Dig.*, pp. 73–74, May 1980.

94. H. C. Yen and K. Chang, "A 63-W W-Band Injection-Locking Pulsed Solid-State Transmitter," *IEEE Trans. Microwave Theory Tech.*, Vol. MTT-29, pp. 1292–1297, December 1981.

95. R. E. Lee, U. H. Gysel, and D. Parker, "High-Power C-Band Multiple-IMPATT-Diode Amplifiers," *IEEE Trans. Microwave Theory Tech.*, Vol. MTT-24, pp. 249–253, May 1976.

96. J. G. Josenhans, "Diamond as an Insulating Heat Sink for a Series Combination of IMPATT Diodes," *Proc. IEEE*, Vol. 56, pp. 762–763, April 1968.

97. C. T. Rucker, G. N. Hill, N. W. Cox, and J. W. Amoss, "Symmetry Experiments with Four-Mesa IMPATT Diodes," *IEEE Trans. Microwave Theory Tech.*, Vol. MTT-25, pp. 75–76, January 1977.

98. C. T. Rucker et al., "Series-Connected GaAs and Si Diode Chips: Some New Results," *Electron. Lett.*, Vol. 13, No. 11, pp. 331–332, May 26, 1977.

99. C. T. Rucker, J. W. Amoss, G. N. Hill, and N. W. Cox, "Multichip IMPATT Power Combining, a Summary with New Analytical and Experimental Results," *IEEE Trans. Microwave Theory Tech.*, Vol. MTT-27, pp. 951–957, December 1979.

100. C. T. Rucker, J. W. Amoss, and G. N. Hill, "Chip-Level IMPATT Combining at 40 GHz," *IEEE MTT-S Inst. Microwave Symp. Dig.*, pp. 347–348, June 1981.

101. C. B. Swan, T. Misawa, and L. Marinaccio, "Composite Avalanche Diode Structures for Increased Power Capability," *IEEE Trans. Electron Devices*, Vol. ED-14, pp. 584–589, September 1967.

102. M. F. Durkin, "35 GHz Active Aperture," *IEEE MTT-S Int. Microwave Symp. Dig.*, pp. 425–427, 1981.

103. R. J. Dinger, D. J. White, and D. Bowling, "A 10 GHz Space Power Combiner with Parasitic Injection-Locking," *IEEE MTT-S Int. Microwave Symp. Digest*, pp. 163–166, 1986.

104. K. A. Hummer and K. Chang, "Spatial Power Combining Using Active Microstrip Patch Antennas," *Microwave Opt. Technol. Lett.*, Vol. 1, pp. 8–9, March 1988.

105. J. M. Mink, "Quasi-optical Power Combining of Solid-State Millimeter-Wave Sources," *IEEE Trans. Microwave Theory Tech.*, Vol. MTT-34, pp. 273–279, February 1986.

106. L. Wandinger and V. Nalbandian, "Millimeter-Wave Power Combiner Using Quasi-optical Techniques," *IEEE Trans. Microwave Theory Tech.*, Vol. MTT-31, pp. 189–193, February 1983.

107. S. Young and K. D. Stephan, "Stabilization and Power Combining of Planar Microwave Oscillators with an Open Resonator," *IEEE MTT-S Int. Microwave Symp. Dig.*, pp. 185–188, 1987.

108. P. Rolland, J.L. Vaterkowski, E. Constant, and G. Salmer, "New Modes of Operation for Avalanche Diodes: Frequency Multiplication and Upconversion," *IEEE Trans. Microwave Theory Tech.*, Vol. MTT-24, pp. 768–775, November 1976.

109. R. E. Collin, *Foundations for Microwave Engineering*, McGraw-Hill, New York, 1966.

110. W. W. Gray, L. Kikushima, N. P. Morenc, and R. J. Wagner, "Applying IMPATT Sources to Modern Microwave Systems," IEEE-ISSCC Meeting, 1969.

111. K. Chang, M. Morishita, and C. Sun, "High Frequency Silicon Impact Ionization Avalanche Transit-Time (IMPATT) Sources beyond 100 GHz," *Soc. Photo-Opt. Instrum. Engi. Meet.*, Vol. 259, pp. 46–50.

112. C. A. Brackett, "The Elimination of Tuning-Induced Burnout and Bias-Circuit Oscillations in IMPATT Oscillators," *Bell Syst. Tech. J.*, Vol. 52, pp. 271–306, March 1973.

113. Y. Hirachi, T. Nakagami, Y. Toyama, and Y. Fukukawa, "High-Power 50 GHz Double-Drift-Region IMPATT Oscillators with Improved Bias Circuits for Eliminating Low-Frequency Instabilities," *IEEE Trans. Microwave Theory Tech.*, Vol. MTT-24, pp. 731–737, November 1976.

114. M. E. Hines, "Large-Signal Noise, Frequency Conversion, and Parametric Instabilities in IMPATT Diode Networks," *Proc. IEEE*, Vol. 60, pp. 1534–1548, December 1972.

115. W. E. Schroeder, "Spurious Parametric Oscillations in IMPATT Diode Circuits," *Bell Syst. Tech. J.*, Vol. 53, pp. 1187–1209, September 1974.

116. D. J. Coleman, Jr., and S. M. Sze, "A Low-Noise Metal-Semiconductor-Metal (MSM) Microwave Oscillator," *Bell Syst. Tech. J.*, Vol. 50, pp. 1695–1699, May/June 1971.

117. C. P. Snapp and P. Weissglas, "On the Microwave Activity of Punch-Through Injection Transit-Time Structure," *IEEE Trans. Electron Devices*, Vol. ED-19, pp. 1109–1118, October 1972.

118. D. J. Coleman, Jr., "Transit-Time Oscillations in BARITT Diodes," *J. Appl. Phys.*, Vol. 43, pp. 1812–1819, April 1972.

119. G. T. Wright, "Small-Signal Characteristics of Semiconductor Punch-Through Injection and Transit-Time Diodes," *Solid-State Electron.*, Vol. 16, pp. 903–912, August 1973.

120. J. R. East, H. Nguyen-Ba, and G. I. Haddad, "Design, Fabrication, and Evaluation of BARITT Devices for Doppler System Applications," *IEEE Trans. Microwave Theory Tech.*, Vol. MTT-24, pp. 943–948, December 1976.

121. J. Chen, J. R. East, and G. I. Haddad, *Development of Innovative Millimeter-Wave Broadband Mixers*, Final report to Naval Research Laboratory, N 00173-80-C-0118, 1983.

122. O. Eknoyan, S. M. Sze, and E. S. Yang, "Microwave BARITT Diode with Retarding Field—An Investigation," *Solid State Electron.*, Vol. 19, p. 795, 1976.

123. H. J. Prager, K. K. N. Chang, and S. Weisbrod, "High-Power, High-Efficiency Silicon Avalanche Diodes at Ultra High Frequencies," *Proc. IEEE*, Vol. 55, pp. 586–587, April 1967.

124. B. C. DeLoach, Jr., and D. L. Scharfetter, "Device Physics of TRAPATT Oscillators," *IEEE Trans. Electron Devices*, Vol. ED-17, pp. 9–21, January 1970.

125. W. J. Evans, "CW TRAPATT Amplification," *IEEE Trans. Microwave Theory Tech.*, Vol. MTT-18, pp. 986–988, November 1970.

126. W. J. Evans, "Circuits for High-Efficiency Avalanche-Diode Oscillators," *IEEE Trans. Microwave Theory Tech.*, Vol. MTT-17, pp. 1060–1067, December 1969.

127. W. J. Evans, "Computer Experiments on TRAPATT Diodes," *IEEE Trans. Microwave Theory Tech.*, Vol. MTT-18, pp. 862–871, November 1970.

128. J. R. East, N. A. Masnari, and G. I. Haddad, "Experimental Investigation of TRAPATT Diode Trigger Conditions," *IEEE J. Solid-State Circuits*. Vol. SC-12, pp. 14–20, February 1977.

129. R. J. Trew, N. A. Masnari, and G. I. Haddad, "Optimization of S-band TRAPATT Oscillators," *IEEE Trans. Microwave Theory Tech.*, Vol. MTT-22, pp. 1166–1170, December 1974.

130. S. G. Liu and J. J. Risko, "Fabrication and Performance of Kilowatt L-Band Avalanche Diodes," *RCA Rev.*, Vol. 31, p. 3, 1970.

131. D. F. Kostichack, "UHF Avalanche Diode Oscillator Providing 400 Watts Peak Power and 75 Percent Efficiency," *Proc. IEEE*, Vol. 58, p. 1282, 1970.

132. N. W. Cox, C. T. Rucker, G. N. Hill, and K. E. Gsteiger, "X-Band TRAPATT Amplifier," *IEEE Trans. Microwave Theory Tech.*, Vol. MTT-22, pp. 1325–1328, December 1974.

133. J. Nishizawa, K. Motoya, and Y. Okuno, "200 GHz TUNNETT Diodes," *Jpn. J. Appl. Phys.*, Vol. 17, pp. 167–172, 1978.

134. J. Nishizawa, K. Motoya, and Y. Okuno, "The GaAs TUNNETT Diodes," *IEEE-MTT Microwave Symp. Dig. Tech. Pap.*, pp. 159–161, 1978.

135. M. E. Elta, H. R. Fetterman, W. V. Macropoulos, and J. J. Lambert, "150 GHz GaAs MITATT Sources," *IEEE Electron Device Lett.*, Vol. EDL-1, pp. 115–116, June 1980.

136. J. Nishizawa, K. Motoya, and Y. Okuno, "The GaAs TUNNETT Diodes," *IEEE MTT-S Int. Microwave Symp. Dig.*, pp. 159–161, 1978.

137. M. E. Elta and G. I. Haddad, "Mixed Tunneling and Avalanche Mechanisms in *p-n* Juntions and Their Effects on Microwave Transit-Time Devices," *IEEE Trans. Electron Devices*, Vol. ED-25, pp. 694–702, June 1978.

138. M. E. Elta and G. I. Haddad, "High-Frequency Limitations of IMPATT, MITATT, and TUNNETT Mode Devices," *IEEE Trans. Microwave Theory Tech.*, Vol. 27, pp. 442–449 May 1979.

139. S. P. Kwok and G. I. Haddad, "Effects of Tunneling on an IMPATT Oscillator," *J. Appl. Phys.*, Vol. 43, pp. 3824–3830, September 1972.

140. W. A. Lukaszek, A. Van Der Diel, and E. R. Chenett, "Investigation of the Transition from Tunneling to Impact Ionization Multiplication in Silicon *p-n* Junctions," *Solid-State Electron.*, Vol. 19, pp. 57–71, January 1976.

141. T. Ishibasi et al., "400 GHz Band Operation of Cooled Silicon IMPATT Diodes," *Jpn. J. Appl. Phys.*, Vol. 17, 1978. pp. 173–178.

142. F. B. Fank, J. D. Crowley, and J. J. Berenz, "InP Material and Device Development for Millimeter-Waves," *Microwave J.*, Vol. 22, pp. 86–91, June 1979.

143. T. A. Midfors and H. C. Bowers, "A Two-Port IMPATT Diode Travelling Wave Amplifier," *Proc. IEEE*, Vol. 56, pp. 1724–1725, October 1968.

144. K. G. Hambleton and P. N. Robson, "Design Considerations for Resonant Traveling Wave IMPATT Oscillators," *Int. J. Electron.*, Vol. 35, pp. 225–244, 1973.

145. M. Franz and J. B. Beyer, "The Traveling-Wave IMPATT Mode," *IEEE Trans. Microwave Theory Tech.*, Vol. MTT-26, pp. 861–865, November 1978.

146. M. Franz and J. B. Beyer, "The Traveling-Wave IMPATT Mode. Part II. The Effective Wave Impedance and Equivalent Transmission Line," *IEEE Trans. Microwave Theory Tech.*, Vol. MTT-28, pp. 215–218, March 1980.

147. Y. Fukuoka and T. Itoh, "Field Analysis of a Millimeter-Wave GaAs Double-Drift IMPATT Diode in the Traveling-Wave Mode," *IEEE Trans. Microwave Theory Tech.*, Vol. MTT-33, pp. 216–222, March 1985.

148. R. K. Mains and G. I. Haddad, "Traveling-Wave IMPATT Amplifiers and Oscillators," *IEEE Trans. Microwave Theory Tech.*, Vol. MTT-34, pp. 965–971, September 1986.

149. B. Bayraktaroglu and H. D. Shih, "Millimeter-Wave GaAs Distributed IMPATT Diodes," *IEEE Electron Device Lett.*, Vol, EDL-4, pp. 393–395, November 1983.

150. J. F. Luy, K. M. Strohm, and J. Buechler, "Monolithically Integrated Coplanar 75 GHz Silicon IMPATT Oscillator," *Microwave Opt. Technol. Lett.*, Vol. 1, pp. 117–119, June 1988.

151. B. Bayraktaroglu and H. Shih, "High Efficiency Millimeter-Wave Monolithic IMPATT Oscillator," *IEEE MTT-S Int. Microwave Symp. Dig.*, pp. 124–127, 1985.

152. N. S. Dogan, J. R. East, M. E. Elta, and G. I. Haddad, "Millimeter-Wave Hecterojunction MITATT Devices." *IEEE Trans. Microwave Theory Tech.*, Vol. MTT-35, pp. 1308–1316, December 1987.

153. T. O. Perkins III, "Active Microstrip Circular Patch Antenna," *Microwave J.*, Vol. 30, pp. 109–117, March 1987.

154. N. Camilleri and B. Bayraktaroglu, "Monolithic Millimeter-Wave IMPATT Oscillator and Active Antenna," *IEEE MTT-S Int. Microwave Symp. Dig.*, pp. 955–958, 1988.

155. H. J. Thomas, D. L. Fudge, and G. Morris, "Gunn Source Integrated with a Microstrip Patch," *Microwave RF*, Vol. 24, pp. 87–89, February 1985.

156. K. A. Hummer and K. Chang, "Microstrip Active Antennas and Arrays," *IEEE MTT-S Int. Microwave Symp. Dig.*, pp. 963–966, 1988.

8

MICROWAVE SILICON BIPOLAR TRANSISTORS AND MONOLITHIC INTEGRATED CIRCUITS

Craig P. Snapp

Avantek, Inc.
Microwave Semiconductor Group
Advanced Bipolar Products
Newark, California

8.1 INTRODUCTION

It has been nearly 40 years since William Shockley fabricated the first bipolar junction transistor. In the ensuing decades bipolar transistor technology has evolved steadily in both performance and practical level of integration. Significant application of this technology to microwave systems first began in the late 1960s with the commercial availability of discrete *npn* silicon transistors with maximum frequencies of oscillation (f_{max}) in excess of 10 GHz [1]. GaAs MESFET technology, however, has emerged in the last 10 years to take a dominant front-end position in most microwave systems primarily because of superior noise, gain, and power performance in linear amplifiers.

Steady and significant progress in silicon process technology has nevertheless continued to improve discrete bipolar transistor performance, dramatically reduce component cost, and make silicon monolithic microwave integrated circuits (MMICs) and gigibit/second digital ICs a commercial reality [2–10]. Discrete transistors with sub-half-micrometer critical dimensions are now practical for low-phase-noise oscillator applications at frequencies as high as 20 GHz. Silicon bipolar MMICs and gigibit/second digital ICs have become established as the most cost-effective microwave semiconductor components for many high-volume and high-performance system applications even at frequencies above 4 GHz. The frequency ranges for some of the most common applications and markets for silicon bipolar transistors and MMICs are summarized in Figs. 8.1 and 8.2.

The intrinsic advantages of GaAs MESFET technology are well known to engineers in the microwave industry. The electron mobility and effective saturated drift velocity are significantly higher in GaAs than in silicon, and the semi-insulating nature of GaAs permits easy electrical isolation of devices and the use of transmission-line circuit elements in MMICs. In addition, the relatively high impedances of GaAs MESFETs make circuit matching easier in many applications. It is germane to the topic of this chapter, however, to review the fundamental advantages of silicon bipolar technology which have

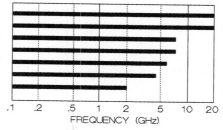

Figure 8.1 Maximum frequency range for common applications of silicon bipolar transistors and MMICs.

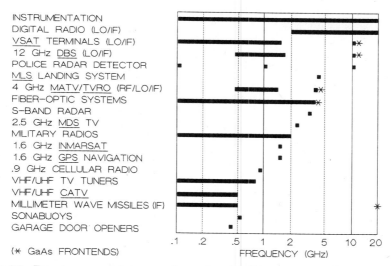

Figure 8.2 Frequency range of representative markets for silicon bipolar transistors and MMICs.

contributed to its staying power during the past 40 years and which ensure its continued contribution to improving the performance and reducing the cost of future microwave and high-speed fiber-optic systems.

Table 8.1 summarizes some of the major intrinsic performance and cost advantages of silicon bipolar technology. Of particular note is the high and strictly current-dependent transconductance of bipolar transistors, which leads to their ability to deliver more current per unit area than GaAs MESFETs and interface to other circuits and loads with less performance degradation. The transconductance of discrete silicon bipolar transistors biased for peak f_t versus emitter length, and GaAlAs HEMT and GaAs MESFET versus gate width are compared in Fig. 8.3. Another important performance advantage stems from the low $1/f$ noise characteristics of silicon bipolar transistors, which leads to lower up-converted phase noise in oscillator applications.

The 50-Ω S_{21} gain, maximum available gain (MAG), and minimum noise figure versus frequency of a current state-of-the art unpackaged GaAs MESFET and a similar-sized

TABLE 8.1 Advantage of Silicon Bipolar Transistors and Technology

Performance	Cost
1. High Transconductance: $g_m = qI/kT = I_c/V_t$ strictly current dependent	1. Low starting-material cost
	2. High manufacturing productivity, due to ease of automation
2. Excellent threshold uniformity: $V_{be} = V_t \ln(I_c/I_b)$ Bandgap determined	3. Ease of Evolution to larger wafers
	4. Negligible wafer breakage
3. Low $1/F$ noise: Stable oxide interface Low trap density Buried junctions	5. Process predictability
	6. Excellent dc–RF performance correlation
4. Ease of device scaling: Vertical profile control Lack of short-channel effects	7. Performance tolerance to lithographic variations

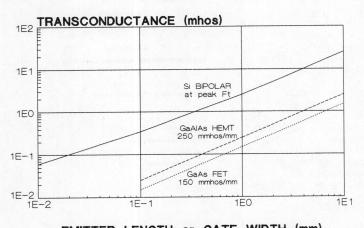

Figure 8.3 Transconductance of silicon bipolar transistors biased for peak f_t versus emitter length, and GaAlAs HEMT and GaAs FET versus gate width.

silicon bipolar transistor are compared in Fig. 8.4. In this comparison the GaAs MESFET has a 0.25-μm gate with a width of 750 μm and the silicon bipolar transistor has a 0.5-μm emitter with 2-μm emitter–emitter pitch and is fabricated with a 10-GHz f_t process. The significantly higher untuned S_{21} gain of the bipolar transistor at the lower microwave frequencies is a direct consequence of its higher transconductance. A further increase in S_{21} gain can easily be achieved with properly sized bipolar transistors by using the merged two-transistor Darlington configuration.

For many microwave applications the increase in device junction temperature over ambient is of critical importance for maximizing performance, maintaining performance stability over temperature, and ensuring acceptable reliability. Minimizing junction temperature rise is obviously critical for reliable power amplification but can also be a major factor in more subtle performance parameters such as post-tuning drift in varactor-tuned oscillators. The higher thermal conductivity (Fig. 8.5) of silicon compared to GaAs is therefore an important intrinsic material advantage.

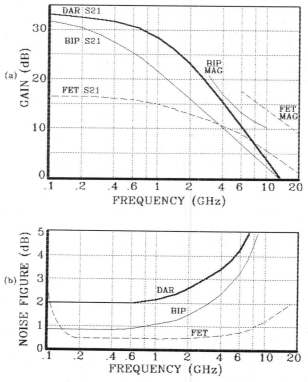

Figure 8.4 Performance comparison of unpackaged 750-μm-gate-width GaAs FET with 2-μm emitter–emitter pitch silicon bipolar discrete and monolithic Darlington transistors: (*a*) S_{21} and maximum available gain (MAG) versus frequency; (*b*) minimum noise figure versus frequency.

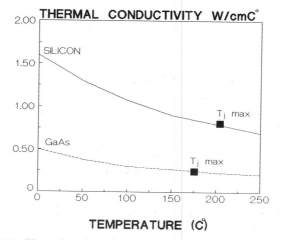

Figure 8.5 Thermal conductivity of silicon and GaAs versus temperature.

Technology factors other than basic material and device properties which affect the potential performance contribution of any high-speed semiconductor in practical production systems include circuit design sophistication and the ability to interconnect to the external world without signal loss or distortion. It is also essential to be able to fabricate the desired device predictably in a production environment with a minimum of unidentified process variables. In these respects silicon bipolar transistors and MMICs are generally superior to alternative high-speed devices.

In the remaining sections of this chapter we review briefly the basis bipolar process structures, together with small- and large-signal modeling approaches (Section 8.2) and then provide an update on the present capabilities and performance of discrete microwave transistors (Section 8.3) and MMICs (Section 8.4). In Section 8.5 we make projections for future performance and practical levels of integration.

8.2   BIPOLAR TRANSISTOR STRUCTURE AND MODELING

The basic device physics and fundamental performance limitations for microwave silicon bipolar transistors are well established and have been thoroughly reviewed [1–4]. Increases in microwave performance are continuing to occur through the straightforward scaling of critical lateral geometries to reduce parasitic capacitances and resistances, and by the more complex scaling of the emitter-to-collector impurity profiles to reduce the fundamental time delays that are the ultimate limitation on performance.

The time delays between emitter and collector determine the frequency where the common-emitter current gain becomes unity. This frequency, f_t, is one of the basic bipolar transistor figures of merit, particularly for digital or current switching applications. A figure of merit that is more relevant for hybrid microwave circuits is $f_{\max}$, defined as the frequency where the unilateral power gain (UG) goes to unity. A simple analytical approximation for $f_{\max}$ includes f_t, total base resistance (R_b), and total collector–base capacitance (C_{cb}):

$$f_{\max} = \sqrt{\frac{f_t}{8\pi R_b C_{cb}}} \tag{8.1}$$

Another, more complex figure of merit has recently been proposed for bipolar transistors in integrated circuits where external matching circuit networks are usually not possible [11].

8.2.1   Structure and Process Technology

Improvements in the intrinsic silicon bipolar transistor figures of merit have occurred and are continuing to occur as the result of dramatic advances in semiconductor process technology. The process architectures and techniques used to fabricate modern microwave silicon bipolar transistors and integrated circuits with submicrometer critical dimension bear little resemblance to those used to manufacture the first microwave transistors with $f_{\max}$ greater than 10 GHz. Some of the features of modern silicon bipolar processes include:

a. Simple process architectures with as few as four masks required to fabricate 25-GHz $f_{\max}$ discrete transistors
b. Either nitride self-aligned or polysilicon self-aligned emitter and base

c. Extensive use of dry etching, including sputter etching, plasma ethcing, and reactive ion etching
d. Fully ion implanted with arsenic emitter
e. Recessed oxide or trench isolation for integrated circuits
f. Thin-film polysilicon resistors
g. Gold-based metal systems

The cross sections of discrete interdigitated silicon bipolar transistors fabricated with nitride self-aligned and polysilicon self-aligned emitter processes are shown in Fig. 8.6. Key features of a nitride self-aligned process are:

a. Simplest process architecture with minimum number of masks
b. One critical mask alignment step
c. Minimum emitter and base contact resistance
d. Maximum performance achieved for multifinger transistors

Key features of a polysilicon self-aligned process are:

a. No critical alignment steps
b. Highest dc current gain
c. Susceptible to erratic emitter or base contact resistance
d. Maximum performance achieved for single-finger transistors

Discrete bipolar transistors generally have a performance advantage over integrated transistors in part because of the lower parasitics and greater epi-layer flexibility inherent in a structure having the collector contact on the bottom of the chip. Most monolithic

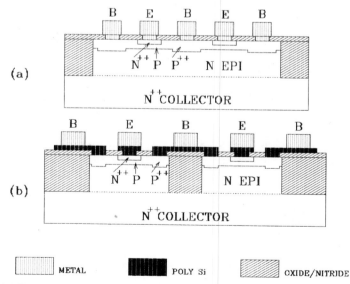

Figure 8.6 Cross sections of modern discrete silicon bipolar transistors: (*a*) nitride self-aligned emitter process; (*b*) polysilicon self-aligned emitter process.

integrated circuits require the complete electrical isolation of adjacent transistors, which usually cannot be accomplished without some negative impact on performance. Two approaches that accomplish isolation with a minimum of performance sacrifice are recessed oxide isolation and deep trench isolation. Cross sections of nitride self-aligned transistors isolated with these two different approaches are shown in Fig. 8.7.

Advances in bipolar process technology have greatly improved the ability to scale and control the very shallow emitter-to-collector impurity profiles required to achieve high f_t. Ion implantation, diffusion from polysilicon, and rapid thermal annealing are used extensively in the fabrication of microwave transistors and integrated circuits. An example of a typical emitter-to-collector vertical impurity profile is shown in Fig. 8.8 for a 10-GHz f_t small-signal discrete transistor. Key features include a very highly doped arsenic emitter, base width less than 0,1 μm, and collector doping and thickness selected to optimize f_{max}.

Future innovations in basic bipolar structures and materials are also possible beyond the evolutionary scaling of vertical and horizontal dimensions. Structures can be conceived that would dramatically increase f_{max} by entirely eliminating the parasitic base resistances and capacitances that are not associated with the intrinsic transistor directly under the emitter contact. Heterojunction structures with emitter and base having different bandgaps are also being explored.

8.2.2   Modeling and Equivalent Circuits

The ability to model the characteristics of individual transistors accurately, quickly, and inexpensively is crucial for circuit designers to exploit the full potential of any technology. Silicon bipolar transistors have well-established models and equivalent circuits that have been proven effective for both small-signal and large-signal operation at microwave frequencies.

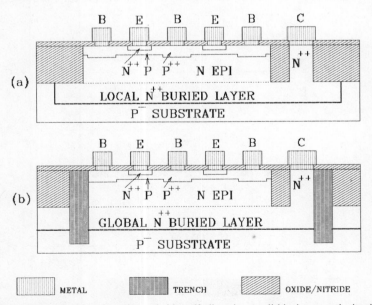

Figure 8.7   Cross sections of modern nitride self-aligned monolithic integrated circuit silicon bipolar transistors: (*a*) recessed oxide-isolated process with local n^{++} buried layer; (*b*) trench-isolated process with global n^{++} buried layer.

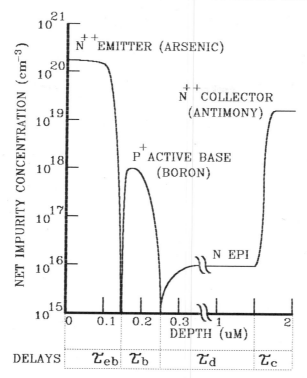

Figure 8.8 Emitter-to-collector doping profile of 10-GHz f_t discrete silicon bipolar transistor versus depth from the emitter surface. Approximate physical location of major time delays are shown.

The T-equivalent circuit shown in Fig. 8.9 is based on a regional physical model of the bipolar transistor chip structure and has been found to be very effective in modeling small-signal performance for fixed-bias conditions [2, 3, 5]. Accurate microwave noise modeling capability can be achieved using this equivalent circuit if both the fixed-base and current-dependent emitter delays are properly accounted for [12, 13]. Definitions

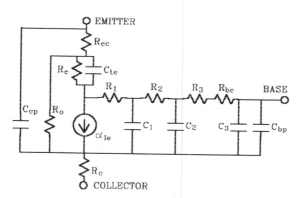

Figure 8.9 Small-signal equivalent circuit of microwave bipolar transistor chip excluding bond wire inductances and package parasitics.

TABLE 8.2 Definition of Small-Signal Equivalent-Circuit Elements for Microwave Bipolar Transistors

Symbol	Definition	Units
C_{ep}	Emitter bond pad capacitance	pF
C_{bp}	Base bond pad capacitance	pF
R_{ec}	Emitter contact resistance	Ω
R_{bc}	Base contact resistance	Ω
R_0	Early effect resistance $$R_0 = \frac{V_e}{I_c}$$	Ω
R_c	Collector resistance	Ω
$\left.\begin{array}{l} R_1 \\ R_2 \\ R_3 \end{array}\right\}$	Distributed base resistance	Ω
$\left.\begin{array}{l} C_1 \\ C_2 \\ C_3 \end{array}\right\}$	Distributed base resistance	Ω
R_e	Emitter resistance $$R_e = \frac{kT}{qI_e}$$	Ω
C_{te}	Emitter–base junction capacitance	pF
α	Common-base current gain $$\alpha = \frac{\alpha_0}{1 + jf/f_b \exp(-j2\pi f \tau_d)}$$	
α_0	Low frequency common-base current gain	
τ_d	Collector depletion region delay time	ps
τ_b	Base region delay time	ps
f_b	Base cutoff frequency $$f_b = \frac{1}{2\pi \tau_b}$$	GHz
f	Operating frequency	GHz

of the circuit elements, time delays, and single-pole current generator for the T-equivalent circuit are given in Table 8.2.

S-parameters calculated using this equivalent circuit can agree very closely with measured results even up to 20 GHz when correct element values are used and unavoidable chip bond wire and carrier parasitics are properly accounted for. For discrete microwave transistors it is particularly important to provide an accurate equivalent circuit for the package or hybrid carrier. Examples of low-parasitic packages suitable for small-signal bipolar transistors and MMICs are shown in Fig. 8.10a. The package equivalent circuit shown in Fig. 8.10b accounts for parasitic lead inductance and transmission-line effects as well as bond wire inductances and shunt capacitances.

For comprehensive large-signal modeling it is necessary to account for device characteristics over the full operation ranges of current, voltage, frequency, and temperature. The complexity of this task can partly be appreciated by considering the generic common-emitter bipolar transistor current–voltage characteristic shown in Fig. 8.11a. The central linear region is bounded by non linear regions which must be properly accounted for in order to accurately simulate large-signal operation. Normal bipolar transistor

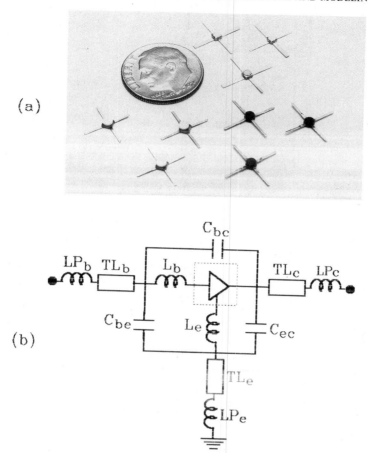

Figure 8.10 Minimum-size packages suitable for small-signal bipolar transistors and MMICs: (a) photographs of hermetic high-reliability 70-mil gold ceramic, hermetically glass-sealed 85-mil ceramic, and low-cost 85-mil plastic microstrip packages; (b) equivalent circuit that can be used effectively to model small-signal bipolar transistor packages.

turn-on non linearities occur at low currents and voltages (regions 2 and 3). The Kirk effect [3] limits operation at high currents (region 4). Other nonlinearities occur at high collector–emitter voltages due to avalanche breakdown (region 1) and at high bias power dissipation due to temperature effects such as thermal runaway (region 5).

The quiescent bias point, signal amplitude, and frequency of operation determine the degree of nonlinearity and signal distortion. The major modes for bipolar transistors are defined by the alternative bias point shown in Fig. 8.11b. Class A bias results in the most linear amplification, lowest noise figure, and the largest dynamic range; class C bias provides the best efficiency and power with the most distortion. Classes AB and B constitute intermediate modes of operation.

Powerful simulation programs such as Berkeley SPICE and its many variations have fortunately proven very accurate and effective for modeling bipolar transistors and circuits over a wide range of large-signal conditions even at microwave frequencies. A basic SPICE equivalent circuit of a microwave bipolar transistor chip excluding bond wire

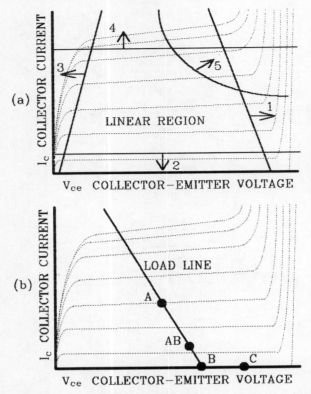

Figure 8.11   Bipolar transistor current–voltage characteristic: (*a*) linear region and non-linear regions (1–5) of operation; (*b*) bias points for class A, AB, B, and C modes of operation.

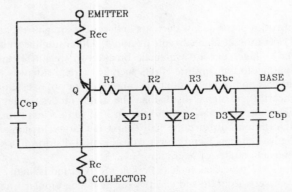

Figure 8.12   Large-signal SPICE equivalent circuit of microwave bipolar transistor chip, excluding bond wire inductances and package parasitics.

TABLE 8.3 Definition of Large-Signal Spice Model Parameters for Microwave Bipolar Transistors

Symbol	Definition	Units
	Transistor (Q)	
IS	Junction saturation current	A
BF	Maximum forward beta	
NF	Current emission coefficient	
VA	Early voltage	V
IK	Corner for high current beta roll-off	
ISE	B-E leakage saturation current	A
NE	B-E leakage emission coefficient	
CJE	B-E 0-bias junction capacitance	F
PE	B-E built-in potential	V
ME	B-E junction grading factor	
FC	Forward-bias depletion capacitance coefficient	
TF	Ideal forward transit time	s
XTF	TF bias dependence coefficient	
VTF	TF dependency on V_{bc}	V
ITF	TF dependency on I_c	A
PTF	Excess phase at $1/(2\pi TF)$	deg
	Diodes (D_1, D_2, D_3)	
IS	Saturation current	A
CJO	0-bias junction capacitance	F
VJ	Junction potential	V
M	Junction grading coefficient	
FC	Forward-bias depletion capacitance coefficient	
BV	Reverse breakdown voltage	V
IBV	Reverse breakdown current	A

inductances and package parasitics is shown in Fig. 8.12. The definition of a minimal set of microwave bipolar transistor SPICE model parameters is given in Table 8.3 for the intrinsic transistor (Q) and the parasitic base–collector diodes (D_1, D_2, D_3).

8.3 DISCRETE TRANSISTOR DESIGN AND PERFORMANCE

8.3.1 Basic Transistor Designs

The optimum and most flexible device for achieving the highest microwave performance from silicon bipolar technology is generally the discrete *npn* transistor. Discrete transistors have been successfully designed, optimized, and made commercially available for a variety of standard microwave components, including:

a. Low-cost wide-band feedback amplifiers

b. Low-noise amplifiers

c. Low-phase-noise oscillators

d. Linear power amplifiers

e. High-power class C CW amplifiers

f. High-power class C pulsed amplifiers

The multifinger interdigitated emitter–base configuration has been established as a very effective basic design for all of these applications [1, 3]. Design optimization for different applications entails proper selection of the emitter–base doping profile, collector doping and thickness, emitter width, emitter length, emitter–emitter pitch, total number of emitter fingers, degree of emitter resistor ballasting, and bonding pad and finger layout configuration. Device physics trade-offs, however, make it impossible to optimize simultaneously for noise, gain, linear power, and class C power performance.

Families of microwave bipolar transistors have been developed and made readily available by a number of manufacturers to address the entire range of standard applications for frequencies up 4 GHz. The SEM photographs shown in Fig. 8.13 are of representative transistor chips designed for low-noise and low-gain (Q414), linear power (Q640), and linear or pulsed class C power (Q6140) applications. The design details, major equivalent circuit element values, figure-of-merit parameters, and performance of transistors optimized for micro-power, low-noise, linear power, and linear or pulsed high-power applications are summarized in Table 8.4.

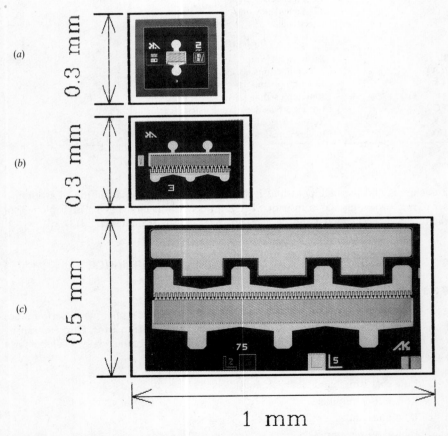

Figure 8.13 SEM photographs of interdigitated silicon bipolar transistor chips designed for (*a*) low noise and gain up to 6 GHz (Q414); (*b*) linear power up to 4 GHz (Q640); (*c*) pulsed class C power up to 4 GHz (Q6140).

TABLE 8.4 Summary of Design, Equivalent Circuit, Figure-of-Merit, and Performance Parameters of Microwave Bipolar Transistor Family

				Transistor			
Parameter	Units	Q220 20-GHz osc.	Q401 Micro-Power	Q405 Low Noise	Q414 Low Noise	Q640 Class A Power CW	Q6140 Class C Power DS = 10%
Design							
Emitters	μo.	20	1	5	14	40	140
Emitter—emitter pitch	μm	2	4	4	4	6	6
Emitter width	μm	0.5	0.7	0.7	0.7	1	1
Emitter length	μm	15	25	25	25	30	75
Emitter periphery	μm	600	50	250	700	2400	21,000
Base area	μm $\times$ μm	720	260	730	1750	8000	70,000
Die size	mm $\times$ mm	0.3 $\times$ 0.3	0.3 $\times$ 0.3	0.3 $\times$ 0.3	0.3 $\times$ 0.3	0.3 $\times$ 0.4	0.5 $\times$ 1
Bias V	V	8	3	8	8	18	45
Bias I_C	mA	18	2	8	25	125	400
Equivalent Circuit							
R_{ec}	Ω	0.2	3.5	0.70	0.30	0.05	0.01
R_{ballest}	Ω	0	0	0	0	2.0	1.0
R_{bc}	Ω	0.60	2.8	0.90	0.40	0.10	0.01
R_1	Ω	0.40	9.0	1.8	0.60	0.30	0.03
R_2	Ω	2.0	35	7.0	2.5	1.0	0.12
R_3	Ω	1.8	28	5.7	2.0	0.80	0.09
R_0	Ω	1100	10,000	2000	700	160	50
R_c	Ω	5	50	9	3	0.8	0.3
C_{ep}	pF	0.020	0.022	0.022	0.022	0.077	0.79
C_1	pF	0.020	0.0019	0.0093	0.026	0.075	0.30
C_2	pF	0.015	0.0041	0.020	0.057	0.20	0.79
C_3	pF	0.030	0.014	0.025	0.051	0.13	0.48
C_{bp}	pF	0.010	0.020	0.020	0.020	0.044	0.59
R_e	Ω	1.6	13	2.7	1.0	0.24	0.07
C_{te}	pF	1.5	0.14	0.72	2.0	8.7	76
τ_d	ps	8	8	8	8	11	23
f_b	GHz	25	23	23	23	23	23
Figure of Merit							
R_b total	Ω	4.8	75	15	5.5	2.2	0.25
C_{cb} total	pF	0.075	0.040	0.074	0.15	0.44	2.2
EP/BA	μm^{-1}	0.83	0.19	0.34	0.40	0.30	0.30
f_t	GHz	9	9	9	9	7	4
$f_{\max}$	GHz	35	11	18	21	17	17
Performance (2 GHz)							
S_{21} gain	dB	16	11	13	12	6	—
50-Ω NF	dB	1.6	5.0	2.0	1.7	6.5	—
minimum NF	dB	1.5	2.2	1.8	1.6	4.5	—
Gain at NF	dB	18	10	12	15	11	—
Power out	W	—	—	—	0.1	1	12
Power gain	dB	—	—	—	16	13	7
Efficiency	%	—	—	—	45	40	55

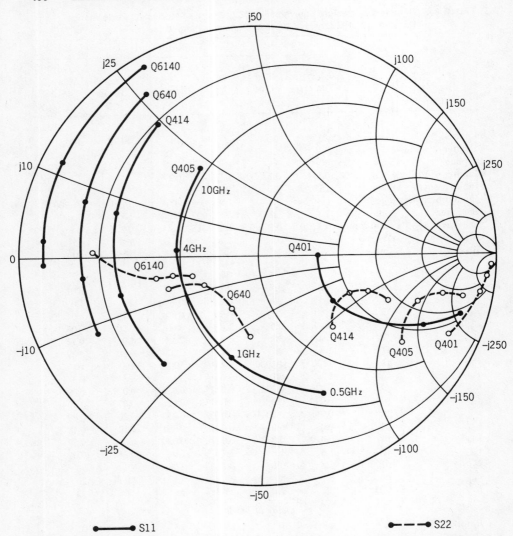

Figure 8.14   Smith chart showing common-emitter S_{11} and S_{22} versus frequency for interdigitated transistor types Q401, Q405, Q414, Q640, and Q6140.

The Smith chart shown in Fig. 8.14 of small-signal common-emitter S_{11} and S_{22} versus frequency for Q401, Q405, Q414, Q640, and Q6140 transistors chips illustrates the wide range of impedances characteristic of different-sized transistors. The common-emitter S_{21} gain at 1 GHz versus current curves of the same transistors are shown in Fig. 8.15 to illustrate the fundamental effect of bias current on performance. Minimum noise figure, maximum gain, and maximum $P-1$ dB bias points are indicated for Q414. The optimum bias current for minimum noise amplification is about 35% of the current that results in maximum S_{21} gain. The optimum bias for linear power amplification is at a current where the S_{21} gain is about $\frac{1}{2}$ dB less than the peak value.

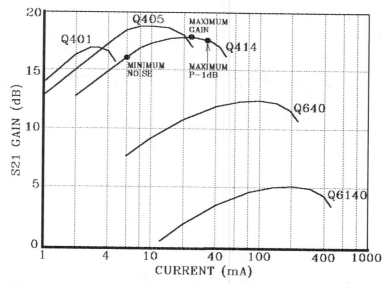

Figure 8.15 Common-emitter S_{21} gain at 1 GHz versus current for transistors Q401, Q405, Q414, Q640, and Q6140. Minimum noise figure, maximum gain, and maximum P–1 dB bias points are indicated for Q414.

8.3.2 High-Gain and Low-Noise Transistors

Dramatic improvements in the small-signal amplifier gain and noise performance of interdigitated silicon bipolar transistors can result from the straightforward scaling of emitter–emitter pitch and reduction of the critical emitter width to submicrometer dimensions. Proportional increases in the ratio of emitter periphery to base area (EP/BA) lead directly to increases in f_{max} and reductions in the minimum achievable noise figure.

Arsenic emitter silicon bipolar transistors with a 0.5-μm emitter width and 2.0-μm emitter–emitter pitch have been successfully fabricated using nitride self-aligned processes, thick planarized field oxide, ion implantation, dry etching, and advanced contact photolithography [14]. A photograph of three such state-of-the-art 2-μm pitch transistors (Q210, Q220, Q240) fabricated on a single 0.3-mm^2 chip is shown in Fig. 8.16a. The SEM photograph in Fig. 8.16b shows the interdigitated 0.5-μm-wide gold contact fingers. This aggressive scaling represents more than a factor of 2 reduction in the critical emitter–emitter pitch compared to previously reported fine-geometry bipolar transistors [15, 16]. The curve tracer photograph shown in Fig. 8.16c reveals a completely normal bipolar transistor current–voltage characteristic despite the very small emitter widths.

The design details, major equivalent-circuit element values, figure-of-merit, and performance parameters of the Q220 transistor chip biased at 8 V, 18 mA are summarized in Table 8.4. The modeled common-emitter, common-base, and common-collector S-parameters of a Q220 chip mounted on a low-parasitic microstrip test carrier with 0.05 μH ground and 0.20 μH input bond wire inductances are listed in Table 8.5. The magnitude of these modeled S-parameters are generally within 10% of those obtained from automatic network analyzer measurements, even at frequencies as high as 20 GHz.

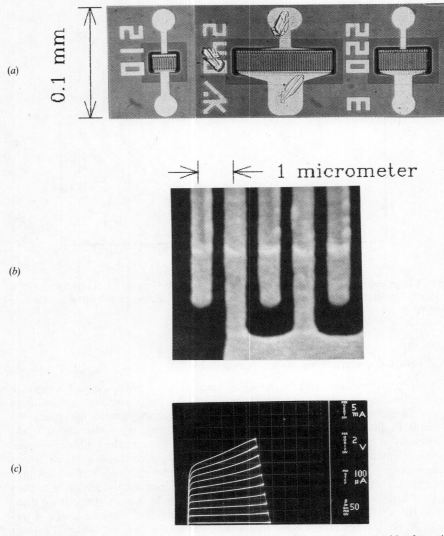

Figure 8.16   Two-micrometer emitter–emitter pitch silicon bipolar transistors capable of practical low-phase-noise oscillator applications up to 20 GHz: (*a*) detailed photograph of Q210, Q220, and Q240 transistors—overall chip size is 0.3 mm²; (*b*) SEM photograph of 0.5-μm-wide interdigitated emitter and base fingers; (*c*) curve tracer current–voltage characteristic of Q220.

The modeled curves of common-emitter S_{21} gain, G_{nf} gain at minimum noise figure, maximum available gain (MAG), maximum stable gain (MSG), and unilateral gain (UG) of the Q220 chip versus frequency are shown in Fig. 8.17*a*. The f_{max} of the Q220 chip, defined as the frequency where UG = 0 dB, is approximately 50 GHz. Extrapolating the MAG curve at 6 dB/ octave to 0 dB results in a more conservative value of 35 GHz

TABLE 8.5   Modeled Q220 Bipolar Transistor S-Parameters

F (GHz)	S_{11} (mag.)	S_{11} (ang.)	S_{21} (dB)	S_{21} (ang.)	S_{12} (dB)	S_{12} (ang.)	S_{22} (mag.)	S_{22} (ang.)
				Common Emitter				
0.1	0.60	−41	31.7	161	−49.1	73	0.90	−6
1	0.75	−153	21.5	102	−38.6	40	0.66	−10
4	0.76	174	9.9	73	−32.9	62	0.63	−12
10	0.77	150	2.1	41	−25.9	74	0.65	−25
15	0.80	133	−1.3	17	−22.4	73	0.68	−37
20	0.82	119	−4.1	−5	−19.8	70	0.72	−49
30	0.88	96	−10.1	−48	−15.9	67	0.82	−74
50	0.90	64	−13.0	87	−11.0	44	0.97	−131
				Common Base				
0.1	0.93	180	5.6	−1	−50.9	4	1.00	0
1	0.93	176	5.6	−10	−51.9	53	1.00	−3
4	1.00	162	5.7	−42	−32.8	143	1.02	−15
10	1.20	121	4.8	−119	−16.0	122	1.01	−45
15	1.07	87	2.0	174	−11.0	95	0.80	−69
20	0.86	67	−0.1	123	−8.7	78	0.65	−83
30	0.65	51	−1.3	65	−5.9	58	0.54	−109
50	0.61	48	−2.2	14	−2.8	22	0.57	179
				Common Collector				
0.1	0.98	−2	5.6	−1	−30.9	47	0.91	178
1	0.96	−15	5.5	−10	−13.6	72	0.90	163
4	0.74	−56	3.7	−32	−3.5	42	0.76	121
10	0.33	−116	−0.3	−49	−0.4	4	0.56	71
15	0.20	−173	−2.3	−53	0	−15	0.49	49
20	0.21	130	−3.4	−55	0.1	−30	0.44	34
30	0.30	71	−3.7	−62	0.1	−58	0.36	14
50	0.35	179	−2.5	−132	−1.5	−148	0.55	82

for f_{max}. The minimum noise figure of the Q220 and the associated 50 Ω system input reflection coefficient versus frequency is shown in Fig. 8.17b. One of the advantages of silicon bipolar transistors for low-noise amplification is that low-noise designs similar in size to the Q220 or Q414 have a minimum noise input match close to 50 Ω for frequencies below 2 GHz.

8.3.3   Oscillator Transistors

Bipolar transistors are in general ideal for fixed-frequency or wide-band oscillator applications. They have inherently lower and more reproducible $1/f$ noise at low frequencies than those of GaAs FETs. Lower $1/f$ noise leads directly to lower upconverted phase noise near the carrier for equally loaded resonator Q_s in YIG-, varactor-, and cavity-tuned oscillators [17, 18].

Bipolar transistors also have advantages over FETs because they can exhibit very wide negative resistance bandwidths, particularly in common-base circuit configurations. As a direct consequence, they can easily be tuned over more than two octaves in simple fundamental YIG-tuned oscillators [19]. Very wide-band hyperabrupt varactor-tuned oscillators have also been demonstrated with upper frequencies exceeding 18 GHz using

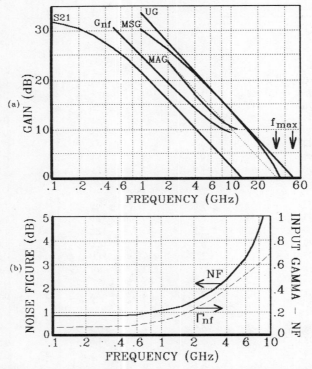

Figure 8.17   Common-emitter performance of 2-μm pitch silicon bipolar transistor chip (Q220): (a) S_{21} gain, G_{nf} gain at minimum noise figure, maximum available gain (MAG), maximum stable gain (MSG), and unilateral gain (UG) versus frequency; (b) minimum noise figure and associated input reflection coefficient versus frequency.

push-push frequency-doubling circuit techniques [20]. Simplified schematics of common hybrid MIC oscillator circuit configurations utilizing silicon bipolar transistors are shown in Fig. 8.18.

The availability of 2-μm emitter–emitter pitch silicon bipolar transistors such as the Q220 has extended the range for practical low-phase-noise fundamental signal generation to over 20 GHz in YIG-tuned [14] or varactor-tuned circuits [21]. The output power performance of state-of-the-art 18-GHz YIG- and varactor-tuned oscillators utilizing Q220 transistor chips and GaAs MMIC output buffer amplifiers is shown in Fig. 8.19a. A comparison of the phase noise performance of 18 GHz YTOs based on the Q220 silicon bipolar transistor and a typical GaAs FET is shown in Fig. 8.19b. The bipolar-based YIG oscillator covers a one-octave-larger bandwidth and is approximately 10 dB quieter than the FET-based version.

8.3.4   Power Transistors

Well-designed microwave bipolar transistors have been established as the highest performance and most cost-effective devices for many applications requiring CW and pulsed power amplification [22–24]. Bipolar transistors designed for power amplifica-

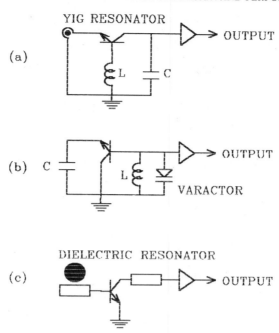

Figure 8.18 Examples of hybrid MIC circuits utilizing silicon bipolar transistors for low-phase-noise oscillators: (*a*) common-base configuration for wide-band YIG-tuned oscillators (YTO); (*b*) common-collector configuration for varactor-tuned oscillators (VTO); (*c*) common-emitter configuration for fixed-frequency dielectric resonator oscillators (DRO).

tion have large emitter areas distributed in a multicell configuration, require epitaxial layer specifications that result in high breakdwon voltages, and have emitter ballasting resistors to prevent hot spots due to current hogging. Significant attention must be paid to packaging and die attach uniformity in order to obtain the low thermal resistance required for reliable operation.

Large power transistors also require input-matching and in some cases output-matching networks inside the packages to increase the chip impedances to levels that can easily be transformed to 50 Ω [23]. Bonding wire inductors and discrete chip capacitors can provide effective high-Q internal matching elements. A typical schematic of an input-matched 50-WCW class C transistor for 900-MHz cellular radio transmitters is shown in Fig. 8.20*a*, and SEM photograph of the multicell and LCL input matching configuration is shown in Fig. 8.20*b*. The schematic of a pulsed class C S-band radar transistor with LCL input matching and monolithic capacitor for output matching is shown in Fig. 8.20*c*.

Discrete bipolar transistors are well suited for class A highly linear power amplification because their distortion characteristics are very predictable and repeatable over a wide range of input powers. An example is shown in Fig. 8.21 of the output versus input power transfer curves of the fundamental and second- and third-order two-tone intermodulation distortion products of a hypothetical 10-dB gain amplifier with $P-1$ dB of + 20 dBm.

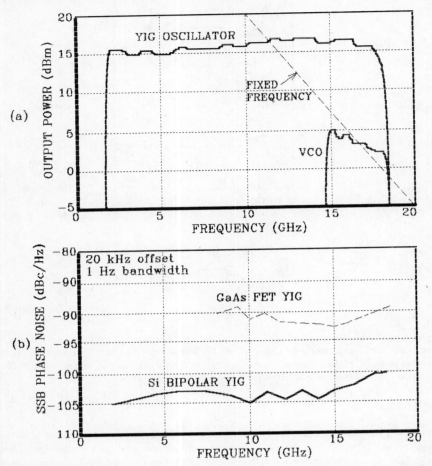

Figure 8.19 Performance of oscillators utilizing 2-μm pitch silicon bipolar transistor chips (Q220) with GaAs MMIC output buffer amplifiers. (*a*) Output power of YIG-tuned and varactor-tuned oscillators versus frequency. Dashed line indicates oscillator power obtainable from single transistors at a fixed frequency. (*b*) Phase noise of silicon bipolar transistor and typical GaAs FET based YTOs versus frequency.

The current-controlled nature of bipolar transistors and their very uniform exponential threshold characteristic allow them to be used effectively in large-signal class B or C modes as well as in the linear class A bias configuration. The maximum CW power that can be obtained from transistors designed for class C operation is about three times the P–1 dB that can easily be obtained from higher-gain linear power transistors. Practical power-added efficiencies as high as 60% can be achieved in class C operation, compared to only about 40% in class A operation. The class C mode is also attractive for pulsed power applications since more than an order of magnitude more peak power can be obtained from transistors designed for low-duty-cycle (DS) operation.

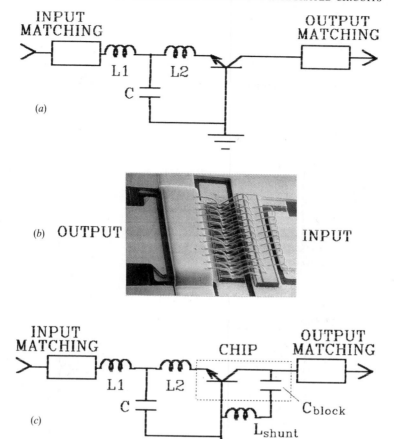

Figure 8.20 Bond wire inductor and chip capacitor matching elements internal to class C power bipolar transistor packages: (a) schematic of input-matched 50-W CW class C transistor for 900-MHz cellular radio transmitters; (b) SEM photograph of multicell and LCL input matching configuration internal to packaged 900-MHz 50-W transistor; (c) schematic of pulsed class C S-band radar transistor with LCL input matching and monolithic capacitor for output matching.

The typical output power versus frequency capabilities of the highest-performance commercially available common-emitter class A and common-base class C power bipolar transistors are shown in Fig. 8.22. The highest power and efficiency is achieved with low-duty-cycle (DS) class C operation. Reliable and cost-effective power amplification using silicon bipolar transistors extends to frequencies as high as 5 GHz.

8.4 MONOLITHIC MICROWAVE INTEGRATED CIRCUITS

The economic and competitive pressures to develop high-speed building blocks that are lower in unit cost, more uniform, more reliable, smaller, of wider bandwidth, and more complex has lead to the microwave and fiber-optic industries' determined search for

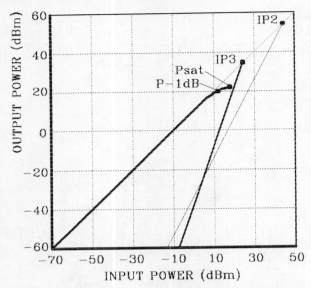

Figure 8.21 Output versus input power transfer curves of fundamental and second- and third-order two-tone intermodulation distortion products of hypothetical 10-dB-gain class A amplifier with P–1 dB of +20 dBm.

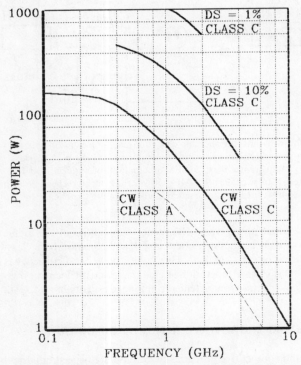

Figure 8.22 Approximate power output of commercial packaged multicell common-emitter class A and common-base class C power bipolar transistors versus frequency. The highest power and efficiency is achieved with low-duty-cycle (DS) class C operation.

manufacturable monolithic microwave integrated circuits (MMICs) that are truly cost-effective alternatives to conventional hybrid microwave integrated circuits (MICs). A broad definition of MMIC is taken here to include both analog and digital circuits as long as they perform IF, RF, or microwave signal conditioning and processing functions and require monolithic microwave semiconductor technology to achieve the desired performance.

Research and development of MMICs has focused predominately on GaAs technology with circuit designs based heavily on reactive matching techniques. MMICs based on well-established bipolar circuit designs and fabricated using aggressively scaled silicon bipolar IC processes with f_t between 10 and 20 GHz have proven, however, to be the optimum approach for many applications from dc to over 10 GHz. This section will review the present capabilities of silicon bipolar technology for amplifier, frequency-converter, frequency-divider, fiber-optic, and semicustom MMICs.

8.4.1 Amplifiers

Signal amplifications, which constitutes the most ubiquitous general function in all microwave and fiber-optic systems, has the highest priority for monolithic integration. The wide range of gain, noise, power, dynamic range, bandwidth, and bias specifications required to meet the diversity of system requirements has lead to the need for compatible families of standard product MMIC amplifiers. Fixed-gain cascadable wide-band MMIC amplifiers in particular have become generally available cost-effective alternatives to hybrid MIC amplifiers for a wide range of applications.

The very high transconductance and S_{21} gain of microwave silicon bipolar transistors makes them well suited for inherently wide-band resistive feedback amplifiers designed to be cascadable in 50- or 75-Ω system [25–28]. Typical examples of single- and two-stage fixed-gain feedback amplifier circuits are shown in Fig. 8.23. These very-small-scale-integrated (VSSI) MMIC amplifiers make use of the very high open-loop-gain. Darlington transistor configuration and can achieve gain–bandwidths as high as 10 GHz using a 10-GHz f_t nitride self-aligned process. SEM photographs of a typical single-stage MMIC amplifier chip and its package configuration are shown in Fig. 8.24.

The use of low-parasitic thin-film feedback resistors combined with the ability to optimize the size of individual transistors leads to considerable design flexibility using the same basic resistive feedback circuit. It is possible to design a diverse family of MMIC amplifiers to meet gain, noise, power, or bandwidth specifications using the single-stage feedback circuits of Fig. 8.23. Individual amplifiers can easily be designed for particular gains, and optimized for either power, noise figure, input VSWR, or gain flatness.

The wide range of possible performance characteristics is illustrated by the curves versus frequency and bias current shown in Fig. 8.25 for an 11-chip MMIC family. Single-stage gains can be as high as 30 dB at 100-MHz, usable bandwidths as wide as 6 GHz, noise figures less than 3 dB, or linear powers as high as 30 dBm. The ripple-free gain roll-of at higher frequencies is a general characteristic of resistive feedback MMIC amplifiers that contributes to easy cascadability. Other attractive features of these MMICs include short group delays on the order of 100 ps, and intrinsic phase match from unit to unit of better than a few degrees.

Although very attractive for wide-band small-signal amplification, the inherent low efficiency of resistive feedback circuits and the relatively low impedances of bipolar transistors results in severe limitations when designing for power applications. One effective approach that can be used to extend the applicability of silicon bipolar MMICs to higher linear powers with improved even-order harmonic distortion performance is to design

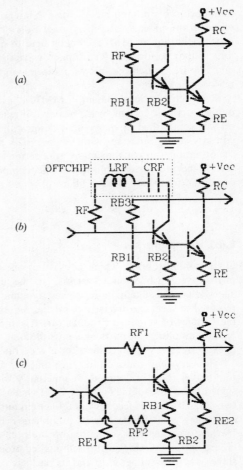

Figure 8.23 Schematics of simple silicon bipolar MMIC resistive feedback amplifiers: (*a*) single-stage amplifier with merged RF feedback and dc bias; (*b*) single-stage amplifier with separate RF feedback and dc bias; (*c*) low-noise two-stage amplifier.

an individual fixed-gain MMIC for 25-Ω impedances and assemble two gain blocks in a push-pull configuration [29]. Figure 8.26 shows the schematic and single-package configuration of such an amplifier using two balanced 25-Ω MSA-10 MMIC chips and external baluns to transform to the unbalanced 50-Ω source and load. This push-pull MMIC amplifier configuration can deliver P–1 dB greater than 1 W and IP3 greater than 42 dBm at 1 GHz and has usable performance even up to 2 GHz.

Many microwave and fiber-optic systems require wide-band amplifiers with special characteristics, such as automatic gain control (AGC), logarithmic gain, or output-power-limiting characteristics. Classical differential circuits based on emitter-coupled pairs of bipolar transistors are particularly well suited for such requirements. AGC amplifiers of 1 GHz bandwidth are commercially available and 2-GHz-bandwidth AGC and equalizing amplifiers optimized for fiber-optic systems have been reported [30, 31]. Differential limiting amplifiers have also been demonstrated for 4-Gbit/s fiber-optic systems [32], and AGC amplifiers with practical 4-GHz bandwidths should be possible in the near future.

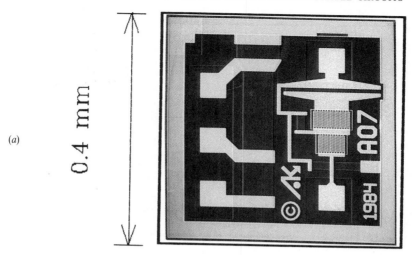

(a)

0.4 mm

(b)

Figure 8.24 SEM photographs of typical silicon bipolar single-stage fixed-gain MMIC feedback amplifier: (*a*) MSA-07 MMIC chip with gain greater than 10 dB up to 3 GHz; (*b*) MMIC chip assembled into 70-mil ceramic microstrip package.

The schematic of a generic bipolar AGC amplifier is shown in Fig. 8.27 as an example of a classical differential bipolar circuit that can be scaled for operation at microwave frequencies. It is based on the principle of "current steering" between transistors Q1 and Q2 which have independent emitter fingers each connected to a different collector of the gain control differential pair. Variations of the AGC voltage cause the current to change in the respective emitters of Q1 and Q2, which causes the gain to vary. This circuit can be optimized for a wide range of target specifications and can be designed to maintain a constant output $P-1$ dB compression point throughout its entire AGC range.

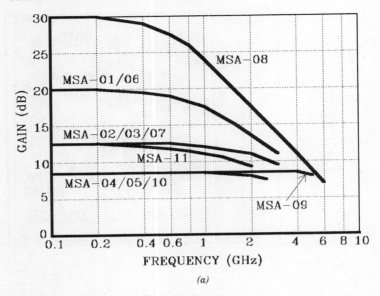

(a)

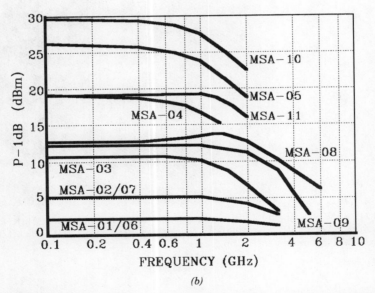

(b)

Figure 8.25 Performance of silicon bipolar single-stage fixed-gain MMIC feedback amplifier family: (*a*) small-signal gain versus frequency; (*b*) *P*–1 dB versus frequency; (*c*) noise figure versus frequency; (*d*) gain at 500 MHz versus bias current; (*e*) *P*–1 dB at 500 MHz versus bias current; (*f*) MMIC bias voltage versus current.

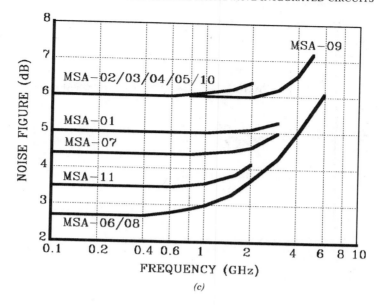

(c)

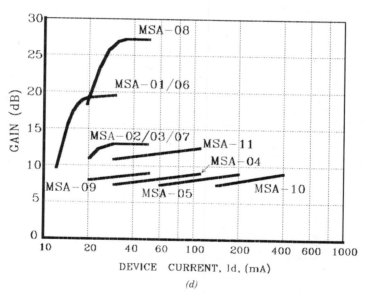

(d)

Figure 8.25   (*Continued*)

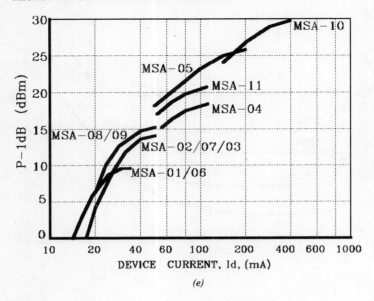

(e)

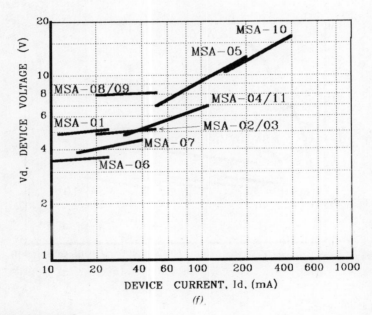

(f)

Figure 8.25 *(Continued)*

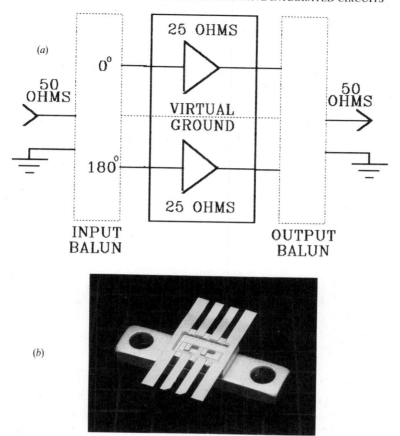

Figure 8.26 Configuration of MSA-10 MMIC-based feedback amplifier designed for medium power push-pull linear amplification up to 2 GHz: (*a*) general schematic of push-pull amplifier usings baluns for interface to single-ended 50-Ω source and load; (*b*) photograph of two 25-Ω MSA-10 MMIC chips assembled into package designed for push-pull operation.

8.4.2 Frequency Converters

Frequency conversion and phase detection are usually accomplished in microwave systems using passive hybrid MIC mixer circuits based on beam-lead Schottky-barrier diodes. These mixers are very effective for up- or down-frequency conversion over extremely wide bandwidths but always feature a conversion loss of from 5 to 8 dB and have an output dynamic range that is a function of the external LO drive level. Much smaller active bipolar MMIC mixers can provide a feasible alternative to diode mixers for applications below 10 GHz and allow easy integration with other MMIC functions, such as low-noise amplifiers, AGC amplifiers, and frequency synthesizers.

The schematic and photograph shown in Fig. 8.28 are an example of an active MMIC mixer based on the well-known Gilbert analog multiplier circuit. This circuit provides true double-balanced operation with excellent isolation between all ports. It can be designed to achieve conversion gain instead of the loss characteristic of diode mixers

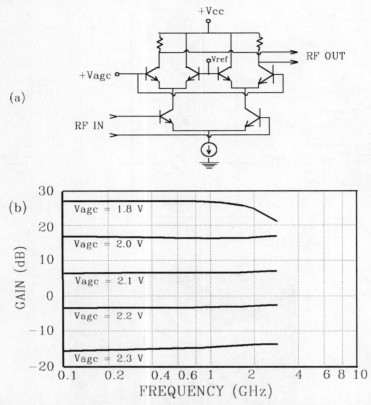

Figure 8.27   Generic silicon bipolar AGC integrated circuit amplifier suitable for microwave applications: (*a*) simplified schematic; (*b*) gain versus frequency with control voltage as a parameter for a design usable to 3 GHz.

and can function successfully with LO powers as low as -10 dBm. Although widely used at lower frequencies, silicon bipolar Gilbert multiplier mixers have only recently been developed for applications above 1 GHz [33]. In general, this class of active mixer has the capability of operating effectively at frequencies up to approximately 40% of f_t. Figure 8.28*c* shows the hypothetical RF, LO, and IF output signal levels versus frequency for a 10-GHz f_t Gilbert multiplier mixer acting as a block down-converter for five satellite transponder signals in the TVRO band at 3.7 to 4.2 GHz. In this example the 3.25-GHz LO input signal power is only -8 dBm and the conversion loss is 0 dB.

Another approach for very low-cost frequency conversion applications is to use a weakly fed-back Darlington MMIC amplifier configuration as the heart of an unbalanced two-port active mixer [34]. The schematic and functional block diagram of a commercially available MMIC family optimized for very high conversion gains with low LO power is shown in Fig. 8.29. The two-port configuration of this MMIC requires that external filters be provided to achieve RF, LO, and IF signal isolation and reduce spurious signals to an acceptable level. The conversion gain versus RF frequency of two different MMIC designs is shown in Fig. 8.29*c* for both 70-MHz and 1-GHz IF frequencies. The dc power consumption for the MSF-88 is 300 mW and the required LO power

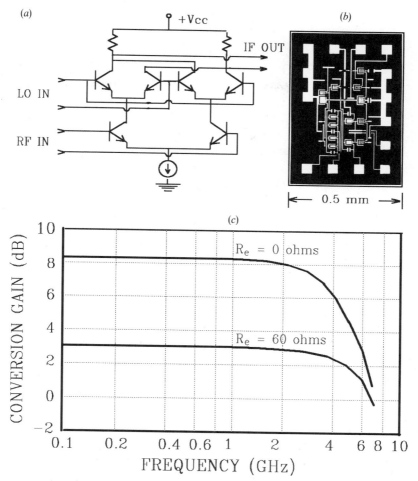

Figure 8.28   Active double-balanced mixer based on Gilbert multiplier circuit: (*a*) schematic of basic Gilbert multiplier circuit; (*b*) photograph of silicon bipolar active mixer MMIC chip; (*c*) conversion gain versus frequency with a 70-MHz IF frequency for basic design and with additional 60 Ω of R_e added to increase flat-gain bandwidth.

is less than +5 dBm. The dc power consumption for the MSF-86 is only 50 mW and the required LO power is only −5 dBm. RF–IF conversion gains as high as 20 dB at 1 GHz and 5 dB at 10 GHz can be achieved using the MSF-88 with a 70-MHz IF.

These two-port frequency converted MMICs can also be used as self-oscillating mixers (SOMs) if an external LO resonator is connected between the input and output ports and the Barkhausen criteria for oscillation are satisfied [34]. The schematic and functional block diagram of an MMIC in a SOM configuration using a ceramic dielectric puck resonator are shown in Fig. 8.30. This merging of the mixing, LO, and IF amplification functions using a very simple bipolar MMIC can lead to dramatic size and cost reductions for certain microwave subsystem applications. Figure 8.30*c* shows the unfiltered RF, LO, and IF output signal levels versus frequency for an MSF-88 SOM

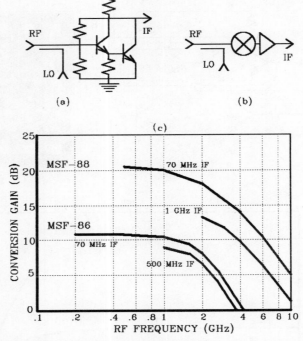

Figure 8.29   Silicon bipolar MMIC optimized for application as a low-cost two-port unbalanced active mixer using a Darlington transistor configuration and minimum feedback: (*a*) schematic of MMIC with injected LO at input; (*b*) functional block diagram of MMIC; (*c*) conversion gain versus RF frequency with IF frequency as a parameter for two active mixer MMICs.

acting as a block down converter for five satellite transponder signals in the TVRO band at 3.7 to 4.2 GHz. The conversion gain to the band at 0.45 to 0.95 GHz is approximately 10 dB.

8.4.3   Frequency Dividers

More stable and lower-phase-noise microwave sources are essential for many military and commercial microwave systems. One of the most attractive techniques for improving the stability and reducing the phase noise of microwave oscillators is to phase lock them to a stable low-frequency reference oscillator. Figure 8.31 shows the schematic of a generic phase-locked-loop (PLL) frequency-stabilized microwave local oscillator using fixed and programable modulus dividers to lock the voltage-controlled LO to a low-phase-noise and temperature-stable low-frequency reference oscillator. The variable modulus divider enables digital control of the LO frequency, while the use of a fixed divider allows operation at frequencies above the practical limits of programable dividers.

Both fixed-and variable-modulus digital dividers are readily implemented in bipolar technologies using current-mode-logic circuits based on the master-slave D flip-flop. Aggressive scaling of silicon bipolar process technologies has enabled fixed-frequency dividers optimized for speed to reach clock frequencies in excess of 10 GHz [35, 36]. In any given technology the maximum clock frequency of a digital circuit can also be

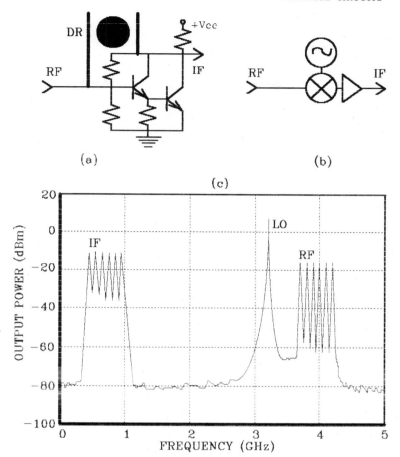

Figure 8.30 Application of silicon bipolar active mixer MMIC as a self-oscillating mixer (SOM): (*a*) schematic of MMIC with dielectric resonator between input and output to establish LO; (*b*) functional block diagram of MMIC and dielectric resonator; (*c*) RF, LO, and IF output signal levels versus frequency with 3.25-GHz dielectric resonator and five TVRO input signals between 3.7 and 4.2 GHz.

traded for lower power consumption. As an example, Fig. 8.32 shows the block diagram, circuit schematic, die photograph, and input/output waveforms of an aggressively scaled static divider capable of dividing by 4 up to 6 GHz with a single clock input and power consumption of only 125 mW [37]. In this case the small 0.5 by 0.75 mm die allows the divider to be packaged in a 100-mil² four-leaded surface-mountable ceramic package. A larger-scale 128/129 variable-modulus CML divider with a power consumption of 125 mW has also been demonstrated for high-volume portable radio applications up to 1 GHz [38].

Higher division frequencies can be achieved using non-digital circuit approaches. A block diagram of the classical regenerative frequency divider using mixing and reactive feedback is shown in Fig. 8.33*a*. A 1.5 to 7.3-GHz divider with this architecture has been developed using a standard bipolar process technology and a fast double-balanced mixer circuit [39]. The Darlington transistor frequency converter MMIC described in Section

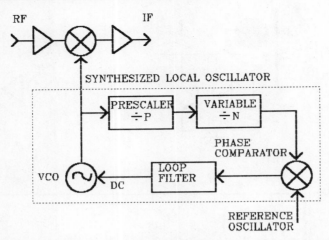

Figure 8.31 Schematic of generic phase-locked-loop (PLL) frequency-stabilized microwave local oscillator using fixed and programable modulus dividers to lock the LO signal to a low-frequency reference oscillator.

8.4.2 can be configured for regenerative frequency division up to even higher frequencies by resonating the device capacitance with a bond wire inductor between the input and output [40]. The schematic of this very simple MMIC divider, which can operate up to 20 GHz, is shown in Fig. 8.33*b*. The input threshold for division versus frequency with the internal feedback optimized for operation between 14.7 and 17.7 GHz is shown in Fig. 8.33*c*.

8.4.4 System Application of Packaged MMICs

The availability of single- and multifunction MMIC packaged products with uniform and guaranteed performance can have a major impact on many RF and microwave systems by providing feasible cost-effective alternatives to thin-film hybrid components. The compatibility of small silicon bipolar MMIC chips with industry standard plastic and ceramic surface-mount packages enables relatively low-cost components to be manufactured with predictable performance and without compromising the inherent size advantage of MMICs.

The block diagram of a hypothetical low-noise RF or microwave receiver front end is shown in Fig. 8.34 as an example of how a frequency-down-converter subsystem might be implemented using generic silicon bipolar MMICs in standard surface-mount packages. Inherently wide-band bipolar MMICs can often be adapted for specific receiver or transmitter subsystems simply by selecting the appropriate filters between various components. The requirements of such subsystems for high-Q signal filtering to achieve desired system performance tends to favor the use of relatively simple standard-function components over technically feasible but more expensive larger scales of MMIC integration.

8.4.5 MMICs for Fiber-Optic Systems

Fiber-optic systems for the high-speed long-distance transmission of digital information have taken a central role in modern telecommunications. Systems are already in operation with data rates of 2.4 Gbit/s, and the need exists for even higher rates in the future.

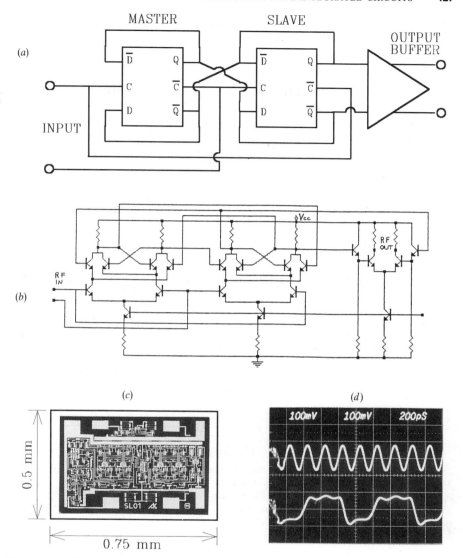

Figure 8.32 Silicon bipolar digital frequency divider: (*a*) generic block diagram of master-slave D flip-flop digital divider and buffer amplifier; (*b*) schematic of digital divider circuit implemented using scaled bipolar emitter-coupled logic (ECL); (*c*) photograph of small silicon bipolar chip capable of dividing by 4 up to 6 GHz with a single clock input and power consumption of only 125 mW; (*d*) sampling oscilloscope photograph showing 5-GHz input signal and output response waveforms of chip shown in (*c*).

Very wide-band microwave amplifiers and high-speed digital circuits are required in the data multiplexers, transmitters, regenerators, receivers, and demultiplexers of these systems. Generic fiber-optic receivers generally include a linear amplification channel, clock recovery, decision circuit, and demultiplexer. Silicon bipolar ICs are capable of addressing all of these functions, even for data rates greater than 4 Gbit/s.

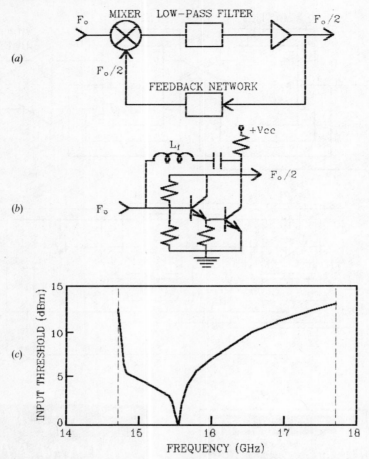

Figure 8.33 Regenerative frequency divider using reactive feedback and silicon bipolar Darlington MMIC optimized for frequency conversion: (a) block diagram of generic regenerative frequency divider ($\frac{1}{2}$); (b) schematic of regenerative divider using silicon bipolar MMIC; (c) input threshold for frequency division versus frequency for MMIC in 70-mil microstrip package with internal feedback optimized for operation between 14.7 and 17.7 GHz.

The linear amplification channel generally consists of a low-noise photodiode current amplifier and a high-gain AGC amplifier with good output-amplitude-limiting properties. The low-noise amplifier usually has a transimpedance circuit configuration using a GaAs FET input transistor for increased sensitivity. The superior sensitivity of GaAs FET versus silicon bipolar transimpedance amplifiers tends to diminish, however, as the data rate increases. Silicon bipolar transimpedance amplifiers have already been shown to be effective for data rates as high as 2 Gbit/s [30]. The schematic of a typical bipolar transimpedance amplifier is shown in Fig. 8.35a. Linear amplification channels based entirely on silicon bipolar technology are serious contenders for future fiber-optic systems for data rates in excess of 4 Gbits/s [32].

Fiber-optic system functions that require high-speed digital components include the data regenerator decision and clock recovery circuits, multiplexers, and demultiplexers.

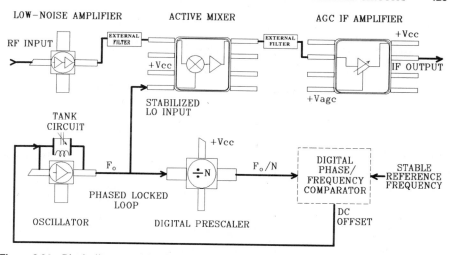

Figure 8.34 Block diagram of hypothetical low-noise RF or microwave receiver front-end sub-system implemented using generic silicon bipolar MMICs in standard surface-mount packages.

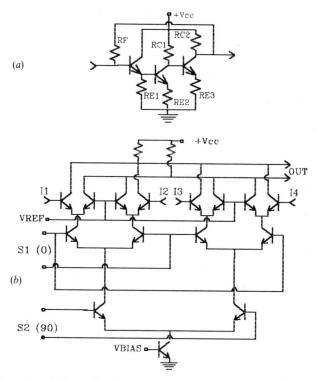

Figure 8.35 Examples of silicon bipolar ICs suitable for fiber-optic systems requiring gigabit/second data rates: (*a*) schematic of transimpedance amplifier circuit; (*b*) schematic of 4:1 digital multiplexer circuit.

The feasibility of using silicon bipolar technology for these primarily digital circuits has been demonstrated for data rates as high as 6 Gbit/s [41, 42]. The schematic of a 4:1 multiplexer is shown in Fig. 8.35b as an example of a standard bipolar circuit that can effectively be scaled for operation in high-speed fiber-optic systems. Future advances in circuit design and process technology are expected to extend the practical use of such digital circuits to data rates above 10 Gbits/s [42].

8.4.6 Semicustom MMICs

In general, the lowest-cost MMICs will be those that are successfully designed by a chip manufacturer to address a large number of general applications and are subse-

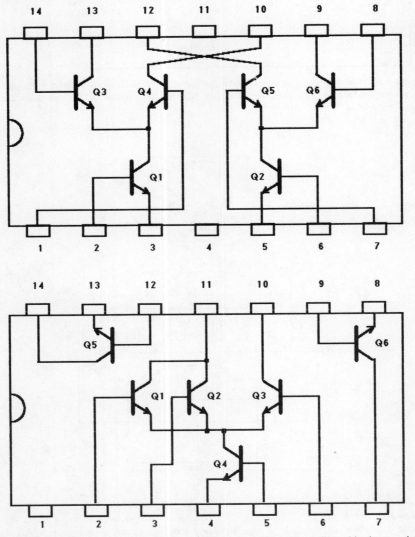

Figure 8.36 Schematics of typical packaged MMIC arrays of high-speed silicon bipolar transistors for semicustom circuit applications.

quently built in the highest possible volume with the greatest degree of standardization. In certain cases it may be justified to use custom or semicustom circuit designs that are highly specific to a particular application. Full-custom MMIC design is possible using established GaAs or silicon bipolar foundry services, but the nonrecurring design and fabrication costs can be prohibitive.

One alternative to expensive full-custom design for simple circuits is to use the high-speed transistor arrays that are available from a number of manufacturers. Two typical examples of 10 GHz f_t silicon bipolar transistor arrays sold in standard 14-pin packages are shown in Fig. 8.36. Custom circuit configuration is easily accomplished by interconnecting the terminals with appropriate external components, but significant performance degradation and variation due to interconnection delays and parasitics is usually unavoidable.

Higher-performance, more complex semicustom circuits can be fabricated using single-chip MMIC arrays of initially disconnected transistors, diodes, resistors, and capacitors. An example is given in Fig. 8.37, which shows the uncommitted schematic of a simple small-scale MMIC chip array of 15-GHz f_t silicon bipolar transistors and thin-film resistors that can be customized by design of the second-metal interconnect

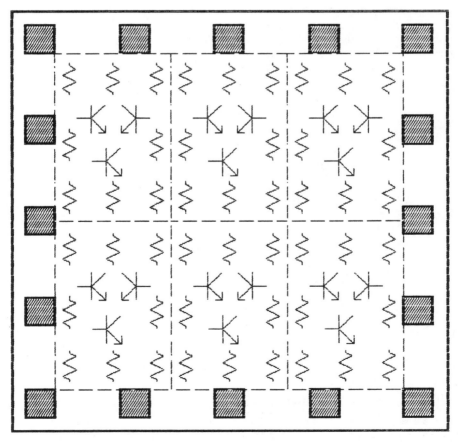

Figure 8.37 Schematic of small-scale MMIC chip array of uncommitted silicon bipolar transistors and thin-film resistors that can be customized by design of metal interconnect patterns.

pattern. More complicated 6.5-GHz f_t bipolar transistor arrays are commercially available which can be used to combine hundreds of high-speed digital ECL gates and analog functions on a single chip with provision for laser trimming of precision thin-film resistors [43]. The demonstration of an 86,000 component VLSI ECL masterslice array fabricated with a 10-GHz f_t polysilicon self-aligned and trench-isolated process [44] suggests that much more complicated semicustom circuits integrating both digital and MMIC functions may already be technologically feasible.

8.5 PROJECTIONS

Future prospects for expanding the contributions of silicon bipolar technology to microwave and high-speed fiber-optic systems are dependent upon the ability to continue the evolution to higher-performance single transistors and more complex levels of integration. The significant advances that have been reviewed in this chapter suggest that rumors of the obsolescence of silicon bipolar technology are highly exaggerated. Projections of future advances can be made with confidence.

Scaling to smaller emitter–emitter pitch is straightforward in principle because of the absence of deleterious "short-channel effects" and will result in significant increases in the f_{max} of small-signal integrated as well as discrete transistors. The experimental and projected f_{max} of optimized discrete silicon bipolar transistors versus emitter–emitter pitch is shown in Fig. 8.38. An f_{max} of over 100 GHz can be predicted for 1-μm pitch transistors when vertical emitter–base impurity profiles are scaled to achieve 15-GHz f_t and extrinsic parasitic resistances and capacitances are reduced significantly.

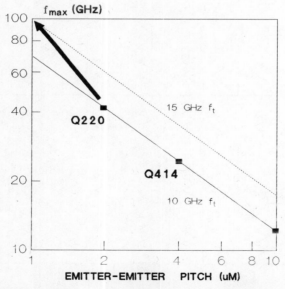

Figure 8.38 Optimum f_{max} of discrete silicon bipolar transistors versus emitter–emitter pitch. A value of 100 GHz f_{max} is projected for 1-μm pitch transistors when emitter–base impurity profile is scaled to achieve 15 GHz f_t.

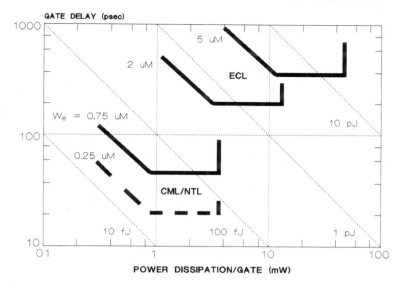

Figure 8.39 High-speed silicon bipolar digital gate delay versus gate power dissipation with emitter width as a parameter. Ten-gigahertz digital dividers and 1.5-GHz VLSI masterslice circuits have been achieved using submicrometer, 15-GHz f_t ECL/CML technology.

The propagation delay and power dissipation of high-speed digital ECL and CML gates can also be reduced dramatically by aggressive scaling of both horizontal critical dimensions and vertical impurity profiles and reduction of internal gate logic levels to 200 mV and below. Figure 8.39 illustrates the effect of scaling silicon bipolar gates on the speed–power product. For logic levels below 200 mV, the propagation delay of ECL/CML gates is expected to be less than 20 ps per gate when the critical emitter width is reduced to 0.25 μm and vertical impurity profiles are scaled to achieve f_t in excess of 30 GHz. MSI digital prescalers fabricated with this technology will have maximum clock rates up to 20 GHz, and true VLSI gigibit/second logic chips may also be practical with design rules relaxed in favor of high yield.

The intrinsic speed advantages of bipolar transistors for high-speed digital circuits have also been demonstrated by digital dividers fabricated with novel AlGaAs/GaAs heterojunction bipolar transistor (HBT) IC processes [45, 46]. These HBT circuits presently have less than a 2:1 speed advantage over comparable silicon bipolar circuits. This modest speed advantage is not expected to widen in the future because of expected parallel advances in both digital III–V and silicon bipolar technologies.

The practical level of integration and useful frequency range for silicon bipolar MMICs designed to perform traditional microwave functions will advance steadily along with process technology evolution. The proposed definitions of silicon bipolar MMIC level of integration given in Fig. 8.40 range from present commercially available VSSI and SSI building-block families to future LSI subsystems such as a complete receiver front end on a single chip, including frequency-synthesized local oscillator and down conversion to baseband. The practical upper frequency limit for standard SSI silicon bipolar MMIC products will extend to as high as 20 GHz for special applications, but cost-effective LSI MMICs will probably be limited to frequencies below 6 GHz. Silicon bipolar technology is also expected to become available to the microwave and

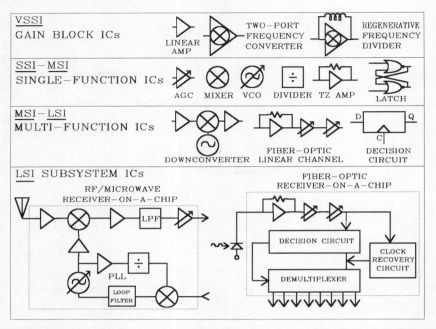

Figure 8.40 Major levels of integration and representative functions for silicon bipolar MMICs range from VSSI fixed-gain blocks to LSI microwave or fiber-optic receiver subsystems on a single chip for applications up to 6 GHz.

fiber-optic industries through full-custom and semicustom foundry services based on extensive libraries of standard cells for the design of application-specific MMICs.

In conclusion, silicon bipolar transistors, MMICs, and gigibit/second digital ICs will continue to make increasingly important performance and cost contributions to microwave and high-speed fiber-optic systems for the foreseeable future. Silicon bipolar and GaAs MMIC components will continue to coexist happily in these systems, and silicon will not be relegated to the role of just an *n*-type impurity for high-speed semiconductors.

Acknowledgments. The author would like to thank all of his colleagues at Avantek, whose dedication to advancing the performance of silicon bipolar processes, transistors, and circuits has helped to ensure that silicon bipolar will continue to improve the high-speed semiconductor technology of the future. Special thanks are due to Jose Kukielka for many valuable discussions and for reviewing the manuscript, to Jim Wholey for permission to include his unpublished power transistor results, and to my family, Barbara, Robyn, and Amy, for their continuing support and understanding.

REFERENCES

1. H. F. Cooke, "Microwave Transistors: Theory and Design," *Proc. IEEE*, Vol. 59, pp. 1163–1181, August 1971.

2. E. D. Graham and C. W. Gwyn, Eds., *Microwave Transistors*, Artech House, Dedham, MA, 1975.

3. J. S. Lamming, "Microwave Transistors," in M. J. Howes and D. V. Morgan, Eds., *Microwave Devices*, Wiley, New York, 1976.

4. S. Y. Liao, *Microwave Solid-State Devices*, Prentice-Hall, Englewood Cliffs, NJ, 1985, pp. 51–70.

5. C. P. Snapp, "Silicon Bipolar Transistors and Integrated Circuits Continue to Grow," *Microwave Syst. News*, Vol. 13, pp. 32–41, November 1983.

6. C. P. Snapp, "Advanced Silicon Bipolar Technology Yields Usable Monolithic RF and Microwave ICs," *Microwave J.* Vol. 26, pp. 93–103, August 1983.

7. S. G. Knorr, "The Potential of Bipolar Devices in LSI Gigibit Logic," *IEEE Circuits Syst. Mag.*, Vol. 3, pp. 2–6, January 1981.

8. S. Konaka et al., "A 30-ps Si Bipolar IC Using Super Self-Aligned Process Technology," *IEEE Trans. Electron Devices*, Vol. ED-33, pp. 526–531, April 1986.

9. Y. Tamaki et al., "A Fine Emitter Transistor Fabricated by Electron-Beam Lithography for High-Speed Bipolar LSIs," *IEEE Electron Device Lett.*, Vol. EDL-7, pp. 425–427, July 1986.

10. H. K. Park et al., "High-Speed Polysilicon Emitter-Base Bipolar Transistor," *IEEE Electron Device Lett.*, Vol. EDL-7, pp. 658–660, December 1986.

11. G. W. Taylor and J. G. Simmons, "Figure of Merit for Integrated Bipolar Transistors," *Solid-State Electron.*, Vol. 29, pp. 941–946, September 1986.

12. R. J. Hawkins, "Limitations of Nielsen's and Related Noise Equations Applied to Microwave Bipolar Transistors, and a New Expression for the Frequency and Current Dependent Noise Figure," *Solid-State Electron.*, Vol. 20, pp. 191–196, March 1977.

13. U. L. Rohde, "Modeling Noise in Microwave Bipolar Transistors," *Microwave Syst. News*, Vol. 17, pp. 91–93, February 1987.

14. C. C. Leung et al., "A 0.5 μm Silicon Bipolar Transistor for Low-Phase-Noise Oscillator Applications up to 20 GHz," *IEEE MTT-S Int. Microwave Symp. Dig.*, pp. 383–386, 1985.

15. T.-H. Hsu and C. P. Snapp, "Low Noise Microwave Bipolar Transistor with Sub-Half-Micrometer Emitter Width," *IEEE Trans. Electron Devices*, Vol. ED-25, pp. 723–730, June 1978.

16. H.-T. Yuan et al., "A 2-Watt X-Band Silicon Power Transistor," *IEEE Trans. Electron Devices*, Vol. ED-25, pp. 731–736, June 1978.

17. J. H. Lepoff and P. Ramratan, "FET vs. Bipolar: Which Oscillator Is Quieter?" *Microwaves*, Vol. 19, pp. 82–83, November 1980.

18. C. Ansorge, "Bipolar Transistor Ku-Band Oscillators with Low Phase-Noise," *IEEE MTT-S Int. Microwave Symp. Dig.*, pp. 91–94, 1986.

19. G. R. Basawapatna and R. B. Stancliff, "A Unified Approach to the Design of Wide-Band Microwave Solid-State Oscillators," *IEEE Trans. Microwave Theory Tech.*, Vol. MTT-27, pp. 379–385, May 1979.

20. R. G. Winch, "Wide-Band Varactor-Tuned Oscillators," *IEEE J. Solid-State Circuits*, Vol. SC-17, pp. 1214–1219, December 1982.

21. A. P. S. Khanna, "Fast-Settling, Low-Noise Ku Band Fundamental Bipolar VCO," *IEEE MTT-S Int. Microwave Symp. Dig.*, pp. 579–581, 1987.

22. J. T. C. Chen and C. P. Snapp, "Bipolar Microwave Linear Power Transistor Design," *IEEE Trans. Microwave Theory Tech.*, Vol. MTT-27, pp. 423–430, May 1979.

23. R. Allison, "Silicon Bipolar Microwave Power Transistors," *IEEE Trans. Microwave Theory Tech.*, Vol. MTT-27, pp. 415–422, May 1979.

24. W. E. Poole, "S-Band Transistors for Radar Applications," *Microwave J.*, Vol. 26, pp. 85–90, March 1983.

25. R. G. Meyer and R. A. Blauschild, "A 4-Terminal Wide-Band Monolithic Amplifier," *IEEE J. Solid-State Circuits*, Vol. SC-16, pp. 634–638, December 1981.

26. J. F. Kukielka and C. P. Snapp, "Wideband Monolithic Cascadable Feedback Amplifiers Using Silicon Bipolar Technology," *IEEE Microwave Millim.-Wave Monolithic Circuits Symp.* June 1982.

27. C. P. Snapp et al., "Practical Silicon MMICs Challenge Hybrids," *Microwaves RF*, Vol. 21, pp. 93–99, November 1982.

28. T. Nakata et al., "0.5–2.6 GHz Si Monolithic Wideband Amplifier IC," *IEEE Microwave Millim.-Wave Monolithic Circuits Symp. Dig.*, pp. 58–62, 1985.

29. J. Wholey and S. Taylor, "Silicon MMIC Amps Push 1 Watt at 1 GHz," *Microwaves* Vol. 26, pp. 197–206, March 1987.

30. M. Ohara et al., "High Gain Equalizing Amplifier Integrated Circuits for a Gigabit Optical Repeater," *IEEE J. Solid-State Circuits*, Vol. SC-20, pp. 703–707, June 1985.

31. T. Kinoshita et al., "Wideband Variable Peaking AGC Amplifier for High-Speed Lightwave Digital Transmission," *Electron. Lett.*, pp. 23–24, January 2, 1987.

32. R. Reimann and H.-M. Rein, "A 40 Gbit/s Amplifier for Optical Systems," *IEEE Int. Solid-State Circuits Conf. Dig.*, 1987.

33. M. Umehira et al., "High-Speed and Precise Monolithic Multiplier with Radiation Hardness Using Silicon Bipolar SST," *Electron. Lett.*, Vol. 22, pp. 744–746, July 3, 1986.

34. I. Kipnis and A. P. S. Khanna, "10 GHz Frequency Converter Silicon Bipolar MMIC," *Electron. Lett.*, Vol. 22, pp. 1270–1271, November 6, 1986.

35. T. Sakai et al., "Prospects of SST Technology for High-Speed LSI," *IEEE Int. Electron Devices Meet. Tech. Dig.*, pp. 18–21, 1985.

36. M. C. Wilson et al., "10.7 GHz Frequency Divider Using Double Layer Silicon Bipolar Process Technology," *Electron. Lett.*, Vol. 24, pp. 920–922, July 21, 1988.

37. I. Kipnis et al., "A Wideband, Low-Power, High-Sensitivity and Small-Size 5.5 GHz Static Frequency Divider IC," *IEEE Bipolar Circuits Technol. Meet*, September 1988.

38. A. Akazawa et al., "Low Power 1 GHz Frequency Synthesizer LSI's," *IEEE J. Solid-State Circuits*, Vol. SC-18, pp. 115–121, February 1983.

39. R. H. Derksen and H.-M. Rein, "7.3 GHz Dynamic Frequency Dividers Monolithically Integrated in a Standard Bipolar Technology," *IEEE Trans. Microwave Theory Tech.*, Vol. MTT-36, pp. 537–541, March 1988.

40. I. Kipnis, "20 GHz Frequency-Divider Silicon Bipolar MMIC," *Electron. Lett.*, Vol. 23, pp. 1085–1086, September 24, 1987.

41. M. Suzuki et al., "A Bipolar Monolithic Multigigabit/s Decision Circuit," *IEEE J. Solid-State Circuits*, Vol. SC-19, pp. 462–467, August 1984.

42. H.-M. Rein, "Multi-Gibibit-per-Second Silicon Bipolar IC's for Future Optical-Fiber Transmission System," *IEEE J. Solid-State Circuits*, Vol. SC-23, pp. 664–675, June 1988.

43. R. Sparkes, "QuickChip 4: A Methodology for the Prototyping of Mixed Analog-Digital Designs," *IEEE Custom Integrated Circuits Conf. Proc.*, 1986, pp. 244–248.

44. M. Suzuki et al., "An 86K Component Bipolar VLSI Masterslice with a 290-ps Loaded Gate Delay," *IEEE J. Solid-State Circuits*, Vol. SC-22, pp. 41–46, February 1987.

45. K. C. Wang et al., "A 20 GHz Frequency Divider Implemented with Heterojunction Bipolar Transistors," *IEEE Electron Device Lett.*, Vol. EDL-8, pp. 383–385, September 1987.

46. Y. Yamauchi et al., "22 GHz 1/4 Frequency Divider Using AlGaAs/GaAs HBTs," *Electron. Lett.*, Vol. 23, pp. 881–882, August 1987.

9

HIGH-ELECTRON-MOBILITY TRANSISTORS: PRINCIPLES AND APPLICATIONS

Jacques Zimmermann and Georges Salmer

Centre Hyperfréquences et Semiconducteurs
Université des Sciences et Techniques de Lille-Flandres-Artois
Villeneuve d'Ascq, France

9.1 INTRODUCTION

The high-electron-mobility transistor (HEMT) was first proposed in the early 1980s as an alternative device which in several respects was considered superior to the conventional MESFET [1, 2]. The development of heterojunction FETs was a lengthy one, made possible thanks to definitive and considerable progress in technological processes of crystal growth and pattern lithography. In the normal MESFET, the electrons move in the active layer, which is a doped N-GaAs layer (in the range of several 10^{17} cm^{-3}). As a result, the electron mobility is degraded due to strong ionized impurity scatterings. This fact becomes exceedingly important in submicrometer structures (gate length $L_g < 1$ μm) since as L_g decreases, the doped layer must be made thinner, and the doping density made higher, in such a way that good control by the gate be preserved, the threshold voltage remaining nearly the same. The possibility of separating the moving carriers from their parent impurities using one or more heterojunctions had been proposed as a solution for achieving very high mobility in III–V materials (see Ref. 3 for a review). This is obtained by doping only the large-bandgap material. This technique is now commonly called doping modulation, from which the acronym MODFET derives for "modulation doped field-effect transistor." This device is strictly equivalent to the HEMT; it is also sometimes called TEGFET: two-dimensional electron gas FET. The typical structure of the HEMT is shown in Fig. 9.1. The main difference with the normal MESFET is the presence of a heterojunction formed at the interface of an AlGaAs doped alloy $[Al_{x_{Al}}Ga_{1-x_{Al}}As = x_{Al}(AlAs) + (1 - x_{Al})GaAs]$ grown on an undoped GaAs layer. AlAs and GaAs have nearly the same lattice constant; as a result, any combination of them will provide good lattice match at the interface ($\Delta a/a < 5 \times 10^{-4}$). Since 1980, the HEMT has made considerable progress, and at present submicrometer HEMTs for microwave applications with high cutoff frequencies ($f_c > 50$ GHz) are available commercially from several companies at a very reasonable cost. The superiority of the HEMT over a MESFET with the same geometry is twofold. First, due to the confinement of the electron channel layer in the undoped GaAs side of the heterojunction, the two

437

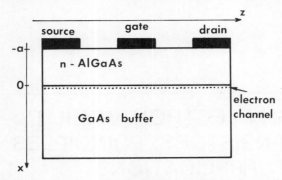

Figure 9.1 Elementary structure of the AlGaAs/GaAs high-electron-mobility transistor.

parameters forming the current (i.e., the electron density and mobility) are optimized such that at a given drain voltage, the transconductance in the HEMT is higher than in the MESFET, thus providing a better cutoff frequency. Second, a number of experimental data clearly show that HEMTs have a better noise figure than MESFETs. This may be due partially to a better source resistance, which is lower in the HEMT because of the better electron mobility in the channel near the source and drain contacts.

The HEMT is also potentially better with regard to logic applications. Actually, since high saturation drain currents can easily be obtained, delay time per gate is minimized. It can be shown that this time is shorter when transconductance and average electron drift velocity under the gate are higher. We will see later that this is realized with the HEMT. Switching delay times per gate of less than 10 ps are obtained, dropping to 5 ps at low temperature. These figures are very competitive with those of other technologies: MOSFET or ECL bipolar. Indeed, one finds from ring oscillator measurements that a 1-μm-gate HEMT is as fast as a $\frac{1}{2}$-μm-gate standard MESFET [4, 5]. Thus integration of HEMTs could be easier and the process could yield better than with MESFETs. Apparent average electron velocities approaching 2×10^7 cm/s are found in HEMTs (see Section 9.3). Concerning dissipation in relation to switching delay times, values as low as 10 to 20 fJ are estimated for the dissipated energy, making HEMT logic an attractive low-consuming technology for integrated circuits [6].

In the rest of this chapter we review first the most elementary theoretical approach of HEMT functioning, the small-signal equivalent circuit useful in microwave applications, and some technological process principles and improvements for the near future. Then a more physical description of the device is presented. Following that, practical applications are reviewed concerning low-temperature behavior, low-noise amplification, and high-power use. We finish the chapter with a summary of the new structures deriving from the basic HEMT structure, using multiple heterolayers and/or new III–V semiconductor compounds, including pseudomorphic structures.

9.2 PRINCIPLE AND MODEL OF THE HEMT

The most elementary general structure of a HEMT is sketched in Fig. 9.1. A more elaborate structure of a real device is shown in Fig. 9.3. Looking at Fig. 9.1, one sees that a HEMT is formed by a thin layer of doped AlGaAs epitaxially grown on a GaAs undoped buffer layer deposited over a GaAs semi-insulating substrate. At not too low temperatures, one expects that most of donor states will deliver a free electron which

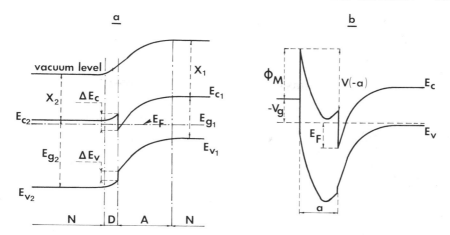

Figure 9.2 Band scheme of the heterojunction: (*a*) without contact, where *N* is the neutral zone, *A* the accumulation zone, and *D* the depleted zone; (*b*) with a biased Schottky contact at a distance *a* from the interface.

then diffuse freely in the doped material. We know from Anderson's theory that the empty electron states in the undoped GaAs layer are much lower in energy than the electron states in the AlGaAs layer. We will see in the next paragraph that the difference in energy is in practice in the range 0.2 to 0.3 eV. In this process a large number of electrons accumulate in the GaAs layer and cannot transfer back to the AlGaAs layer, due to the barrier at the heterojunction. At the same time, since the local neutrality is destroyed, a space-charge reaction tends to cause the electrons to accumulate against the interface between the two materials. The situation obtained is depicted in Fig. 9.2*a*, where the band scheme is shown as a function of the distance *x* of Fig. 9.1. In Fig. 9.2*a* we make the distinguish among the neutral zones, the depleted zone, and the accumulation zone.

Now, we suppose that a metallic (Schottky) contact has been made over the AlGaAs layer (the gate) and a bias voltage is applied to this contact, the other contacts being at zero. This situation is depicted in Fig. 9.2*b*. If the AlGaAs layer is not too wide (in practice, about 500 Å when $N_D \sim 10^{18}$ cm^{-3}), a moderate negative bias will deplete the totality of the AlGaAs electrons into the GaAs layer, such that the areal electron density n_s in GaAs may be in the range of about 10^{12} cm^{-2}. One then obtains a large amount of electrons moving in a high-quality material, and the areal density n_s can easily be controlled by the gate bias voltage. Next, if one of the ohmic contacts is positively biased (drain) while the other (source) is grounded (technologically, one tries to have direct access to the channel electron layer by minimizing access contact resistances), the electrons will move, subject to the electric field between source and drain, under control of the gate bias. A field-effect transistor is obtained. As in the conventional FET, normally-on or normally-off structures can be obtained depending on whether or not the AlGaAs layer is thick. The barrier height of the metal contact is about 1 V (increasing slightly with the Al mole fraction x_{Al}), and for the total depletion voltage given by $V_p = qN_D a^2/2\epsilon$ at $N_D = 10^{18}$ cm^{-3} we find $a \simeq 400$ Å using $\epsilon = 12.9\epsilon_0$.

Figure 9.3 shows a real structure of the HEMT. Compared to Fig. 9.1, two important details appear which indeed improve substantially the functioning of the device in current applications. First, to improve the quality of the source and drain access zones and make the access resistances to the electron channel as low as possible, a doped GaAs

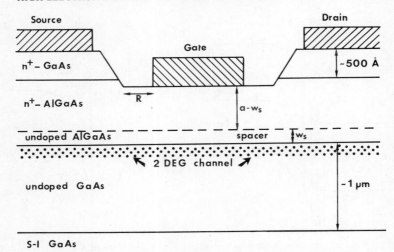

Figure 9.3 Typical structure of a real HEMT. Note that the gate metal is deposited in the recessed doped GaAs layer at the top; also, note the position of the undoped spacer layer next the heterojunction.

layer ($N_D \sim 3 \times 10^{18}$ cm^{-3}) is grown over the AlGaAs layer. A second improvement comes from the presence of the undoped spacer layer next to the interface. As the electrons move in the channel, they stay in close contact with the heterojunction interface, and thus in the vicinity (although outside) of the positive ionic layer. If one wishes to obtain a very good electron mobility (and/or drift velocity), it is necessary to separate the electron layer from the Si ions. It is well known that the coulombic field due to ions on the electrons strongly degrades the mobility by deviating the trajectories of the electrons continuously as they move along the channel. It is obvious that the deviations are smaller when the electrons are farther away from the ions. Unfortunately, at the same time, if the spacer is too wide, it will prevent electrons from transferring correctly to the GaAs layer. A number of microscopic simulations and experiments have shown that a good trade-off between these two conflicting effects is a spacer in the range 20 to 40 Å in usual structures. Now we turn to a more quantitative, although still elementary analysis of the HEMT.

9.2.1 Anderson Theory

The Anderson theory [7], which was originally developed for Ge/GaAs heterojunctions, applies readily to GaAs/AlGaAs heterojunctions. This theory postulates that the vacuum level is continuous between the two materials and that no interface defects exist. Then the Fermi levels on each side line up when no current is flowing through the interface. Thus, depending on the type of doping of each material (n or p), several configurations for the band scheme may occur. Here, with the HEMT, we are more interested in a p^--GaAs/n^+-AlGaAs system. Indeed, experience shows that undoped GaAs, as grown by molecular beam epitaxy, for instance, most often is of p^- type, due to the presence of various contaminents which are not all well identified. The electron affinity χ is defined as the energy to be spent to extract an electron from the conduction band edge to the vacuum level; thus

$$\Delta E_c = \chi_1 - \chi_2, \qquad \Delta E_v = \chi_2 - \chi_1 + E_{g_2} - E_{g_1} \qquad (9.1)$$

where 1 stands for GaAs and 2 for AlGaAs. Experience shows that $E_{g_1} < E_{g_2}$ and $\chi_1 > \chi_2$; taking into account that the positions of the Fermi levels in the two materials may be very different because of the types, according to the Anderson theory the band scheme of Fig. 9.2a is obtained, in which the conduction band states are lower in GaAs than in AlGaAs. When the types are reversed, one can obtain an accumulation of holes in GaAs rather than of electrons. Since the bandgaps are easily measured with photoluminescence experiments in AlGaAs as a function of the Al content x_{Al}, one needs a formula for ΔE_c and ΔE_v in terms of the bandgaps rather than the affinities. This is not an easy problem, but here again, a number of experiments show that the most probable values are

$$\Delta E_c(x_{Al}) = 0.62(E_{g_2} - E_{g_1}), \qquad \Delta E_v = 0.38(E_{g_2} - E_{g_1}) \tag{9.2}$$

where $E_{g_2}(x_{Al})$ and E_{g_1} are now well known [8]. x_{Al}, in practice, lies in the range 0.2 to 0.4. As a result, the band offset ΔE_c lies in the range 0.2 to 0.3 eV.

9.2.2 Charge Control Law by the Gate

This is one of the most important point in the analysis of the HEMT mode of functioning [9]. Charge control law by the gate determines for the most part the performance to be expected from a device. We refer to Fig. 9.2b as the basis for the calculations presented here. We assume that the bias applied to the gate is sufficient to totally deplete the doped AlGaAs layer. Let N_D and a be the doping density and the thickness of this layer, respectively. The origin $x = 0$ is taken at the interface. Assuming that the impurities are fully ionized, one can readily integrate Poisson's equation between 0 and $-a$ and obtain

$$V(-a) = -E_s a + \frac{q}{\epsilon_2} \int_0^{-a} dx \int_0^x N_D(x') \, dx' = -E_s a + \frac{q N_D a^2}{2\epsilon_2} = -E_s a + V_p \tag{9.3}$$

where E_s is the electric field at the interface on side 2. Assuming that the doping concentration in GaAs is negligible, we also have

$$E_s = \frac{q n_s}{\epsilon_2} = \frac{Q_s}{\epsilon_2} \tag{9.4}$$

Next, from Fig. 9.2b we note that $V(-a)$ is also given by

$$V(-a) = \Phi_M - V_g + E_F - \Delta E_c \tag{9.5}$$

where Φ_M is the metal barrier height. Combining Eqs. (9.3) and (9.5) gives

$$\epsilon_2 E_s = \frac{\epsilon_2}{a}(V_p - \Phi_M - E_F + \Delta E_c + V_g) \tag{9.6}$$

Using Eq. (9.4) for E_s, we then have

$$q n_s = \frac{\epsilon_2}{a}(V_g + \Delta E_c + V_p - \Phi_M - E_F) \tag{9.7}$$

As we will see later, the Fermi level E_F is a complicated function of n_s. Worse, this dependence can only be evaluated in an electron system at thermal equilibrium. Then

Eq. (9.7) is transcendental and cannot be solved analytically. However, in most cases E_F is small compared with the other quantities, and we can write

$$qn_s \simeq \frac{\epsilon_2}{a}(V_g - V_T), \quad \text{where} \quad V_T = \Phi_M - \Delta E_c - \frac{qN_D a^2}{2\epsilon_2} \tag{9.8}$$

is the threshold voltage, which depends only on technological parameters. When $V_g = V_T$ one obtains the flat conduction band condition in which the electron accumulation no longer exists. When there is a spacer layer of width w_s in the AlGaAs layer at the interface, a must be replaced by $a - w_s$ in Eq. (9.8). If an interface state charge Q_{ss} is also present, the threshold voltage takes the form

$$V'_T = V_T - \frac{a}{\epsilon_2}Q_{ss} \tag{9.9}$$

Equation (9.8) predicts a linear dependence to hold between the accumulated charge n_s and the gate voltage V_g. Although this law is approximate, it is a good basis for the calculation of the HEMT characteristics together with the gradual channel approximation. The result obtained is generally in good agreement with experiment. Equation (9.8) also predicts that the source–gate capacitance (per unit area) $C_{gs} = d(qn_s)/dV_g$ is a constant. In practice, however, this is never the case because the charge control by the gate is more complicated than described here. Together with the free-electron contribution in the source–drain channel, there is also a contribution from the free electrons remaining in the AlGaAs layer (total depletion is not always realized), and also a contribution from the ionic charge because the Si ions create several donor levels (some of

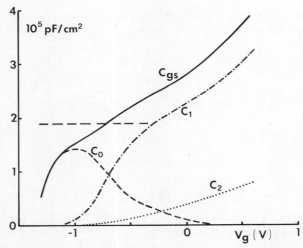

Figure 9.4 Gate–source capacitance as a function of gate bias voltage. The horizontal dashed line is the result of Eq. (9.8). The solid line is the total capacitance $C_{gs} = C_0 + C_1 + C_2$, where C_0 is the contribution from the electrons in the channel, C_1 the contribution from the electrons in the AlGaAs layer, and C_2 the contribution of the electrons trapped on the deep donor level in AlGaAs. $x_{Al} = 0.3$, $a = 600$ Å, $N_D = 10^{18}$ cm^{-3}, $T = 300$ K.

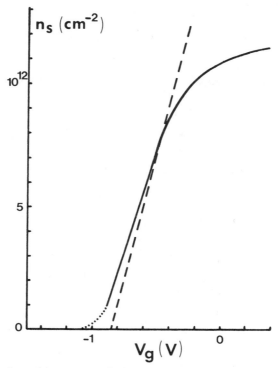

Figure 9.5 Comparison of an accurate calculation of $n_s(V_g)$ (solid line) with the prediction of Eq. (9.8) (dashed line). $x_{Al} = 0.28$, $a = 370$ Å, $w_s = 40$ Å, $N_D = 2 \times 10^{18}$ cm^{-3}, $T = 300$ K.

them are deep DX centers) which interact with the conduction electron states. In Fig. 9.4, for the same structure, we compare the results of an accurate calculation of C_{gs} with the value predicted by Eq. (9.8). Of course, the difference is significant. At present in submicrometer HEMTs, gate capacitances as low as 0.2 to 0.5 pF/mm are commonly realized. In Fig. 9.5 we compare the results of a careful analysis of $n_s(V_g)$ with the result of Eq. (9.8). One notes that there exists a nearly quadratic part when $V_g \simeq V_T$, which indeed must be considered in low-current, low-noise applications, and at $V_g \gg V_T$ a sublinear part, a trend to saturation of $n_s(V_g)$, which indeed is a principal limitation for power applications. The linear part of $n_s(V_g)$ is observed between these two extremes. It is important to note, however, that Eq. (9.8) is extremely useful and is used extensively in the elaboration of CAD models of the HEMT.

9.2.3 Effect of a Source–Drain Bias Voltage

When a drain voltage V_d is applied, the potential (and the electric field) applied to the electrons changes all along the channel. The source–drain abcissa is referred to here as z. If the potential at z is $V(z)$, the charge in the channel can be written as

$$Q_s(z) = q n_s(z) = \frac{\epsilon_2}{a} [V_g - V(z) - V_T] \qquad \text{C/m}^2 \qquad (9.10)$$

and the drain current is given, using the gradual channel approximation, by

$$I_d = q n_s(z) v_d(z) Z = \text{constant} \tag{9.11}$$

where Z is the width of the source and drain contacts and $v_d(z)$ is the electron drift velocity due to the electric field at z. Equation (9.11) assumes that no gate current is flowing. We assume a $v_d(E)$ dependence of the form

$$v_d(E) = \mu(E)E = \frac{\mu_0 E}{1 + E/E_c} \tag{9.12}$$

where E_c is a critical field and μ_0 is the ohmic (low-field) mobility. Using Eq. (9.12), to be justified later, we have

$$I_d \left(1 + \frac{dV/dz}{E_c} \right) = \frac{Z \epsilon_2 \mu_0}{a} \left[V_g - V(z) - V_T \right] \frac{dV}{dz} \tag{9.13}$$

which can readily be integrated between $V(0) = R_s I_d$ and $V(L_g) = V_d - R_d I_d$, where L_g is the gate length and R_s and R_d are the source and drain access resistances, respectively. Assuming that $R_s = R_d$, we can express the drain current as a function of V_d and V_g as

$$I_d = \frac{-B - (B^2 - 4AC)^{1/2}}{2A} \quad \text{with } A = \frac{2R_s}{E_c}$$

$$B = -\left[L_g + \frac{V_d}{E_c} + \frac{2Z\epsilon_2\mu_0 R_s (V_g - V_T - V_d/2)}{a} \right] \tag{9.14}$$

$$C = \frac{Z\epsilon_2\mu_0(V_g - V_T - V_d/2)V_d}{a}$$

From Eq. (9.14) expressions for the transconductance $g_m(V_d, V_g) = dI_d/dV_g$ and the output drain conductance $G_d(V_d, V_g) = dI_d/dV_d$ are easily obtained. These, together with C_{gs}, are essential to the development of an equivalent electric circuit of the HEMT.

Next, the integration of Eq. (9.13) over z from 0 to z gives an expression for $V(z)$ that can be put in Eq. (9.10) to obtain the space dependence of $n_s(z)$ as a function of bias voltages. Interested readers can derive this expression for themselves and will observe that $n_s(z)$ decreases monotonously from the source to the drain and as a consequence the electron drift velocity and the electric field strength increase from the source to the drain accordingly. Equation (9.14) describes the so-called ohmic part of the static characteristics. Drain current saturation can be treated using other assumptions. If the current saturation is the consequence of drift velocity saturation, due to hot electron effects, as soon as $E(z) \geqslant E_c$ in the channel, one assumes that $v_d(E \geqslant E_c) = v_s = \mu_0 E_c/2$. The point $z = L$ where $E(z) = E_c$ can be called the saturation point where velocity saturation begins. Accordingly, the saturation drain current is obtained by saying that at a given V_g, $E(L_g) = E_c$, which also gives the value of the corresponding V_d. Beyond this point, the n_s remains equal to $n_s(L)$ and the saturation zone can be treated using Poisson's equation. If access resistances are considered, one observes a very slow rise of the saturation drain current as V_d increases. An example of application and a comparison with experimental data are shown in Fig. 9.6. The two fitting parameters needed are μ_0 and E_c. Although the values used for these parameters are physically justified, the resulting saturation drift velocity v_s is not ($v_s \sim 1.2 \times 10^7$ cm/s). In micrometer and submicrometer structures, the

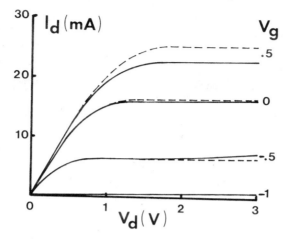

Figure 9.6 Static characteristics of a large-gate-length HEMT: $L_a = 5$ μm. Solid lines are the experimental data, dashed lines are the results of Eq. (9.14) with $\mu_0 = 6200$ cm^2/V/s, $E_c = 4000$ V/cm, $a = 450$ Å, $w_s = 40$ Å, $N_D = 8 \times 10^{17}$ cm^{-3}, $R_s = R_d = 10$ Ω, $Z = 150$ μm, $T = 300$ K.

situation can be even worse since v_s as much as 2 to 2.5 $\times 10^7$ cm/s is necessary to obtain the fit, values in total contradiction to the values obtained for microscopic physical theories or from experiment ($v_s \sim 0.8 \times 10^7$ cm/s). At any rate, a fitting of the experimental $I_d(V_d, V_g)$ with the results of the present model is by no means able to deliver a correct evaluation of v_s. This is particularly true in submicrometer structures. Here v_s is simply taken as an empirical parameter and must not be considered as a physical parameter.

9.3 SMALL-SIGNAL EQUIVALENT CIRCUIT

The small-signal equivalent circuit (SSEC) of a HEMT closely resembles that of the conventional MESFET. It is described at the top of Fig. 9.7. Only the intrinsic device circuit can be studied using physical models of the HEMT [10]. This is represented in the dashed rectangle of Fig. 9.7. The parasitic access elements are represented outside the dashed rectangle; they do not depend, as a first assumption, on the bias voltages applied to the device. The experimental analysis of the SSEC can be accomplished by adjusting the various lumped-element values in order to fit closely the otherwise measured broad-band S-matrix parameters of the transistor in the fixture or through wafer-probing systems. These are illustrated at the bottom of Fig. 9.7, where we display the transconductance, the gate–source capacitance, the output conductance, and the cutoff frequency, all as functions of the gate-to-source voltage. One notes, for instance, that the transconductance, which increases initially with the gate voltage as a result of the increase of n_s, at positive gate voltage begins to drop as a result of the appearance of electrons in the AlGaAs layer when the gate voltage is so high that the conduction band there approaches the Fermi level. As these electrons move in a highly doped layer, where the mobility is very low, the drain current cannot continue to increase; thus the transconductance drops rapidly. This phenomenon is sometimes called the "parallel MESFET."

A new method for the determination of the elements of the SSEC has recently been developed [11] which no longer necessitates broad-band S-parameter measurements. It

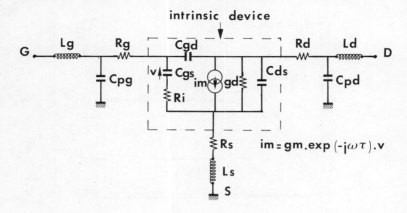

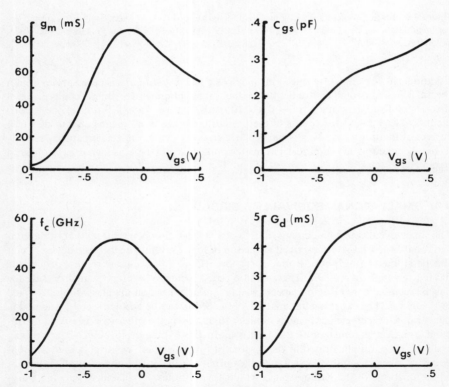

Figure 9.7 Small-signal equivalent circuit of the HEMT (top of figure), including the extrinsic access elements represented outside the dashed rectangle. Below we represent typical experimentally measured values of the circuit: transconductance, gate capacitance, and output conductance of a 0.5-μm-gate HEMT ($Z = 200$ μm) at room temperature as functions of the gate voltage at $V_d = 2$ V. We also represent the variation of the cutoff frequency calculated from Eq. (9.15).

has already been applied with success to the SSEC of conventional FETs; it is also usable for the determination of the SSEC of HEMTs. To abstract the Z-matrix parameters of the intrinsic FET, the parasitic elements of the extrinsic FET must be eliminated. Using S-matrix measurements at $V_d = 0$ and forward gate bias V_g on one side, and $V_d = 0$ and subthreshold V_g on the other side, all eight parasitic elements can be calculated from the S-matrix data. Then, using successive transformations from Z to Y matrices and conversely, the Z-matrix of the intrinsic transistor is arrived at for any values of V_d and V_g. An interesting point is that all the parasitic elements can be obtained with measurements at relatively low frequencies only ($F \leqslant 5$ GHz).

Of principal importance is the determination of the intrinsic cutoff frequency of the transistor, which is defined as

$$f_c = \frac{g_m}{2\pi C_{gs}} \tag{9.15}$$

and can be related to the electron transit time under the gate $\tau = L_g/\langle v_d \rangle$, where $\langle \cdots \rangle$ indicates the average value along the channel. If we write $I_d = qZ\langle n_s \rangle\langle v_d \rangle$, we have $g_m \simeq (\epsilon_2/a)Z\langle v_d \rangle$ and $C_{gs} = ZL_g d(qn_s)/dV_g = ZL_g(\epsilon_2/a)$, so that $C_{gs}/g_m \simeq L_g/\langle v_d \rangle = \tau$, or $f_c = 1/2\pi\tau$. The measurement of f_c thus gives the average electron drift velocity during transit under the gate. In submicrometer HEMTs, $\langle v_d \rangle$ values in excess of 1.5×10^7 cm/s are commonly obtained at ambient temperature, corresponding to cutoff frequencies higher than 50 GHz. This is the case for the transistor illustrated in Fig. 9.7, where the maximum f_c is 54 GHz and the average velocity is 1.7×10^7 cm/s, since $L_g = 0.5$ μm. These average velocities can be substantially higher than those obtained in a MESFET having the same geometry. With regard to G_d, the output drain conductance, it is essentially linked to injection conditions at the contacts and to the short-channel effect in very short structures. Its numerical value depends strongly on the technology and on the geometry of the device. In general, it goes through a maximum when the gate forward regime is approached.

9.4 TECHNOLOGY OF THE HEMT

The fabrication of good-quality heterojunctions was rendered possible recently with the development of very sophisticated techniques for growing epitaxial monocrystal layers. At present, the two techniques used most often are molecular beam epitaxy (MBE) and metalorganic vapor-phase epitaxy (MOVPE) [12]. Of course, the quality of the epitaxial layers is strongly linked to the quality of the semi-insulating substrates usually obtained from Czochralski or Bridgeman grown ingots. Semi-insulating GaAs substrates still contain rather large densities of localized and extended defects, making large-scale integration more difficult than in silicon technology. However, at present, the improvement in quality of substrates allows processing of 3-in. wafers with good uniformity of threshold voltages and yield, at least for discrete microwave transistors. The problem of the influence of defects at the surface of semi-insulating substrates is resolved in part by careful cleaning and etching, and by growing a buffer layer about 1 μm thick before fabrication of the active layers, which are much thinner. This, of course, requires more time for the growing process and the cost of devices is therefore higher. AlGaAs doping, usually in the range of 10^{18} cm^{-3}, is generally done with silicon atoms in very well controlled conditions, such that the atoms are in the position of donor states. Moreover, silicon does not induce serious pollution in the growth chambers and can be handled

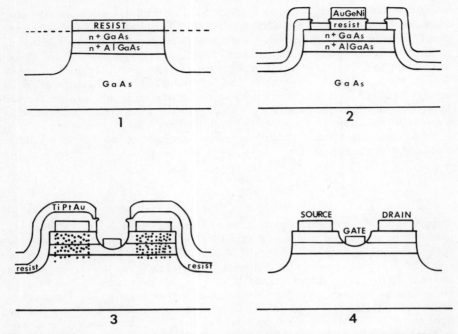

Figure 9.8 Elementary HEMT technological process. (1) Insulation by mesa chemical etching (or ion implantation). (2) Ohmic contacts patterns and metallization; lift-off and annealing. (3) Recess and gate pattern; recess etching and gate metallization. (4) Gate lift-off and annealing. Surface passivation and contact pads plating are not represented.

easily in the technological process of the material. With regard to device processing, gates as short as 0.3 μm are currently fabricated using electron-beam lithography. More and more in use now are plasma-enhanced etching and deposition techniques, which might lead to much better results than chemical processes because of better accuracy in the control of the process kinetics and more regular delineation of the various patterns. This kind of technological process might become an important interesting alternative for etching in the near future. A current process for the fabrication of a HEMT is described in Fig. 9.8. On doped GaAs layers, very high quality ohmic contacts can be made with the present technology. They are much more difficult to make directly on a AlGaAs layer, especially when the Al content is high. Thus, before gate deposition, the GaAs layer must be recessed such that the gate metal is against (or very near) the AlGaAs layer. The problem of access resistances is extremely important in these devices, because of their effect on microwave and noise performance, as we will see later.

The trend in HEMT technology at present is toward the development of transistors with cutoff frequencies higher than 100 GHz, to obtain the largest possible bandwidths for telecommunications applications, using 94- and 140-GHz atmospheric windows, for instance. It is clear that in this domain very short (i.e., submicrometer) gate lengths are needed. But to keep good control of the current channel over so small distances, the thickness a of the active layers must be diminished in proportion, and the width of the transistor must be such that impedance matching to propagation networks is still possible. At the same time, as a result of a small a, doping concentrations must be increased to or above 10^{18} cm^{-3}. With high dopings in the AlGaAs layer, severe limitations arise, because highly doped layers where $x_{Al} > 0.2$ contain deep donor levels, limiting the areal density n_s of the electron channel. These deep levels are called DX centers [13, 14].

Another problem resulting from shrinking the gate length is known as the short-channel effect. In the calculations of Section 9.2 we supposed that the driving field applied to the electrons was essentially directed along the source–drain axis. If L_g is very small, this is no longer true. Even near the ohmic contacts the electric field may have a significant component in the direction transverse to the heterojunction plane, and complete simulation of the transistor in two dimensions becomes necessary. We study this question in more detail in the next section. Moreover, at pinch-off in particular, the channel electrons may diffuse into the substrate near the drain-end side of the gate, where the electrons accelerated by the field can be very energetic, thus increasing the output drain conductance. To overcome these problems, several solutions are possible. For instance, by growing a p-GaAs layer beneath the undoped layer where the electrons move, one can prevent them from diffusing into the substrate since they are repulsed by the diffusion barrier of the p layer. This solution has already been applied with success to the case of MESFETs [15]. An alternative solution can also be tried by growing an AlGaAs barrier beneath the undoped layer or use a so-called inverted structure [16, 17]. Unfortunately, this solution, which is much simpler than the previous one, is more difficult to apply because the quality of the interface obtained in this way is not very good compared with that at the top heterojunction.

The existence of a gate access resistance R_g and source access resistance R_s also degrade the transconductance and the available gain. For instance, when $R_s \neq 0$, g_m takes on the value

$$g'_m = \frac{g_m}{2\pi C_{gs}(1 + g_m R_s)} \tag{9.16}$$

At very high frequencies, however, the structure of the source access zone is more complicated than a simple ohmic resistance, since displacement currents through the heterojunction layers (via a R-C transmission-line scheme) must be taken into account [10]. As a result, one can show that the real part R_s of the access impedance Z_s may drop by a factor of 3 between the dc regime and the millimeter wave range. At any rate, it also depends on the sheet resistance of the doped contact layer, on the ohmic contact resistivity, and on the distance between the contact metal and the edge of the gate.

We have shown previously how a n^+ cap layer may improve the quality of the ohmic contacts. However, it is clear from Fig. 9.3 that the electrons coming from the source metal do not have direct access to the channel under the gate. To minimize the distance between the source contact and the gate edge, self-aligned deeply ion-implanted contacts can be made (managing for gaps, such that the gate cannot short the source–drain channel), a process in which the metal of the gate serves as a protecting mask for the ion implantation. Unfortunately, ion implantation is always followed by high-temperature annealing for crystal reconstruction and ion activation, and refractory metals must be employed for constructing the gate. This may lead in turn to a degradation of the gate resistance which is rather high with refractory metals (W or Mo). This solution is, however, good for logic applications where the gate width is generally small ($Z \simeq 10\ \mu m$), whereas in microwave applications $Z \simeq 100\ \mu m$. The cross-sectional shape of submicrometer metal gates is triangular rather than rectangular, making the gate access resistance higher that expected. To improve gate access resistance, mushroom-shaped or multidigit gates must be made.

9.5 PHYSICAL ANALYSIS OF THE HEMT

Despite its apparent simplicity, the HEMT is the seat of many interesting physical phenomena. A good understanding of the essential physical phenomena occurring in this device is necessary if we wish to develop accurate models in order to undertake the optimization of a transistor. In this section we describe several of them. We finish with a brief description of some sophisticated methods used for accurate modeling of HEMTs. A more detailed account and an abundant bibliography giving the present state of the art on all these questions are given in Refs. 6 and 10.

9.5.1 Low-Field Mobility and Charge Control Law by the Gate

Figure 9.9 shows the temperature dependence of the experimental low-field electron Hall mobility in a high-quality AlGaAs/GaAs heterojunction [18] compared with that measured in undoped bulk GaAs [19]. If the bulk GaAs were doped at 10^{17} cm^{-3}, the mobility would drop to the range 2500 to 3500 cm^2/V/s between 77 and 300 K. The

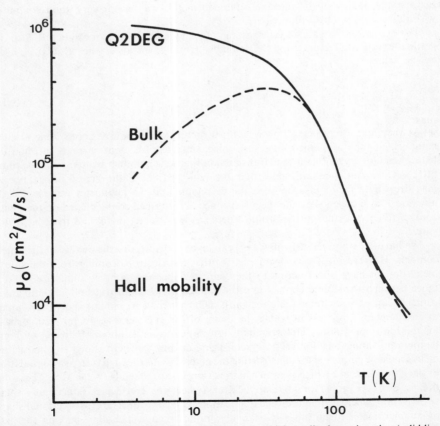

Figure 9.9 Experimental low-field mobility measured in a high-quality heterojunction (solid line) and in undoped bulk GaAs (dashed line), as a function of temperature.

difference between the two curves of Fig. 9.9 is striking at low temperature. Here the spacer of the heterolayer (about 300 Å) completely prevents ionized impurity scattering, while even in undoped bulk GaAs, the residual background impurities are sufficiently effective for the mobility to drop strongly below 40 K. At low temperature, mobility is dominated by lattice phonon interactions whose probability evolves as $T^\alpha (\alpha \geqslant 1)$, α depending on the type of interaction. On the contrary, ionized impurity scattering probability evolves as $T^{-3/2}$. As a result, very high electron mobility can be found, in excess of 10^6 cm^2/V/s in heterojunctions, as shown in Fig. 9.9. Values as high as 3×10^6 are commonly reported now at 4 K. This high mobility makes heterojunction devices very attractive for low-temperature applications (see Section 9.6).

Next, as mentioned earlier, the electron space charge in the heterojunction is usually very thin. As a result, the conduction band is also very narrow and is called a "well." Accurate calculations of the band structure based on quantum mechanical treatments show that the width of the well near the Fermi level is on the order of 100 Å, which is also the thickness of the electron layer. This number is comparable (even at high temperature) with the thermal wavelength of the electrons perpendicular to the heterojunction. The motion is thus impeded along x from the point of view of quantum mechanics. As a consequence, the electrons are free to move along the plane of the heterojunction (y and z) only. The electron system, which is then called a quasi two-dimensional electron gas (or simply Q2DEG or 2DEG), evolves on a ladder of energy subbands of which only the lowest few are occupied. The conduction band of a 2DEG then takes the form

$$E_c = E_{n_x} + \frac{\hbar^2}{2m^*(k_y^2 + k_z^2)} \tag{9.17}$$

where the series of E_{n_x} (n_x is a subband index) can be calculated using Schrödinger's equation (and Poisson's equation for a self-consistent treatment), and k_y and k_z are the quasi continuous-wave vector components in the other directions. A detailed account of this question is given in Refs. 20 and 21. At the same time the envelope wave functions in each subband for a given n_s in the well, the Fermi level E_F, and the conduction band are calculated (via Poisson's equation), from which the potential at the gate can be deduced. Thus the control law $n_s(V_g)$ can be calculated exactly. An example was illustrated in Fig. 9.5. In this process the ion charge density N_D^+ in the AlGaAs layers can be introduced in the calculation since E_F and the donor levels are known. Of course, the calculation proceeds iteratively. In this way the effect of the doping density N_D and x_{Al} in the AlGaAs layers can be taken into account. In particular, the areal density n_{s_0}, the maximum n_s that can be obtained in the heterojunction and its dependence with N_D and x_{Al}, is one of the essential parameters entering into optimization of the device.

9.5.2 Hot-Electron Effects

From the point of view of the transistor, high low-field mobility is a good criterion only when the material is being tested before device processing. As a matter of fact, in the device the electric field is rather high and of complicated geometry. As a consequence, knowledge of hot-electron drift velocity as a function of field is more important than ohmic mobility in order to study the physics of electron transit under the gate. Figure 9.10 compares the electron drift velocity $v_d(E)$ in bulk undoped GaAs [22] with that of a 2DEG in a heterojunction based on Monte Carlo simulations [23], and experimental results [24]. At room temperature, the difference between the curves is not exceedingly large. This may not be the case at lower temperatures.

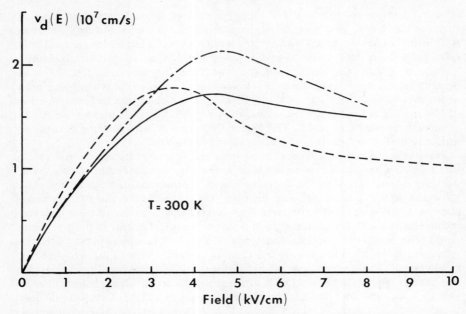

Figure 9.10   Experimental heterojunction drift velocity as a function of the electric field (solid line) with $x_{Al} = 0.3$ [24]. Simulation results (Monte Carlo technique) in undoped bulk GaAs (dashed–dotted line) [22] and in a heterojunction with $x_{Al} = 0.3$ (dashed line) [23].

The physics of charge transport in heterojunctions is more complicated than in bulk GaAs. As the driving field increases, electrons absorb energy, and their equivalent temperature may increase well above the lattice temperature. In these conditions higher-energy subbands can be occupied. This enhances thermal diffusion, and the carriers tend to disperse more easily into space than they do at equilibrium. As a consequence, the change in shape of the electron layer all along the channel, as the electrons move along the gate, in turn leads to a change in the conduction band from place to place. One can then wonder to what extent the two-dimensional character of the electron channel layer is preserved in the case of a high field. Moreover, the energy absorbed by the electrons may be large enough for them to transfer back to the AlGaAs layer much more easily than they do at equilibrium. This phenomenon, known as "real space transfer," may lead to transconductance degradation because the electron mobility in AlGaAs is low due to the ionized impurity scattering. As a consequence, a high value of ΔE_c is desirable, which is obtained with a high value of x_{Al}. Unfortunately, several phenomena appear when x_{Al} is too high: Γ-L-X valley crossover in AlGaAs when $x_{Al} > 0.4$; enhancement of deep levels in the doped AlGaAs, which strongly limits the n_{s_0}; and degradation in the quality of interface stoichiometry.

As in the normal MESFET, nonstationary transport can be obtained in submicrometer HEMTs. Roughly speaking, the high field zone exists only under the gate. If the gate length is comparable with the carrier mean free path (nearly 1000 Å at room temperature), the carriers coming from the source near equilibrium may not have enough time to establish an equilibrium velocity or energy distribution in the presence of the field changing over very short distances. Then the transit is accomplished in essentially transient conditions before the carriers reach the drain. In that case, transient drift velocities can be achieved which are much higher than the usual maximum velocities

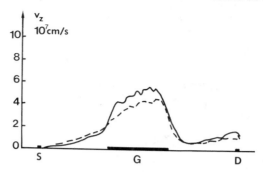

Figure 9.11   Monte Carlo simulations: influence of the lattice temperature on the space variation of the electron drift velocity in the HEMT channel under the gate with $L_g = 0.3$ μm. Solid line, $T = 77$ K; dashed line, $T = 300$ K.

attainable in bulk GaAs. In other words, if the transit time under the gate is too short, the electrons do not have time to reach the energy of about 0.3 eV that is necessary for them to transfer to a sattelite valley of GaAs, even though they drift in a high electric field. This situation is illustrated in Fig. 9.11, in which results of HEMT static simulations are displayed. One notes that the drift velocity may approach 4×10^7 cm/s, and the average velocity under the gate $\langle v_d \rangle$ is about 3×10^7 cm/s. We have shown earlier that the latter quantity is directly related to the cutoff frequency. Although there is no direct experimental observation of this phenomenon (called "velocity overshoot"), it is striking that simulation results are in fair agreement with evaluations of velocities from f_c measurements. As a result, in submicrometer structures, the drift velocity is no longer dependent on the local electric field, but rather, on the energies of the particles, or to first order on the average energy of the electron system. This idea is the source of a number of submicrometer HEMT simulations. Thus the $v_d(E)$ of Eq. (9.12) is probably wrong for a submicrometer structure. It is not surprising that the good agreement obtained between the $I_d(V_d, V_g)$ derived in Section 9.2 and experimental data of submicrometer structures leads to a $v_d(E)$ that is not physical [e.g., this dependence is silicon-like, although the true $v_d(E)$ has a negative differential mobility region]. Actually, Eq. (9.12) incorporates in an empirical manner a number of physical phenomena, the most important being velocity overshoot.

9.5.3   HEMT Physical Simulations

Several kinds of simulation models have been developed. Here we mention the Monte Carlo model and the solution of fundamental semiconductor equations. The basic principle of Monte Carlo simulations [25] is to follow the motion of carriers representative points in reciprocal and direct spaces, taking into account both the deterministic effect of the driving electric field according to Newton's law and the stochastic effect of the various types of scattering events the carriers may undergo in the course of time. The bias voltages applied at the contacts are introduced by means of boundary conditions, and Poisson's equation can be solved in two dimensions since the distribution of the carriers in space is known at all times. From this one gets a map of the potential everywhere in the simulated structure. Then external currents and device characteristics can be computed. Figure 9.12a represents the carrier equidensity contours in a submicrometer gate HEMT. This clearly shows, for instance, carrier injection into the buffer layer near the exit of the gate.

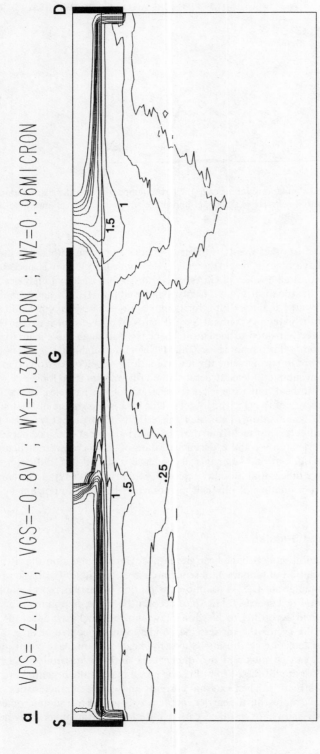

a VDS= 2.0V ; VGS=-0.8V ; WY=0.32MICRON ; WZ=0.96MICRON

Figure 9.12 (a) Electron equidensity contours in a planar HEMT simulated with a Monte Carlo model. A typical Monte Carlo uncertainty is clearly shown. $x_{Al} = 0.3$, $a = 40$ Å, $w_s = 0$, $N_D = 10^{18}$ cm^{-3}, $L_g = 0.3$ μm (all distances are at scale). All densities are in 10^{17} cm^{-3}. External $V_g = 0.2$ V, $V_d = 2$ V. (b) Electron equidensity contours (top) and the equi-energy lines (bottom) in a planar HEMT calculated after the solution of the semiconductor balance equations. They are labeled in 10^{16} cm^{-3} and eV, respectively. The horizontal distances are in micrometers, the vertical in angstrom. A spacer of 40 Å is included here. $N_D = 10^{18}$ cm^{-3}, $x_{Al} = 0.3$. A depleted zone extending on each side of the gate appears because a surface potential has been included in the simulations. External $V_g = 0.4$ V, $V_d = 2.5$ V.

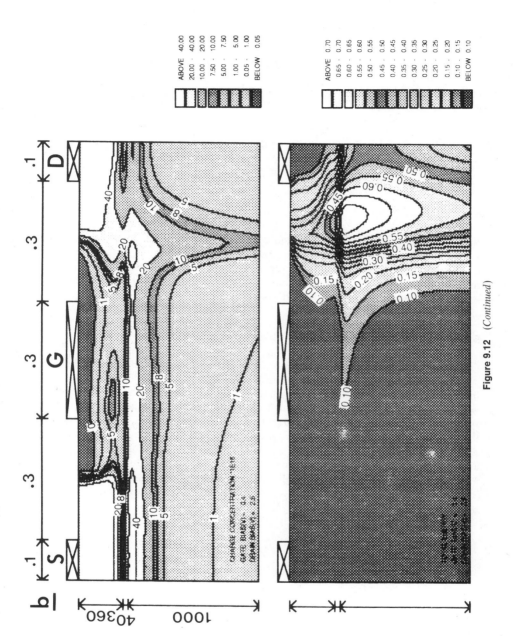

Figure 9.12 (*Continued*)

Another method is based on the solution of balance equations of the carrier mean energy and momentum [26]. These equations are deduced from the moments of the Boltzmann transport equation and use the relaxation time approximation. The basic postulate is to assume that all the moments of the distribution function are a function of the average energy only, which depends on space and time. The equations are

$$\mathbf{J} = \mu(W) \cdot [qn\mathbf{E} - \vec{\nabla}(nkT(W))] \tag{9.18}$$

$$\frac{qd(nW)}{dt} = \mathbf{J} \cdot \mathbf{E} - \vec{\nabla}(\mathbf{J} \cdot W) - \vec{\nabla}(\mathbf{J}kT(W)) - \frac{n(W - W_0)}{\tau} \tag{9.19}$$

plus the usual current continuity and Poisson equations. n is the electron density, $\mathbf{J}$ the current density, μ an equivalent mobility, $\mathbf{E}$ the electric field, W the average electron energy, T the electron temperature, and τ an energy relaxation time. These quantities are all dependent on x and z. k is Boltzmann's constant. A typical result of this technique is shown in Fig. 9.12b in case of a submicrometer HEMT. Of course, these methods of simulation are computer time consuming, but they are able to predict in detail what should occur in a device with a high degree of accuracy.

9.6 LOW-TEMPERATURE BEHAVIOR

One of the most important advantage of HEMTs is that the 2DEG mobility increases strongly as temperature decreases. Figure 9.9 showed the typical variation of low-field mobility with temperature compared with that observed in bulk GaAs. As a consequense, we can expect a significant improvement of device characteristics at low temperature such as transconductance g_m, and current gain cutoff frequency f_c, and as a result of the main device performance, such as propagation time, noise figure, and associated gain. However, the g_m enhancement observed when the temperature drops from the ambient down to 77 K as predicted, for example, by Monte Carlo simulations (see Fig. 9.13) is less important than what could be expected from the mobility rise of Fig. 9.9. This is due to the fact that g_m and so on are roughly proportional to the average velocity under the gate rather than to the low-field mobility.

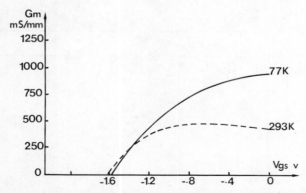

Figure 9.13 Monte Carlo theoretical prediction of the transconductance as a function of gate voltage at room temperature (dashed line) and 77 K (solid line), in a 0.3-μm-gate HEMT.

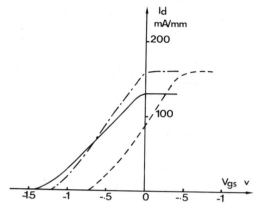

Figure 9.14   Typical transfer characteristics of a HEMT. Solid line, at 300 K; dashed line, at 77 K in dark; dashed-dotted line, at 77 K with light illumination.

However, it is not easy to profit from the potential advantages of HEMTs at low temperature, due to the existence of parasitic phenomena which change the expected behavior considerably. First, at low temperature, the HEMT static characteristics is very sensitive to light illumination [27, 28]. Even under low-intensity light, the $I_d(V_d, V_g)$ are excellent and exhibit no anomaly. On the contrary, in the dark they are strongly degraded; one observes a shift of the threshold voltage and a collapse of the curves at low drain voltage. When the HEMTs are cooled down in the dark, in most cases, the shift δV_T of V_T is positive (see Fig. 9.14). With normally-on devices, the shift is much larger than with normally-off devices. With light illumination, the shift can be reduced considerably.

The typical degradation of the static characteristics, so-called collapse, is illustrated in Fig. 9.15. These characteristics are measured under light and in the dark after cooling down to 77 K. They are obtained using ordinary test measurement setups. This degradation, however, was found to be critically dependent on the V_d and V_g bias ranges of

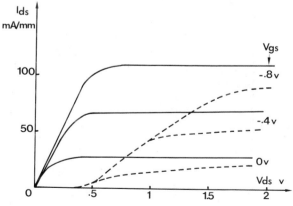

Figure 9.15   HEMT static characteristics at 77 K. Dashed lines, in dark; solid lines, with light illumination.

operation, as shown by Kastalsky et al. [29]. For instance, a pronounced degradation is observed only if V_d and V_g are higher than 1 V and 0.5 V, respectively. In addition, it is observed that once degradation has begun, it persists even though V_d is reduced below 1 V. This degradation is characterized by a strong increase in the access drain resistance, reaching values as high as several kilohms [30]. Note that although light illumination has an instantaneous effect, relaxation is very long once light has been switched off.

This rather complicated behavior can be explained by considering the existence of deep trap levels in the AlGaAs layer, called DX centers. A controversy still exists as to the exact nature of the DX centers, but their influence on phenomena occuring in HEMTs at low temperature is now generally admitted. If we represent the position of the DX center energy level as a function of x_{Al} (see Fig. 9.16) with respect to the positions of the Γ-L-X band edges, we can observe that the DX center level approximately follows the L band edge and becomes resonant with the Γ valley at $x_{Al} < 0.22$ [31]. As a consequence, at $x_{Al} > 0.2$ a good part of the electrons are trapped in DX centers and cannot take part in the drain current. This proportion increases quickly with doping concentration and may reach 20 and 95% for x_{Al} equal to 0.2 and 0.3, respectively, at N_D near 10^{18} cm^{-3} [32]. This scheme explains why the potential barrier height for emission from a deep level is independent of x_{Al} and why it decreases with x_{Al} for capture into it.

The shift of V_T can also be explained by electron trapping in deep DX centers in the undepleted part of the AlGaAs layer. Because of the exceedingly long emission and capture times at low temperature, equilibrium cannot be reached after the perturbation has been applied to the system, unless a very long time has elapsed. Collapse can also be explained by hot electron trapping in the AlGaAs layer between gate and drain. The result is a degradation of the drain access resistance, due to the reduction of the free electron density in both the AlGaAs layer and the channel quantum well.

Several authors have shown that it is possible to suppress the collapse with a proper design of the recessed zone, depending on the gate length. For instance, collapse disappears if the distance between the end of the gate metal and the edge of the recessed zone is made shorter than 0.4 μm. This effect can be explained by considering the localization of the high-field domain between gate and drain and the electron transfer from

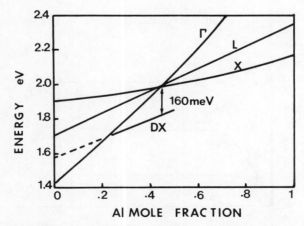

Figure 9.16   Lower-conduction band edges of Al$_{x_{Al}}$Ga$_{1-x_{Al}}$As as functions of the Al mole fraction x_{Al}. Note the band crossover at nearly 0.45. The DX center deep level is also represented.

the channel to the drain contact through the AlGaAs layer. However, it is always possible to aviod these parasitic effects simply by using x_{Al} lower than 0.15 to 0.17 in normal, uniform doping levels (10^8 cm^{-3}). Unfortunately, in this case, the height of the potential barrier ΔE_c and the available n_{so} are strongly reduced compared with conventional compositions ($x_{Al} > 0.2$). For these reasons, specific devices for low-temperature applications have been conceived and realized. They are described in Section 9.9.

9.7 LOW-NOISE AMPLIFICATION

HEMTs have demonstrated better noise performance than conventional MESFETs. For instance, Duh, Pospiezalski et al. [33] report noise performance of 0.25-μm HEMTs and MESFETs fabricated at General Electric Laboratory, with the same gate topology. They are compared in Fig. 9.17 in the range 8 to 60 GHz at room temperature. In the entire frequency range, the HEMT has a lower noise figure, making this device very attractive for low-noise applications, up to millimeter wave range. Now, we will describe briefly some of the physical aspects of HEMT noise behavior and describe how devices and circuits can be optimized.

9.7.1 Physical Aspects of Noise

In semiconductor devices, four kinds of noise can be observed: thermal or diffusion noise; shot noise; generation–recombination noise; $1/f$ noise. The last three appear only in amplifiers at low frequency, and as a consequence, they do not have an influence on microwave low-noise amplification. Diffusion noise is due to carrier velocity fluctuations, increasing with electronic temperature. From a macroscopic point of view, this results in local fluctuations of the conduction current. These fluctuations occur not only in the channel, but also in the access zones and at the electrodes. As to their influence on the microwave signal, they are considered to generate thermal noise and their contributions to the total noise are easily evaluated. Each can be represented as a noise generator with the amplitude

$$\langle e_R^2 \rangle = 4kTR\,\Delta f \tag{9.20}$$

Source and drain access resistances have the largest influence on noise performance because they are located in series with the input of the amplifier.

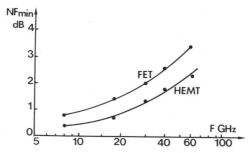

Figure 9.17 Comparison of noise figures for 0.25-μm-gate MESFETs and HEMTs at room temperature between 8 and 60 GHz. (After Ref. 33.)

In the channel, the noise source is a bit more complicated. Fluctuations in the conduction current can be represented by uncorrelated current noise sources distributed in series all along the channel. If we assume a one-dimensional model of the device along the source–drain axis, we can divide the channel into elementary sections of equal length Δz. The mean-square value of the noise current source between z and $z + \Delta z$ is given by

$$\langle i^2(z) \rangle = \frac{q \langle v_{||}^2 \rangle Q_s(z) Z}{\Delta z} \tag{9.21}$$

where $\langle v_{||}^2 \rangle$ is the average quadratic fluctuation of the drift velocity component parallel to the interface. The noise generators can also be characterized in terms of current spectral density given by

$$S_i = \frac{4q D_{||} Q_s Z}{\Delta z} \tag{9.22}$$

where $D_{||}$ is the component of the diffusion coefficient parallel to the axis of the channel. These coefficients are usually considered as dependent on the electric field, but in fact, it is more realistic to assume that they are related to the average energy of the carrier, in analogy to what was said at the end of Section 9.5. Although $D_{||}$ is now relatively well known in bulk GaAs, the field dependence in the case of a 2DEG is still rather unclear. Some results (obtained with Monte Carlo simulations) are presented in Fig. 9.18, indicating that the $D_{||}(E)$ obtained in the 2DEG are nearly similar to the bulk GaAs. This is not the case with the diffusion coefficient perpendicular to the interface, which is much smaller than in the bulk. This results from the fact that velocity fluctuations in this direction are reduced due to the existence of a potential barrier at the interface. This might partially explain why noise is reduced in HEMTs.

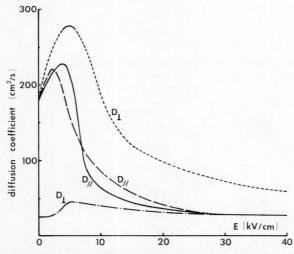

Figure 9.18   Evolution of diffusion coefficients as functions of the electric field at 300 K. Solid line, $D_{||}(E)$ in bulk GaAs; dashed line, $D_{||}(E)$ in a AlGaAs/GaAs heterojunction; dotted line, $D_{\perp}(E)$ in bulk GaAs; dashed–dotted line, $D_{\perp}(E)$ in a AlGaAs/GaAs heterojunction.

Gate and drain current noise fluctuations provoke gate and drain voltage fluctuations by coupling through the gate and drain external circuits. This process of noise generation is somewhat complicated, and as a result, it is not easy to say what part of the device really contributes the main part of the total noise. First, as the $D_{\parallel}$ decreases almost monotonously with the average energy of the carriers, Eq. (9.22) shows that the amplitude of the noise source decreases from the input to the output of the electron channel. Second, the influence coefficients relating the noise sources to the corresponding contributions to the gate and drain noise currents change all along the source–drain axis. These coefficients can be calculated by using the impedance field method [34–36], consisting in a calculation of the gate and drain current variations resulting from a vanishingly small current fluctuation in a section of the channel. The equivalent gate and drain noise sources can be calculated by a summation of the quadratic fluctuations as

$$\langle V_{do}^2 \rangle = 4q^2 Z \, \Delta f \int_0^{L_g} Q_s(z) D_{\parallel}(z) \left| \frac{dZ(z,\omega)}{dz} \right|^2 dz \qquad (9.23)$$

$$\langle i_{go}^2 \rangle = 4q^2 Z \, \Delta f \int_0^{L_g} Q_s(z) D_{\parallel}(z) \left| \frac{dA}{dz} \right|^2 dz \qquad (9.24)$$

The terms $Z(z, \omega)$ and A are determined by application of the impedance field method. $dZ(z, \omega)/dz$ is the impedance field [37] and A is an influence coefficient relating the noise currents in two adjacent sections of the channel [36].

Moreover, as gate and drain noise fluctuations are correlated, the correlation coefficient entering the expression of the noise figure (as we will see later) is calculated assuming that the transit time under the gate is negligible compared with the period of the microwave signal. In this case, the correlation coefficient is purely imaginary:

$$C_0 = jC = \frac{\langle i_{go} V_{do}^* \rangle}{(\langle i_{go}^2 \rangle \langle V_{do}^2 \rangle)^{1/2}} \qquad (9.25)$$

9.7.2 Calculation of the Noise Figure

The noise figure can be evaluated by a method similar to that of Rothe and Dahlke [38]. All the noise sources are referenced at the input of the FET equivalent circuit and represented by three elements: voltage and current sources which are uncorrelated, and a correlation impedance $Z_{cor} = R_{cor} + jX_{cor}$ (see Fig. 9.19). All the elements are related in terms of drain and gate noise sources [35] correlation coefficients, and gate and source access resistances. Finally, the noise figure F can be expressed in terms of a noise resistance R_n, a noise conductance G_n, and the correlation impedance Z_{cor}, which depends only on the dc bias voltages:

$$F = 1 + \frac{1}{R_0} (R_n + G_n |Z_0 + Z_{cor}|^2) \qquad (9.26)$$

Here $R_n = \langle U_n^2 \rangle / 4kT\Delta f$, $G_n = \langle i_n^2 \rangle / 4kT\Delta f$, and $Z_0 = R_0 + jX_0$ is the internal impedance of the "equivalent generator" connected to the FET.

Equation (9.26) shows that a proper matching can minimize the noise figure. The matching conditions are

$$X_0 = X_m = -X_{cor}, \qquad R_0 = R_m = \left(R_{cor}^2 + \frac{R_n}{G_n} \right)^{1/2} \qquad (9.27)$$

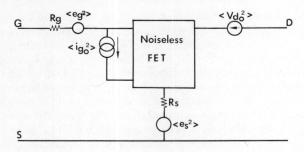

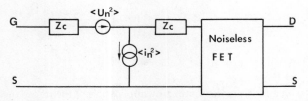

Figure 9.19 Representation of a noisy FET as a noiseless FET in series with a noise equivalent network at the input.

The minimum noise figure is given by

$$F_{\min} = 1 + 2G_n(R_{\text{cor}} + R_m) \tag{9.28}$$

and the dependence with the generator impedance Z_0 is given by

$$F = F_{\min} + \frac{G_n}{R_g}|Z_0 - Z_m|^2 \tag{9.29}$$

9.7.3 Gain Calculation and Main Dependences

For low noise as well as power amplification, the available power gain is an important device characteristics. It can easily be calculated from the equivalent circuit elements as

$$G = \left(\frac{f_c}{f}\right)^2 \frac{1}{[4G_d(R_s + R_i + R_g + L_s\omega_c/2) + 2C_{gd}\omega_c(R_s + R_i + 2R_g + L_s\omega_c)]} \tag{9.30}$$

where L_s is the parasitic access source inductance, and $\omega_c = 2\pi f_c$. Note the role of f_c, which directly determines the frequency and amplitude dependences of the available power gain. The gain is directly proportional to f_c^2, and a maximum value of f_c must be researched in device optimization. On the other hand, in the denominator of Eq. (9.30), we note the influence of the intrinsic elements G_d and C_{gd} and the access resistances R_s and R_g which must be reduced all together as much as possible in order to improve the available power gain. However, the most important parameter remains the current gain cutoff frequency shown in Fig. 9.20, representing the variations of G and f_c as functions of the dc bias current in a typical submicrometer HEMT. Note the important degradation of both G and f_c at low or high current. The maximum is obtained at medium current near 25 mA in the HEMT illustrated in the figure, a value not too far from $I_{\text{dss}}/2$.

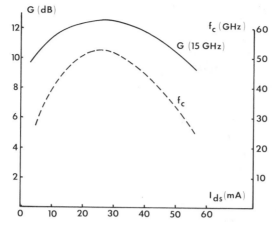

Figure 9.20 Evolutions of the cutoff frequency f_c and the available gain as functions of the drain current. The structure is the same as in Fig. 9.7.

9.7.4 Main Dependences of Noise Figure

We have shown that the noise figure depends strongly on the generator impedance and matching circuit. Moreover, its minimum value also depends on operating conditions (dc biases and frequency) and on technological parameters of the device.

Figure 9.21 represents typical variations of the minimum noise figure as a function of the dc bias current. We observe a minimum value of noise figure for drain currents

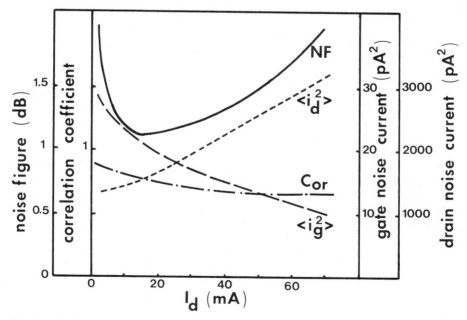

Figure 9.21 Typical evolutions of drain and gate noise currents, correlation factor, and noise figure NF as functions of the dc drain current ($F = 10$ GHz, $L_g = 0.5$ μm, $Z = 300$ μm). (After Ref. 35.)

in the range 40 to 60 mA/mm. This dependence can be explained by considering the corresponding variations of the noise sources $\langle i_d^2 \rangle$, $\langle i_g^2 \rangle$ and the correlation coefficient with the drain current (see Fig. 9.21). Since the noise figure decreases with $\langle i_d^2 \rangle$ (while C increases slightly), it is obvious that F decreases with I_d. At low I_d, g_m and f_c decreases, involving a strong increase in the noise figure. The minimum of F is less pronounced as the gate length is shorter [39]. In a 0.25-μm gate length the variation of F with I_d is very weak and related only to the presence of source and drain resistances. If they can be ignored, one finds that the noise figure becomes independent of I_d. This is very interesting because the associated gain strongly decreases near pinch-off regime. We can also note that the optimum value of the drain current (the one at which F is a minimum) is an increasing function of the frequency. For instance, with $L_g = 0.25\ \mu$m, this optimum value increases from 40 mA/mm at 10 GHz to about 250 mA/mm at 60 GHz. This result, which has been observed both experimentally and theoretically, is very interesting because it allows a minimum of F and a maximum of the associated gain to be obtained simultaneously at millimeter waves. This clearly shows the potential interest of HEMTs in the millimeter wave range.

The dependence of the noise figure on frequency and device parameters, however, remains a subject of controversy. Figure 9.17 represented typical noise figures, depending almost linearly on frequency in a large range. This is in good agreement with the well-known Fukui equation [40].

$$F = 1 + k_F \frac{f}{f_c} \left[g_m (R_s + R_g) \right]^{1/2} \tag{9.31}$$

where k_F is the so-called Fukui factor. Several authors have found that k_F is close to 1.6 in HEMTs while it is 2.5 in MESFETs. This shows the potential superiority of HEMTs over MESFETs from the point of view of noise. More recently, Cappy et al. [35] have proposed an approximate formula giving a frequency dependence similar to that of Fukui resulting from the preceding theory by using the low-frequency assumption ($\omega \to 0$) and neglecting the influence of noise gate current:

$$F = 1 + 4\pi L_g f \left[(R_s + R_g) \frac{1}{\langle v_d \rangle} (\alpha Z + \beta I_d) \right]^{1/2} \tag{9.32}$$

where α and β are two numerical coefficients not depending on the device architecture ($\alpha = 2 \times 10^5$ pF/cm^2 and $\beta = 125$ pF/mA·cm).

Equation (9.32) clearly shows the important parameters for noise figure improvement. L_g, R_s, and R_g must be made as low as possible. To reduce R_s, recessed gate structures and n^+-GaAs cap layers are now commonly made (see Fig. 9.3). Gate-to-source distances are also reduced as much as possible, but the gate metal must not be in contact with the edge of the recess; this would create additional parasitic capacitances. Today, L_g of nearly 0.25 μm are used in the best HEMTs. It looks advantageous to reduce L_g down to 0.1 μm for even better results. Nevertheless, technological efforts are still necessary to solve two problems related to very small gate lengths: first, the fact that R_g changes as $1/L_g^2$, due to the triangular form of the gate metallization; and second, the fact that higher G_d may limit the power gain [see Eq. (9.30)], due to carrier injection into the buffer or the substrate. To solve these problems, new technologies are being developed: mushroom gates, allowing a reduction of R_g by a factor of about 5, and the introduction of a p-GaAs buffer layer, which may reduce G_d. Considering Eq. (9.32) and the

TABLE 9.1

Frequency (GHz)	F (dB)	G (dB)	F_∞ (dB)
8	0.4	15.2	0.41
18	0.7	13.8	0.73
32	1.2	10.0	1.31
60	1.8	6.4	2.22

corresponding expression for k_F

$$k_F = 2\left(\frac{a}{\epsilon Z}\,\frac{\alpha Z + \beta I_d}{1 + C_p a/\epsilon Z L_g}\right)^{1/2}$$

where a is the epilayer thickness and Z the width of the transistor, it is possible to understand why HEMTs exhibit better noise figures than MESFETs. The epilayer thickness is usually smaller in HEMTs than in MESFETs, making k_F weaker. The average carrier velocity in the channel of HEMTs is higher (up to twice), again making noise figures smaller. We can also note that the term L_g introduced in Eq. (9.32) actually represents the electric gate length incorporating fringing effects. As the epilayer thickness a is smaller, fringing effects are smaller and the electric gate length tends to the real metallic L_g.

For all these reasons, the HEMT remains the most attractive device for low-noise operation at millimeter waves. Table 9.1 shows the best results obtained in 1988 [41] with 0.25-μm AlGaAs/GaAs HEMTs. The table gives not only the noise figure per stage, but also the associated gain and the noise figure F_∞ of an infinite chain of cascaded single-stage amplifiers. These figures show the potential interest of HEMTs, which are at present the most promising FET technology for millimeter wave applications.

9.8 POWER AMPLIFICATION

The development of microwave systems often requires three terminal power devices. Power MESFETs allow interesting power performance to be obtained (Chapter 10). However, the available output powers remain lower than 0.5 to 0.8 W/mm. Moreover, they can presently be used only below 30 GHz, despite important possibilities of applications in the millimeter wave range. It is thus interesting to have a look at the HEMT capability for power amplification.

For this kind of application, we must consider four different requirements: available power gain, linearity, efficiency, and maximum output power.

9.8.1 Gain

The formula for available power gain was given in the preceding section. The most important parameter controlling the power is the current gain cutoff frequency f_c. Since in HEMTs f_c is higher than in MESFETs (with the same gate length) by about 20 to 30%, the power gain should be higher in HEMTs by about the same factor. Of course, the output conductance plays a role in power gain. But presently, it is not possible to draw coherent and definitive conclusions concerning this gain in comparison with MESFETs. However, any technology progress tending to reduce carrier injection in the buffer layer and the output conductance will allow improvement of power gain.

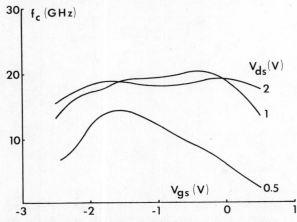

Figure 9.22 Comparison of performance obtained in MESFETs and HEMTs, as a function of gate length. For equal length, HEMTs have larger cutoff frequencies than MESFETs. (After Ref. 42.)

9.8.2   Linearity

This property is related to the variation of gain with the input and output levels. It is characterized both by the power at -1 dB gain compression and by the third-order intermodulation intercept point. Up to now, no result has been published concerning this performance for HEMTs. However, we can have an idea of the possible linearity of HEMTs by looking at the variations of f_c and G as functions of I_{ds} (see Fig. 9.20). From this point of view, the standard HEMT is still not as good as the MESFET. On the contary, the linearity might be greatly improved in multichannel heterojunction transistors, as illustrated in Fig. 9.22, in which the cutoff frequency as a function of V_{gs} is represented. At high drain voltage, the case of interest here, the linearity is better than at low drain voltage.

9.8.3   Efficiency

Power efficiency is an important quantity for systems applications. Experiments have shown that its value is an increasing function of the ratio f_c/f. From this point of view again, HEMTs appear very promising.

9.8.4   Maximum Output Power

Under large-signal conditions, the power provided by the FET to the load at the fundamental frequency is given by

$$P_0 = \tfrac{1}{2}\Delta I_d\,\Delta V_d \cos \psi \tag{9.33}$$

where ΔI_d and ΔV_d are the maximum amplitudes of the drain current and the drain-to-source voltage, respectively, and ψ is the phase angle between I_d and V_d at the fundamental frequency. Assuming the load impedance to be the complex conjugate of the FET output impedance, $\cos \psi$ can be expressed as

$$\cos \psi = \frac{1}{(1 + \omega^2 C_p^2 R_p^2)^{1/2}} \tag{9.34}$$

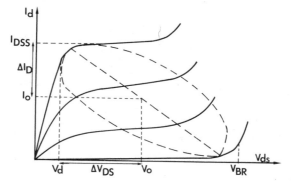

Figure 9.23 Typical static characteristics of a HEMT with breakdown at high drain voltage, and time representative point swing as a dashed ellipse.

where C_p and R_p are the FET output impedance parallel components at the frequency of operation. It is obvious that obtaining a maximum output power requires ΔI_d, ΔV_d, and cos ψ to be maxima.

Drain Current Swing ΔI_d. The drain current swing is limited by the maximum permissible value of drain saturation current I_{dss} (see Fig. 9.23). Usually, the maximum is close to $I_{dss}/2$. The drain maximum current I_{dss} is proportional to the average carrier velocity under the gate and the maximum free carrier density in the channel. On one hand, the latter is higher in HEMTs than in MESFETs. On the other hand, the maximum free charge density in the channel n_{so} is always smaller than 10^{12} cm^{-2} in conventional one-channel AlGaAs HEMTs, although it may reach 2×10^{12} in power MESFETs. For these reasons, the maximum drain current I_{dss} is always smaller than 300 mA/mm instead of 400 mA/mm in power MESFETs. From this point of view, power HEMTs require n_s to be as high as possible; therefore, high doping levels ($> 10^{18}$ cm^{-3}) and high x_{Al} are needed.

Drain–Source Voltage Swing ΔV_d. As shown in Fig. 9.23, the drain–source voltage swing is limited by the breakdown voltage V_{br}. Of course, in power FETs, V_{br} is a function of V_{gs} [43, 44]. But to simplify our discussion, we consider only breakdown voltages near pinch-off. As an example, Fig. 9.24 shows typical V_{br} near pinch-off conditions in conventional single-channel HEMTs [45]. It can be noted that V_{br} is a decreasing function of the AlGaAs doping level N_D and becomes very low for doping levels above 10^{18} cm^{-3}. Nevertheless, substantial improvement can be achieved by using pulsed-doped structures where a thin undoped layer separates the gate from the normally doped AlGaAs layer. As shown in Fig. 9.25, this arrangement can give a breakdown voltage of about 15 V with only a slight dependence on the undoped layer thickness. This value, however, is slightly smaller than those usually obtained (20 to 25 V) in power MESFETs. On the other hand, it seems possible to improve V_{br} and take advantage of the small AlGaAs ionization coefficients by using low doping levels in the AlGaAs layer. Thus it is not possible to satisfy all these requirements at the same time: high I_{dss} (and high doping levels), with high breakdown voltages (and low doping levels). As a consequence, the product $\Delta I_d \Delta V_d$ in HEMTs remains smaller than in power MESFETs by about 30 to 40%.

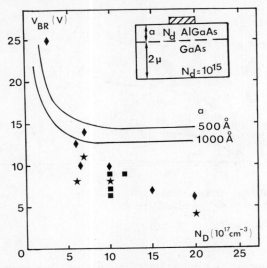

Figure 9.24   Calculated breakdown voltage (solid lines) of a conventional HEMT as a function of AlGaAs layer doping density for two thicknesses, and comparison with experimental data of transistors in diode configuration (stars) and transistor configuration at pinch-off (rhombus and squares). (After Ref. 45.)

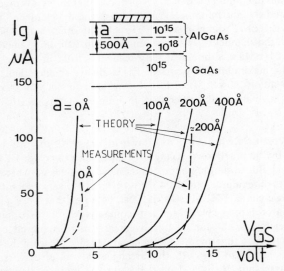

Figure 9.25   Static transfer characteristics of Schottky gate MIS HEMTs, for various thicknesses of the undoped AlGaAs layer, with a comparison between theory (solid lines) considering tunnel effect only, and experiments (dashed lines).

Phase Angle ψ. The phase angles ψ mainly depend on the output impedance of the device; they are comparable in MESFETs and HEMTs.

9.8.5 First Conclusion

If we consider the various requirements (gain, linearity, efficiency) for power applications, HEMTs appear to be very promising devices. However, their main weakness is a low maximum output power. Several research works have been undertaken in order to find new structures allowing for both higher I_{dss} and higher V_{br}. They are presented in the next section.

9.9 NEW HEMT STRUCTURES

A great number of new structures have been proposed and realized in order to overcome a number of limitations inherent to the standard structure made of AlGaAs/GaAs. They were mentioned in previous sections. Here we describe these new structures briefly and discuss their potential advantages and drawbacks.

9.9.1 Inverted HEMTs

Originally [2, 16], inverted structures with undoped GaAs as the top layer just under the gate and placed over the doped AlGaAs layer were proposed (see Fig. 9.26). This structure was thought to be very advantageous: the electron channel is placed much closer to the gate than in the conventional HEMT, and carrier injection into the substrate is naturally reduced. Thus both higher transconductance and lower output conductance were expected. The first practical realizations, however, indicated a bad low-field mobility, probably due to an outdiffusion of silicon impurities through the heterojunction. Moreover, the conduction in the embedded AlGaAs layer, unless completely depleted, led to undesirable parasitic conductance. More recently, optimized growth conditions and use of a AlAs/GaAs superlattice instead of n^+-AlGaAs allowed better carrier mobility to be obtained in the channel. Cirillo et al. [17] have obtained the highest transconductance ever reported for a 1-μm-gate FET (1810 mS/mm at 77 K and 1180 mS/mm at 300 K), despite a high source access resistance due to surface effects. The main limitation seems to come from the gate current in this structure. Fujishiro [46] has improved the results obtained with this structure by putting n and n^+ layers at the top of the undoped GaAs in order to minimize the influence of surface effects and thus reduce access resistances. He showed the capability of this device for digital high-speed integrated circuits: 11.8-ps/gate propagation delay and 33-mV threshold voltage standard deviation on a wafer. A very small output conductance was observed in short-gate-length devices ($G_d = 2$ mS/mm with a 0.3-μm gate), indicating better confinement of the

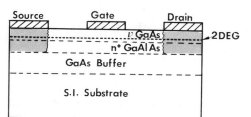

Figure 9.26 Schematic cross-sectional view of an inverted HEMT.

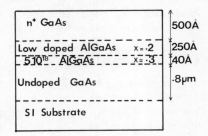

Figure 9.27 Example of a structure in which low-temperature parasitic effects can be eliminated.

carriers and reduced short-channel effects. The optimization of these devices, however, is still a difficult problem, and up to now, no convincing result has been published for microwave low noise and power applications.

9.9.2 Devices for Low-Temperature Applications

To overcome parasitic effects resulting from low-temperature conditions as described previously, a structure can be conceived in which the majority of DX centers will be empty of electrons whatever the conditions. This can be achieved by using nonuniform N_D and x_{Al} in the AlGaAs layer [30], as shown in Fig. 9.27. In this case, trapping on deep levels occurs only in the thin, highly doped part of the AlGaAs, which always remains fully depleted. As a consequence, no persistent photoconductivity is observed in this device. However, such a structure requires gate self-aligned ion implantation if a small source access resistance is desired.

More complicated solutions have been proposed by several authors (e.g., by Hiyamizu et al. [47]). A delta planar doped AlAs/GaAs/AlAs quantum well or superlattice is used as the electron supplier instead of the single doped AlGaAs layer. As the donor atoms are located in GaAs, no trapping on DX centers can take place and no parasitic effect at 77 K can occur. Another advantage of this device is that the 2DEG density can be higher than in normal HEMTs having the same spacer width. Moreover, very thin AlGaAs layers between the gate and the channel can be realized, making high transconductance possible. Once again, practical realization of this device requires a very accurate control of epitaxial growths parameters.

9.9.3 SISFETs, HIGFETs, and MIS-Like FETs

The basic idea [48–50] is to try to improve the threshold voltage uniformity by using an undoped AlGaAs layer instead of a doped one. This layer plays the role of an insulator and thus a device is obtained that closely resembles the MOSFET. A schematic cross-sectional view of this device is shown in Fig. 9.28. In the present case, the threshold voltage is independent of the characteristic of the AlGaAs layer and depends only on the material used for the gate electrode. First realized was the so-called SISFET (standing for "semiconductor insulator semiconductor FET") with a gate made of a n^+-GaAs ohmic contact. A 2DEG is formed only at positive gate bias voltages with a threshold voltage very close to zero. More recently, WSi/n^+-Ge gates have been proposed giving uniform threshold voltages in the range 0.1 to 0.4 V.

Other structures, called HIGFET (for "heterojunction isolated gate FET") or MIS-like-FET, with a Schottky WSi/undoped AlGaAs gate have threshold voltages close to 0.8 V. As shown in Fig. 9.29, n channels or p channels (using a 2D hole gas) have been proposed, with $V_T \simeq -0.5$ V for the p channel. Complementary logic can be realized

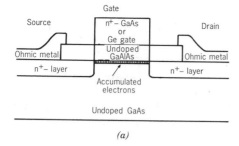

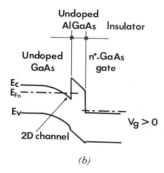

Figure 9.28 (*a*) Schematic structure of a GaAs SISFET (After Ref. 48); (*b*) Band structure of the SISFET with a positive bias applied to the n^+-GaAs gate.

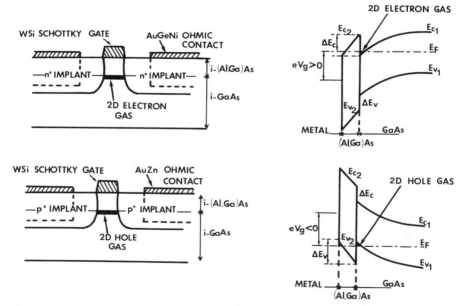

Figure 9.29 Schematic cross-sectional view of *n*-channel and *p*-channel MIS-like FETs and their corresponding band structures on the right.

with HIGFETs. This may be very interesting for digital applications. For instance, p-channel MIS-like FETs and n-channel SISFETs can be fabricated on the same wafer. Very encouraging results have been obtained. Mizutani et al. [51] reported V_T standard deviations of 13 mV on a 2-in. wafer with Ge-gate SISFETs and a transconductance as high as 470 mS/mm for a 0.6-μm gate with $f_c \simeq 54$ GHz. A frequency divider was realized based on a 0.9-μm-gate technology; a maximum toggle frequency of 16 GHz was achieved at room temperature. On the other hand, p-channel MIS-like FETs have transconductances higher than 100 mS/mm and might reach 200 mS/mm in the near future. In these devices, the charge control law is very linear and no parasitic effects due to deep levels are to be feared at low temperature. However, two main problems limit their capability. The gate current is generally very high. It can reach 10^2 to 10^3 A/cm^2 for a gate-to-source voltage of 0.8 V above threshold. Moreover, the gate access resistance may be 100 times higher in SISFETs than in conventional HEMTs. Techonological efforts are still necessary to solve the latter troublesome problem.

9.9.4 AlGaAs/GaAs Multilayer Structures

The simplest way to obtain very high drain currents I_{dss} for power applications is by use of multichannel structures instead of single channels (see Fig. 9.30). Two, three, four, or even six parallel channels [52–54] are feasible to obtain higher drain currents and output powers: 600 to 800 mA/mm instead of 250 to 300 mA/mm in normal HEMTs. Unfortunately, it is not easy to make good contact with all the 2DEG channels with the same low access resistance. The transconductance can be very weak when the gate voltage is small and the deepest 2DEG channel is controlled. However, with proper design and good theoretical optimization of the structure, a very good linearity can be obtained, as shown recently by Crosnier et al. [45] in a three-channel structure (see

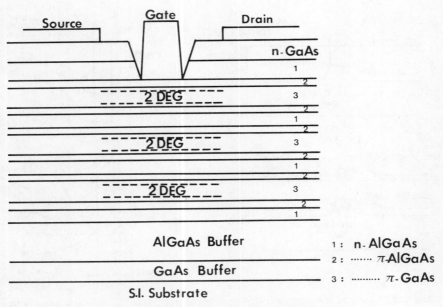

Figure 9.30 Six-channel multilayer HEMT.

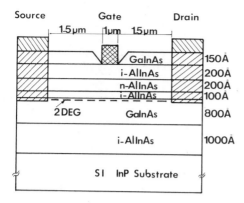

Figure 9.31 AlInAs/GaInAs HEMT grown on a SI InP substrate with a AlInAs buffer layer. (After Ref. 56.)

Fig. 9.22). Very interesting results have also been obtained in the millimeter wave range by Hikosaka et al. [54]. With a six-2DEG-channel HEMT, the maximum output power reaches 1.2 W at 30 GHz with a 2.4-μm-gate periphery device. However, this structure is very difficult to realize in practice because it requires very good control of the epitaxial growth conditions.

9.9.5 InAlAs/InGaAs Devices

Ternary compounds $In_{0.53}Ga_{0.47}As$ lattice match to InP appears to be a very attractive material for digital and microwave applications because of its high electron mobility (13,800 cm^2/V/s at 300 K) and its high electron drift velocity as observed in submicrometer structures. But as it is not possible to construct good Schottky barriers on n^+-InP, use of $In_{0.52}Al_{0.48}As$ as the wide-gap n^+ material instead of InP was proposed [55] (see Fig. 9.31). InAlAs/InGaAs now appears to be a very interesting heterojunction because of the large conduction band offset $\Delta E_c = 0.53$ eV at the interface, compared with 0.2 to 0.3 eV for AlGaAs/GaAs. As a result, the 2DEG density n_{so} may reach 3.7×10^{12} cm^{-2}, whereas only $\simeq 1 \times 10^{12}$ can be obtained in normal HEMTs with the same doping densities. Also, transport properties in AlInAs are quite comparable with GaAs. Thus the deleterious effect of parasitics when the free electrons of the top layer are controlled by the gate voltage are less pronounced that in AlGaAs/GaAs heterojunctions. Of course, a number of problems remain as to the quality of SI InP substrates, the epitaxial growth of InAlAs, and the Schottky barrier height, which is still rather low on this material ($\phi_M \simeq 0.5$ eV). The performance of this device, however, is very attractive. For instance, on a 0.2-μm-gate [57] a transconductance of 700 mS/mm with a f_c of 130 GHz and 6 ps/gate propagation delay time at 300 K (4.8 ps at 77 K) have been reported. This result, therefore, seems to be among the best ever reported in HEMT technology. Note that inverted structures and MIS-like FETs have also been proposed and realized with the same kind of heterojunction.

9.9.6 Pseudomorphic HEMTs

As SI GaAs substrates are of better quality and less expensive than SI InP substrates, keeping in mind the good prospects of InGaAs, it seems interesting to join together these two materials, although lattice match is not realized [58]. In this heterojunction

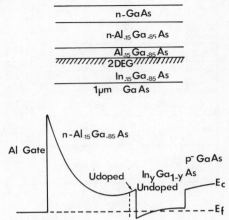

Figure 9.32 Example of a pseudomorphic InGaAs/GaAs HEMT.

(see Fig. 9.32), in fact, the interface is able to accommodate the mismatch as an elastic surface strain, making the interface almost free from dislocations, although pseudomorphic. It has been shown that the critical thickness of the strained layer is a decreasing function of the In mole fraction y_{In}, a typical value being 200 Å at $y_{In} = 0.15$. As the carrier transport properties are better as y_{In} increases, a satisfying trade-off is found near $y_{In} = 0.2$. Due to the higher conduction band offsets ($\Delta E_c \geqslant 0.3$ eV at the interface with $Al_{0.15}Ga_{0.85}As$), the carrier density n_s can be higher than in normal HEMTs. Thus injection into the buffer layer is also weaker and the output conductance is reduced due to the potential barrier between the strained layer and the GaAs buffer. Although transport properties of electrons and holes at low and high fields still are not very well known, it has been shown that the hole effective mass is low and close to $0.1m_0$. For this reason, p-channel pseudomorphic structures appear very attractive for digital applications.

Based on this heterojunction principle a large number of new devices have been proposed and realized: n single-channel HEMTs, n planar-doped double quantum wells [59], single and multiple p-channel quantum wells [60, 61], MIS-like FETs, and so on. Pseudomorphic HEMTs have shown the best performance reported today. For instance, a 0.1-μm-gate pseudomorphic ($y_{In} = 0.22$) HEMT can have a maximum frequency of oscillation of 350 GHz and a gain of 15 dB at 60 GHz. With a 0.18-μm gate, a maximum output power of 0.67 W/mm has been reached at 60 GHz with 22% efficiency. Power amplification is possible up to 92 GHz with a 0.25-μm-gate pseudomorphic device with a maximum output power of 180 mW/mm.

A last family of devices has been invented, based on the use of pseudomorphic InGaAs with $y_{In} > 0.5$ to take advantage of a better carrier drift velocity. This is possible by using SI InP substrates and $In_{0.52}Al_{0.48}As$ lattice matched to InP, as shown in Fig. 9.33. Using a mole fraction $y_{In} \geq 0.53$ in the quantum well (matched to InP) allows the barrier height and carrier density to increase substantially. The electron confinement, the output conductance, and the electron dynamics are improved. It has also been shown that both carrier mobility and concentration increase when y_{In} is brought from 0.53 to 0.65: a 40% better transconductance may result. These first results are very encouraging [62]. For instance, a drain saturation current of 1 A/mm and a cutoff frequency of 64 GHz with a 0.5-μm-gate pseudomorphic HEMT with $y_{In} \simeq 0.6$ can be

n^+ In$_{.53}$Ga$_{.47}$As	$3\,10^{18}$	200 Å
i In$_{52}$Al$_{.48}$As		300 Å
n In$_{.52}$Al$_{.48}$As	$3\,10^{18}$	200 Å
i In$_{.52}$Al$_{.48}$As		100 Å
i In$_x$Ga$_{1-x}$ As		150 Å
i In$_{.53}$Ga$_{.47}$As		400 Å
i In$_{.52}$Al$_{.48}$As		4000 Å
i InAlAs/GaInAs	Superlattice	
S.I. InP		

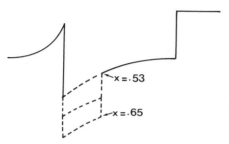

x = .53

x = .65

Figure 9.33 Multilayer AlInAs/InGaAs structure matching semi-insulating InP substrate with a superlattice as the buffer layer.

achieved. However, here, too, technological problems arise with this device, which are quite similar to those encountered with InGaAs compounds matching SI InP substrates.

9.10 CONCLUSION

We have shown that the advent of new structures proposed for heterojunction FET technology indeed has improved the performance of HEMTs impressively in less than 8 years. The present situation is probably temporary, and a breakthrough can be expected in the next few years. Among all the structures described here, those having undoped layers only are particularly well suited for logic applications because they allow good uniformity of threshold voltages. InGaAs seems to be better for low noise as well as power applications. Considering the variety of heterojunction FET structures at present, some of them giving very interesting results, it is almost impossible to say which will emerge as the standard device for the future: pseudomorphic heterojunctions on SI GaAs substrates, or lattice matched on SI InP. According to some authors [63, 64], the latter system seems to have the most promising capabilities (see Fig. 9.34).

Acknowledgments. The authors wish to acknowledge several colleagues for their interest and encouragement: Profs. E. Constant, Y. Crosnier, and R. Fauquembergue, and Dr. A. Cappy. Very special thanks are due to Dr. J. L. Thobel, Mr. T. Shawki, and Mr. A. Belache, who provided us with some of their unpublished work.

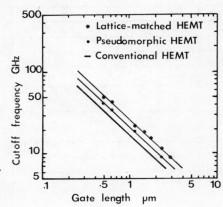

Figure 9.34 Comparison of various kinds of HEMTs showing the cutoff frequency as a function of gate length L_g for different material options.

REFERENCES

1. T. Mimura, S. Hiyamizu, T. Fujii, and K. Nanbu, *Jpn. J. Appl. Phys.*, Vol. 19, Pt. 2, p. L 225, 1980.

2. D. Delagebeaudeuf, P. Delescluse, P. Etienne, M. Laviron, J. Chaplart, and N. T. Linh, *Electron. Lett.* Vol. 16, p. 667, 1980.

3. E. E. Mendez, *IEEE J. Quantum Electron.*, Vol. QE-22, p. 1720, 1986.

4. N. J. Shah, S. S. Pei, C. W. Tu, and R. C. Tiberio, *IEEE Trans. Electron Devices*, Vol. ED-33, p. 543, 1986.

5. D. Fritzsche, *Solid State Electron.*, Vol. 30, p. 1183, 1987.

6. T. J. Drummond, W. T. Masselink, and H. Morkoç, *Proc. IEEE*, Vol. 74, p. 773, 1986, and references cited therein.

7. R. L. Anderson, *IBM J. Res. Dev.* Vol. 4, p. 283, 1960.

8. S. Adachi, *J. Appl. Phys.*, Vol. 58, p. R1, 1985.

9. D. Delagebeaudeuf and N. T. Linh, *IEEE Trans. Electron Devices*, Vol. ED-29, p. 955, 1982.

10. G. Salmer, J. Zimmermann, and R. Fauquembergue, *IEEE Trans. Microwave Theory Tech.*, Vol. MTT-36, p. 1124, 1988.

11. G. Dambrine, A. Cappy, F. Heliodore, and E. Playez, *IEEE Trans. Microwave Theory Tech.*, Vol. MTT-36, p. 1151, 1988.

12. S. M. Sze, 1985, *Semiconductor Devices: Physics and Technology*, Wiley New York, Chap. 8.

13. Land, D. V., R. A. Logan, and M. Jaros, *Phys. Rev.*, Vol. B19, p. 1015, 1979.

14. M. I. Nathan, *Solid State Electron.*, Vol. 29, p. 167, 1986.

15. B. J. Van Zeghbroeck, W. Patrick, H. Meier, and P. Vettiger, *IEEE Electron Device Lett.*, Vol. EDL-8, p. 118, 1987.

16. D. Delagebeaudeuf and N. T. Linh, *IEEE Trans. Electron Devices*, Vol. ED-28, p. 790, 1981.

17. N. C. Cirillo, M. S. Shur, and J. K. Abrokwah, *IEEE Electron Device Lett.*, Vol. EDL-7, p. 71, 1986.

18. M. Heiblum, E. E. Mendez, and F. Stern, *Appl. Phys. Lett.*, Vol., 44. p. 1064, 1984.

19. D. L. Rode, "Low Field Electron Transport," in R. K. Willardson and A. C. Beer, Eds., *Semiconductors and Semimetals*, Vol. 10, Academic Press, New York, 1975, Chap. 1.

20. T. Ando, A. B. Fowler, and F. Stern, *Rev. Mod. Phys.*, Vol. 54, p. 437, 1982.

21. F. Stern and S. Das Sarma, *Phys. Rev.*, Vol. B30, p. 840, 1984.

22. J. G. Ruch and W. Fawcett, *J. Appl. Phys.* Vol. 41, p. 3843, 1970.

23. K. Yokoyama and K. Hess, *Phys. Rev.*, Vol. B33, p. 5595, 1986.

24. W. T. Masselink, N. Braslau, D. LaTulipe, W. I. Wang, and S. L. Wright, *Solid State Electron.*, Vol. 31, p. 337, 1988.

25. P. J. Price, "Monte Carlo Calculation of Electron Transport in Solids," in R. K. Willardson and A. C. Beer, Eds., *Semiconductors and Semimetals*, Vol. 14, Academic Press, New York, 1979, Chap. 4.

26. O. El Sayed, S. El Ghazali, G. Salmer, and M. Lefebvre, *Solid State Electron.*, Vol. 30, p. 643, 1987.

27. J. F. Rochelle, P. Delescluse, M. Laviron, D. Delagebeaudeuf, and N. T. Linh, *Int. Symp. GaAs Related Compounds*, Albuquerque, NM., 1982.

28. R. Fischer, T. J. Drummond, J. Klem, W. Kopp, T. S. Henderson, D. Perrachione, and H. Morkoç, *IEEE Trans. Electron Devices*, Vol. ED-31, p. 1028, 1984.

29. A. Kastalsky and R. A. Kiehl, *IEEE Trans. Electron Devices*, Vol. EDL-33, p. 414, 1986.

30. A. Belache, A. Vanoverschelde, G. Dambrine, and M. Wolny, "Proceedings 18th European Solid State Device Research Conference, Montpellier France," *J. Phys.* Vol. 49, Coll. C4, Suppl. 9, p. 709, 1988.

31. T. N. Theis, *Int. Symp. GaAs Related Compounds*, Heraklion, Greece, 1987.

32. T. Ishikawa, T. Yamamoto, K. Kondo, J. Komeno, and A. Shibatomi, *Int. Symp. GaAs Related Compounds*, Las Vegas, Ne, 1986.

33. K. H. G. Duh, M. W. Pospieszalski, W. F. Kopp, P. Hu, A. A. Jabra, P. C. Chao, P. M. Smith, J. F. Lester, J. M. Ballingall, and S. Weinreb, *IEEE Trans. Electron Devices* Vol. ED-35, p. 249, 1988.

34. J. P. Nougier, J. C. Vaissiere, and D. Gasquet, *Proc. 6th Int. Conf. Noise Phys. Syst.*, NBS Washington, DC, 1981, p. 42.

35. A. Cappy, A. Vanoverschelde, M. Schortgen, C. Versnaeyen, and G. Salmer, *IEEE Trans. Electron Devices*, Vol. ED-32, p. 2787, 1985.

36. A. Cappy and W. Heinrich, *IEEE Trans. Electron Devices*, Vol. ED-36, p. 403, 1989.

37. W. Shockley, J. A. Copeland, and R. P. James, "The Impedance Field Method of Noise Calculation in Active Semiconductor Devices," in P. O. Löwdin, Ed., *Quantum Theory of Atoms, Molecules and the Solid State*, Academic Press, New York, 1966, p. 537.

38. H. Rothe and W. Dahlke, *Proc. IRE*, Vol. 44, p. 811, 1956.

39. B. Carnez, A. Cappy, R. Fauquembergue, E. Constant, and G. Salmer, *IEEE Trans. Electron Devices* Vol. ED-28, p. 784, 1981.

40. H. Fukui, *IEEE Trans. Electron Devices*, Vol. ED-26, p. 1032, 1979.

41. K. H. G. Duh, P. C. Chao, P. M. Smith, L. F. Lester, B. R. Lee, J. M. Ballingall, and M. Y. Kao, *Proc. IEEE Microwave Theory Tech. Symp.*, New York, 1988, p. 923.

42. C. A. Liechti, *Proc. Eur. Microwave Conf.*, Paris France, 1986, p. 21.

43. S. U. Wemple, W. C. Nichlaus, H. M. Cox, J. V. DiLorenzo, and W. O. Schlasser, *IEEE Trans. Electron. Devices*, Vol. ED-27, p. 1013, 1980.

44. R. Wroblewski, G. Salmer, and Y. Crosnier, *IEEE Trans. Electron. Devices*, Vol. ED-30, p. 154, 1983.

45. Y. Crosnier, F. Temcamani, D. Lippens, and G. Salmer, "Proceedings 16th European Solid State Device Research Conference Montpellier, France," *J. Phys.* Vol. 49, Coll. C4, Suppl. 9, p. 563, 1988.

46. H. Fujishiro, *14th Int. Symp. GaAs Related Compounds*, Heraklion, Greece, 1987.

47. S. Hiyamizu, S. Sasa, T. Ishikawa, K. Kondo, and H. Hishikawa, *Jpn. J. Appl. Phys.*, Vol. 24, Pt. 2, p. L431, 1985.

48. T. Wada, K. Matsumoto, M. Ogura, K. Shida, T. Yao, I, Igarashi, N. Hashizume, and Y. Hayashi, *Jpn. J. Appl. Phys.*, Vol. 24, Pt. 2, p. L213, 1985.

49. K. Arai, T. Mizutani, and F. Yanagawa, *Jpn. J. Appl. Phys.* Vol. 24, Pt. 2, p. L623, 1985.

50. K. Maezawa, T. Mizutani, K. Arai, and F. Yanagawa, *IEEE Electron Device Lett.*, Vol. EDL-7, p. 454, 1986.

51. T. Mizutani, M. Hirano, S. Fujita, and K. Maezawa, *Int. Electron Device Meet. Dig.*, p. 27.3, 1987.

52. P. Saunier and J. W. Lee, *IEEE Electron Device Lett.*, Vol. EDL-7, p. 503, 1986.

53. E. Sovero, A. K. Gupta, J. A. Higgins, and W. A. Hill, *IEEE Trans. Electron Devices*, Vol. ED-33, p. 1434, 1986.

54. K. Hikosaka, N. Hidaka, Y. Hirachi, and M. Abe, *IEEE Electron Device Lett.*, Vol. EDL-8, p. 521, 1987.

55. M. Kamada, et al., *Int. Symp. GaAs Related Compounds*, Las Vegas, NE., 1986.

56. L. F. Palmeteer, P. J. Tasker, T. Itoh, A. S. Brown, G. Wicks, and L. F. Eastman, *Electron. Lett.*, Vol. 23, p. 53, 1987.

57. C. K. Peng, et al., *IEEE Electron Device Lett.*, Vol. EDL-8, p, 24, 1987.

58. A. Ketterson, M. Moloney, W. T. Masselink, J. Klem, R. Fischer, W. Kopp, and H. Morkoç, *IEEE Electron Device Lett.*, Vol. EDL-6, p. 628, 1985.

59. P. M. Smith, et al., *IEEE Microwave Theory Tech. Symp.*, New York, 1988.

60. T. E. Zipperian et al., *Appl. Phys. Lett.*, Vol. 49, p. 461, 1986.

61. T. E. Zipperian et al., *Appl. Phys. Lett.*, Vol. 52, p. 975, 1988.

62. M. A. Fathimulla, H. Hier, and J. Abrahams, *Electron. Lett.*, Vol. 24, p. 93, 1988.

63. K. Hikosaka, S. Sasa, N. Harada, and S. Kuroda, *IEEE Electron Device Lett.*, Vol. EDL-9, p. 241, 1988.

64. L. Eastman, *Proc. IEEE Microwave Theory Tech. Symp.*, New York, 1988.

10

FETs: POWER APPLICATIONS

Hing-Loi A. Hung, Thane Smith, and Ho-Chung Huang

COMSAT Laboratories
Clarksburg, Maryland

10.1 INTRODUCTION

Advances in gallium arsenide (GaAs) technology have resulted in the development of microwave and millimeter wave field-effect transistors (FETs) for use in both hybrid and monolithic microwave integrated circuits (HMICs and MMICs). These components are gaining in importance for applications such as satellite communications, electronic warfare, and phased-array systems. Power metal-semiconductor FETs (MESFETs) in particular are increasingly being used instead of traveling-wave tubes (TWTs) for many reasons. MESFETs require simpler, lower-voltage power supplies for biases, and less warm-up time. Use of transistors also reduces the size and weight required to produce a given amount of RF power. MESFETs are also more reliable than TWTs, particularly in applications where long periods of operation without bias and circuit adjustment are required. Although TWT amplifiers (TWTAs) still provide excellent dc-to-RF power conversion efficiency, the performance of MESFET power amplifiers is constantly improving, to provide better linearity. For operating frequencies higher than a few gigahertz, GaAs FETs are used exclusively in power applications instead of silicon bipolar junction transistors (BJTs) because of the lower electron mobility in silicon.

In this chapter we present basic principles for the computer-aided design (CAD) of both discrete GaAs power FETs and their associated matching circuits. The analysis of the device is based on device physics, FET structure, device geometry, and GaAs material parameters. The dc and RF equivalent-circuit model of the power FET is discussed, and the optimal load impedance for the device to achieve maximum output power or power-added efficiency is derived. The optimization of the amplifier circuit design, as well as other considerations to obtain higher power from a single device and multiple devices, are discussed.

Recently, there has been an important trend toward the use of MMICs. As an example, a FET power amplifier with several stages can be formed on a single GaAs chip, and large batches of these chips can be manufactured at one time on a GaAs wafer. This approach can lead to considerable savings in assembly costs and potential advantages in terms of reliability. It also gives designers the opportunity to custom-design power FETs for particular applications. Therefore, the development of monolithic circuits will also be emphasized.

The information presented here can be used as a guide to both power FET designs and their applications. The principle of the design approaches to be discussed has been demonstrated in practical circuits ranging in frequency from 2 to 60 GHz. The discussion centers on MESFETs. Two special circuits using the dual-gate power MESFET for variable-power application, as well as high-voltage FET for improving system power conversion efficiency, are described, and two examples of device and circuit fabrication are given. The related subjects of device and circuit characterization techniques, thermal considerations, packaging designs, and radiation effects are also discussed. The projected performance of emerging high electron mobility transistor (HEMT) and heterojunction bipolar transistor (HBT) devices is then presented.

10.2   STATUS OF POWER FETs AND AMPLIFIERS

10.2.1   Discrete Power Devices

For frequencies below 20 GHz, FET chips, package FETs, and internally matched (IM) power devices on carriers that provide excellent RF performance are commercially available [1]. At C-band (6 GHz), IM FET devices (with a partial matching circuit of FET to 50-Ω input/output impedance) exhibit an output power of over 17 W, a gain of 6.5 dB, and a power-added efficiency of 28%. X-band (10-GHz) FETs are capable of providing 7 W of output power, 5 dB of gain, and 22% power-added efficiency. IM devices with an output power of 3.5 W, with 4.5 dB of associated gain, and 21% efficiency, are available in the 14- to 14.5-GHz communications band. At 18 GHz, FET chips provide 29 dBm (800 mW) of output power, 5 dB of gain, and 23% power-added efficiency. With packaged devices, output power is reduced significantly, to 22.5 dBm (178 mW). More recently, Toshiba has developed IM GaAs FETs with output power exceeding 12 W, gain of 5 to 5.8 dB, and efficiency of 21 to 25% at 14.25 and 11.2 GHz [2]. A summary of the current status of power FET performance is given in Fig. 10-1.

Recently, low-power-level performance has been demonstrated using conventional AlGaAs HEMT devices with respectable power-added efficiency. At 60 GHz, a maximum output of 50 mW with 3 dB of gain and 11% efficiency was reported [3].

10.2.2   HMIC Power Amplifiers

Power amplifiers with FETs that use the HMIC matching technique and with bias networks have demonstrated remarkable performance. Power devices operating in the class B mode have shown a power-added efficiency of 65% at 4 GHz with an output power of 1 W [4]. A 20-W GaAs FET amplifier at 5 GHz has been developed for a microwave landing system [5]. Using a TM-mode cavity eight-way power divider/combiner, an 80-W, 26-dB gain, GaAs FET amplifier with 18% efficiency has been developed for the 6-GHz communications band application [6]. An amplifier with harmonic tuning has achieved 5 W, 26% power-added efficiency, and 6 dB of gain at 10 GHz [7]. In the same frequency band, slightly better performance (6 W, 6.7 dB of gain, and a 34.8% efficiency) [8] has also been demonstrated by using a novel molecular-beam epitaxy (MBE) GaAs FET.

An X-band control gain amplifier providing more than 12 W continuous wave (CW) (16 W pulsed) with 60 dB of associated gain and a 10% bandwidth has been reported [9]. The power-added efficiency is 15%. Higher-frequency amplifier results include an output power of 1 W and a power-added efficiency of 30% at 20 GHz, and 0.5 W and 27% at

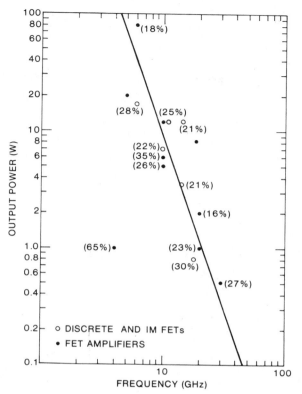

FIGURE 10.1   Status of discrete power FET and HMIC amplifier performance. (Numbers in parentheses are power-added efficiencies, in percent.)

30 GHz [10, 11]. With the same devices, an output power of 2 W at 20 GHz has also been achieved using a divider/combiner in a balanced amplifier configuration [11]. A higher-power-level amplifier with 8.2 W of output power and 36 dB of gain has also been demonstrated over a 17.7- to 19.1-GHz band [12]. A summary of the results for HMIC amplifiers is also given in Fig. 10.1.

10.2.3   MMIC Power Amplifiers

In the MMIC area, a C-band single-chip GaAs MMIC amplifier using a refractory self-aligned gate process has demonstrated 10 W of output power, with an associated gain of 5 dB and a power-added efficiency of 36% at 5.5 GHz [13]. A 2-W monolithic amplifier has been obtained at 18 GHz [14]. A single-ended amplifier at K-band with an output power greater than 0.5 W, a gain of 5.2 dB, and a power-added efficiency of 27% has also been demonstrated [15, 16]. A multistage amplifier has achieved 2 W of output power and 35 dB of gain [16]. At K_a-band (28 to 30 GHz), monolithic amplifiers are capable of delivering output power at the 1-W level [17–19], while a power-added efficiency of 21% has been obtained at an output of 0.5 W [17]. More recently, in the 44-GHz satellite band, an output of 135 mW and a power gain of 4 dB were obtained at 41 GHz [20], and a saturated output

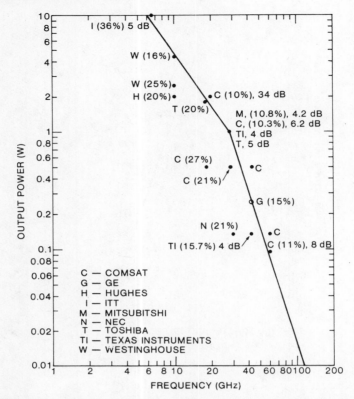

FIGURE 10.2 Status of MMIC power FET amplifier performance. (Numbers in parentheses are power-added efficiencies, in percent.)

of 0.5 W with a linear gain of 15.5 dB at 42.5 GHz [21] was achieved based on MBE technology. Excellent accomplishments have also been demonstrated at V-band, with the achievement of 136 mW of output power and 7.5 dB of linear gain in a balanced two-stage amplifier [22, 23]. Figure 10.2 summarizes the significant performance of the power MMICs over the frequency range 2 to 60 GHz.

10.3   POWER FET MODELING

In this subsection we discuss the physics and mathematics of the theory and device modeling of power FETs. This information will be useful in the design of power FETs and the prediction of power FET performance. Examples are given to illustrate the modeling approaches and results presented. The discussion focuses on optimizing the device to achieve the maximum linear output power and associated power gain and efficiency. The results can readily be applied to circuit designs without elaborate three-dimensional or time-domain computation. This quasi-linear modeling approach has been demonstrated in a number of designs for power amplifier circuits to millimeter wave frequencies. Other, more elaborate device modeling and analysis based on a time-domain nonlinear model of the FET device have been reported [24–26].

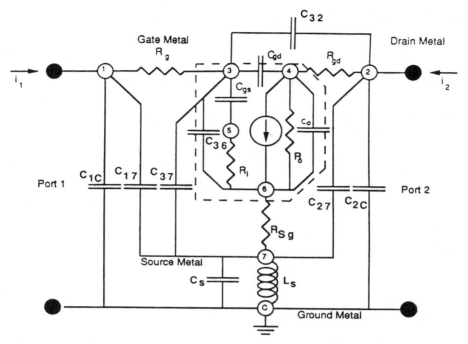

FIGURE 10.3   Example of a lumped-element equivalent circuit model for a FET.

10.3.1 Lumped-Element Equivalent Circuit

The basis for computing the RF performance of any electronic device is the lumped-element equivalent-circuit (LEEC) model. Figure 10.3 shows an eight-node (numbers 1–7, C) LEEC that could be used to represent a power FET. The three terminals of the FET (source, gate, and drain) are usually configured to have the source common to both imput and output currents, as in an amplification stage. In the figure, node C is the common source, the signal input at the gate is node 1, and the output at the drain terminal is node 2.

It is possible to calculate all the node voltages of an LEEC by using the "method of nodes," based on knowledge of all input currents and the voltage at one node. Since the common terminal of a FET is usually grounded, node C in Fig. 10.3 is assigned a node voltage of zero. As mentioned above, the input current (i_1) and output current (i_2) must be known in order to solve for the other node voltages. These currents are represented by complex numbers that contain both amplitude and phase information.

The first step toward achieving a solution is to express the unknown currents between pairs of nodes in terms of the two unknown node voltages and the known complex admittance through which each pair of nodes is connected. In addition, there is always a current generator in the LEEC of a FET, which generates a current, i_{ds}, proportional to the voltage difference ($v_3 - v_5$ or v_{35})* between the nodes across the gate-to-source

---

* v_{mn} represents the difference between the voltage at node m and that of node n.

capacitor given by

$$i_{ds} = g_m e^{(j\omega\tau)} v_{35} \qquad (10.1)$$

where g_m is the intrinsic transconductance, ω the angular frequency of the ac input current, and τ the time delay associated with the current generator.

Application of Kirchhoff's current law to all nodes except the common (grounded) node results in N simultaneous equations. These may be solved for the N unknown complex node voltages. For the LEEC of Fig. 10.3 N is 7. Two examples illustrating the method of nodes are given in Appendix A.

10.3.2 Matrix Methods

Matrix notation provides a very concise means of expressing the effect of a power FET (represented as an LEEC) on an RF signal. Matrix representation is particularly useful when a computer is used because of the availability of standard subroutines for computations involving matrices. In fact, some programing languages allow variable names to be declared as matrices. Such variables are automatically processed using the rules of matrix algebra. An example of how matrixes are set up and used in the method of nodes is given in Appendix B.

10.3.3 Physical Basis for the Equivalent Circuit

Figure 10.4 shows a scanning electron micrograph of the top surface of an example of a power FET structure [11]. The device features several parallel gate fingers between the source and drain fingers. The source fingers cross over gate metal to connect to the

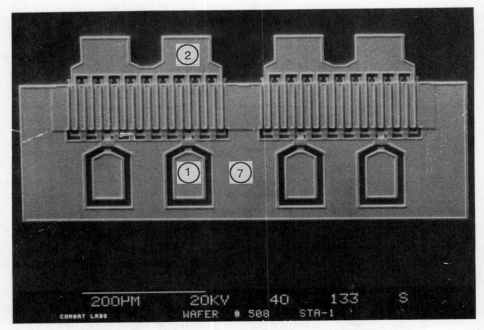

FIGURE 10.4 Scanning electron micrograph of the top view of a power FET.

source-pad areas, which are connected to the metallized bottom (not shown) of the GaAs chip by metallized through-substrate via-holes. High power parallel-gate structures with dimensional and process variations have also been reported [27].

The circled numbers in Figure 10.4 correspond to the node numbers in Fig. 10.3. Node 1 is the gate-pad metal, node 2 is the drain metal, and node 7 is the source metal. Figure 10.5 is a cross section through the FET channel which shows nodes C and 2, and the physical locations of nodes 3, 4, and 6. The semiconductor volume directly under the gate and above the buffer layer contains the portion of the LEEC of Fig. 10.3 within the dashed rectangle. This part of the FET is called the intrinsic FET. The rest of the circuit elements are extrinsic, causing parasitic effects on the device performance.

In the following discussion, frequent reference is made to the test structures to determine certain parameters used to calculate element values. These structures are created in the semiconductor wafer along with the FETs.

Extrinsic Circuit Elements. In Fig. 10.3 the resistor connected between nodes 1 and 3 represents the power dissipation in the gate metal. This is usually dominated by the gate fingers because of their small cross-sectional area. The gate finger can be most accurately represented as a transmission line with a resistance and capacitance per unit gate width. The inductance of the gate finger is usually small and can be neglected. The term "gate

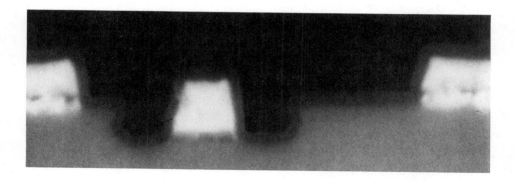

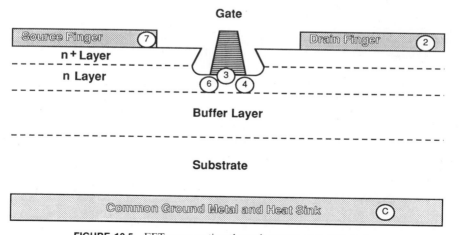

FIGURE 10.5 FET cross section through source, gate, and drain.

width" (W_g) is used to denote the length of the transmission line and is distinguished from the gate length, L_g.

When the gate width of each finger is much shorter than a quarter wavelength of the RF signal, it can be assumed that the shunt ac current is uniformly distributed. In this case, and taking $x = 0$ to be at the end of the finger farthest from its connection to the input signal, the rms RF current at distance x, $i(x)$, is given by $i(x = W_u)/W_u$ times x, where W_u is the unit gate width. The resistance of an element, dR, of gate width dx is $(R_{ee}/W_u)\,dx$, where R_{ee} is the end-to-end resistance of the unit gate finger. The RF power dissipated in dx is the square of $i(x)$ times dR. Integration from $x = 0$ to $x = W_u$ gives the total RF power dissipated in the finger as $R_{ee}i(x = W_u)^2/3$.

In the LEEC model of the FET, the series resistance of the gate finger is assumed to be $R_{ee}/3$ [28] because of the distributive effect of the RF signal through the fingers. Thus the correct power dissipation, and therefore the correct gain, can be obtained. The resistor connected between nodes 1 and 3 is given by the parallel combination of all the gate finger resistances, plus the resistance of the gate metal connecting the gate pad to the gate fingers.

The capacitor between nodes 1 and C is a parallel-plate capacitor with GaAs (relative dielectric permittivity of 12.9) between the gate pad metal and the ground plane metal on the bottom of the chip. The capacitance, neglecting fringing, is calculated based on the area of the gate pad metal, the chip thickness, and the GaAs dielectric constant. Similarly, the capacitor (C_s) between nodes 7 and C is calculated using the area of the source metal, and the capacitor between nodes 2 and C is calculated using the drain metal area.

The actual distributed structure of the source, ground plane, and via-holes forms a parallel-plate resonator. This is represented in the LEEC by the parallel combination of a capacitor, C_s, and an inductor, L_s, between nodes 7 and C. The inductance of L_s can be calculated if the resonant frequency and C_s are known. In principle, the resonant frequency can be calculated from the geometry. The capacitance is usually ignored, since its admittance is small compared to that of the via-hole inductance. Commercial circuit modeling programs, such as SUPERCOMPACT [29], contain via-hole models that can be used to calculate the inductance from geometric parameters. This inductance is usually on the order of 0.005 to 0.01 nH for a 90-μm-thick substrate.

The resistor connected between nodes 6 and 7 is known as the source-gate resistor, R_{sg}. It is the resistance to current flowing between the source metal and the point under the source end of the gate where the carriers reach their maximum velocity. This point, the physical location of node 6, is indicated in Fig. 10.5. For the purpose of calculating the value of R_{sg}, it is reasonable to locate node 6 directly under the edge of the gate metal closest to the source.

R_{sg} is the sum of three components. One component is the contact resistance, R_c, between the source metal and a point in the semiconductor directly under the source-metal edge nearest the gate. The usual method of measuring R_c is to use a test structure based on the transmission-line model (TLM) originally proposed by Shockley [30]. This structure consists of ohmic contact metal having several gaps with different distances between the metal edges. An example of a TLM test structure is shown in Fig. 10.6. The conducting layer of the semiconductor is the same width (measured parallel to the metal edges) for each gap. The total resistance of each gap is plotted versus gap length. Extrapolation to a gap length of zero gives $2R_c$. Multiplying R_c by the contact width gives a normalized contact resistance which can be used to find R_c for any other width. For example, if a FET has ohmic contacts with a normalized contact resistance of 0.1 Ω-mm and one contact is 0.075-mm wide, the contact resistance is 0.1 Ω-mm divided by 0.075 mm, or 1.33 Ω.

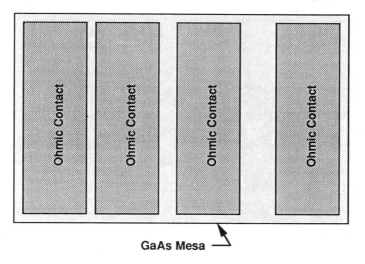

GaAs Mesa

FIGURE 10.6 TLM test structure for measuring contact resistance.

The next component of R_{sg} is the resistance through the unrecessed portion of the GaAs. It is given by the sheet resistance (R_{su}) of the combined n- and n^+- layers, times the distance (L_{su}) between the metal and the recess edges, divided by the gate width, W_g. The sheet resistance, in $\Omega/\square$, can be calculated from the TLM test data structure used to determine R_c. It consists of the slope of a plot of gap resistance versus gap length multiplied by the gap width. A somewhat more accurate method of finding sheet resistance involves measurements taken on a Van der Paw pattern test structure [31].

The final component of R_{sg} is the resistance from the recess edge to the gate metal edge, a distance denoted as L_{sr}. The sheet resistance, R_{sr}, in this area is higher than R_{su} because the n^+-layer and some of the n-layer have been removed in this region. The value of R_{sr} can be computed if the carrier profile and mobility of the n layer under the gate are known. To obtain the correct value, the effect of the bare surface depletion layer must be accounted for. In most cases it can be assumed that this layer has the same thickness as the zero-bias depletion depth under a Schottky barrier. Sheet carrier concentration is calculated by integrating the carrier profile from the surface depletion depth to the semi-insulating substrate. To obtain the sheet conductivity, the sheet carrier concentration can be multiplied by the mobility and charge of an electron. Sheet resistance is the reciprocal of sheet conductivity. A typical R_{su} value for the n^+/n epitaxial layers used for power FETs is $80\ \Omega/\square$. A typical R_{sr} value of the remaining n-layer, after recessing the gate, is $500\ \Omega/\square$. The total value of R_{sg} is given by

$$R_{sg} = R_c + (L_{su}R_{su} + L_{sr}R_{sr})/W_g \tag{10.2}$$

The resistance between nodes 4 and 2, R_{gd}, is computed in the same way as R_{sg}. Its value may differ from R_{sg} because the geometric details on the drain side may be different from those on the source side. For example, the gate is often deliberately placed closer to the source than to the drain. Several capacitors shown in Fig. 10.3 resulted from close proximity between metal areas on the top surface of the FET.

The capacitor between nodes 1 and 7 is designated C_{17} and represents the capacitance between the gate-pad metal and the source metal. C_{37} is the capacitor between the gate fingers and the source metal, C_{32} between the gate fingers and the drain metal, and C_{27}

between the drain metal and the source metal. Clearly, these capacitors will have values that are highly dependent on the geometric layout of the FET. For example, the layout might resemble an interdigitated capacitor, or perhaps a coplanar waveguide [32].

Intrinsic Circuit Elements. The intrinsic circuit elements are associated with the movement of the depletion layer boundary under the gate, which is caused by a change in the gate–source voltage, V_{gs}. The current generator results from the fact that the depletion-layer configuration limits the drain–source current, I_{ds}, to a value that depends on V_{gs}. The intrinsic resistors account for the input and output power dissipation associated with the current generation mechanism. When V_{gs} changes, the depletion layer near the gate expands or contracts. This changes the total charge in the depletion layer. The fixed charge density, ρ, in a depletion layer is equal to $q(N_d - N_a)$, where q is the elementary charge, N_d the donor impurity density, and N_a the acceptor impurity density. In undepleted regions with an electric field in the constant mobility range, free carriers exactly balance the fixed charge, resulting in zero space charge. Thus a change in depletion-layer volume caused by a change voltage results in a change in the total charge associated with the depletion layer. This may be represented by a capacitor, C_{gs}.

Saturated source-drain current is determined by the depletion-layer positions that confine it. This is the physical basis of the current generator. Figure 10.7a shows a cross section of a FET channel in the vicinity of the gate. Both the upper and lower boundaries of the current-carrying layer are depletion-layer edges. The upper boundary position is controlled by V_{gs}; the lower boundary is caused by the depletion layer in the low-doped buffer layer. This layer results from the 0.7-eV shift in the Fermi energy at the junction between the semi-insulating substrate and the n-type buffer layer (back junction). The entire buffer layer should be depleted to avoid excessive output conductance. For example, based on device physics analysis, a buffer-layer thickness of 1 μm should not have a carrier concentration of more than 1.0×10^{15} cm^{-3}. For a 6-μm-thick buffer layer, the maximum carrier concentration is 2.8×10^{13} cm^{-3}.

In Fig. 10.7a, two boundaries are shown for the upper depletion layer. One is for an open-circuited (floating) gate and represents the depletion-layer configuration near the upper limit of drain current. Because of the drain–source voltage, this boundary is closer to the gate than to the zero-bias depletion depth near the source end of the gate and is deeper than the depth near the drain end. Outside the gate, this boundary becomes the boundary of the bare-surface depletion depth. The other boundary shown is for a V_{gs} such that the depletion layer under the drain edge of the gate is just beginning to merge with the back junction depletion layer. Vertical dashed lines on the figure show the edges of the zone of carrier transport at saturated velocity. The locations of the edges can change slightly with both gate and drain voltages. The current generator is physically located in this zone where the source–drain current is controlled.

The calculation of source–drain current from depletion-layer boundaries is shown in a Speakeasy program, FLATIDSS, which calculates the maximum drain current ($I_{ds,\text{max}}$), the grounded gate drain current (I_{dss}), and the pinch-off voltage (V_{po}) for a FET with a flat carrier profile under the gate. A Speakeasy session in which FLATIDSS is used is reproduced in Appendix C.

Figure 10.7b shows the equivalent-circuit elements associated with semiconductor regions near the gate. The extrinsic (also called "parasitic") elements, R_{sg} and R_{gd}, are also shown for reference. The physical origins for the remaining circuit elements are described in the following paragraphs.

Two capacitors, C_{36} and C_{gd}, included as intrinsic, are also parasitic in the sense that their nonzero values lead to degradation of high-frequency performance. C_{36} is the result of the change in depletion-layer charge versus V_{gs} caused by gate depletion-layer move-

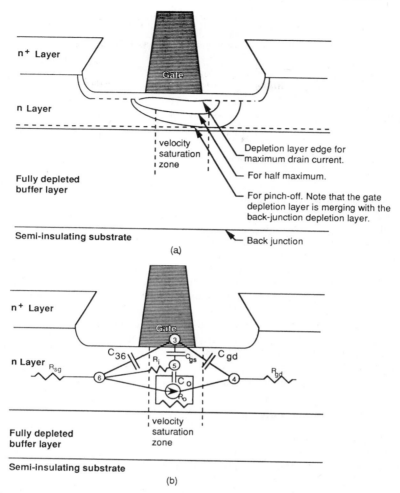

FIGURE 10.7   Cross section of a FET channel: (*a*) variation of depletion layer with biases; (*b*) physical location of the intrinsic circuit elements.

ment in the region outside the velocity saturation zone and between gate and source. C_{gd} is associated with depletion-layer charge versus V_{gs} outside the velocity saturation zone and between gate and drain. These depletion-layer boundaries are illustrated in Fig. 10.7*a*.

The circuit elements (C_{gs} and the current generator) shown in the velocity saturation zone of Fig. 10.7*b* are very closely interrelated. C_{gs} is given by the change in depletion-layer charge in this zone, divided by the change in the voltage across the zone. The I_{ds} change resulting from the same change in V_{gs} is proportional to the same change in depletion-layer charge. Therefore, g_m (equal to dI_{ds}/dV_{ds}) is proportional to C_{gs}. [The constant of proportionality depends on gate length (L_g).] The effect of the gate voltage on I_{ds} is given by Eq. (10.1). The phase delay of the current generator, τ, from that of v_{35} is given by the length of the velocity saturation zone, divided by the carrier saturation velocity (v_s). For the FET model, discussed in Section 10.3.4, τ is calculated using the simple formula C_{gs}/g_m. The resistor R_i is used to account for the power dissipated in changing the position of the depletion-layer edge. According to Pucel et al. [33], R_i times C_{gs} is expected to be proportional to τ. The constant of proportionality is usually taken to be $\frac{1}{2}$ [33], in which case $R_i = 1/(2g_m)$.

The remaining circuit elements associated with the velocity saturation zone are a resistor (R_0) and a capacitor in parallel with it. The capacitor accounts for the enlargement of the gate depletion layer toward the drain as the voltage difference, v_{46}, increases. R_0 accounts for the power dissipated by the carriers flowing under the gate at saturated velocity.

There are several models for calculating R_0 [33], [34]. The usual procedure is to begin with the classical model for constant mobility current flow across merged depletion layers, and modify it to include a saturated velocity region in addition to the constant-mobility region. However, although their parameters of these models can be adjusted to fit dc I–V characteristics reasonably well, the predicted R_0 values do not necessarily agree with values derived from RF measurements. Also, the fitting parameters (saturation velocity and saturation field) required to fit the dc I–V data typically do not agree well with direct measurements of velocity versus field. The dc value of R_0 can be measured easily as the reciprocal of the slope of the saturated drain current versus drain voltage. If the gate voltage is adjusted to give a saturated current curve that extrapolates to $I_{ds} = 0$ at $V_{ds} = 0$, the slope yields a reasonable estimate of the large-signal RF value of R_0 (at least for a properly passivated FET).

It is desirable to have R_0 as large as possible, since low values can limit both maximum output power and gain. From the theoretical models [33, 34], R_0 should be proportional to the hyperbolic sine of $\pi/2$, times the ratio of gate length to channel height (L_g/H). The optimum L_g/H ratio for power FETs is in the range 4 to 5. Long L_g reduces gain, and small H reduces the maximum possible drain current.

10.3.4 Physical Limitations on Output Amplitude

A two-dimensional model of a FET can provide a reasonable prediction of power FET performance. The onset of current saturation is not caused by depletion of the channel, as it would be if mobility were constant, but to the carriers reaching velocity saturation at the point where the undepleted channel is thinnest. If the drain voltage, V_{ds}, is increased, the drain-to-gate voltage, V_{gd}, is also increased. This would increase the thickness of the gate depletion layer, except that the increased V_{ds} lowers the potential barrier presented to the carriers by the depletion layer. This allows the carriers into regions that would otherwise be depleted. The net result is that even though the carrier velocity is constant, an increase in V_{ds} still causes some increase in the I_{ds}. The present model will determine the maximum value of I_{ds} for the case where the drain voltage is just sufficient to cause velocity saturation under most of the L_g. This is the situation at the high-current end of the load line (the locus of drain I, V points passed through in one RF cycle) for a typical power FET amplifier. Even though the current will increase slightly at higher V_{ds}, it is the current at the knee of the $I_{ds,\,max}$ versus V_{ds} curve that determines the maximum RF voltage amplitude.

Based on the discussion above, the dc value of I_{ds} is taken to be constant at qNv_shW, where N is the carrier concentration in the channel, v_s is 1.2×10^7 cm/s, h is the undepleted channel height at the point under the gate where velocity saturation begins, and W is the total gate width. If the carrier concentration varies with depth under the gate, the product of Nh is replaced by the integral of N over depth, starting from the depletion-layer edge. The upper limit of the integration is the edge of the depletion layer caused by the back junction. In the model, $I_{ds,\max}$ and and associated source–drain voltage [the onset (or "knee") voltage for velocity saturation at $I_{ds,\text{nax}}$] are needed. All load lines will be limited to lie between zero and the maximum current, and between the onset voltage and a maximum voltage (the voltage at which drain-to-gate breakdown occurs).

To obtain $I_{ds,\max}$, a positive V_{gs} is required to move the gate depletion layer up to the point where current is limited by the bare surface depletion layer. This is exactly the saturated source–drain current (I_{dsu}, where u stands for "ungated") obtained if the gate metal is removed. In cases where the bare surface potential, ϕ_s, equals the Schottky barrier built-in voltage, ϕ, $I_{ds,\max}$ is nearly equal to the saturated source–drain current, I_{dsf}, obtained with an open-circuited (floating) gate. I_{dsu} and I_{dsf} are always larger than I_{dss}. This is because, with V_{gs} equal to 0, the voltage (v_{35}) across the source end of the gate depletion layer is $-I_{dss}R_{sg}$.

The discussion above leads to the model's current-generator characteristics. Figure 10.8a shows the external characteristics I_{ds} versus V_{gs}. The figure is for the case of $\phi_s = \phi$. For modeling purposes, the curve is simplified, making it linear toward zero current at $-V_p$. Figure 10.8b shows the same current as a function of the internal node voltage difference, v_{35}, across the source end of the depletion layer. It is seen that the limits of linearity are $-V_{po}$ and zero. The voltage v_{35}, including dc bias, is required to fall within this range to avoid clipping of the output current waveform.

Two other node voltage differences (see Fig. 10.7), v_{43} and v_{46}, place limits on linearity. The voltage responsible for avalanche breakdown on the drain end of the gate is v_{43}, the voltage across C_{gd}. The avalanche breakdown voltage, $V_{dg,\text{br}}$, is geometry dependent. For the case where the depletion-layer edge is everywhere parallel to a planar junction, the "bulk" breakdown voltage across a Schottky barrier depletion layer as a function of carrier concentration is given by Kressel et al. [35, 36]. The geometric configuration at the drain edge of the gate is far from being a planar configuration. Because the gate depletion layer merges with the back junction depletion layer well below the avalanche breakdown voltage, the breakdown voltage can be higher than bulk value. A simple model [37] for determining $V_{dg,\text{br}}$ has been presented which is valid when the recessed area between the gate and drain extends a sufficient distance toward the drain. If $I_{ds,\max}$ is below 500 mA/mm, the product of $V_{dg,\text{br}}$ and $I_{ds,\max}$ is a constant. (This "constant" can vary due to geometric details at the gate edges.)

The limit on v_{43} discussed above places an upper bound on the drain voltage. A lower bound is supplied by the condition that the electric field in the velocity saturation zone must be above the saturation field, E_s. The value of E_s can be taken as 4000 V/cm. This is the field at which electrons in GaAs that have carrier concentrations typical of FETs reach their peak velocity. (Various physical effects prevent I_{ds} from decreasing at higher field where v_s is actually below its peak value.) For the present model, the length of the velocity saturation zone is taken to be equal to L_g. A lower bound on drain voltage is provided by the requirement that V_{43} not go below the value required for velocity saturation, $E_s L_g$. For example, for a gate length of 1 μm, the minimum voltage difference is 0.4 V.

In computing the maximum RF power for linear FET operation, a dc drain voltage ($V_{2\,\text{dc}}$) and a dc drain current ($I_{2\,\text{dc}}$) which allow the maximum possible RF voltage amplitudes within the limits discussed above must be determined. This can be done once a load impedance, Z_L, is specified. If this computation is repeated for many different loads (varied in some systematic way), the optimum load can be selected. This optimum could be based on maximum output power, maximum power added ($P_0 - P_i$), maximum power-added efficiency, ($P_0 - P_i$)/P_{dc}, or some other criterion, such as power gain, determined by the intended application. This procedure of varying the load impedance to find an optimum is called "load pull."

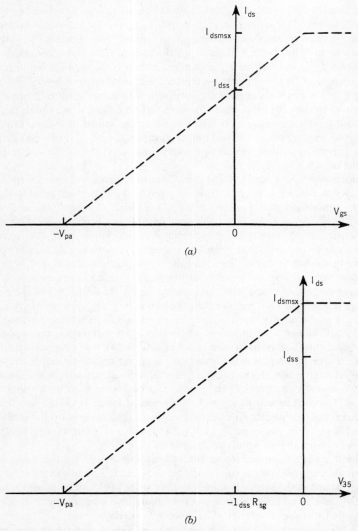

FIGURE 10.8   Examples of a transfer curve: (*a*) nominal I_{ds} versus V_{gs} for FETs; (*b*) I_{ds} versus V_{35}.

10.3.5   Mathematical Load-Pull Model

In this subsection we discuss the mathematical procedures used to compute power FET performance. The computation of an M-port impedance matrix, given the $N \times N$ **Y**-matrix representing the LEEC at a given signal frequency, is discussed in Appendix B. The reciprocal of this is an M-port admittance matrix (Y-matrix). Since the physical limits discussed in the preceding subsection involve nodes 3 through 6, and the RF performance is determined by the voltages and currents at nodes 1 and 2, $M = 6$ will be discussed here, as

$$\mathbf{Y}_6 \begin{bmatrix} v_1 \\ v_2 \\ v_3 \\ \vdots \\ v_6 \end{bmatrix} = \begin{bmatrix} i_1 \\ i_2 \\ 0 \\ \vdots \\ 0 \end{bmatrix} \tag{10.3}$$

This is the form required by the simultaneous equation subroutine SEQ discussed in Appendix B. Here a node current matrix with many columns is needed. All i_1 values will be 1, all i_2 values will be different, and all external currents for nodes 3 through 6 will be zero. Thus

$$\mathbf{V}_6 = \text{SEQ}\left[\mathbf{Y}_6 \begin{bmatrix} 1 & 1 & \cdots & 1 & \cdots \\ i_{21} & i_{22} & \cdots & i_{2n} & \cdots \\ 0 & 0 & \cdots & 0 & \cdots \\ 0 & 0 & \cdots & 0 & \cdots \\ 0 & 0 & \cdots & 0 & \cdots \\ 0 & 0 & \cdots & 0 & \cdots \end{bmatrix}\right] \tag{10.4}$$

The SEQ subroutine computes the complex node voltages in $\mathbf{V}_6$. Each column of $\mathbf{V}_6$ is the solution for the node currents in the same column of the node current matrix shown in Eq. (10.4). The load impedance corresponding to the nth i_2 value (i_{2n}) is $Z_n = -v_{2n}/i_{2n}$. The present power FET has a current generator that is linear within certain operating limits (see Fig. 10.8). Linearity implies that any column, n, of the node current matrix can be multiplied by a factor, f_n, and the new solution is given by multiplying the corresponding column in $\mathbf{V}_6$ by the same factor.

All i_2 values not requiring the load to generate power (which is physically impossible) fall within a circle in the complex i_2 plane shown in Fig. 10.9. All i_2 values on this circle result in zero output power. Normalizing parameters are computed [38] so that the positive output power circle in the normalized current plane (x-plane) has radius 1 and is centered at $1 + j0$. They are derived from the two-port $\mathbf{Z}$-matrix discussed in Appendix B. The normalized coordinate system (the x-plane) is shown in Fig. 10.9. To select the i_2 values for Eq. (10.4), begin by choosing points inside the radius 1 circle in the x-plane. For example, a hexagonal array of points with each point 0.1 from its six nearest neighbors has 331 points inside the unit circle. These 331 points are plotted inside the circle in the figure. The complex number represented by each point, x_n, in the x-plane can be converted to i_2 values as follows. First use two-port $\mathbf{Z}$-matrix elements to calculate a phase angle, ψ, in radians, as

$$\psi = \text{ANGLE}[(-Z_{12}Z_{11})^*] \tag{10.5}$$

where the function ANGLE gives the phase angle of a complex number, and the asterisk denotes complex conjugation. Next, a stability factor, K, is computed from

$$K = \frac{[1 - 2\,\text{Re}(Z_{11})\,\text{Re}(Z_{22})]\cos\psi}{\text{Re}(Z_{12}Z_{21})} \tag{10.6}$$

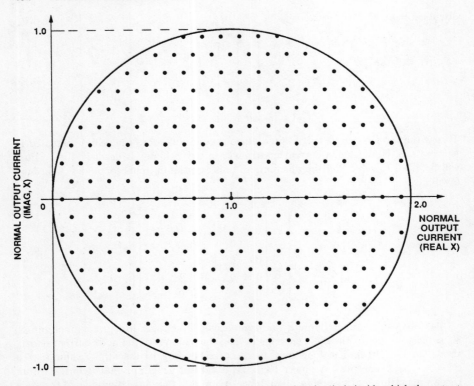

FIGURE 10.9   Complex plane of the output current, i_2, showing the circle inside which the output power is positive.

where Re denotes the real part of the variable. If K is less than 1 and negative, some load impedances will result in oscillation. The x_n values corresponding to these impedances can be eliminated by removing all that have a real part less than or equal to $-K$. The remaining x_n values are converted to i_{2n} values for use in Eq. (10.4), using the equation

$$i_{2n} = -Z_{21}(1 + x_n e^{j\psi})/[2 \text{ Re}(Z_{22})] \tag{10.7}$$

This set of i_2 values can now be used in Eq. (10.4) to compute the $6 \times N$-node voltage matrix, $\mathbf{V}_6$, discussed above. N is the total number of i_2 values under consideration.

10.3.6   Maximum Amplitudes for Linearity

Once the node-voltage matrix $\mathbf{V}_6$ has been computed, each column and each corresponding i_1 and i_2 value may be multiplied by a factor, f_n, to be determined (the corresponding i_1 becomes equal to f_n). For each column (n) of $\mathbf{V}_6$, the maximum f_n will be found such that none of the node voltages exceed the limits discussed in Section 10.3.4. Appropriate dc bias values will also be determined.

In the following discussion of how to find each of the f_n values, the subscript n will be omitted to simplify notation. The ac voltage amplitude for node i, for maximum linear power, will be denoted as v_i, from Eq. (10.4). For FETs, only two dc bias parameters can be chosen independently; the drain voltage ($V_{2\,dc}$) and the drain current ($I_{2\,dc}$). Once $I_{2\,dc}$ is selected, the dc gate voltage, $V_{1\,dc}$, is determined (see Fig. 10.8). To resolve ambiguity in the choice of dc bias parameters, they are chosen according to the following rules, which minimize the dc power required while maximizing f:

• *Rule 1.* $I_{2\,dc} - f|i_{gs}| = 0$, where i_{gs} is given by Eq. (10.1). This rule implies class A operation and gives priority to minimizing dc power. An alternative "rule" is $I_{2\,dc} + f|i_{gs}| = i_{dsu}$, which in some cases gives higher RF power but lower drain efficiency. (Drain efficiency is RF output power divided by dc power.)

• *Rule 2.* $V_{2\,dc}$ must be as low as possible while maximizing the ac amplitude and staying within the limits of linearity.

Once f has been determined, $I_{2\,dc}$ may be calculated using rule 1, as

$$I_{2\,dc} = f|i_{gs}| \tag{10.8}$$

All dc node voltages can now be calculated in terms of $I_{2\,dc}$. Figure 10.8 shows that g_m is given by I_{dsu}/V_{po}, and that the external transconductance, g_{mx}, is given by $i_{dsu}/(V_{po} + i_{dsu}R_{sg})$. Thus

$$V_{3\,dc} = I_{2\,dc}/g_{mx} - V_{po} \tag{10.9}$$

$$V_{1\,dc} = V_{3\,dc} \tag{10.10}$$

$$V_{6\,dc} = I_{2\,dc}R_{sg} \tag{10.11}$$

$$V_{5\,dc} = V_{6\,dc} \tag{10.12}$$

Rule 2 requires that $V_{4\,dc} - V_{6\,dc} = E_sL_g + f|v_{46}|$. If $V_{4\,dc}$ were any lower, $V_{4\,dc} - V_{6\,dc} + f|v_{46}|$ would fall below the voltage required for velocity saturation, and thus below the lower limit on linearity. That is,

$$V_{4\,dc} = V_{6\,dc} + E_sL_g + f|v_{46}| \tag{10.13}$$

and

$$V_{2\,dc} = V_{4\,dc} + I_{2\,dc}R_{gd} \tag{10.14}$$

For sufficiently low values of load impedance, Z_L, the amplitude is limited by the current limits. The ratio of $|i_2|$ to I_{dsu} cannot be used to find f because there are many ac current paths through the LEEC. Instead, it is noted that the current from the current generator can only vary when v_{53} is between V_{po} and the value at which the surface depletion layer limits further increase in current. This value of v_{53} is $\phi_s - \phi$, where ϕ_s and ϕ are the potentials discussed above in Section 10.3.4. The maximum f value that allows operations within these current generator voltage limits, f_c, is given by the following equation, which says that twice the amplitude limit of the ac potential difference, v_{35}, is given by the voltage range between the two limits on current generator linearity:

$$f_c = (V_{po} + \phi - \phi_s)/(2|v_{35}|) \tag{10.15}$$

For example, let one of the node voltage solutions to Eq. (10.4) have $v_3 = 6 + 4j$ and $v_5 = 2 + 1j$, also let $V_{po} = 3$ V and $\phi_s - \phi = 0.7$ V. The value of v_{35} is $4 + 3j$, which has a magnitude of 5 V. Inserting these values into Eq. (10.15) give an f_c factor of 0.3.

For higher values of Z_L, amplitude will be limited by the onset of avalanche break-down between nodes 3 and 5. The f value, f_v, which gives the maximum linear amplitude in the voltage-limited case, will be required to satisfy the following conditions simultaneously:

$$V_{4 \text{ dc}} - V_{3 \text{ dc}} + f_v|v_{43}| = V_{dg,\text{br}} \qquad (10.16)$$

$$V_{4 \text{ dc}} - V_{6 \text{ dc}} - f_v|v_{46}| = E_s L_g \qquad (10.17)$$

Note that Eq. (10.17) is derived from Eq. (10.13). Using Eq. (10.8), (10.9), (10.11), and (10.14), and subtracting Eq. (10.17) from Eq. (10.16), allows f_v to be calculated as

$$f_v = \frac{V_{dg,\text{br}} - V_{\text{po}} - E_s L_g}{|v_{46}| + |v_{43}| + |i_2|(R_{sg} - 1/g_{mx})} \qquad (10.18)$$

An f_c and f_v for every column in V_6 may be calculated using the i_2 from the same column in the node current matrix of Eq. (10.4). The correct f value for each column is the lower of the two.

The following equations give the RF and dc performance parameters for the FET model and for the case of maximum linear amplitude to the load. The subscript n is included in these equations to emphasize that each parameter is associated with the nth column of the node-voltage matrix $\mathbf{v}_6$ given by Eq. (10.4).

$$P_{0,n} = f_n^2 v_{2n}(-i_{2n}^*) \qquad (10.19)$$

$$P_{i,n} = f_n^2 v_{1n}(-i_{1n}^*) \qquad (10.20)$$

$$P_{\text{dc},n} = f_n^2 V_{2 \text{ dc}} I_{2 \text{ dc}} \qquad (10.21)$$

$$\eta_{pa,n} = (P_{0,n} - P_{i,n})/P_{\text{dc},n}. \qquad (10.22)$$

$$G_n(\text{dB}) = 10 \log_{10}(P_{0,n}/P_{i,n}) \qquad (10.23)$$

$$Z_{L,n} = v_{2n}/(-i_{2n}) \qquad (10.24)$$

$$Z_{in,n} = v_{1n}/i_{1n} \qquad (10.25)$$

In the equations above, recall that all i_{1n} have the value of 1, and all i_{2n} were selected as explained above. Once all n sets of f values and node voltages are computed, the optimum set can be selected according to any chosen criteria, for example, maximum P_0. The procedure above can be iterated for a finer mesh of x-plane points near the optimum point of the original set.

10.3.7 Input Signal Amplitude Resulting in Gain Compression

The derivation above results in performance parameters at the limits of linearity of the FET model. Power FETs can be overdriven to obtain more power at reduced (compressed) gain. Data sheets for commercial power FETs usually specify a maximum power output at 1-dB gain compression. This can be calculated from the present model exactly if $f_n = f_c$ [see Eq. (10.15)], and approximately otherwise. In the first case, the i_2 waveform from Fig. 10.8 is clipped at zero and $I_{ds,\text{max}}$ when linearity limits are exceeded. The dc bias point selection rules lead to symmetrical clipping of the sinusoidal i_2 waveform.

For the case of $f_n = f_{v,n}$ [see Eq. (10.18)], the discussion so far does not provide a quantitative determination of what happens when the voltage limits [Eqs. (10.16) and

(10.17)] are exceeded. The model assumes that the current waveform is simply clipped for larger amplitudes of v_{43}. The following qualitative description shows why this might be a reasonably good approximation. Near the maximum drain current and minimum drain voltage, g_m must decrease rapidly because the ohmic portion of the drain I–V characteristics are being entered. Near the minimum drain current and maximum drain voltage, avalanche breakdown current is added in a direction opposite that of the ac component of current from the current generator. This reduces, and could even reverse, the sign of g_m during this portion of the ac cycle. The net result, in most cases, is a small increase in dc drain current, while the amplitude of the ac waveform cannot greatly exceed $f_v|i_2|$. Furthermore, only a portion of any amplitude difference from $f_v|i_2|$ will appear in the signal-frequency Fourier component.

Consider an ac since wave of amplitude A greater than 1, clipped at 1 and -1. The first Fourier amplitude, A_1, is given by

$$A_1 = [A \arcsin(1/A) + \mathrm{sqrt}(1 - 1/A^2)](2/\pi) \tag{10.26}$$

Power is proportional to the square of the amplitude, so the gain compression in decibels is $20 \log_{10}(A/A_1)$. This amount of gain compression will result for the present model when the input power is increased by a factor of A^2 from the value given by (10.19). The resulting output power is A_1^2 times the value given by Eq. (10.20). Because the clipping is symmetric, no change in dc power results. To find the resulting power-added efficiency, substitute the new values for P_i and P_0 in Eq. (10.22). In the present example, the values of A and A_1 associated with 1 dB of gain compression are 1.25 and 1.12, respectively, and A_1^2 is 1.26. This means that for this model, the output power at 1-dB gain compression is 1.26 times the power calculated in Eq. (10.19). For a real FET, this factor will give the power at gain compression greater than 1 dB, because of the non-linearity of the current generator.

10.4 POWER FET DESIGNS

The design of FETs for any purpose usually requires careful optimization of the physical design parameters. Typically, changing a physical parameter to optimize a particular circuit element in the LEEC will adversely affect the other elements. Thus a model such as the one described above which can relate physical device parameters to power performance can be very useful in the design of power FETs. Several examples are discussed in this section.

The design parameters that have the greatest influence on power FET performance are the carrier concentration profile under the gate, gate length, channel height, and certain geometric details that affect breakdown voltage. High breakdown voltage requires a region on the drain side of the gate in which the channel height is not much greater than it is under the gate metal [37]. It is also important to minimize the parasitic resistances R_{sg} and R_g, as well as the values of the capacitors connecting LEEC nodes 3 and 4 and nodes 3 and 6. The parasitic resistance values need only be made small compared to R_i, since further reduction will have little effect on the input power required for a given output power (i.e., on gain). Similarly, the parasitic capacitors C_{34} and C_{36} need only be made small compared to C_{gs}. Figure 10.7 shows the physical location of the LEEC components of most interest. This area near the gate is also where geometric details are most critical.

10.4.1 Output Power and Efficiency

A computer model based on the discussion in Section 10.3.5 is now presented. The program takes its input from another program (subprogram), which specifies a particular FET model with certain parameters specifiable by the user. This subprogram provides a menu for user input (and default values for all parameters in the menu). It returns to the main program a 6×6 **Y**-matrix representing the FET at a particular frequency of interest, and calculates and returns the current and voltage limits. The main program then finds the load required to give the maximum linear power. The maximum power at 1 dB of gain compression, associated gain, and power-added efficiency are determined.

A description of the subprograms used to generate the data to be used in the main program is given in Appendix D. To generate the sample data for the performance curves of interest, two parameters were varied: channel height H and channel layer carrier concentration, N. W_g has been fixed at 1 mm, L_g at 1 μm, and the frequency at 10 GHz. Figure 10.10 shows the assumed variation of output resistance, R_0, versus channel height, H. The model's results can be scaled as follows to obtain the same power per millimeter, and the same gain and efficiency at other frequencies. The geometric parameters (L_g, W_g, and H) are multiplied by a factor of $10/f$, where f is the desired operating frequency in gigahertz. N is multiplied by $0.01f^2$. This choice of N gives the

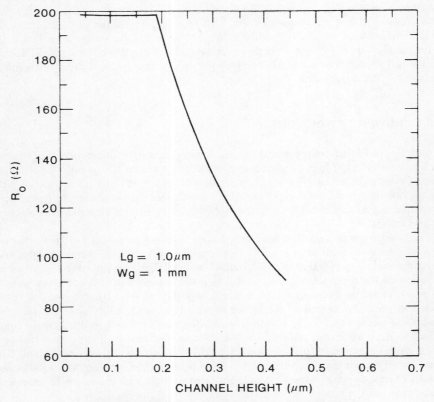

FIGURE 10.10 Output resistance versus channel height.

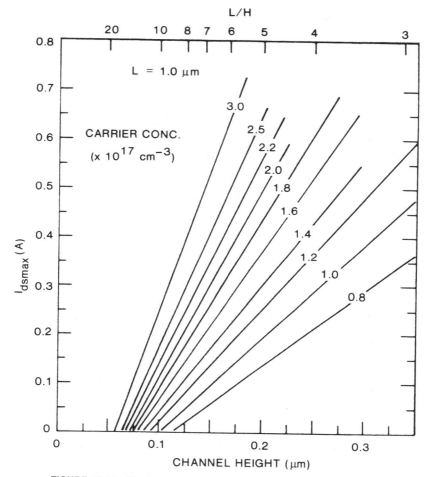

FIGURE 10.11 Maximum drain–source current versus channel height.

same current per millimeter gate width, and the same V_{po}, and compensates for the smaller H needed to maintain the L_g/H ratio.

Even though the H was the parameter varied to generate the performance curves, it is more useful to plot performance parameters versus the more easily measured parameter, $I_{ds,max}$. This parameter can be determined by increasing the forward V_{gs} until g_m is zero, and then measuring the value of I_{ds} at the knee of the I_{ds} versus V_{ds} curve. Figure 10.11 relates $I_{ds,max}$ to H. The ratio of L_g/H is shown at the top of the figure. The $V_{dg,br}$ used at each H value is also shown.

Figure 10.12 shows that the value of $I_{ds,max}$ at which the power is maximized is nearly independent of N. However, the maximum power increases with N. As N increases, the V_{po} associated with a given value of $I_{ds,max}$ decreases (Fig. 10.13). Since the maximum drain voltage, $V_{ds,max}$, is $V_{dg,br}$ minus V_{po}, a greater amplitude of RF drain voltage is allowed, and hence more power is obtained.

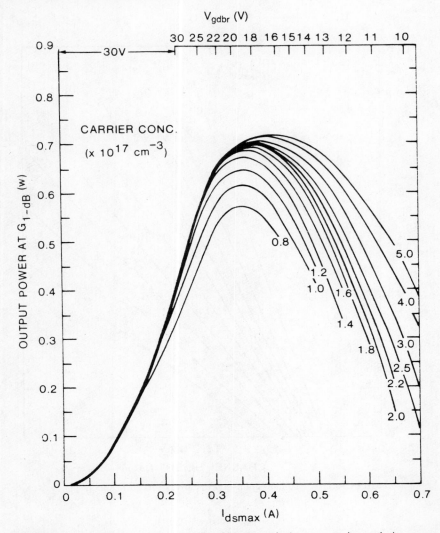

FIGURE 10.12   RF power output (at 1-dB gain compression) versus maximum drain current.

Figure 10.14 shows power-added efficiency versus $I_{ds,max}$. These are the values obtained when the load is chosen for maximum power. At low values of $I_{ds,max}$, upper limits have been arbitrarily assigned to R_0 and $V_{dg,br}$, resulting in the plateau evident in the curves. At higher values of $I_{ds,max}$, the maximum power-added efficiency (like the maximum power) increases with increasing N. Higher N allows a higher RF voltage amplitude to be obtained for a given $I_{ds,max}$. The knee voltage of the $I_{ds,max}$ versus V_{ds} does not change. The net result is higher drain efficiency, because the ratio of RF v_{ds} amplitude to the optimum dc drain voltage is increased. In addition, since the H is lower, the g_m is increased. As shown in Fig. 10.15, this is associated with some increase in the power gain, further increasing power-added efficiency.

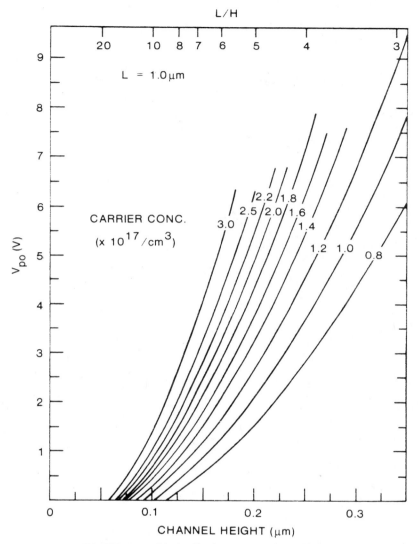

FIGURE 10.13 Pinch-off voltage versus channel height.

The results discussed are for the optimum class A dc biases. These allow the maximum RF voltage amplitude on the gate before the output current waveform is clipped. Class B operation is sometimes of interest. In class B, the dc gate bias is chosen so that there is zero drain current when no RF signal is present. This results in a half-wave rectified drain current waveform. If the load is configured to provide a short circuit at all the even harmonics of the signal frequency, the power to the load still increases linearly with input power. The drain efficiency can be greatly increased because the dc drain current is significantly less than the class A value. There is also some reduction in power

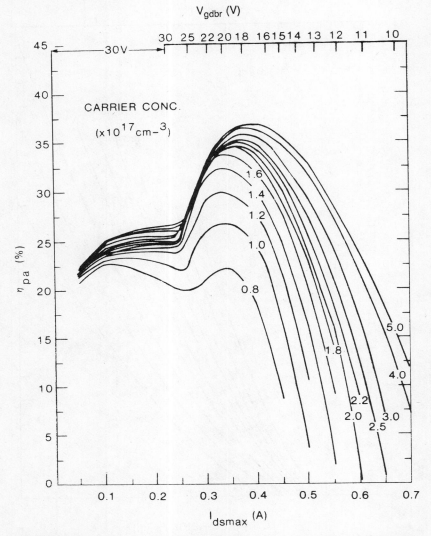

FIGURE 10.14   Associated RF power-added efficiency versus maximum drain current.

and gain. When the FET has a high gain in class A, and the pinch-off voltage for maximum power is low, these penalties are often acceptable in a situation where efficiency is important. The optimum operating condition is often class AB, with the gate dc-biased somewhere between the class B value and the optimum class A value.

Class B modeling is quite similar to the class A modeling already described. Among the necessary changes are modification of the dc bias selection rule 1, discussed above (Section 10.3.6). Rule 1 for class B is $I_{2\,dc} = 0$. This leads, in a straightforward way, to changes in the definitions of g_m and C_{gs}, and to some of eqs. (10.8) through (10.18).

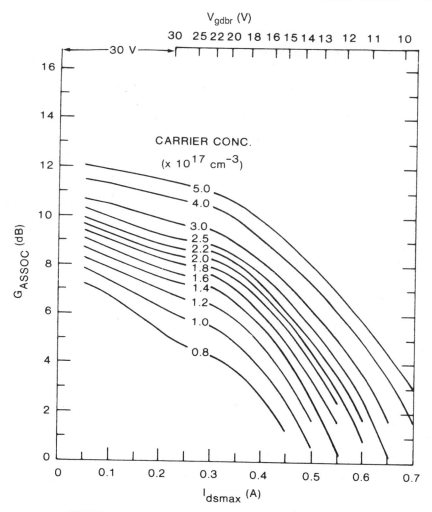

FIGURE 10.15 Associated gain versus maximum drain current.

10.5 MATCHING CIRCUIT DESIGNS

The MESFET structure must be determined accurately in order to optimize its performance in an embedded amplifier circuit. The CAD programs described in previous sections can be used to generate the dc $I–V$ characteristics and equivalent circuits of a MESFET as a function of bias. Figure 10.16a shows an example of the predicted and measured drain characteristics of a power FET device in a common-source configuration. A comparison between the predicted and measured S-parameters of a 1.5-mm-gate-width power FET is shown in Fig. 10.16b. The load-pull computer simulator allows the designer to determine the optimal load for the device to be matched, depending on

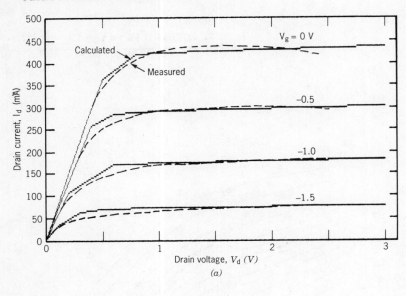

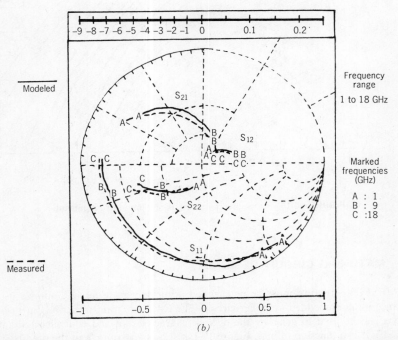

FIGURE 10.16 Comparison of the predicted and measured characteristics of a power FET: (*a*) dc results; (*b*) S-parameters.

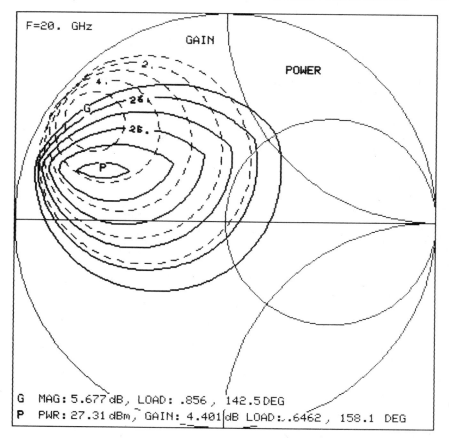

F=20. GHz

GAIN

POWER

G MAG: 5.677 dB, LOAD: .856 , 142.5 DEG
P PWR: 27.31 dBm, GAIN: 4.401 dB LOAD: .6462 , 158.1 DEG

FIGURE 10.17   Example of the theoretical load-pull plot of a power FET at 20 GHz.

the design requirements of power-added efficiency, output power, or power gain. An example of the theoretical load-pull plot on a Smith chart for a FET at 20 GHz is shown in Fig. 10.17. These customized CAD programs, together with the SUPERCOMPACT and Touchstone commercial frequency-domain CAD programs, can be used to provide a complete sensitivity and parametric study of the performance of the design circuit as a function of device parameters such as gate length or width, device geometry, and doping profile.

The circuit design begins with the output power and power-added efficiency requirements, which determine the total gate width for the unit cell of the FET device. Assuming a value for the figure of merit of power per unit gate width, the circuit design is then optimized with respect to the FET structure and circuit approaches to provide the best trade-off among output power, linear gain, bandwidth, and yield of usable devices and/or circuits. Figure 10.18 is a flowchart describing the sequence for designing an amplifier circuit using either HMIC or MMIC technology. This design approach allows a parametric study of the circuit performance as a function of device parameter variation. Thus the power device can be optimized to achieve a certain circuit performance to

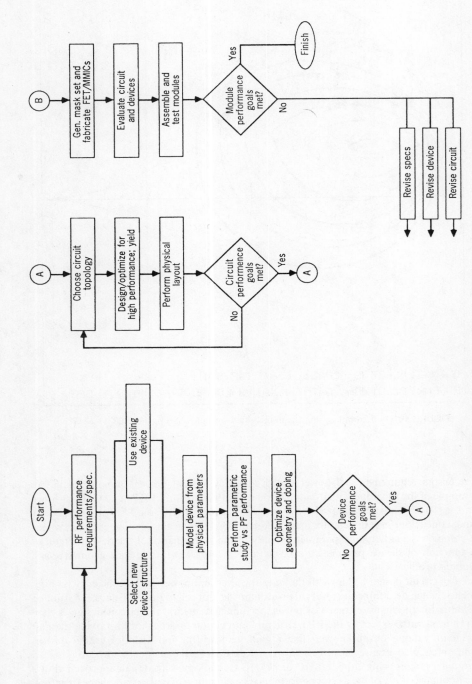

FIGURE 10.18 Flowchart of power amplifier development.

satisfy a particular system requirement. A number of HMIC and MMIC power amplifiers have been designed successfully, with excellent performance, based on a single design cycle [9, 13, 14, 19, 20].

The general approach to designing the power amplifier circuit is as follows. The impedance-matching circuit for an amplifier is to have a complex conjugate match at the input to the FET, whose drain-to-source port is terminated with an optimal load for maximum output power or power-added efficiency. The load circuit is designed using the method described in Section 10.3.5. The amplifier circuit (in particular the input) is then optimized for minimum input return loss and flat gain response across the required frequency band by using microwave circuit analysis programs such as SUPERCOM-PACT or Touchstone. As the result of possible changes in device parameters and circuit dimensions due to fabrication process variation, design uncertainty, and temperature changes, an RF performance sensitivity analysis is usually conducted by varying the device parameters. A number of network topologies used in circuit design can be chosen based on microwave network synthesis and analysis principles [39–43].

10.5.1 Lumped Elements

Impedance-matching circuits for applications at the lower microwave frequencies are usually designed with lumped elements. The use of lumped elements for the realization of HMICs or MMICs has been described extensively in the literature [44–47]. These resistor–inductor–capacitor ($R–L–C$) elements exhibit highly predictable RF behavior in the frequency range below 20 GHz. They also have a size advantage over the larger distributed elements with comparable RF circuit loss. Presently, high-Q (quality factor) lumped capacitors are physically small enough to be well modeled as a lumped circuit elements up to lower K_a-band. Q factors from 50 to 120 at X-band are readily realizable, even with GaAs substrates, and the values can improve to 300 with lower RF-loss substrate material such as fused silica and alumina. Other advantages of matching networks that employ lumped elements are the ability to match to low impedance levels and circuit compactness. Lumped or semilumped circuit elements are also important in monolithic applications because of their real-estate-saving attributes. Inductors are implemented with single or multiturn thin transmission line [48, 49], while capacitors are fabricated with metal–insulator–metal (MIM) or interdigited [50, 51] structures. Figure 10.19 shows a 4-GHz MMIC power amplifier circuit using a lumped-element design.

As frequency increases, the physical dimensions of the matching circuit elements decrease. Since the elements are normally designed to be less than a quarter-wavelength

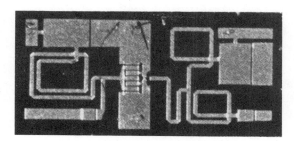

FIGURE 10.19 Power amplifier using a lumped-element design.

in dimension to avoid any spurious resonance in the amplifier response, the use of distributed-element matching is more favorable. In addition, circuit reproducibility is better in networks that employ mostly distributed elements.

10.5.2 Distributed Elements

Distributed-element matching circuits are commonly realized in microstrip-line structures [52]. Slotline [52] is an alternative transmission structure that can be included in microstrip circuits by etching the slotline circuit in the ground plane of the substrate. This type of hybrid combination allows flexibility in the design of distributed elements in microwave circuits [53]. New types of circuits, such as a hybrid branchline directional coupler, can be realized. Also, some of the circuit elements, which are not readily realizable in microstrip configuration, can be incorporated in the slotline section of the circuit. These elements could be short circuits, high-impedance lines, or series stubs. Recently, coplaner waveguide (CPW) [32] structures have also been employed in a number of amplifiers or subsystem designs. CPW provides a convenient means of grounding shunt elements; however, CPW circuits suffer higher RF loss than does microstrip line, especially at millimeter wave frequencies. Finline [54] and suspended-substrate structures [55] have also been used in V- and W-band (50 to 100 GHz) amplifier circuits because of their relatively low RF-loss characteristics.

In the amplifier circuit implementation, narrow transmission lines are avoided wherever possible. Instead, radial and low-impedance shunt elements [56] can be used to minimize RF circuit loss. In general, transmission lines of characteristic impedance

FIGURE 10.20   20-GHz HMIC 1-W power amplifier.

between 35 and 50 Ω provide minimum RF loss. This design consideration is especially important in millimeter wave applications because the available gain of the FET device decreases as the frequency of operation increases. Series input and output capacitors are employed for dc blocking. This allows the direct cascade of several amplifier circuits to achieve a multistage amplifier with useful power gain for system applications. The MIM capacitor is also modeled as a transmission element in the CAD analysis, to account for any parasitic effect caused by discontinuities. Bias networks consisting of a low-pass network design are commonly used with added resistive elements to stabilize the amplifier. Such a resistor should be located in the input (whenever possible) and in shunt with the matching element, since any resistive element in the output circuit affects the output power. RF grounding with low parasitic inductance can be achieved by using via-holes in both HMICs and MMICs. Examples of power amplifier circuits using the distributed-elements approach [11, 17, 23] are shown in Fig. 10.20 to 10.22.

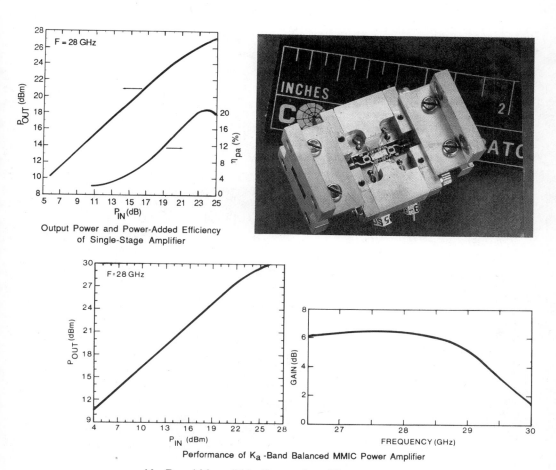

Output Power and Power-Added Efficiency
of Single-Stage Amplifier

Performance of K$_a$-Band Balanced MMIC Power Amplifier

K$_a$-Band Monolithic Power Amplifiers

FIGURE 10.21   K$_a$-band monolithic power amplifiers.

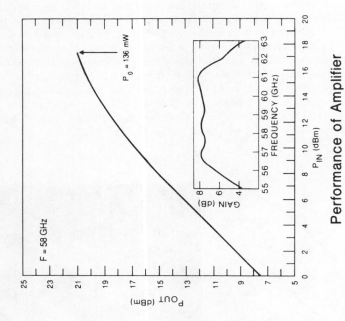

Performance of Amplifier

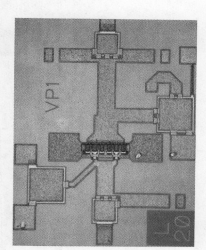

FIGURE 10.22 60-GHz MMIC power amplifier.

10.5.3 Power-Combining Techniques

The output power capability of a FET is directly proportional to the size (total gate width) of the device. However, the size of a FET chip or MMIC is limited by thermal impedance, matching circuit bandwidth, and manufacturing yield considerations. The spacings among the heat-generating elements and drain–gate–source regions must be of sufficient dimension to achieve adequate thermal dissipation to the heat sink. This will ensure a FET junction temperature at which the required mean-time-to-failure (MTTF) is obtainable. The wider the device, the lower the matching impedance of the FET and the more difficult it will be to achieve broadband matching. Also, the larger the chip size, the lower will be the dc and RF yield of the processed devices or MMICs per wafer. In addition, the mechanical handling of the thin chips (normally 50- and 100-μm-thick for power FETs and MMICs, respectively) will be more difficult. Therefore, power combining circuits are used for IM power FET modules and/or MMICs to achieve amplifiers with high power beyond the output level attainable with an individual module.

Various techniques in power dividing and combining have been reported in the literature [57–59]. Figure 10.23 summarizes the various approaches. A single-combining approach can be appropriately selected and applied individually to devices. Alternatively, a mixture of two or three techniques can be used in a multiple-combining scheme, depending on the power-level requirement, the type of transmission-line medium, whether a planar circuit or waveguide structure is used, the size limitation, and the input/output impedance match requirement of the amplifier.

Chip-level combining can be achieved by designing the complete matching circuits of two individual FET cells using elements of characteristic impedance values that are half of that used for the input/output impedance. Thus direct combining of the two

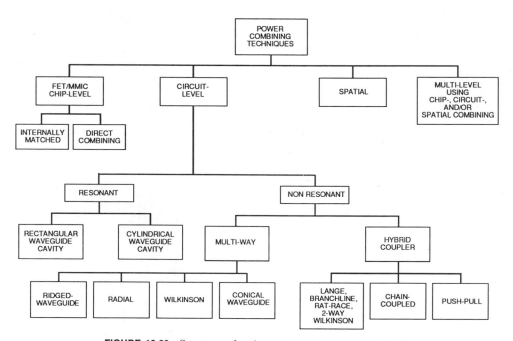

FIGURE 10.23 Summary of various power combining techniques.

circuits can be accomplished. Although the circuit size of such an approach can be reduced using this approach (especially in an MMIC implementation), the need to use
higher impedance-matching elements implies higher RF loss. Also, resistive elements
may have to be added to improve the electrical isolation between the amplifier units.

Another approach to chip-level combining is to perform a partial matching of the
FET devices using small chip capacitors and MIC matching circuits to 50-Ω impedance.
Stabilization and bias networks are added externally. Most commercially available high-
power devices are supplied in this configuration.

The different topologies used for the common types of planar 3-dB couplers are the
Lange coupler, Wilkinson combiner, branch-line coupler, and rat-rice coupler. Figure
10.24 shows examples of two of these types of combiner/dividers. These topologies are
readily implemented on hybrid microstrip and GaAs MMICs. Because of its requirement for small spacing between the thin coupling fingers and, thus higher RF loss at
high frequencies, the Lange coupler is generally restricted to applications below K_a-band.
These 3-dB couplers share the characteristics of wide bandwidth potential and high
port-to-port isolation, thus minimizing interactions among devices. They can also be
cascaded to form a multiway combiner. However, the performance of these couplers
will become limited as the number of combined units increases. In addition to the RF
loss and phase problem, the multiway combiner can become relatively large compared
to other approaches, especially at millimeter wave frequencies. The Wilkinson combiner
is a simpler network with relatively lower RF loss; while the other circuits, such as the
Lange and branchline coupler, can provide improved input/output VSWR to amplifier
modules of poor I/O match that are to be combined.

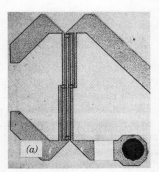

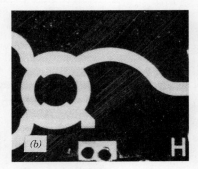

FIGURE 10.24 Examples of divider/combiner circuits:
(a) Lange coupler with air bridges; (b) Branchline
couplers.

N-way multiport combining using radial lines has been demonstrated with excellent results [12, 60, 61]. The configuration is usually implemented with an isolation resistor connected between two neighboring lines. Eight-way and 30-way power combiners have been reported to provide FET amplifiers with an output power of 8.2 W at 20 GHz [12] and 26 W at 11 GHz [60], respectively. Multiport traveling-wave combiners have also been demonstrated in HMICs on alumina substrates [62] and in MMICs on GaAs substrates [63]. Resistive elements are generally required between each combining/dividing lines to achieve proper isolation. Pulsed output of 16 W has been obtained with this power-combining approach at X-band [9].

Another divider/combiner configuration is the resonant-cavity type. Broadbanding techniques using waveguide TM_{omo}-mode double cavities have been demonstrated for an eight-way divider/combiner with insertion loss of 0.2 dB and a bandwidth of 600 MHz at 6 GHz. Output power as high as 80 W has been obtained by combining GaAs FET amplifier modules [6].

Spatial combiners utilize the proper phase relationship of many radiating elements to combine power in space. The combined power can either be collected by another antenna or simply reflected off a target in a radar system. This type of structure has been demonstrated in an active array at K_a-band [64] and more recently analyzed [65]. The radiating elements can be implemented by using solid-state sources amplified by FET power amplifiers. The spatial combiner is better suited for millimeter wave applications.

10.5.4 High-Voltage FET Power Amplifiers

One circuit design approach to achieving the high bias voltage applied to FETs involves connecting the dc bias of a number of FETs in series (drain to source), while combining the RF power of each FET in parallel to achieve high output power [66]. For certain system applications such as satellite communications and phased-array radars, this device/circuit configuration can allow dc bias to the drain circuits at a much higher voltage (e.g., 40 V) than the device normally would allow, as limited by avalanche breakdown at the schottky-to-semiconductor junction. The potential prime power saved by using FET amplifiers with higher bias voltages can be significant in these applications.

Figure 10.25 shows the high-voltage FET (HVFET) topology for four cells. The gate bias design of the individual FET cell can be implemented by using a resistive network and a self-bias scheme for the last cell in the dc bias series. Thus a single high-voltage bias will be the only supply required.

In the amplifier design, two technical issues must be considered. First, because the FET sources are RF-grounded through capacitors, a large capacitive reactance is presented to the FET source below the frequency of operation, and at very low frequencies the capacitor appears electrically open: The stability of the amplifier must be carefully examined to prevent oscillation. Therefore, the capacitance–inductance network has to be included in the device design and optimization. Second, because the FETs are dc-biased in series, any variation in the FET characteristics will produce an imbalance in dc voltage distribution, resulting in premature gate–drain breakdown, inefficient RF power combining, or both. Because they enjoy good uniformity in the electrical characteristics of adjacent FETs, MMICs are highly suitable for the high-voltage circuit configuration approach.

The total gate width of the FET in the high-voltage amplifier is usually chosen to be the same as that of a low-voltage amplifier of comparable output power. However, because the gate width is divided into smaller units, several advantages are realized in amplifier design. The large gate-width devices normally used to achieve higher power

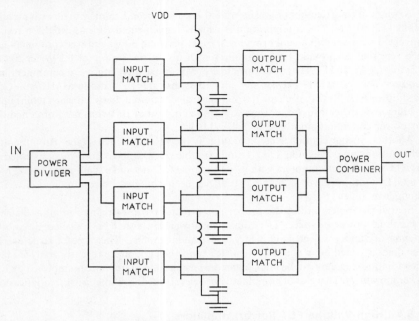

FIGURE 10.25   HVFET amplifier topology.

are difficult to match to the appropriate impedances because of their relatively low impedances. Higher matching circuit loss also results. The smaller FET cells in HVFETs can be more efficiently matched for a given bandwidth, and because total gate width is the same as in the conventional FET, device process yields are also expected to be the same. Furthermore, because heat dissipation is distributed across several HVFET cells, lower channel temperatures and extended lifetime are also expected. The HVFET amplifier circuits can be realized by using either the lumped-element or distributed-element approach.

Two-cell microwave integrated-circuit amplifiers using the HVFET approach with discrete power FETs at C- and X-band have previously been reported [66]. More recently, HVFET designs and measured microwave results for two-cell and four-cell MMIC amplifiers were demonstrated [67]. Figure 10.26 shows a photograph of the four-cell MMICs and their performance. Bias voltage of 40 V, output power of 1 W, and power-added efficiency of 20% were achieved in the 11-GHz satellite communications band.

The HVFET amplifiers will lead to improved dc-to-RF conversion efficiency, as well as reduction in prime power for both satellite and phased-array applications. As an example, assuming a spacecraft with 24 8.5-W transponders and amplifiers with 30% power-added efficiency, 170 W of dc power may be saved for a regulated satellite bus, instead of using electronic power conditioners (EPCs) to convert the higher bus voltage to that of the power FET. The nominal efficiency of the EPC is 80%. In phased-array applications with a large number of active devices, the I^2R power loss in the dc distribution network is quite high. Increasing the dc power supply voltage from 8 to 32 V, for example, will reduce the dc distribution loss by a factor of 16. For a large phased-array system consisting of a thousand 10-W elements with a distribution cable having a resistance of only 0.001 Ω, power distribution losses will drop from more than 15% to less than 1% of the dc power supplied.

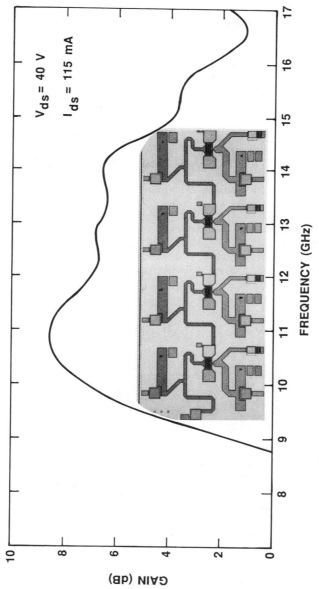

FIGURE 10.26 MMIC HVFET power amplifier: circuit and performance.

10.5.5 Dual-Gate Power Amplifiers

Another power circuit approach using a dual-gate MESFET (DGMF) is now presented. In phased-array systems for communications satellite and radar applications, it may be desirable to change output power with minimum variation in insertion phase or power-added efficiency. The GaAs DGMF is well suited for such applications. This type of device has two parallel gate fingers between each pair of source and drain pads; one of the gates can be used as a control terminal. The DGMFs have been employed in a number of small-signal control circuits for gain control in the amplifier chain, and as modulating and switching elements. A small-signal gain control monolithic amplifier up to the 28-GHz band has been demonstrated [68]. More recently, DGMF circuits have been designed for power applications, resulting in a dual-gate power MMIC module with 3-W output power at 18 GHz [69]. Another MMIC using a DGMF for variable-power application in the 33-GHz band has also been developed [70].

The design of a DGMF power amplifier can use the same device/circuit approach presented in preceding sections. The DGMF structure shown in Fig. 10.27a is similar to that of the parallel single-gate MESFET described in Fig. 10.4 and Ref. 17. It consists of twelve 0.3-μm recessed dual-gate fingers, providing a total gate width of 800 μm. The device/circuit modeling program has been extended to obtain the dc characteristics and an equivalent-circuit model at different gate biases. Thus the dc and small-signal characteristics of the DGMF modeled as two single-gate devices in a cascode configuration (Fig. 10.27b) can be studied at various biases of the two gates. To perform the power and associated gain analysis of the device, the existing load-pull subroutine [11] has been modified for the DGMF. The contours of output impedance for constant power and gain at different frequencies for a given device structure can be plotted and evaluated. Figure 10.27c shows an example of the calculated load impedance, optimized for maximum linear output power at 33 GHz.

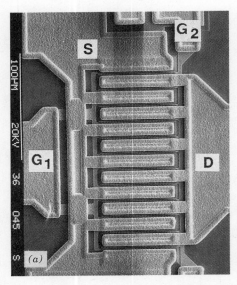

FIGURE 10.27 Dual-gate power MESFET: (a) MESFET structure; (b) equivalent circuit of a power DGMF (800-μm gate width); (b) contours of output impedance for constant output power and gain.

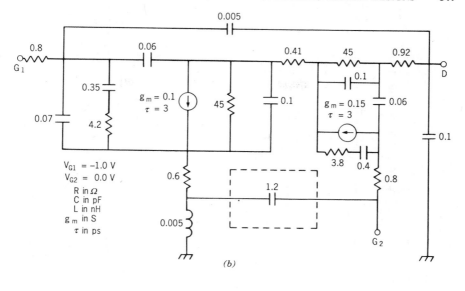

(b)

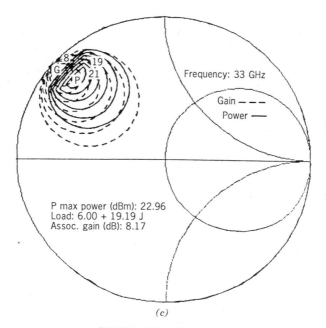

P max power (dBm): 22.96
Load: 6.00 + 19.19 J
Assoc. gain (dB): 8.17

Frequency: 33 GHz

Gain — — —
Power ——

(c)

FIGURE 10.27 (*Continued*)

In the power amplifier design, the input impedance is conjugately matched to the input of the device, while the output was terminated with the optimal load impedance. The second gate is RF-grounded through two MIM capacitors whose values are selected to ensure the stability of the circuit and to minimize phase variation as amplifier gain is changed. In the MMIC implementation, these capacitors can be grounded through

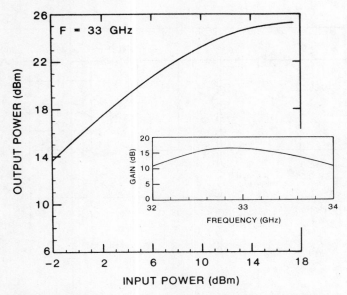

FIGURE 10.28   Performance of a DGMF MMIC power amplifier.

the same via-holes used for the DGMF source. Integrated bias networks (low-pass-filter type) and dc-blocking capacitors should be incorporated in the design to allow direct cascading of the individual amplifier stages. Figure 10.28 shows the dynamic range and power performance of a variable-power DGMF circuit that can be achieved in an MMIC with epitaxial/electron-beam technology at K_a-band [70]. The single-stage amplifier demonstrated a linear gain of 9.1 dB and a dynamic range of 30 dB. A two-stage balanced amplifier provided a linear gain of 16.5 dB and maximum output power of 25.3 dBm at 33 GHz. Minimal variations in the DGMF as a function of second gate control were also demonstrated.

10.6   FABRICATION PROCESSES

10.6.1   Materials

The quality of the gallium arsenide material will strongly influence the resultant performance of discrete power FETs and MMICs, as well as device uniformity and yield. Active layers for FET device and circuit development can be obtained using a number of different fabrication techniques: ion implantation, or epitaxial growth using vapor-phase epitaxy (VPE) or molecular beam epitaxy (MBE).

Ion implantation onto semi-insulating substrates places limitations on doping profile shapes and abruptness in the transition between different doping concentrations. The resulting material properties depend on the quality and uniformity of the starting material. Halide-transport VPE process can yield the materials for high-power applications; however, this technique does not routinely provide the desired thickness and doping uniformity across 3-in. wafers. MBE is known for its ability to achieve a high degree of growth control, abrupt interfaces, and layers with excellent electrical properties, but at the expense of the system throughput. In general, MBE material provides a narrower transition region between the different layers and a more uniform doping density across the wafer than does VPE material. Furthermore, for critical applications such as higher power, ultra-low-noise devices and high-frequency circuits, GaAs epitaxial layers generally provide a higher level of performance than ion-implanted layers can provide.

A typical MBE process for power MESFET application is as follows. The device or MMICs can be fabricated from epitaxial GaAs n^+/n/buffer layers deposited by MBE fabrication equipment such as a Riber 2300P or Varian Gen II machine [21, 71]. The layers are deposited onto undoped, semi-insulating, liquid-encapsulated Czochralski (LEC) grown substrate, which is oriented 2° off the (100) plane toward the nearest (110) plane. Epitaxial deposition using elemental arsenic and gallium sources can be performed at 580°C, with an arsenic-to-gallium flux ratio of 20. Three layers are deposited sequentially on the substrate: an unintentionally doped p^- buffer layer, an n active layer doped with silicon to 1 to 5×10^{17} cm^{-3} (depending on the frequency of operation for which the device is designed), and an n^+-layer. The MBE-grown active/buffer layers provide both highly uniform doping across the wafer and improved active channel definition with reduced leakage current.

10.6.2 Lithography

The power MESFET or MMIC can be fabricated using optical lithography to define all mask levels. The gate can be defined by either optical lithography (for gate lengths longer than 0.5 μm) or electron-beam (e-beam) lithography for direct-writing of sub-half-micrometer gates. A mesa or planar (based on selective ion-implantation) process similar to those reported previously ([16, 23, 72, 73] with sequence variation) can be used to fabricate the power device or monolithic amplifiers.

Mesa Process. Figure 10.29 is an example of the major processing steps for a mesa process [23] that has provided excellent results in power FETs and MMICs operating at frequencies up to 60 GHz. The fabrication process begins with mesa isolation for the active devices and resistors. A Au/Ge/Ni/Ag/Au alloy using a furnace heating or rapid-thermal annealing (RTA) technique can be used for the ohmic contact. MIM capacitor base metal can be defined by lifting off Ti/Pt/Au, and the Si$_3$N$_4$ dielectric for the MIM capacitor (Fig. 10.30) can be deposited using the plasma-enhanced chemical vapor deposition (PECVD) technique to a thickness of 2500 to 5000 Å. An evaporated Ti/Pt/Au metallization can be employed for the gates. Figure 10.31 shows the detailed gate structure of a power MESFET. The air bridge structures and transmission lines can be fabricated using two mask levels: plating-via and plating. The top plate metallization can be Ti/Au at a thickness of 2 μm, depending on the frequency of operation and ease of fabrication. To minimize RF circuit loss, the metal thickness for the transmission line should be greater than twice the skin depth.

Planar Process. Another approach to the fabrication of power MESFETs and MMICs uses the refractory metal, multifunctional self-aligned gate (MSAG) process [72, 73]. This planar approach has demonstrated excellent power performance at the 10-W level for MMICs operating at C-band (5.5 GHz). The application has thus far been limited to the frequency range below 18 GHz.

The process flow diagram is shown in Fig. 10.32. The fully planar process employs selective ion implantation into undoped LEC substrates. The substrates are first passivated with 85 nm of plasma-deposited oxy-nitride (SiON) followed by ^{29}Si implantation for the FET active channel. The other process steps include Au/Ge/Ni metallization for

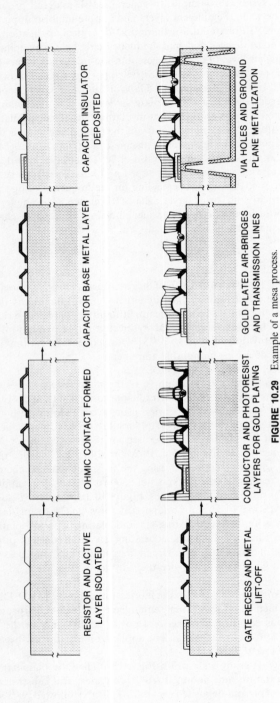

RESISTOR AND ACTIVE
LAYER ISOLATED

OHMIC CONTACT FORMED

CAPACITOR BASE METAL LAYER

CAPACITOR INSULATOR
DEPOSITED

GATE RECESS AND METAL
LIFT-OFF

CONDUCTOR AND PHOTORESIST
LAYERS FOR GOLD PLATING

GOLD PLATED AIR-BRIDGES
AND TRANSMISSION LINES

VIA HOLES AND GROUND
PLANE METALIZATION

FIGURE 10.29 Example of a mesa process.

FIGURE 10.30 Example of a MIM capacitor.

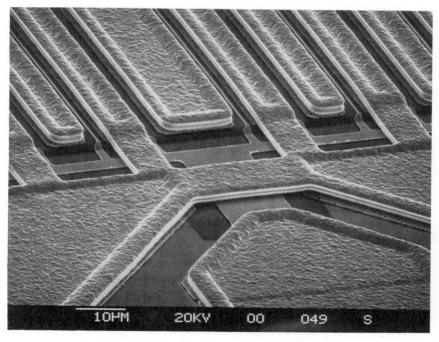

FIGURE 10.31 Detailed gate structure of a power MESFET.

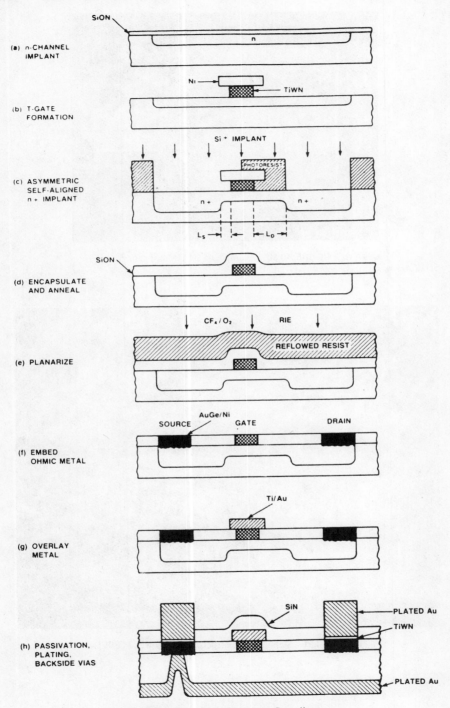

FIGURE 10.32 MSAG process flow diagram.

the ohmic contacts, 0.5-μm TiWN Schottky barrier "T-gates" and ion-implanted resistors. The 0.5-μm TiWN gates are covered by a 0.5-μm Ti/Au overlay by evaporation and lift-off after planarization. Silicon nitride is used for both capacitors and passivation. The air bridges and microstrip lines are gold plated to a thickness of 5 μm.

10.6.3   Back-Side Processes

After front-side processing is completed, the wafer is thinned to an appropriate thickness. In general, for discrete power devices, the substrate is lapped down to about 35 μm and then metallized to similar thickness. The major reason for the process is to achieve the lowest thermal impedance for the device, resulting in maximum output power. Because the width of the microstrip line circuit in MMICs is proportional to the substrate thickness, the thinner transmission line (corresponding to a thinner substrate layer) will result in higher microwave circuit loss; therefore, the nominal substrate thickness of a completed MMIC is usually not reduced to less than 75 to 100 μm. For high-frequency FETs or circuits, via-hole structures in conjunction with capacitors of appropriate value are used to provide low-inductance grounding for the source of the FET or shunt elements that require an RF ground. Through-substrate via-holes can be attained by infrared alignment and spray or dry-etch techniques. Via-hole size is nominally 50 × 80 μm using the wet-etch process and is smaller with the dry plasma-etch technique.

After the back-side processes are complete, the wafer can be mounted front side up on a stretched tape, and RF on-wafer testing of discrete devices or MMICs can be performed to identify the working units before they are finally diced. Each device or circuit can be serialized (using dielectric or metallic patterns) during an earlier process step to facilitate the separation of good dies from those that have failed.

10.7   MEASUREMENT TECHNIQUES

10.7.1   DC Characterization

The characterization of power devices commences with on-wafer dc measurements. These include the saturated drain current (I_{dss}), pinch-off voltage (V_{po}), reverse breakdown voltage (V_{bk}), forward voltage (V_f), and transconductance (g_m). Other resistance measurements allow determination of the parasitic resistances for drain, gate, and source terminals.

Recently, a device characterization technique using pulse transient measurement [74, 75] has been demonstrated to provide useful information for predicting the existence of surface states in the channel region of power FETs. Thus the RF performance of the device in terms of output power capabilities can be estimated prior to actual microwave measurements.

10.7.2   Microwave Characterization

The two common methods of introducing a microwave or millimeter wave signal into the power FET device or circuits are through coaxial launchers to a microstrip-line or CPW structure, or via waveguide-to-microstrip transitions. These components are used with the commercially available automated network analyzer to perform small-signal characterization of the power device or circuits. Other customized load-pull measurement setups and automated systems have been established at various laboratories and companies to evaluate large-signal characteristics. New optical techniques that can provide very broadband characterization and time-domain waveform analysis of MMICs have recently been demonstrated.

Coaxial Launchers. In the frequency range up to K_u-band, power FET devices and circuits can be evaluated by using coaxial launchers such as the Wiltron k, or APC-3.5. In general, these devices provide good microwave transitions from the measurement system to microstrip-line or CPW structures. More recently, the APC-2.4 mm connectors have demonstrated return loss equal to or better than 20 dB up to 40 GHz. A coaxial launcher to microstrip line using this type of connector has yet to be proven to perform equally well at millimeter wave frequencies.

Waveguide-to-Microstrip Transitions. At frequencies above 20 GHz, the signal gain of MESFET power devices begins to decrease rapidly (approaching 6 dB or more per octave frequency band). Therefore, microwave launchers for the characterization of discrete devices and amplifier circuits must be of low RF loss and minimal mismatch. To evaluate the devices accurately, high-performance waveguide-to-microstrip transitions at different waveguide bands can be used. A ridged-waveguide approach can be used up to the 60-GHz frequency band. Beyond this band, the dimensions required for the physical machining of the waveguide structure would be too small to achieve adequate accuracy for the values designed. The rigid structure also lends itself to space applications.

Typical insertion losses per waveguide-to-microstrip transition are 0.1, 0.15, and 0.3 dB in the 20-, 30-, and 44-GHz frequency bands, respectively [16, 17, 21]. The return losses are better than 30, 26, and 25 dB for the three frequency bands, respectively. For the 60- and 94-GHz bands, waveguide-to-microstrip line transitions with finline structures [54] are preferable. Excellent performance (0.5-dB insertion loss and 18-dB return loss for the 60-GHz band [23] and 0.75 dB and 15 dB, respectively, for the 94-GHz band (76)] has been achieved. Figure 10.33 shows a waveguide-to-microstrip transition at 60 GHz.

Other waveguide-to-microstrip transition design approaches for millimeter wave applications employ a waveguide-to-microstrip cross-junction for suspended stripline on dielectric substrates. These designs offer only satisfactory RF performance with a limited bandwidth of 20% [77, 78]. Recently, full waveguide bandwidth (40%) transitions have been demonstrated by using an E-plane metallic probe on 10-mil RT-5880/Duroid substrates [79]. In this design a metal strip supported by the substrate is used as a probe to couple electromagnetic (EM) energy from the waveguide to the microstrip line, as shown in Fig. 10.34. A quarter-wave impedance transformer matches the input impedance of the probe to a 50-Ω microstrip line. One disadvantage of such a design approach is that the input/output signals are not in the same plane with the microstrip line circuit, making circuit evaluation inconvenient.

Network Analyzer Technique. Characterizations of power FET devices, HMIC amplifiers, and MMICs can be grouped into two categories. The first set consists of small-signal parameters, including amplifier linear gain and I/O return loss. These parameters can be readily obtained using the Hewlett-Packard (HP) 8510B or Wiltron automatic network analyzer system. The HP system has recently been extended to the range of measurements from 45 MHz to 40 GHz using coaxial connectors with a single connection, and to 100 GHz using different waveguide-band attachments. The system makes error-corrected measurements of magnitude, phase, and group delay and removes predictable system errors such as directivity, source mismatch, load mismatch, crosstalk, and frequency response at the measurement reference plane selected. In addition to measuring

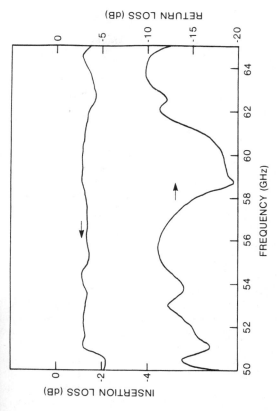

FIGURE 10.33 Waveguide-to-microstrip transition with finline structure at V-band.

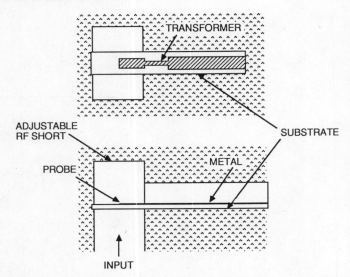

FIGURE 10.34 Cross section of a waveguide-to-microstrip transition using an E-plane probe design.

the traditional network parameters versus frequency, the system also offers the capability of viewing the network's time-domain response (for transmission-line faults or test fixture discontinuities) in real time. The nature of the impedance changes (R, L, C) can be identified and selectively "removed" from the time-domain response and then corrected in the frequency mode.

The HP 8510B system can be extended to achieve pulsed-RF network analysis from 2 to 20 GHz. It allows evaluation of the dynamic pulsed-RF characteristics, relative magnitude, and phase shift of the components as a function of time (with an equivalent bandwidth of 2.5 MHz) for pulse widths down to 1 μs. These types of measurements are essential to the characterization of the transmit/receive (T/R) module for radar systems.

The second set of characterizations includes the nonlinear characteristics associated with the large-signal operation of the power amplifier. Power circuits are characterized for parameters such as output power, power-added efficiency, third-order intermodulation product, AM-to-PM coefficient, and noise-power ratio (NPR) for multicarrier operation. Other parameters, such as pulse droop and phase droop, must also be characterized and are significant for pulse signal operation.

Discrete power FETs can be characterized for optimal output impedance by using a load-pull measurement system. To minimize the input return loss, the input of the assembled FET on a carrier can be tuned using a simple shunt element (a gold ribbon on a microstrip line at the frequency of test). The tuners in the load-pull setup should be capable of covering the entire Smith chart; otherwise, an active load (using an amplifier) will have to be used.

In amplifier measurement, computer-controlled measurement systems are essential for obtaining accurate results and well-documented data. Such a system can also assist in diagnosing the causes of failure during power amplifier development. An example of such a system is the AMPAC (automated microwave power amplifier characterization) system shown in Fig. 10.35, which has been ussed to measure and document the linear and nonlinear parameters of solid-state power amplifiers qualified for space applications.

FIGURE 10.35 Automated microwave power amplifier characterization (AMPAC) system. (From Ref. 80)

10.7.3 Optical Techniques

Current FET device and MMIC testing techniques are proving to be inadequate, especially at millimeter wave frequencies. The use of metal waveguide test fixture components offers a low-loss medium for device characterization; however, this approach is limited by the required transition from a waveguide to a planar interface on the FET or MMIC under test. Additionally, because the waveguide measurement system is bandwidth-limited, multiple waveguide sizes are needed for broadband measurements. Also, at frequencies below the waveguide cutoff, a high reflection coefficient is presented to the device or MMIC, which may result in circuit oscillations. On-wafer contacting CPW probes are routinely used up to 26.5 GHz, and have recently been extended to 40 GHz. However, it is difficult to achieve a low-loss, impedance-matched probe, as frequencies are increased into the millimeter wave range, and the thin wafer may be damaged by the inevitable mechanical contact between the probe tips and the FETs or MMICs.

An elegant approach to providing a potentially low-cost, noncontacting test methodology that is compatible with on-wafer MMIC measurements employs optical techniques for both signal generation and sampling. The use of photoconductive switches to generate ultra-short-duration pulse signals along a transmission line is well known [81]. Sufficiently fast-switching photoconductive switches and a picosecond laser source will create pulses with high millimeter wave frequency content. The application of these pulses to devices enables the broadband characterization of FETs and MMICs. By launching the pulses along a terminated quasi-TEM transmission line, an inherently excellent broadband source match is provided to the MMIC, thus eliminating the discontinuities of a waveguide-to-microstrip transition.

Laser excitation of the photoconductive switch is also capable of performing signal sampling, with a response function limited only by the photoconductive switch dynamics.

The alternatives are to take advantage of the electro-optic nature of GaAs, or to use an electro-optic sampling probe. When a laser beam passes through a GaAs substrate that has electric fields set up by transmission-line voltages, the optical birefringence causes a polarization change proportional to the sampled voltage. Electro-optic sampling also allows probing of the internal voltages of the chip. By comparing the Fourier transforms of the sampled incident and reflected or transmitted waveforms, the complex two-port S-parameters can be determined.

Photoconductive switch reflection measurements of passive elements such as resistors [82] has been demonstrated. Experimental results for a power MMIC in which both

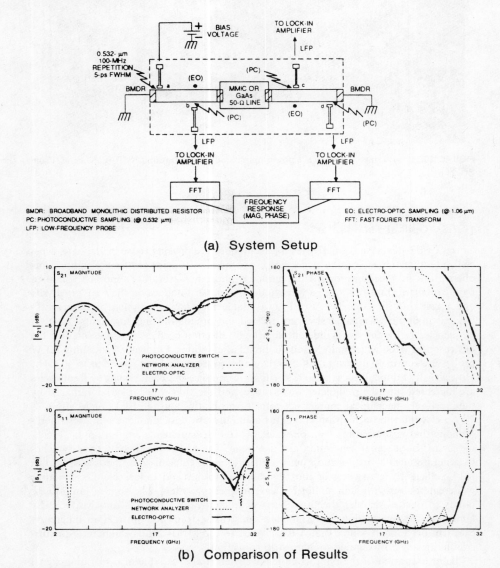

(a) System Setup

(b) Comparison of Results

FIGURE 10.36   Optical measurement system: (*a*) system setup; (*b*) comparison of measured results.

complex reflection (S_{11}) and transmission (S_{21}) measurements obtained by photoconductive switch and electro-optic sampling techniques have been compared directly to network analyzer measurements with good agreement [83–85]. Figure 10.36a is a schematic of the optical measurement system for pulse generation and sampling. A comparison of the measured results in phase and magnitude is shown in Fig. 10.36b. In addition to offering noncontacting measurements, this optical approach has the potential of characterizing multiport devices such as mixers, as well as evaluating circuit performance at any internal node.

10.8 THERMAL CONSIDERATIONS

Thermal limitations are important considerations in microwave power FET operation, since FET channel temperature affects microwave performance and device reliability. There are three methods by which thermal parametric measurements of power FETs and MMICs can be applied, and parameters such as thermal impedance and device junction temperature can be derived. Each method has its own characteristics, such as ease of measurement, measured spot size, degree of accuracy, and destructive or nondestructive nature [86, 87]. It has been observed that thermal impedance can vary with both temperature and bias conditions [88]. The channel-to-carrier or channel-to-case thermal impedance, θ_j, can be calculated as

$$\theta_j = \frac{T_t - T_c}{P_d} \qquad (10.27)$$

where T_t and T_c are the channel temperature and carrier or case temperature, respectively, and P_d is the dc power plus the net of the RF power between the output and the input.

10.8.1 Analysis

A number of approaches have been employed to analyze the junction temperature of power FETs. One technique [89, 90] takes advantage of the similarity between analyzing strips of heat sources and analyzing microwave microstrip lines. Simple close-form expressions can be obtained to achieve a first-order effect on the channel temperature of the device with respect to source-gate-drain spacing and via-hole location. More elaborate three-dimensional analyses based on computer calculations [91, 92] will generally provide upper limits on the thermal performance of the devices prior to its fabrication. Commercially available thermal analysis computer programs such as SINDA [92], which are based on the finite-difference nodal analysis method, can be applied to the steady-state and transient thermal evaluation of mounted devices. The prediction will assist in optimizing the device, especially for MMIC applications. In the following subsections we discuss three techniques that can be used to obtain thermal impedance values.

10.8.2 Measurements

Electrical Method. The electrical measurement technique is a versatile method that provides average channel temperature rather than peak value. It can be applied to a FET chip on-carrier, or to a packaged device or circuit with directly accessible dc biases. Thus it is useful for the evaluation of fabricated components.

The method involves first obtaining a calibration factor (K-factor) for the FET which relates the temperature characteristics of the forward-biased gate voltage at a fixed gate current [87, 93]. The K-factor is the temperature coefficient for the gate-source junction (in °C/mV). The device is normally terminated with appropriate matched loads at the input and output to prevent oscillation during the measurement. A bias network should be implemented to ensure that true Schottky barrier gate-to-source characteristics are measured.

During the calibration measurement, a pulsed forward gate current from a constant-current supply is applied to the gate-source port, and the drain terminal of the device is unbiased. The forward gate voltage of the FET as a function of temperature can then be measured. Following this, the device is dc-biased (at the drain and gate port) at the normal operating point, with the carrier temperature control set at the desired environmental value. During a short interval (e.g., 10 μs), the drain bias is switched off and the gate bias becomes positive and is biased by a constant-current source at a value similar to that used in the calibration measurement. The measured forward gate voltage can then be used with the calibration data to determine the FET channel temperature. Since the increase in junction temperature caused by the heating power is equal to K times the change in forward voltage, the thermal impedance can be derived by using eq. (10.27).

Infrared Imaging Method. The infrared (IR) imaging technique [94, 95] provides a temperature profile of a power FET by detecting the thermal IR radiation emitted from the surface of the device. The spot size resolution is about 15 μm, and areas close to a drain or via-hole usually have a lower temperature than that in the channel. Thus the thermal impedance value derived may tend to yield a lower number than the actual value, especially for small devices for millimeter wave applications.

The essential features of this measurement system are given in Ref. 95. The visible and IR energy from the device is collected by a reflective objective. An optical filter is used to direct the visible portion to an eyepiece for observation. The IR beam is chopped at a low frequency and received by a detector (e.g., an InSb detector). The signal can then be amplified and demodulated and the result stored in a computer. (In practice, a calibration run is made first.) The thermal image of the entire device can be reconstructed through beam scanning or by moving the stage of the device mount, with the desired operating dc bias on the device. An example of the IR thermal pattern of a power FET is shown in Fig. 10.37. The thermal impedance can be calculated based on the FET channel temperature obtained.

Liquid Crystal Method. The liquid crystal technique [86, 88] depends on the properties of temperature-sensitive nematic crystals and is capable of measuring hot spots in the channel of the device. Above a critical temperature, molecules in nematic crystals transition from an orderly (nematic) state to a randomly oriented (isotropic) state. Liquid crystals that have a transition temperature ranging from 30 to 300°C are available and are readily soluble in a carrier liquid at room temperature.

The thermal impedance of the GaAs FET is measured by applying a thin coating of liquid crystal to the surface of an eutectically mounted device on a carrier or case to which dc bias is applied. The device is then illuminated with polarized light, and the reflected light is examined using a rotating analyzer. The analyzer at the reflected path is first calibrated to produce cross polarization by rotating it with respect to regions without liquid crystal, and the image appears dark. When polarized light is directed onto the regions covered with liquid crystal (which acts as a third polarizer), these areas will not appear dark until the transition temperature, T_t (which is the device channel temperature) is reached and the crystal becomes isotropic. As the bias voltage is gradually

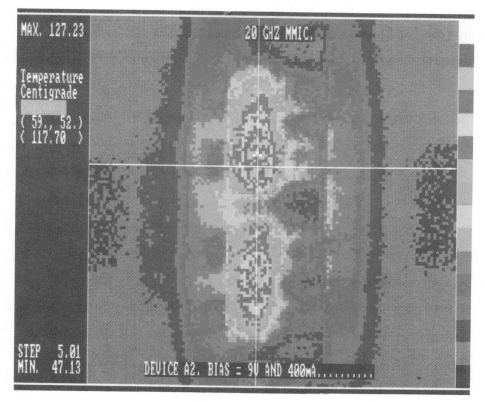

FIGURE 10.37 IR image of a power FET.

increased, at a certain point the reflected image will change from light to dark. By noting the dc power dissipated in the device, P_d, and the carrier temperature, T_c, the thermal impedance of the device from channel to carrier, θ_j, can be calculated. This measurement approach can provide a temperature accuracy of $\pm 0.5°C$ and spatial resolution of 2 μm.

10.9 PACKAGING DESIGN

The rapid progress that has occurred in the development of power FETs and MMICs has not been accompanied by corresponding growth in packaging and interconnect technology. Currently, limited packages that provide satisfactory microwave performance are available up to 20 GHz [96–98]. The performance of a packaged device, compared to that offered by FET or MMIC chips, is generally degraded due to package parasitics. However, efforts are still being made to achieve high-performance packaged devices, especially with MMICs at 30 GHz and up to the 60-GHz frequency band [99, 100]. Analytical studies of the parasitic effects of packages on microwave device performance are also being performed [101].

Packaging design considerations can be grouped into three categories: mechanical, electrical, and thermal. The parameters that govern the mechanical designs are reliability, manufacturing cost, flexible size, and a trend toward surface mount technology. Electromagnetic waves must be transmitted into and out of the enclosure in a controlled manner. Electrical considerations include good return loss (or VSWR), high lead isolation, suppression of the waveguide mode, elimination of resonance, parasitic inductance, and capacitance to ground. In the area of thermal design, the reduction of thermal resistance (as mentioned above) and the provision of adequate heat sinking are major factors.

Consideration must also be given to the fact that excellent electrical materials may be poor mechanical and/or thermal materials. In fact, a package is not just a housing but part of the circuit, especially for MMICs and high-frequency circuits. Because the packaging process has become a part of the design optimization, requiring iterations in both design and fabrication, knowledge of the mechanical, electrical, and thermal disciplines is essential.

10.9.1 Discrete and Internally Matched Devices

Discrete power devices designed to operate at an output power level of less than 1.5 W and in the lower microwave frequency range (below X-band) can be mounted inside a small, hermetically sealed package. Figure 10.38a shows a typical microwave FET package for discrete devices. As the power and frequency requirement increases, partial matching circuits fabricated on alumina or fused silica substrates are incorporated into the device on a copper or thermcom carrier. In general, the device is eutectically bonded to the package or carrier. The common solder used is AuSn (gold-tin), with a eutectic temperature of 285°C. The circuit substrate can first be soldered to the thermcom package or copper carrier using AuGe (gold-germanium) preforms at 356°C, depending on the operating temperature requirements. Figure 10.38b shows an IM power FET on a carrier.

10.9.2 MMICs

Recently, several approaches have been presented for developing a low-cost, high-performance package for a multichip MMIC module. One approach, which uses ceramic packaging techniques, has been shown to function satisfactorily up to 30 GHz [99]. The multilayer ceramic frame consists of metallized layers over most of both surfaces and co-fired. Microstrip line and through-wall microstrips with shielded walls for the RF input/output (I/O) ports are designed for 50-Ω impedance. The insertion of each I/O port is less than 0.5 dB up to 33 GHz, and isolation is in excess of 30 dB. The package also provides an input VSWR of 1.3:1.

Another approach, using the waffleline packaging technique, has been extended to 60 GHz [100]. The body of the EHF waffleline is a waffle-iron-like grid. Teflon-coated wires lying in the waffle channels are used to interconnect circuits mounted on round chip carriers. The wires are quasi-coaxial transmission lines that exhibit an effective characteristic impedance of 50 Ω. The round chip carriers are mounted in cavities within the body structure, using differential thermal expansion properties for mechanical and electrical ground attachments. An insertion loss of less than 5 dB, and a return loss better than 10 dB, have been achieved from 50 to 60 GHz.

MMICs can also be packaged with a conventional waveguide package. An example of a 60-GHz amplifier package with waveguide-to-microstrip transitions is shown in Fig. 10.22. This type of package can be hermetically sealed using quartz windows at the input and output, and still provide excellent millimeter wave performance.

Finally, a new approach to packaging MMICs for low-cost testing uses the frame tape concept [102]. Conventional on-wafer probing can still be performed with the MMIC chip mounted on the tape carrier.

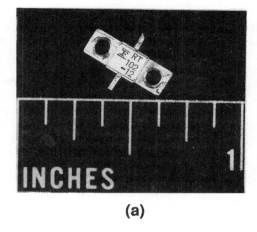

(a)

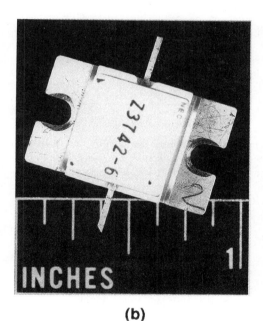

(b)

FIGURE 10.38 FET packaging: (*a*) hermetically sealed package; (*b*) IM FET carrier.

10.10 RELIABILITY AND RADIATION EFFECTS

The reliability of GaAs power FETs has already been proven, and the devices are employed in large commercial and military systems. FET devices are also currently being used on-board a number of satellites, such as SATCOM and INTELSAT, replacing the TWTAs used previously.

The reliability of GaAs power MMICs has recently received increased attention. A number of studies have been performed on the effects of total-dose and transient radiation on the various elements in GaAs MMICs [103, 104]. The results indicate that GaAs MMICs are resistant to radiation levels orders of magnitude beyond that from the natural environment. Therefore, there is normally no need to increase their radiation tolerance. However, for an enhanced radiation environment, hardening techniques may be applied to further increase GaAs MMIC radiation resistance. MIM capacitors using silicon nitride dielectrics showed a change of less than 3% in capacitance value following irradiation of 10^8 rads. This change is small enough to be tolerable in circuits [103].

Epitaxially grown GaAs is less affected by radiation than is ion-implanted material. In monolithic circuits fabricated on epitaxial material, microwave performance is compromised at gamma doses exceeding 10^9 rad (from Co^{60}). MMICs are well behaved in a transient-radiation environment up to 10^{11} rad/s, and with proper design could be made even more radiation resistant [104]. No significant deviation from the total gamma does results was observed from cumulative neutron exposure at 4×10^{13} neutrons/cm^2. However, new synergistic effects on FET and MMIC performance have been observed when combined radiation (instead of separate pulsed neutron and flash x-ray) is applied to simulate a nuclear event [105].

Since dc measurements can be performed on test structures at the chip level, the characterization of radiation-induced changes in HMICs or MMICs has been demonstrated to be useful based on the characterization of the simpler discrete devices [104]. If increased MMIC hardness to total-dose or transient radiation is desired, improvements in circuit design, material properties, device structures, or circuit pretuning have all been shown to be effective.

10.11   CONCLUSIONS

Since monolithic circuits have both an output power and bandwidth advantage over discrete devices as a result of close-to-device impedance matching, the trend in millimeter wave power components development will probably emphasize the MMIC approach. Whenever possible, MMIC power components will replace two-terminal devices such as IMPATTs and Gunn devices in the millimeter frequency range. Although higher power levels can currently be achieved with these devices, MESFET technology offers advantages such as suitability for low-cost, small-size planar integration in HMICs or MMICs. MESFET also provide the potential for improved reliability. Accurate CAD modeling and circuit analysis will be essential to the successful development of microwave and millimeter wave FETs and monolithic circuits and will help to reduce the development cycles.

By reducing the gate length, but maintaining a high gate length channel height ratio, conventional MESFET power MMICs should be able to provide output power of 0.15 W per chip at 60 GHz. Higher output power can be achieved by using external divider/combiner circuits.

Two emerging technologies, the pseudomorphic HEMT and the heterojunction bipolar transistor (HBT), will provide better gain, if not power, performance for applications at millimeter wave frequencies. The pseudomorphic HEMT has demonstrated better gain and power-added efficiency than do conventional HEMT devices in the 60-GHz band. A pseudomorphic HEMT device with a 0.15-μm gate length has demonstrated a projected maximum frequency of oscillation, f_{max}, as high as 350 GHz [106]. With a longer-gate-length device (0.25 μm), higher f_{max} has been reported from a HEMT structure with an InP-base material device [107]. Further development of larger-gate-width HEMTs, pseudomorphic HEMTs, and optimization of device topology can be expected to yield a single device with an output power of 1 W at 35 GHz and 0.25 W in the 60-GHz band. Future advanced millimeter wave components operating in the higher-frequency band (94 GHz) will be largely dependent on these types of devices.

Extremely high f_t (the frequency at which short-circuit forward current transfer becomes unity) has been estimated for fabricated HBT devices. The transistor will be a super-power device for oscillator and pulse power applications, because of its large transconductance, uniform current density over the device cross section, large breakdown voltage, low $1/f$ noise, and higher-frequency capability [108, 109]. Power densities as high as 2.5 W/mm of emitter periphery has been demonstrated at 10 GHz under CW conditions [110, 111]. Device operation under pulsed conditions produced even higher power densities of 5.4 W/mm [111]. Microwave performance of *npn* and *pnp* AlGaAs/ GaAs HBTs have also been studied [112, 113], leading to the potential implementation of complementary microwave circuits. It is projected that HBT will also reach a maximum frequency of oscillation of over 300 GHz. However, major efforts will be required in technology/circuit development before significant power performance can be achieved.

The technology of these types of device structures, and the related MMICs, will definitely be expanded in the future. With these new devices for power applications, not only can performance of existing systems be improved, but also new concepts for the next generation of phased-array, satellite communications, and electronic warfare systems can be realized.

APPENDIX A
Examples of Solving Networks for Node Voltages

EXAMPLE 1: PASSIVE NETWORK

Figure A.1 shows a simple four-node circuit that will be used to illustrate the method of nodes. The nodes are connected to one another through resistors with conductance values of G_{mn} between nodes m and n. Since there are three nodes in addition to the common node, three simultaneous equations will be obtained.

First, consider node 1. The four currents into this node are i_1, $-G_{10}v_1$, $G_{21}v_{21}$, and $G_{31}v_{31}$.[†] The sum of these currents must be zero as shown in the following equation:

$$i_1 - (G_{10} + G_{21} + G_{31})v_1 + G_{21}v_2 + G_{31}v_3 = 0 \tag{A.1}$$

This equation can be rearranged as follows:

$$(G_{10} + G_{21} + G_{31})v_1 - G_{21}v_2 - G_{31}v_3 = i_1 \tag{A.2}$$

Repeating the procedure above for nodes 2 and 3, two more equations are obtained:

$$-G_{21}v_1 + (G_{20} + G_{21} + G_{32})v_2 - G_{32}v_3 = i_2 \tag{A.3}$$

$$-G_{31}v_1 - G_{32}v_2 + (G_{30} + G_{31} + G_{32})v_3 = 0 \tag{A.4}$$

These three simultaneous equations can now be solved for the three node voltages. Their solution by matrix methods is discussed in Appendix B.

[†] v_{mn} represents the difference in voltage between nodes m and n.

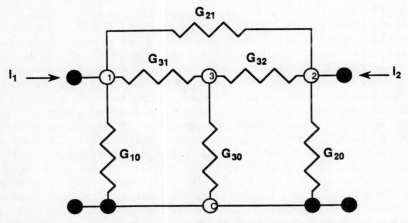

FIGURE A.1   Lumped-element equivalent circuit for Example 1 without current generator.

EXAMPLE 2: NETWORK WITH A CURRENT GENERATOR

Figure A.2 shows that a current generator exists in the circuit between nodes 2 and 3 of the same circuit as Fig. A.1. Let the generated current, i_{23}, be given by the voltages at nodes 1 and 3 as

$$i_{23} = g_m v_{13} \tag{A.5}$$

The right-hand side of Eq. (A.3) becomes $i_2 + i_{23}$, and the right-hand side of Eq. (A.4) becomes $-i_{23}$. Using eq. (A.5) and rearranging the terms, the following two equations

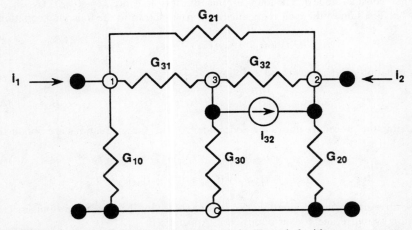

FIGURE A.2   Lumped-element equivalent circuit for Example 2 with current generator.

can be obtained:

$$-(G_{21} + g_m)v_1 + (G_{20} + G_{21} + G_{32})v_2 - (G_{32} - g_m)v_3 = i_2 \qquad \text{(A.6)}$$

$$-(G_{31} - g_m)v_1 - G_{32}v_2 + (G_{30} + G_{31} + G_{32} - g_m)v_3 = 0 \qquad \text{(A.7)}$$

The three node voltages for this case are found by solving the set of simultaneous equations given by eqs. (A.2), (A.6), and (A.7).

APPENDIX B
Examples of Solving the Network Using the Matrix Method

The examples discussed in Appendix A will be used to illustrate how the matrices are set up. First, the simultaneous equations are arranged with the node voltages, and their coefficients, on the left side, and the currents externally input to the nodes on the right side. These external node currents are all zero except for nodes 1 and 2. On the left side, the node voltages are represented in order of increasing node number, from left to right. The equations representing the current sums for each node are also arranged in order of node number, starting with node 1 at the top. The node voltage coefficients become elements of an $N \times$ **Y**-matrix of admittance values. The right-hand side forms an $N \times 1$ matrix.

B.1 EXAMPLES

For the circuit without a current generator (Fig. A.1), the 3×3 voltage coefficient matrix discussed above is found from Eqs. (A.2), (A.3), and (A.4). The matrix is given as

$$\mathbf{Y} = \begin{bmatrix} (G_{10} + G_{21} + G_{31}) & -G_{21} & -G_{31} \\ -G_{21} & (G_{20} + G_{21} + G_{32}) & -G_{32} \\ -G_{31} & -G_{32} & (G_{30} + G_{31} + G_{32}) \end{bmatrix} \qquad \text{(B.1)}$$

Note that this matrix is symmetrical across the main diagonal. This is true only for networks that do not contain generators.

For the network of Fig. A.2, which contains a current generator, the following matrix can be obtained from Eqs. (A.2), (A.6), and (A.7):

$$\mathbf{Y} = \begin{bmatrix} G_{10} + G_{21} + G_{31} & -G_{21} & -G_{31} \\ -(G_{21} + g_m) & G_{20} + G_{21} + G_{32} & -(G_{32} - g_m) \\ -(G_{31} - g_m) & -G_{32} & G_{30} + G_{31} + G_{32} - g_m \end{bmatrix} \qquad \text{(B.2)}$$

Note that this matrix is not symmetrical across the main diagonal.

For both networks, the 3×1 external current matrix is

$$\mathbf{I} = \begin{bmatrix} i_1 \\ i_2 \\ 0 \end{bmatrix} \qquad \text{(B.3)}$$

The simultaneous equations can now be expressed as

$$\mathbf{YV} = \mathbf{I} \tag{B.4}$$

where $\mathbf{V}$ represents the three unknown node voltages (a 3×1 matrix). If a computer language is used that allows matrix arithmetic to be performed on variable names representing matrices, the solution to Eq. (B.4) is obtained by using the statement $\mathbf{V} = (1/\mathbf{Y})\mathbf{I}$. Alternatively, a subroutine such as the SEQ subroutine discussed in Section B.2 could be used.

To solve the two example, let the resistors in Figs. A.1 and A.2 have the following numerical values of conductance (in siemens).

$$G_{10} = 0.002 \text{ S}$$

$$G_{20} = 0.001 \text{ S}$$

$$G_{21} = 0.003 \text{ S}$$

$$G_{30} = 1.0 \text{ S}$$

$$G_{31} = 0.3 \text{ S}$$

$$G_{32} = 0.001 \text{ S}$$

To solve the network of Fig. A.1, let $i_1 = 1$ A and $i_2 = 0.01$ A. If the values above are substituted into Eq. (B.1) to obtain the $\mathbf{Y}$ matrix

$$\mathbf{Y} = \begin{bmatrix} 0.105 & -0.003 & -0.1 \\ -0.003 & 0.005 & -0.001 \\ -0.1 & -0.001 & 1.101 \end{bmatrix} \tag{B.5}$$

then Eq. (B.4) yields

$$\begin{bmatrix} 0.105 & -0.003 & -0.1 \\ -0.003 & 0.005 & -0.001 \\ -2.0 & -0.001 & 1.101 \end{bmatrix} \mathbf{V} = \begin{bmatrix} 1 \\ 0.01 \\ 0 \end{bmatrix} \tag{B.6}$$

The solutions are the matrix for element values of v equal to 10.703, 8.618, and 0.97998, corresponding to voltage at nodes 1, 2 and 3, respectively. From these solutions, the input power at nodes 1 and 2 is found to be 10.703 W and 0.08618 W, respectively.

In the example with current generation (Fig. A.2), let the conductance values be the same as in the previous example, and let $i_1 = 0.01$ A, $i_2 = -4$ A, and $g_m = 10$ S. The solution to this example is again found by substituting the numerical values into Eqs. (B.2) and (B.4), yielding

$$\begin{bmatrix} 0.105 & -0.003 & -0.1 \\ -10.003 & 0.005 & 9.999 \\ 9.9 & -0.001 & -8.899 \end{bmatrix} \mathbf{V} = \begin{bmatrix} 0.01 \\ -4.0 \\ 0.0 \end{bmatrix} \tag{B.7}$$

and the solutions for the matrix element values are v equaling to 1348.1, 7.5251, and -6.6937. Based on this solution, the input power at node 1 is 13.481 W and at node 2 is -30.1 W. The negative sign indicates that the network is generating power at this node; in other words, power is flowing out from node 2. The negative of the voltage-to-current ratio at node 2, 1.88 Ω, is the load impedance into which the power is flowing.

B.2   TWO-PORT MATRICES

Usually, v_1 and v_2 are of interest, since i_1, v_1, i_2, and v_2 completely describe the small-signal performance of a FET. In this case, the two-port network represented by its LEEC can be characterized by a 2×2 complex matrix. Implicit in the use of a two-port matrix to represent a FET mathematically is the assumption that the signal amplitude is small enough so that linearity can be assumed. Methods for making engineering estimates of the large-signal effects that limit FET output power were discussed in Section 10.3.5, the load-pull model.

The two-port impedance matrix, or "Z-matrix," can be obtained from the $N \times N$ Y-matrix discussed above. The method is to solve for the node voltages for two special sets of node currents: i_1 equals 1 and all others zero, and i_2 equals 1 and all others zero. The current values of 1 are chosen so that voltages resulting from the two solutions can be reinterpreted as impedances.

Computer subroutines exist that allow an $N \times M$ matrix to represent M different columns for the simultaneous equations represented by an $N \times N$ coefficient matrix. In the present case, each column of an $N \times 2$ external current matrix, **I**, would represent the two special sets of external currents. This type of subroutine returns an $N \times M$ matrix, with each column representing the node voltages that are the solution for the corresponding column of the input $N \times M$ matrix.

In the following example, "SEQ" represents a hypothetical subroutine that allows complex element values. (Substitute the name of an actual subroutine.) **Y** is the $N \times N$ admittance matrix, and **I** is the $N \times 2$ input current matrix, representing the two special sets of node currents mentioned above. **V** is the $N \times 2$ node voltage matrix computed by SEQ, as follows:

$$\mathbf{V} = \mathrm{SEQ}\left[\mathbf{Y}\begin{bmatrix} 1 & 0 \\ 0 & 1 \\ 0 & 0 \\ \vdots & \vdots \\ 0 & 0 \end{bmatrix}\right] \tag{B.8}$$

The two-port **Z**-matrix is constructed by retaining the first two rows of **V** and discarding the rest. To find v_1 and v_2 given i_1, i_2, and the **Z**-matrix, the matrix analog of Ohm's law ($V = ZI$) is used, where V is a 2×1 matrix (v_1, v_2), Z is the Z-matrix derived above, and I is a 2×1 matrix (i_1, i_2).

An alternative method of obtaining the two-port **Z**-matrix is to first obtain the N-port **Z**-matrix by inverting the $N \times N$ Y-matrix, using standard subroutines. The 2×2 sub-matrix at the upper left (Z_{11}, Z_{12}, Z_{21}, and Z_{22}) is the desired two-port Z-matrix. The methods discussed above may be used to verify that the two-port **Z**-matrix corresponding to the 3×3 **Y**-matrix of Eq. (B.5) is

$$\mathbf{Z} = \begin{bmatrix} 10.638 & 6.577 \\ 6.577 & 204.1 \end{bmatrix} \tag{B.9}$$

As mentioned above, other matrices can be derived from the **Z**-matrix. The two-port admittance matrix, or **Y**-matrix, is the reciprocal of the two-port **Z**-matrix.

Another very useful two-port matrix is the general function matrix, often called the ABCD matrix, which is used to find v_1 and i_1 given v_2 and i_2, as

$$\begin{bmatrix} v_1 \\ i_1 \end{bmatrix} = \begin{bmatrix} A & B \\ C & D \end{bmatrix} \begin{bmatrix} v_2 \\ -i_2 \end{bmatrix} \tag{B.10}$$

$$A = Z_{11}/Z_{21} \tag{B.11}$$

$$B = Z_{11}(Z_{22}/Z_{21}) - Z_{12} \tag{B.12}$$

$$C = 1/Z_{21} \tag{B.13}$$

$$D = Z_{22}/Z_{21} \tag{B.14}$$

It is convenient to use this matrix to find power gain when the load impedance, Z_L, is given. In this case, i_2 can be set to 1 and V_2 to Z_L, and then we can solve for i_1 and i_2. The gain in decibels is given by 10 times the base-10 logarithm of the ratio $Z_L/(v_1 i_1^*)$, where the asterisk (*) denotes the complex conjugate. $ABCD$ two-port matrices also have the convenient property that the product of two of them represents the $ABCD$ matrix for the two two-port networks cascaded in series.

The two-port matrices mentioned above are used when calculating with currents and voltages. Once the values of i_1, i_2, v_1, and v_2 are known, all ac parameters of interest to the circuit designer are easily calculated. The output power, P_0, is $(-i_2)v_2^*$, and the load impedance, Z_L, is $v_2/(-i_2)$. The impedance, Z_{in}, presented to the input signal is v_1/i_1, and the input power, P_i, is $i_1 v_1^*$. The power gain, in decibels, is $10 \log(P_0/P_i)$.

For microwave circuits, it is convenient to treat input and output signals as voltage waveforms in the transmission lines used to connect microwave circuit elements, such as FETs. For this purpose a two-port matrix known as the "scattering" matrix, **S**, is used. The four matrix elements are known as "S-parameters." A matrix element S_{mn} represents the ratio of the amplitude of the signal traveling outward from port m to that of the signal traveling inward to port n, with all ports terminated by the characteristic impedance. In matrix notation, and based on a system of 50-Ω characteristic impedance,

$$\mathbf{S} = 1 - 2/(\mathbf{Z}/50 + 1) \tag{B.15}$$

where **Z** is the two-port Z-matrix. Matrix arithmetic must be used in Eq. (B.15). FET manufacturers typically provide tables of S-parameters versus frequency in their catalogs. To find the **Z**-matrix from the scattering matrix, Eq. (B.15) can be inverted to obtain

$$\mathbf{Z} = -50(1 + 2/(\mathbf{S} - 1)) \tag{B.16}$$

The **Z**-matrix can be used to compute the maximum available gain and stability [38]. The starting point is to set $i_1 = 1$ and consider all complex values of i_2. Input and output power can be computed for every point in the i_2 plane. It turns out that positive output power occurs only inside a circle on this plane. The i_2 plane can be normalized by multiplying by a complex normalization factor so that this circle has a radius of 1 centered at $+1$ on the real axis. In Section 10.3, regarding the load-pull method, the extension of this type of computation to determine the maximum available power from a power FET is illustrated.

When the limitations on the output power of FETs are considered, certain internal node voltages related to the physical limits of the device must be computed. For example, maximum and minimum voltages are required across the gate capacitor ($v_3 - v_5$) in order for the voltage to have any effect on the output of the current generator.

The methods used to compute the two-port $\mathbf{Z}$-matrix can be extended to any number, M, of nodes of interest. To do this, it is convenient to assign the node numbers such that the nodes of interest are numbered consecutively 1 through M. An $M \times M$ impedance matrix, Z_m, can be computed by inverting the $N \times N$ Y-matrix, retaining only the $M \times M$ upper left-hand submatrix of the resulting Z-matrix. Or, the SEQ subroutine can be used with the $N \times 2$ matrix in Eq. (B.8)) by replacing an $N \times M$ matrix with ones on the main diagonal and discarding all rows with a number greater than M from the resulting $N \times M$ matrix, $\mathbf{v}$. Node voltages are found from i_1 and i_2 using the following matrix equation:

$$
\begin{bmatrix} v_1 \\ v_2 \\ v_3 \\ \vdots \\ v_m \end{bmatrix} = Z_m \begin{bmatrix} i_1 \\ i_2 \\ 0 \\ \vdots \\ 0 \end{bmatrix}
\tag{B.17}
$$

All input currents to nodes other than 1 and 2 must be zero. The usefulness of Z_m in computer modeling of power FETs is illustrated in Section 10.3.5 on load-pull modeling.

APPENDIX C
A Sample Run of the FLATIDSS Program

The user supplies the carrier concentration (N), the channel height (H), the channel width (W), and the source–gate resistance (R_{sg}). H is the distance from the gate metal to the edge of the back junction depletion layer. I_{dss} is calculated by expressing it as the undepleted channel height (H minus the gate depletion-layer thickness) times the product of electron charge, saturation velocity, and gate width. The depletion-layer thickness is proportional to the square root of the voltage across it, plus the "built-in" voltage, Phi or ϕ. The voltage across the depletion layer is $I_{dss}R_{sg}$. The resulting expression for I_{dss} is squared, giving a quadratic equation that is solved for I_{dss}. The $I_{ds,\max}$ is determined by the bare surface depletion-layer thickness. For low carrier concentrations, this is determined by a fixed bare surface potential, Phi$_s$ or ϕ_s. The positive charge in the depletion layer is balanced by negative ionization of surface states. At a high carrier concentration, there may not be enough surface states to maintain a surface potential at the fixed value. In this case the FLATIDSS program calculates ϕ_s assuming a surface state density of 1.7×10^{12} cm^2 (this value is used as an example only).

```
        Speakeasy IV Delta   5:40 PM November 18, 1987
  1 PROGRAM FLATIDSS    BY T. SMITH   11/05/87        MOD 11/18/87
  2 $ "FLATIDSS" COMPUTES IDSS AND IDSMAX FOR FLAT PROFILE FETS
  3 $
  4 $ THE USER IS PROMPTED FOR INPUT
  5 $
  6 IF(KIND(GAASCONS).EQ.0) GETDECK GAASCONS
  7 GAASCONS                  $ PROVIDES CONSTANTS OF GAS
  8 ASKDATA: AUTOPRINT
  9 ASK("CARRIER CONCENTRATION (CM**-3):" "N=" "N")
 10 ASK("CHANNEL HEIGHT (CM)            :" "H=" "H")
 11 ASK("CHANNEL WIDTH  (CM)            :" "W=" "W")
```

```
 12 ASK("SOURCE-GATE RESISTANCE (OHMS) :" "RSG=" "RSG")
 13 ENDAUTOPRINT
 14 "ENTER CHANGES OR NULL LINE TO CONTINUE" PAUSE
 15 DOCMS("DVHUTL CLEAR")
 16 C1=W*IDSCON1               $ IDSCON1 IS Q(ELCTRN)*SAT.VELOCITY
 17 C2=C1**2*DEPCON1           $ DEPCON1 IS 2*PERMITIVITY/Q(ELCTRN)
 18 PHIS=MAXSSDEN**2/(N*DEPCON1)  $ MAXSSDEN: SURF. STATE DENSITY
 19 WHERE(PHIS.GT.PHISO) PHIS=PHISO$FERMI LEV. PINNING PHENOMENON
 20 IDSOO=C1*N*H
 21 IDSMAX=IDSOO — C1*N*SQRT(DEPCON1*PHIS/N)
 22 BQF=*2*IDSOO — C3*N*RSG$ BQF,B:(-B+OR-SQRT(B**2-4*A*C)/(2*A)
 23 CQF=IDSOO**2 — C3*N*PHI$ CQF,C: THE SAME "QUADRATIC FORMULA"
 24 X=1-4*CQF/BQF**2          $ X IS SQRT(B**2-4*A*C)/B.     (A=1)
 25 WHERE(X.LE.O) X=O         $ NO IMAGINARY SQUARE ROOTS
 26 IDSS=-BQF*(1-SQRT(X))/2$ SOLUTION TO QUADRATIC EQ. FOR IDSS
 27 WHERE(IDSS.LE.O) IDSS=0$ PINCH-OFF OF IDSS
 28 VPO=N*H**2/DEPCON1 — PHI $ PINCH-OFF VOLTAGE, FLAT PROFILE
 29 TABULATE PHI N PHIS H W IDSMAX RSG IDSS VPO
 30 "ENTER NULL LINE TO CONTINUE" PAUSE
 31 DOCMS("DVHUTL CLEAR")
*32 GOTO ASKDATA
:%end
MANUAL MODE
: flatidss
EXECUTION STARTED


CARRIER CONCENTRATION (CM**-3): (1,2,3,4,5,6,7)*1e17
  N (A 7 COMPONENT ARRAY)
  1E17   2E17   3E17   4E17   5E17   6E17   7E17
CHANNEL HEIGHT (CM)          : sqrt(depconl*(phi+2)/n)$ vpo=2
  H (A 7 COMPONENT ARRAY)
1.962E-5 1.3874E-5 1.1328E-5 9.8102E-6 8.7745E-6 8.01E-6 7.4158E-6
CHANNEL WIDTH (CM)           :W =  .1
SOURCE-GATE RESISTANCE (OHMS) :RSG =  1
"ENTER CHANGES OR NULL LINE TO CONTINUE" PAUSE
:><<< NULL LINE ENTERED >>>
```

PHI	N	PHIS	H	W	IDSMAX	RSG	IDSS	VPO
***	****	******	*********	**	******	***	******	***
.7	1E17	.7	1.962E-5	.1	.18515	1	.16385	2
	2E17	.7	1.3874E-5		.26185		.22177	2
	3E17	.67565	1.1328E-5		.32653		.26314	2
	4E17	.50674	9.8102E-6		.42761		.29619	2
	5E17	.40539	8.7745E-6		.51666		.32403	2
	6E17	.33783	8.01E-6		.59717		.34827	2
	7E17	.28956	7.4158E-6		.67121		.36981	2

```
"ENTER NULL LINE TO CONTINUE" PAUSE
:>idsmax/idss
IDSMAX/IDSS (A 7 COMPONENT ARRAY)
  1.13    1.1807  1.2409  1.4437  1.5945  1.7147  1.815
:>quit
```

APPENDIX D
Description of Computer Subprograms for Generating Device and Performance Parameters

The LEEC used in the following discussion is that of Fig. 10.3. The menu shown in Table D.1 allows the user to specify the FET by using material parameters and geometric parameters associated with the channel. In the menu, N is the channel-layer carrier concentration, NTHK is the thickness of this layer before any recess etching, and NCOMPR is the compensation ratio of this layer. The compensation ratio is used to calculate drift mobility, and hence bulk resistivity, for the given carrier concentration. Similarly, NPLUS, NPTHK, and NPCOMPR are the carrier concentration, thickness, and compensation ratios, respectively, of the N-plus layer, which is grown on top of the channel layer to reduce parasitic resistance.

The program assumes the FET channel cross section shown in Fig. D.1. L_g is the gate length, H is the channel height from the bottom of the channel layer to the gate metal, and W_g is the gate width. It is assumed that the user-specified channel height is achieved by a recess etch process, using the same lift-off mask used for subsequent gate metal deposition. The etch is assumed to undercut the mask horizontally at the same rate that it etches vertically. This assumption allows the source-gate and gate-drain resistances to be computed from the parameters above. The total gate width is assumed to be distributed among 20 unit gate fingers.

The menu also allows specification of RGSHEET, the sheet resistance of the gate metal. The gate resistance, RG, is computed from RGSHEET, L_g, and W_g, with the addition of 2.18 squares of gate metal in series with the parallel combination of all the gate fingers. The computation of output resistance, R_o, is based on a hyperbolic sine

TABLE D.1   User Menu Produced by the Listed Subprogram

VARIABLE NAMES	VALUES	UNITS
*************	*********	*******
N	3E17	CM**-3
NTHK	3E-5	CM
NCOMPR	3	
NPLUS	2E18	CM**-3
NPTHK	1E-5	CM
NPCOMPR	4	
L	5E-5	CM
H	1.1836E-5	CM
LSG	1.2E-4	CM
LSD	3E-4	CM
W	.145	CM
FREQ	20	GHZ
RGSHEET	.12	OHMS/SQ
LHRMAX	6	
MAXVLIMI	18	VOLTS

```
ENTER CHANGES (NULL LINE TO CONTINUE):
PAUSE
:POWER6b>indx=indx+1;h=spech(indx)
INDX = 4  H = 1.2703E-5
:POWER6B><<< NULL LINE ENTERED >>>
```

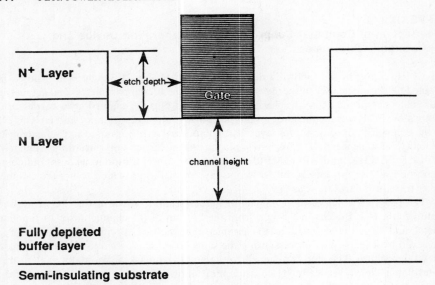

FIGURE D.1   Recess shape assumed in the listed subprogram.

function of the gate length/channel height ratio. This is oversimplified and can give unreasonably high values. The menu parameter ROUTMAX provides an upper limit to the R_o computation. Similarly, MAXVLIMIT provides an upper limit to the gate-to-drain breakdown voltage computation [37]. FREQ is the operating frequency in gigahertz.

Changes can be made to the menu parameters by the user. To indicate "no further changes," the user should press "enter" immediately following the computer's :> prompt for user input. The equivalent-circuit element values and other parameters used in computing RF performance are then computed and printed, and the user is given the opportunity to change any of these. The element values are used to calculate the load conditions and dc bias values to maximize RF power at 1-dB gain compression. Associated values for power gain and power-added efficiency are also computed.

REFERENCES

1. Microwave FET catalogs from Fujitsu, NEC, Avantek, and Mitsubishi.

2. Y. Yamada, H. Kuroda, H. Izumi, T. Soezima, H. Wakamatsu, and S. Hori, "X and Ku Band High Power GaAs FETs," *IEEE MTT-S Int. Microwave Symp. Dig.*, pp. 847–850, May 1988.

3. P. M. Smith, P. C. Chao, K. H. G. Duh, L. F. Lester, B. R. Lee, and J. M. Ballingall, "Advances in HEMT Technology and Applications," *IEEE MTT-S Int. Microwave Symp. Dig.*, pp. 749–752, June 1987.

4. B. D. Geller and P. Goettle, "Quasi-monolithic 4-GHz Power Amplifiers with 65-Percent Power-Added Efficiency," *IEEE MTT-S Int. Microwave Symp. Dig.*, pp. 835–837, May 1988.

5. K. Hirai, H. Takamatsu, S. Morkiawa, and N. Tomita, "5 GHz 20 Watt GaAs FET Amplifier for MLS," *IEEE MTT-S Int. Microwave Symp. Dig.*, pp. 447–450, June 1986.

6. Y. Tokumitsu, T. Saito, N. Okubo, and Y. Kaneko, "A 6-GHz 80-W GaAs FET Amplifier with a TM-Mode Cavity Power Combiner," *IEEE Trans. Microwave Theory Tech.*, Vol. MTT-32, pp. 301–308, March 1984.

7. B. Kopp and D. D. Heston, "High-Efficiency 5-Watt Power Amplifier with Harmonic Tuning," *IEEE MTT-S Int. Microwave Symp. Dig.*, pp. 839–842, May 1988.

8. M. Avasarala, D. S. Day, S. Chan, P. Gregory, and J. R. Basset, "High Efficiency Small Size 6 W Class AB X-Band Power Amplifier Module Using a Novel MBE GaAs FET," *IEEE MTT-S Int. Microwave Symp. Dig.*, pp. 843–846, May 1988.

9. C. Piegnet, Y. Mancuso, G. LeMeur, L. Remiro, A. Bert, J. F. Jouen, and P. Savary, "A 16 W Pulsed X-Band Solid-State Transmitter," *IEEE MTT-S Int. Microwave Symp. Dig.*, pp. 417–420, May 1988.

10. "COMSAT Reports GaAs FET Advance," *Microwave Syst. News*, Vol. 15, p. 20, 1985.

11. H.-L. A. Hung, G. M. Hegazi, K. E. Peterson, and H. C. Huang, "Design and Performance of MESFET Power Amplifiers at K- and K_a-Band," *Microwave J.*, Vol. 31, pp. 177–190, June 1988.

12. J. Goel, "A K-Band GaAs FET Amplifier with 8.2-W Output Power," *IEEE Trans. Microwave Theory Tech.*, Vol. MTT-32, pp. 317–324, March 1984.

13. I. J. Bahl, R. Wang, A. Geissberger, E. Griffin, and C. Andricos, "C-Band 10 Watt MMIC Amplifier Manufactured Using Refractory SAG Process," *IEEE Microwave Millim.-Wave Monolithic Circuits Symp. Tech. Dig.*, pp. 21–24, June 1989.

14. H. M. Macksey, H. O. Tserng, and H. D. Shih, "A 2-W Ku-Band Monolithic GaAs FET Amplifier," *IEEE Microwave Millim.-Wave Monolithic Circuit Symp. Dig.*, pp. 27–29, June 1985.

15. H.-L. A. Hung, A. Ezzeddine, F. R. Phelleps, J. F. Bass, and H. C. Huang, "GaAs MMIC Power FET Amplifiers at K-Band," *17th Eur. Microwave Conf. Dig.*, pp. 255–260, September 1987.

16. H.-L. Hung, F. Phelleps, J. Bass, A. Cornfeld, A. Ezzeddine, and H. Huang, "A 2-W Multistage K-Band GaAs Monolithic FET Amplifier," *IEEE GaAs IC Symp. Tech. Dig.*, pp. 239–242, November 1987.

17. H.-L. A. Hung, A. Ezzeddine, L. B. Holdeman, F. R. Phelleps, J. A. Allison, A. B. Cornfeld, T. Smith, and H. C. Huang, "K_a-Band Monolithic GaAs Power FET Amplifiers," *IEEE Microwave Millim.-Wave Monolithic Circuit Symp. Dig.*, pp. 97–100, June 1987.

18. N. Camilleri, B. Kim, H. O. Tseng, and H. D. Shih, "Ka-Band Monolithic GaAs FET Power Amplifier Modules," *IEEE MTT-S Int. Microwave Symp. Dig.*, pp. 179–182, May 1988.

19. Y. Oda, S. Arai, T. Yoshida, and H. Nakamura, "Ka-Band 1 Watt Power GaAs MMICs," *IEEE MTT-S Int. Microwave Symp. Dig.*, pp. 413–416, May 1988.

20. B. Kim, H. M. Macksey, H. Q. Tserng, H. D. Shih, and N. Camilleri, "Millimeter-Wave Monolithic GaAs Power FET Amplifiers," *IEEE GaAs IC Symp. Tech. Dig.*, pp. 61–63, November 1986.

21. G. Hegazi, H.-L. A. Hung, J. L. Singer, F. R. Phelleps, A. B. Cornfeld, T. Smith, J. F. Bass, H. E. Carlson, and H. C. Huang, "GaAs Molecular Beam Epitaxy Monolithic Power Amplifiers at U-Band," *IEEE Microwave Millim.-Wave Monolithic Circuit Symp. Dig.*, pp. June 1989.

22. G. Hegazi, H.-L. Hung, F. Phelleps, L. Holdeman, A. Cornfeld, T. Smith, J. Allison, and H. Huang, "V-Band Monolithic Power MESFET Amplifiers," *IEEE MTT-S Int. Microwave Symp. Dig.*, pp. 409–412, May 1988.

23. H.-L. Hung, G. Hegazi, T. Lee, F. Phelleps, J. Singer, and H. Huang, "V-Band MMIC Low-Noise and Power Amplifiers," *IEEE Trans. Microwave Theory Tech.*, Vol. MTT-36, pp. 1966–1975, December 1988.

24. V. Rizzoli and A. Neri, "State of the Art and Present Trends in Nonlinear Microwave CAD Techniques," *IEEE Trans. Microwave Theory Tech.*, Vol. MTT-36, pp. 343–365, February 1988.

25. W. R. Curtice, "GaAs MESFET Modeling and Nonlinear CAD," *IEEE Trans. Microwave Theory Tech.*, Vol. MTT-36, pp. 220–230, February 1988.

26. R. H. Jansen, R. G. Arnold, and I. G. Eddison, "A Comprehensive CAD MESFET Approach to the Design of MMIC's Up to MM-Wave Frequencies," *IEEE Trans. Microwave Theory Tech.*, Vol. MTT-36, pp. 208–219, February 1988.

27. M. Fukuta, M. Takashi, S. Hidetake, and S. Katsuhiko, "4-GHz 15-W Power GaAs MESFET," *IEEE Trans. Electron Devices*, Vol. ED-25, pp. 559–562, June 1978.

28. H. Fukui, "Design of Microwave GaAs MESFET's for Broad-Band Low-Noise Amplifiers" *IEEE Trans. Microwave Theory Tech.*, Vol. MTT-27, pp. 643–650, July 1979.

29. SUPERCOMPACT, Compact Software, Inc., Paterson, NJ.

30. W. Shockley, *Research and Investigation of Inverse Epitaxial UHF Power Transistors*, Rep. Al-TOR-64-207, Air Force Atomic Lab., Wright-Patterson Air Force Base, Ohio, September 1964.

31. J. M. David and M. G. Buehler, "A Numerical Analysis of Various Cross Sheet Resistor Test Structures," *Solid-State Electron.*, Vol. 20, p. 539, 1977.

32. C. P. Wen, "Coplanar Waveguide: A Surface Strip Transmission Line Suitable for Nonreciprocal Gyro-Magnetic Device Applications," *IEEE Trans. Microwave Theory Tech.*, Vol. MTT-17, pp. 1087–1090, December 1969.

33. R. A. Pucel, H. A. Haus, and H. Statz, "Signal and Noise Properties of GaAs Field Effect Transistors," *Electron. Electron Phys.*, Vol. 38, pp. 195–265, 1975.

34. J. A. Higgins, "Modeling the Influence of Carrier Profiles on MESFET Characteristics," *IEEE Trans. Electron Devices*, Vol. ED-27, pp. 1066–1073, June 1980.

35. H. Kressel, A. Blecher, and L. H. Gibbons, Jr., "Breakdown Voltage of GaAs Diodes Having Nearly Abrupt Junctions," *Proc. IRE*, Vol. 50, pp. 2493–2493, 1962.

36. H. Kressel and N. Goldsmith, "High Voltage Epitaxial Gallium Arsenide Microwave Diodes," *RCA Rev.*, Vol. 24, pp. 182–198, 1963.

37. S. H. Wemple, W. C. Niehaus, H. M. Cox, J. V. DiLorenzo, and W. O. Schlosser, "Control of Gate-Drain Avalanche in GaAs MESFETs," *IEEE Trans. Electron Devices*, Vol. ED-27, pp. 1013–1018, June 1980.

38. A. J. Cote and J. B. Oakes, *Linear Vacuum Tube and Transistor Circuits*, McGraw-Hill, New York, 1961.

39. K. Kurokawa, *An Introduction to the Theory of Microwave Circuits*, Academic Press, New York, 1969.

40. H. J. Carlin and A. B. Giordano, *Network Theory: An Introduction to Reciprocal and Nonreciprocal Circuits*, Prentice-Hall, Englewood Cliffs, NJ, 1964.

41. G. Matthei, L. Young, and E. M. T. Jones, *Microwave Filters, Impedance-Matching Networks, and Coupling Structures*, Artech House, Norwood, MA, 1980.

42. P. S. Pengelly, *Microwave Field-Effect Transistors: Theory, Design and Applications*, 2nd ed., Wiley, New York, 1986.

43. R. Soares, *GaAs MESFET Circuit Design*, Artech House, Norwood, MA, 1988.

44. M. Caulton, S. P. Knight, and D. A. Daly, "Hybrid Integrated Lumped-Element Microwave Amplifiers," *IEEE Trans. Electron Devices*, Vol. ED-15, pp. 459–466, 1968.

45. M. Caulton, B. Hershenov, S. P. Knight, and R. E. DeBrecht, "Status of Lumped Elements in Microwave Integrated Circuits—Present and Future," *IEEE Trans. Microwave Theory Tech.*, Vol. MTT-19, pp. 588–599, 1971.

46. R. S. Pengelly and D. C. Rickard, "Measurement and Application of Lumped Elements Up to J-Band," *Proc. 7th Eur. Microwave Conf.*, pp. 460–464, September 1977.

47. E. Pettenpaul, H. Kapusta, A. Weisgerber, H. Mampe, J. Luginsland, and I. Wolff, "CAD Models of Lumped Elements on GaAs Up to 18 GHz," *IEEE Trans. Microwave Theory Tech.*, Vol. MTT-36, pp. 294–304, February 1988.

48. F. W. Grover, *Inductance Calculations*, D. Van Nostrand, Princeton, NJ, 1946; reprinted by Dover, Mineola, NY, pp. 17–47, 1962.

49. D. M. Krafcsik and D. E. Dawson, "A Closed-Form Expression for Representing the Distributed Nature of the Spiral Inductor," *IEEE MTT-S Int. Microwave Theory Tech.*, Vol. MTT-24, pp. 87–92, June 1986.

50. G. D. Alley, "Interdigital Capacitors and Their Applications to Lumped-Element Microwave Integrated Circuits," *IEEE Trans. Microwave Theory Tech.*, Vol. MTT-18, pp. 1028–1033, 1970.

51. J. L. Hobdell, "Optimization of Interdigital Capacitors," *IEEE Trans. Microwave Theory Tech.*, Vol. MTT-27, pp. 788–791, 1972.

52. K. C. Gupta, R. Garg, and I. J. Bahl, *Microstrip Lines and Slotlines*, Artech House, Norwood, MA, 1979.

53. H. Y. Yand and N. G. Alexopoulos, "A Dynamic Model for Microstrip-Slotline Transition and Related Structures," *IEEE Trans. Microwave Theory Tech.*, Vol. MTT-36, pp. 286–293, February 1988.

54. B. Bhat and S. K. Koul, *Analysis, Design and Applications of Fin Lines*, Artech House, Norwood, MA, 1987.

55. J. I. Smith, "The Even- and Odd-Mode Capacitance Parameters for Coupled Lines in Suspended Substrates," *IEEE Trans. Microwave Theory Tech.*, Vol. MTT-19, pp. 424–431, May 1971.

56. F. Giannini, C. Paoloni, and M. Ruggieri, "CAD-Oriented Lossy Models for Radial Stubs," *IEEE Trans. Microwave Theory Tech.*, Vol. MTT-36, pp. 305–313, February 1988.

57. K. J. Russell, "Microwave Power Combining Techniques," *IEEE Trans. Microwave Theory Tech.*, Vol. MTT-27, pp. 472–478, May 1979.

58. E. J. Wilkinson, "An *N*-Way Hybrid Power Divider," *IRE Trans. Microwave Theory Tech.*, pp. 116–118, January 1960.

59. K. Chang and C. Sun, "Millimeter-Wave Power-Combining Techniques," *IEEE Trans. Microwave Theory Tech.*, Vol. MTT-31, pp. 97–107, February 1983.

60. E. Belohoubeck, R. Brown, H. Johnson, A. Fathy, D. Bechte, D. Kalotitis, and E. Mykietyn, "30-Way Radial Power Combiner for Miniature GaAs FET Power Amplifiers," *IEEE MTT-S Int. Microwave Symp. Dig.*, pp. 515–519, June 1986.

61. G. W. Swift and D. I. Stones, "A Comprehensive Design Technique for the Radial Wave Power Combiner," *IEEE MTT-S Int. Microwave Symp. Dig.*, pp. 279–281, June 1988.

62. A. G. Bert and D. Kaminsky, "The Traveling-Wave Power Divider/Combiner," *IEEE MTT-S Int. Microwave Symp. Dig.*, pp. 487–489, June 1980.

63. H. Q. Tserng and P. Saunier, "10-30 GHz Monolithic GaAs Travelling-Wave Divider/Combiner," *IEE Electron. Lett.*, Vol. 21, pp. 950–951, October 1985.

64. M. F. Durkin, "35-GHz Active Aperture," *IEEE MTT-S Int. Microwave Symp. Dig.*, pp. 425–427, June, 1981.

65. J. W. Mink, "Quasi-optical Power Combining of Solid-State Millimeter-Wave Sources," *IEEE Trans. Microwave Theory Tech.*, Vol. MTT-34, pp. 273–279, February 1986.

66. A. Ezzeddine, H.-L. A. Hung, and H. C. Huang, "High-Voltage FET Amplifier for Satellite and Phased-Array Applications," *IEEE MTT-S Int. Microwave Symp. Dig.*, pp. 336–339, June 1985.

67. K. E. Peterson, H.-L. Hung, F. R. Phelleps, T. F. Noble, and H. C. Huang, "Monolithic High-Voltage FET Power Amplifiers," *IEEE MTT-S Int. Microwave Symp. Dig.*, pp. 945–948, June 1989.

68. V. Sokolov, J. Geddes, and A. Contolatis, "Two Stage MESFET Monolithic Gain Control Amplifier for Ka-Band," *IEEE Microwave Millim.-Wave Monolithic Circuit Symp. Dig.*, pp. 75–79, June 1987.

69. R. B. Culbertson and D. C. Zimmermann, "A 3-Watt X-Band Monolithic Variable Gain Amplifier," *IEEE Microwave Millim.-Wave Monolithic Circuits Symp. Dig.*, pp. 121–124, May 1988.

70. H.-L. A. Hung, G. Hegazi, K. E. Peterson, F. R. Phelleps, and H. C. Huang, "Monolithic Dual-Gate MESFET Power Amplifiers," *IEEE GaAs IC Symp. Tech. Dig.*, pp. 41–44, November 1988.

71. T. T. Lee, H.-L. A. Hung, A. Cornfeld, F. R. Pelleps, J. L. Singer, J. F. Bass, and H. C. Huang, "GaAs MBE Monolithic Amplifiers for Millimeter-Wave Applications," *GOMAC Dig.*, pp. 369–372, November 1988.

72. A. E. Geissberger, I. J. Bahl, E. L. Griffin, and R. A. Sadler, "A few Refractory Self-Aligned Gate Technology for GaAs Microwave Power FET's and MMIC's," *IEEE Trans. Electron Devices*, Vol. ED-35, pp. 615–623, May 1988.

73. A. E. Geissberger, R. A. Sadler, M. L. Balzan, G. E. Menk, I. J. Bahl, and E. L. Griffin, "High-Efficiency X- and Ku-Band GaAs Power FET's Fabricated Using Refractory SAG Technology," *IEEE GaAs IC Symp. Tech. Dig.*, pp. 309–312, November 1988.

74. COMSAT Laboratories, private communication.

75. M. Paggi, P. H. Williams, and J. M. Borrego, "Nonlinear GaAs MESFET Modeling Using Pulsed Gate Measurements," *IEEE MTT-S Int. Microwave Symp. Dig.*, pp. 229–231, May 1988.

76. COMSAT Laboratories, private communication.

77. R. H. Knerr, "A New Type of Waveguide-to-Stripline Transition," *IEEE Trans. Microwave Theory Tech.*, Vol. MTT-16, pp. 192–194, March 1968.

78. B. Glance and R. Trumbaralo, "A Waveguide-to-Suspended Stripline Transition," *IEEE Trans. Microwave Theory Tech.*, Vol. MTT-21, pp. 117–118, February 1973.

79. Y.-C. Shih, T.-N. Ton, and L. O. Bui, "Waveguide-to-Microstrip Transitions for Millimeter-Wave Applications," *IEEE MTT-S Int. Microwave Symp. Dig.*, pp. 473–475, May 1988.

80. T. T. Lee, H.-L. Hung, and C. E. Mahle, "AMPAC—An Automated Microwave Power Amplifer Characterization System," *IEEE 21st ARFTG Conf. Dig.*, pp. 133–155, Spring 1983.

81. D. A. Austin, "Impulse Response of Photoconductor in Transmission Lines," *IEEE J. Quantum Electron.*, Vol. QE-19, pp. 639–648, April 1983.

82. C. Rauscher, "Picosecond Reflectometry Technique for On-Chip Characterization of Millimeter-Wave Semiconductor Devices," *IEEE MTT-S Int. Microwave Symp. Dig.*, pp. 881–884, May 1987.

83. H.-L. A. Hung, T. T. Lee, P. Polak-Dingels, E. Chauchard, K. Webb, C. H. Lee, and H. C. Huang, "Characterization of GaAs Monolithic Circuits by Optical Techniques," *SPIE Proc., Opt. Technol. Microwave IV*, Vol. 1102, pp. July 1989.

84. H.-L. A. Hung, P. Polak-Dingels, K. J. Webb, T. Smith, C. H. Lee, and H. C. Huang, "Millimeter-Wave Monolithic Integrated Circuit Characterization by a Picosecond Optoelectric Technique," *IEEE Trans. Microwave Theory Tech.*, Vol. MTT-37, pp. 98–106, March 1989.

85. P. Polak-Dingels, H.-L. A. Hung, K. J. Webb, T. T. Lee, T. Smith, and C. H. Lee, "An Optoelectronic Technique for *S*-Parameter Measurements of GaAs Monolithic ICs," *Int. Conf. Infrared Millim.-Waves Conf. Dig.*, pp. 69–70, December 1988.

86. S. H. Wemple and H. Huang, "Thermal Design of Power GaAs FETs," in J. V. Dilorenzo, Ed., *GaAs FET Principles and Technology*, Artech House, Norwood, MA, 1982, pp. 314–347.

87. B. S. Sage, "A Proposed Method for Testing Thermal Resistance of MESFETs," *Microwave Syst. News*, Vol. 7, pp. 66–70, November 1977.

88. M. M. Minot, "Thermal Characterization of Microwave Power FETs Using Nematic Liquid Crystals," *IEEE MTT-S Int. Microwave Symp. Dig.*, pp. 495–498, June 1986.

89. H. F. Cooke, "Precise Technique Finds FET Thermal Resistance," *Microwaves and RF*, Vol. 25, pp. 85–87, August 1986.

90. H. Fukui, "Thermal Resistance of GaAs Field-Effect Transistors," *Int. Electron Devices Meet. Tech. Dig.*, pp. 118–121, December 1980.

91. R. Bar-Gadda, "The Thermal Modeling of Integrated-Circuit Device Packages," *IEEE Trans. Electron Devices*, Vol. ED-34, pp. 1934–1938, September 1987.

92. Network Analysis Associates, Inc., SINDA, a commercial thermal analysis computer program, Network Analysis Associates, Inc., Fountain Valley, CA.

93. *Sage Thermal Resistance Tester Manual*, Model Theta 220, Sage Enterprise, Inc., Palo Alto, CA.

94. F. Sechi, B. S. Perlman, and J. M. Cusack, "Computer Controlled Infrared Microscope for

Thermal Analysis of Microwave Transistors," *IEEE MTT-S Int. Microwave Symp. Dig.*, pp. 143–146, June 1977.

95. EDO Corporation, CompuTherm, thermal analysis system for microelectronics, Barnes Engineering Division, EDO Corporation, Shelton, CT.

96. H. Bierman, "Microwave Packages Meeting GaAs Challenges," *Microwave J.*, Vol. 29, pp. 26–39, November 1986.

97. J. A. Frisco, "High Speed Microwave Packaging of GaAs Integrated Circuits," *IEEE GaAs IC Symp. Tech. Dig.*, pp. 173–176, November 1986.

98. B. A. Ziegner, "High Performance MMIC Hermetic Packaging," *Microwave J.*, Vol. 29, pp. 133–139, November 1986.

99. F. Ishitsuka and N. Sato, "Low Cost, High-Performance Package for a Multi-chip MMIC Module," *IEEE GaAs IC Symp. Tech. Dig.*, pp. 2221–224, November 1988.

100. G. C. Rieder, D. E. Heckaman, J. A. Frisco, R. H. Vought, R. H. Higman, and G. R. Perkins, "Waffleline Packaging Techniques for Millimeterwave Integrated Circuits," *IEEE GaAs IC Symp. Tech. Dig.*, pp. 217–220, November 1988.

101. A. Christ and H. L. Hartnagel, "Three-Dimensional Finite-Difference Method for the Analysis of Microwave-Device Imbedding," *IEEE Trans. Microwave Theory Tech.*, Vol. MTT-35, pp. 688–696, August 1987.

102. R. Esfandiari, D. Yang, S. Chan, S. Lin, and R. K. Ellis, "A Low Cost Packaging/Testing Procedure for Manufacturing GaAs MMIC," *IEEE Microwave Millim.-Wave Monolithic Circuit Symp. Dig.*, pp. 135–137, May 1987.

103. K. Aono, O. Ishihara, K. Nishitani, M. Nakatani, K. Fujikawa, M. Ohtani, and T. Odaka, "Gamma Ray Radiation Effects on MMIC's Elements," *IEEE GaAs IC Symp. Tech. Dig.*, pp. 139–142, November 1984.

104. A. Meulenburg, H.-L. A. Hung, K. E. Peterson, and W. T. Anderson, "Totel Dose and Transient Radiation Effects on GaAs MMICs," *IEEE Trans. Electron Devices*, Vol. ED-35, pp. 2125–2132, December 1988.

105. W. Anderson, R. Harrison, J. Gerdes, J. Beall, and J. Roussos, "Combined Pulsed Neutron and Flash X-Ray Radiation Effects in GaAs MMICs," *IEEE GaAs IC Symp. Tech. Dig.*, pp. 53–56, November 1988.

106. L. F. Lester, P. M. Smith, P. Ho, P. C. Chao, R. C. Tiberio, K. H. G. Duh, and E. D. Wolf, "0.15 μm Gate-Length Double Recess Pseudomorphic HEMT with F_{max} of 350 GHz," *Int. Electron Device Meet. Tech. Dig.*, pp. 172–175, December 1988.

107. P. Ho, P. C. Chao, K. H. G. Duh, A. A. Jabra, J. M. Ballingall, and P. M. Smith, "Extremely High Gain, Low Noise InAlAs/InGaAs HEMTs Grown by Molecular Beam Epitary," *Int. Electron Device Meet. Tech. Dig.*, pp. 184–186, December 1988.

108. M. Madihian, H. Shimawaki, and K. Honjo, "A 20–28 GHz AlGaAs/GaAs HBT Monolithic Oscillator," *IEEE GaAs IC Symp. Tech. Dig.*, pp. 113–116, November 1988.

109. M. Kim, A. Oki, J. Cannon, P. Chow, B. Nelson, D. Smith, J. Canyon, C. Yang, B. Allen, and Dixit, "12–40 GHz Low Harmonic Distortion and Phase Noise Performance of GaAs Heterojunction Bipolar Transistors," *IEEE GaAs IC Symp. Tech. Dig.*, pp. 117–120, November 1988.

110. B. Bayraktaroglou, N. Camilleri, H. D. Shih, and H. Q. Tserng, "AlGaAs/GaAs Heterojunction Bipolar Transistors with 4 W/mm Power Density at X-Band," *IEEE MTT-S Int. Microwave Symp. Dig.*, pp. 969–972, June 1987.

111. N. H. Sheng, M. F. Chang, P. M. Asbeck, K. C. Wang, G. J. Sullivan, D. L. Miller, J. A. Higgins, E. Sovero, and H. Basit, "High Power GaAlAs/GaAs HBTs for Microwave Applications," *Int. Electron Devices Meet. Tech. Dig.*, pp. 95–98, December 1987.

112. D. A. Sunderland and P. D. Dapkus, "Optimizing *n-p-n* and *p-n-p* Heterojunction Bipolar Transistors for Speed," *IEEE Trans. Electron Devices*, Vol. ED-34, p. 367, 1987.

113. B. Bayraktaroglou and N. Camilleri, "Microwave Performance of *npn* and *pnp* AlGaAs/GaAs Heterojunction Bipolar Transistors," *IEEE MTTS-S Int. Microwave Symp. Dig.*, pp. 529–532, May 1988.

11

FETS: LOW-NOISE APPLICATIONS

Thomas A. Midford

Microwave Products Division
Hughes Aircraft Company
Torrance, California

11.1 INTRODUCTION

In 1976, Charles Liechti [1] published a significant review paper summarizing the state-of-the-art of microwave field-effect transistors. A large portion of this review dealt with the first five years of rapid progress in GaAs metal field-effect transistor (MESFET) technology achieved during the first half of the 1970s. In his introduction, Leichti predicted that field-effect transistor technology had a great deal of potential for yet further near-term advances because:

1. A variety of available device structures, in addition to MESFETs, junction field-effect transistors (JFETs), and insulated-gate field-effect transistors (IGFETs), were suitable for microwave amplification.
2. A number of III–V materials including GaAs, InP, InGaAs, and InAsP which have superior majority carrier transport properties to silicon were available, although with relatively primitive materials technologies.
3. Further reduction in the size of critical transistor features to submicrometer dimensions was possible with existing technology.
4. Monolithic integration on semi-insulating substrates promised effective device isolation with low parasitic capacitance, low-loss interconnections, and high packing densities.

Some of these prognoses were later proven to be remarkably accurate. The GaAs field-effect transistor has continued to be the dominant microwave solid-state device for both low-noise and power applications and is still by far the most important active device employed in monolithic microwave integrated circuit (MMIC) technology. During the 1980s, heterojunction field-effect transistors which are called HEMTs, MODFETs, or HFETs [2], which are based on ternary and quaternary III–V materials systems, have made spectacular advances and are described in Chapter 9 of this volume. Throughout the 1980s, continuing improvements in materials, high-resolution lithography, and processing techniques have resulted in major performance improvements for both MESFETs and HEMTs; and during this period, operation of these devices has been extended to millimeter wave frequencies.

This chapter will examine progress in low-noise GaAs MESFET devices and amplifiers over the past 20 years. An overview perspective may be obtained from Fig. 11.1, which shows the noise performance of narrow-band multistage MESFET amplifiers as a function of frequency at three times during this period. Amplifier performance was chosen as a status indicator to avoid the problems inherent in comparing device level results between organizations, for example, how to deal with test fixture corrections and as being more representative of applications.

The 1975 data, taken from a survey paper by Pucel et al. [3], indicate the status of low-noise MESFETs after about five years of intensive development. During this time, basic device technology, which included epitaxial materials and 1-μm gate lengths based on contact lithography, was developed [1]. During approximately the same time period, a comprehensive theoretical model of low-noise MESFET operation evolved which culminated in the papers of Statz et al. [4] and Pucel et al. [5]. These works provide the basis for understanding low-noise MESFET operation to the present day.

The second data line in Fig. 11.1 represents the state of the low-noise MESFET early in the 1980s. During the intervening five years, substantial progress was made in performance improvement as well as extending operation of the device to higher frequencies. Further materials enhancements, including the introduction of molecular beam epitaxy (MBE) [6], the reduction of gate lengths to 0.3 μm [7], and optimized device processing leading to the reduction of source and gate parasitic resistances, contributed to a substantial reduction of noise figure and extended low-noise MESFET operation to near 30 GHz [7].

The third data line shown in Fig. 11.1 represents the current status of low-noise MESFET amplifiers. During the 1980s, major research activities were shifted to the development of MIMIC technology and to heterojunction field-effect transistors. However, as described in Section 11.4, continuing improvements in substrate, buffer, and active layer material quality, the latter including optimized channel profiles such as spike doping, and the reduction of minimum gate length to near 0.1 μm have resulted in the extension of MESFET operation to frequencies greater than 60 GHz.

During the same period, techniques and hardware for the accurate characterization of millimeter wave three-terminal devices, in particular network analysis and on-wafer

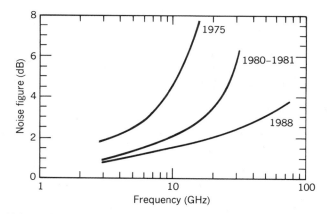

Figure 11.1   Noise performance of low-noise MESFET amplifiers shown as a function of frequency at three times during the period of intensive MESFET development. Data shown for 1975 are from Ref 3.

probing [8], have advanced rapidly and have made a significant contribution to the development of devices with improved performance at higher frequencies.

This chapter is organized in five subsequent sections, each dealing with a major portion of low-noise MESFET technology and applications. Section 11.2 provides a brief review of MESFET device physics including small-signal equivalent circuits and limitations on high-frequency operation. Section 11.3 deals with MESFET noise theory including the noise models of Statz et al. [4] and Pucel et al. [5] and the Fukui empirical noise theory [9]. Section 11.4 describes the design and fabrication of low-noise MESFETs including materials and processing and concludes with a design example of a highly optimized 0.1-μm gate MESFET which has demonstrated state-of-the-art low-noise performance near 60 GHz. Section 11.5 focuses on low-noise MESFET applications including low-noise device characterization and amplifier design and concludes with a second design example of a K-band three-stage low-noise amplifier. Section 11.6 gives the outlook of future development.

11.2 LOW-NOISE MESFET DEVICE PHYSICS

11.2.1 Principles of Operation

Understanding the low-noise properties of the GaAs MESFET requires an understanding of critical details of the device physics, in particular, electron transport in the active region of the device as influenced by the velocity-field characteristics of carriers and the electric field distribution in the vicinity of the gate as determined by the device structure as well as the gate and drain bias voltages.

Velocity Saturation of Carriers. Fundamental to operation of the GaAs MESFET is the carrier velocity–electric field relationship within the active region of the semiconductor. Figure 11.2 illustrates the electron velocity-field dependency in both gallium arsenide and silicon over the range of electric fields encountered in device structures [10]. For small fields both materials behave as linear resistors with carrier velocity proportional to electric field. At larger fields, both materials exhibit velocity saturation and field-independent velocities of approximately 10^7 cm/s at fields greater than a critical field

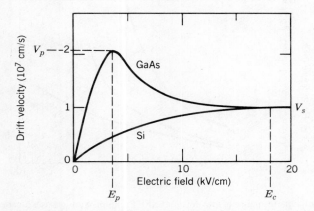

Figure 11.2 Equilibrium electron drift velocity versus electric field in GaAs and silicon (after Ruch [10]).

(E_c) of about 20 kV/cm. However, the velocity-field behavior of GaAs is strikingly different from that of silicon at lower fields. At fields below about 3 kV/cm, the electron velocity in both materials rises rapidly with electric field in the linear (constant-mobility) region. Here the mobility for electrons in doped GaAs is approximately six times that of electrons in silicon. At fields greater than a few hundred volts per centimeter, the electron velocity in GaAs begins to saturate, reaching a peak velocity of near 2×10^7 cm/s at 3 kV/cm. The peak velocity of electrons in GaAs is thus about twice that in silicon. For larger fields, the velocity then decreases with increasing field to the saturated value. In contrast, the electron velocity in silicon increases monotonically and saturates gradually, reaching a peak velocity of near 10^7 cm/s at an electric field value of approximately 20 kV/cm. Carrier transport in the high-electron-field region of a GaAs MESFET is complicated by the velocity-field behavior [10, 11]; in addition, for gate lengths shorter than 2–3 μm, nonequilibrium velocity-field characteristics give rise to velocity overshoot effects which are discussed below [12].

MESFET Device Operation. Figure 11.3a schematically illustrates a gateless GaAs MESFET in which current is carried in a thin n-type channel layer which is supported by an insulating substrate. At the surface of the conducting layer, two ohmic contacts (source and drain) are provided.

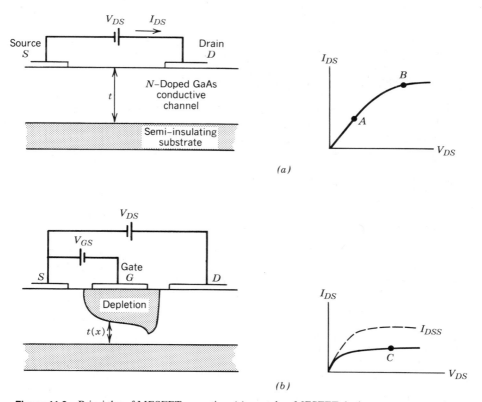

Figure 11.3 Principles of MESFET operation: (a) a gateless MESFET depicts current saturation; (b) voltage applied to the gate of a gated FET controls the extent of channel depletion region and hence the current flowing from source to drain.

Upon applying a positive voltage V_{DS} to the drain, electrons flow from the source to the drain. For small values of V_{DS}, the conductive layer behaves as a linear conductance (region A). For larger voltages, the electron drift velocity increases more slowly than the electric field; thus the relationship between drain current and voltage falls below the linear conductance line. For further increase in V_{DS}, the electric field (in some portions of the source–drain region) exceeds the peak field E_p at which the electrons reach a maximum velocity v_p. At this value of drain voltage, the drain current–voltage exhibits hard saturation (region B).

Figure 11.3b shows a complete MESFET device structure including the Schottky barrier gate (G). The MESFET $I–V$ characteristics are influenced by the voltage applied to the gate. With the gate at the source potential, the Schottky, or "built-in," voltage associated with the gate contact causes a portion of the channel under the gate to be depleted of carriers. The depleted region behaves like an insulator and consequently reduces the cross section of the channel which contributes to current flow. The extent of the depletion region depends on the voltage applied to the gate. For increasingly negative values of gate voltage, the cross section of the depletion region expands, thereby further constricting the conductive cross section and limiting the current. For sufficiently negative gate voltage, the channel over some portion becomes fully depleted of carriers so that drain current is effectively shut off. The value of V_g at which this occurs is the pinchoff voltage v_p and in general v_p varies with the applied drain voltage.

The noise performance of the GaAs MESFET depends critically on the spatial dependence of velocity saturation in the conducting channel. In addition, for device gate lengths below 1–3 μm, nonequilibrium field characteristics also influence noise performance. Current MESFET devices may have gate lengths as short as 0.1 μm.

An early treatment of electron transport in GaAs MESFETs based on equilibrium, or steady-state, electron transport was performed by Himsworth [13]. Figure 11.4 [1] summarizes the qualitative features of the operation of a GaAs MESFET with a 3-μm gate length operated in the saturated current region using data from Ref. 13. For bias voltages $V_{DS} = +3.0$ V and $V_{GS} = -1.0$ V, the narrowest channel cross section is located at the drain end of the gate. The electric field also peaks in this region. The electron velocity rises to its peak value v_p at X_1 near the center of the channel, falls to the minimal saturated value at the gate edge, and then again rises to the peak value at the edge of the depletion region closest to the drain. The drain current flowing from drain to source, I_{DS}, is given by

$$I_{DS} = wqn(x)v(x)t(x) \tag{11.1}$$

where w is the gate width, q the electronic charge, n the density of conduction electrons, v the electron drift velocity, t the conductive layer thickness, and x the coordinate in the direction of electron flow. The electron density n is equal to the donor density as long as the electric field does not exceed the critical field E_c. As the voltage along the channel increases from source to drain, the metal–semiconductor junction formed at the gate becomes more strongly reversed biased, and the depletion layer correspondingly becomes wider on the drain edge of the gate where the conductive channel is consequently the narrowest. In the region between X_1 and X_2, where the channel narrows and the electron velocity slows with increasing X, there is an accumulation of electrons to preserve current continuity as dictated by Eq. (11.1). In the region between X_2 and X_3, the conductive channel widens, the electric field decreases, and the electrons move faster resulting in a partial depletion of electrons. The amount of charge contained in the depletion and accumulation regions is approximately equal. The stationary dipole layer formed by the depletion and accumulation region contains most of the drain vol-

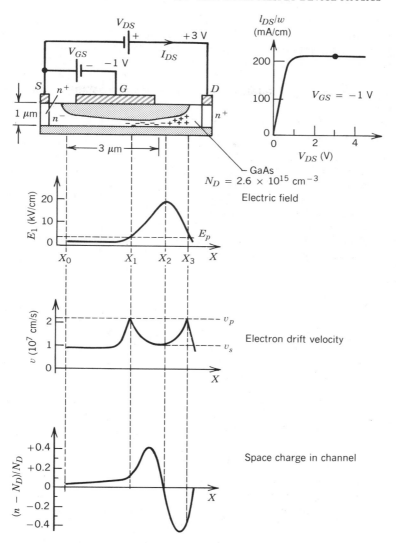

Figure 11.4 Channel cross section, electric field, electron drift velocity, and channel space-charge distribution for a GaAs MESFET operated in the current-saturated region (data from Himsworth [13], after Leichti [11]).

tage drop and is a contributor to the drain-channel capacitance C_{dc} in the intrinsic MESFET equivalent circuit model.

Short-Gate Effects and Velocity Overshoot. In microwave and millimeter wave MESFETs, which have gate lengths significantly below 1 μm, electrons do not remain in the high field of the channel long enough to reach equilibrium velocity conditions. Monte Carlo–based computer simulations have been used to study nonequilibrium velocity-field behavior in GaAs [12–15]. The results of these simulations may be summarized as follows: for values of E below the threshold field E_p, electron transport is governed by equilibrium conditions. However, if electrons enter a region where $E > E_p$, they are accelerated to a higher velocity before relaxing to the equilibrium velocity. Such velocity overshoot may be as much as twice the peak equilibrium velocity v_p [10]. The overshoot decreases the electron transit time through the high-field region and shifts the accumulation region toward the drain [15].

11.2.2 Small-Signal Equivalent Circuits

A number of small-signal equivalent circuits have been developed for low-noise MESFETs [16–21]. Although an accurate equivalent circuit representation should model the channel as a distributed RC network [1], simple lumped-element models provide accurate representations of measured FET S-parameters to greater than 12 GHz and beyond and are actually used to much higher frequencies.

Figure 11.5a shows the equivalent circuit for a MESFET connected in the common source configuration. The physical origin of the circuit elements is shown in Fig. 11.5b. The circuit elements for the intrinsic transistor are located within the dashed lines, while the parasitic or extrinsic elements are outside of this area. The circuit elements are defined as follows:

INTRINSIC ELEMENTS

C_{gs}	gate-to-channel capacitance
C_{dc}	drain-to-channel capacitance
C_{dg}	drain-to-gate feedback capacitance
R_i	channel resistance
R_{ds}	output resistance
g_m	low-frequency transconductance
τ_0	carrier transit time in through the portion of the channel where carrier velocity is saturated ($E > E_p$)

EXTRINSIC ELEMENTS

C_{ds}	drain–source capacitance
R_g	gate resistance
R_s	source-to-channel resistance including contact resistance
R_d	drain-to-channel resistance including contact resistance

In the intrinsic FET model, the total gate-to-channel capacitance is represented by the sum of $C_{dg} + C_{gs}$; R_i and R_{ds} model the channel resistance and i_{ds} defines the voltage-controlled current source related to the RF voltage applied across C_{gs} through the transadmittance Y_m. At microwave frequencies Y_m is related to the low-frequency transconductance g_m and the phase delay τ_0 corresponding to the carrier transit time through the portion of the channel where there is velocity saturation. The gain of the MESFET is embodied in the transconductance g_m, which is defined as the ratio of a small change in drain current produced by a small change in gate voltage with a fixed source–drain voltage. Internal feedback is provided by the capacitances C_{dc} and C_{ds}. The capacitance C_{ds} is associated with the depletion/accumulation dipole layer in the drain end of the channel [1].

The parasitic elements in the extrinsic circuits include the source and drain resistances R_s and R_d, the gate metal resistance R_g, and the substrate capacitance C_{ds}. Other

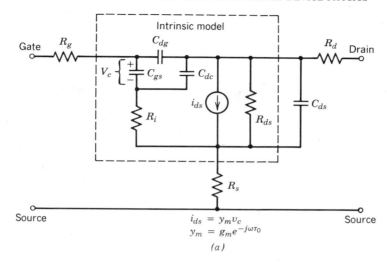

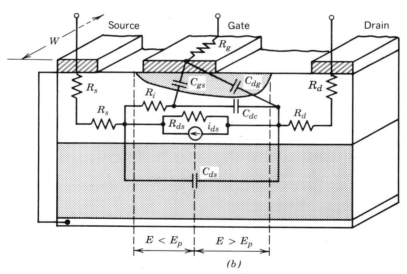

Figure 11.5 (a) Equivalent circuit for the MESFET in the common source configuration and (b) the physical basis of the circuit elements (after Leichti [1]).

parasitic elements not shown include contacting inductances in series with the gate, source and drain resistances, and capacitance values associated with bonding pads [1, 17].

Because of the internal feedback mechanisms, the MESFET is subject to low-frequency instabilities. Analysis of the equivalent circuit leads to the critical frequency f_k above which the MESFET is unconditionally stable. The frequency f_k is given approximately by

$$f_k \approx \frac{1}{2\pi(\tau_0 + \tau_1 + \tau_2)} \tag{11.2}$$

where τ_0 is defined in Fig. 11.5 and

$$\tau_1 = \frac{C_{dg}(2R_g + R_i + R_s)}{C_{dg}/C_{gs} + R_s/R_{ds}} \tag{11.3}$$

$$\tau_2 = \frac{2}{(g_m/C_{gs})(C_{dg}/C_{gs} + R_s/R_{ds})} \frac{R_g + R_i + R_s}{R_{ds}} \tag{11.4}$$

At frequencies below f_k, a MESFET with a matched input port may become unstable because an increasing fraction of the output voltage is fed back to the input through C_{dg} [1].

Other important device figures of merit include:

The Cutoff Frequency f_T. The frequency at which current gain falls to unity, which is given approximately by

$$f_T \approx \frac{1}{2\pi} \frac{g_m}{C_{gs}} \tag{11.5}$$

The Unilateral Gain G_u [22]. This is approximated by

$$G_u \approx \left(\frac{f_{max}}{f}\right)^2 \tag{11.6}$$

The Maximum Frequency of Oscillation f_{max} [17]. This is given by

$$f_{max} \approx \frac{f_T}{2\sqrt{r_1 + f_T\tau_3}} \tag{11.7}$$

where r_1 is a ratio of the input to output resistance:

$$r_1 = \frac{R_g + R_i + R_s}{R_{ds}} \tag{11.8}$$

and $\tau_3 = 2\pi R_g C_{dg}$.

Equation (11.6) implies a gain decrease of 6 dB per octave with increasing frequency. In order to enhance the high-frequency performance of MESFETs, both the intrinsic and parasitic elements must be optimized. In order to increase f_{max}, f_T and the resistance ratio R_{ds}/R must be optimized in the intrinsic MESFET. The parasitic resistances R_g and R_s must be minimized as must the feedback capacitance C_{dg} [1].

11.2.3   MESFET Operation at High Frequencies

The high-frequency limitations of MESFETs result from device geometry as well as material and processing parameters. In device geometry, the gate length L is the most critical parameter. Decreasing L decreases the gate-to-channel capacitance C_{gs} and increases the transconductance g_m. For short gate devices, f_T is proportional to $1/L$ [23]. High-frequency operation is obtained by reducing the gate length to the minimum value possible with the available technology. In 1989, this lower limit on gate length is near

0.1 μm using electron or ion beam lithography. Since decreasing gate length decreases the gate cross section for a given thickness of gate metal, shrinking the gate length will result in higher gate resistance unless special techniques are employed. These are discussed in Section 11.4.

A practical lower limit for the gate length is established when the gate length is approximately equal to the channel thickness. Early theoretical calculations by Drangeid and Summerholder [24] showed that MESFET performance is improved by reduction of the gate length if the channel thickness is decreased correspondingly so that the ratio of gate length to channel thickness is greater than approximately 3. As the gate length is decreased, increased doping density is required to maintain channel current and transconductance. A thin, highly doped channel under the gate also contributes to a high value of g_m. For uniformly doped channels, doping densities of up to 1×10^{18} may be used with both short ($L \leq 0.25$ μm) and long ($L \simeq 1$ μm) devices [25]. For short gate lengths, increasing the doping concentration leads to a substantial increase in f_T, as shown in Fig. 11.6. An upper limit on channel doping density is provided by the avalanche breakdown voltage, which decreases with increasing doping. For experimental devices with uniform doping of 1×10^{18} cm^{-3}, Daembkes et al. [25] have reported gate–drain breakdown voltages of near 6 V, a value great enough for most low-noise applications. An alternative approach to uniform values of channel doping is through the use of a spike doping profile as described in Section 11.4.

Optimization of MESFETs for low-noise operation at high frequencies is described in Section 11.3.

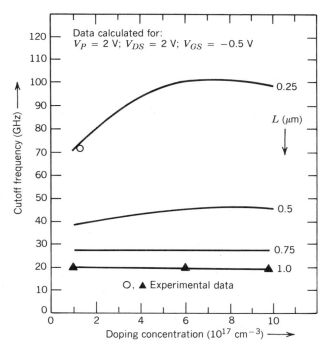

Figure 11.6   Calculated dependence of the cutoff frequency f_T on the channel doping density for MESFETs with various gate lengths (after Daembkes et al. [25]).

11.3 NOISE PERFORMANCE AND MODELS

11.3.1 Summary of MESFET Noise Theory Development

The noise theory of the field-effect transistor was developed between 1960 and 1975. During the mid-1970s a number of comprehensive accounts of MESFET noise were published [4, 5] summarizing the developments of this period. Only a brief account is presented here.

Schockley [26] first described the operation of the field-effect transistor. He assumed a constant mobility throughout the conducting channel region. In a series of pioneering papers using the Schockley model, Van der Ziel [27–29] developed the intrinsic signal and noise properties of the MESFET. This work provided the basis for subsequent refinements of the noise theory and models by later workers. Van der Ziel showed that noise in the intrinsic MESFET is thermal in origin and can be depicted, from an equivalent circuit perspective, by two white noise generators, one in the drain circuit and one in the gate circuit. The two noise sources are partially correlated because the gate noise generator results from noise induced on the gate electrode by thermal noise fluxuations on the drain current. This representation is a specific embodiment of the more general representation that any linear noisy two-port system can represented by a noiseless two-port one with noise current generators connected across the input and output ports [30].

For short gate GaAs MESFETs, the constant-mobility model of Schockley and Van der Ziel is not accurate. These devices, when biased above pinchoff, where drain current is strongly saturated, have a longitudinal electric field much greater than E_p over a substantial portion of the channel region. In this region, conduction is nonohmic and mobility is a function of the electric field. Baechtold [31, 32] corrected for the nonconstant mobility and also revised the Schockley–Van der Ziel models to account for a field-dependent electron temperature. The mobility is assumed to be constant up to the critical field E_p while for fields larger than the critical field, velocity is assumed to be constant. This approach neglects the region above E_p and below E_c where the electron velocity is decreasing with increasing electric field. The region in which electron velocity is saturated can occupy a major part of the total channel length. The operating characteristics of the transistor, including noise, are thus strongly influenced by the region of saturated velocity.

A major advance in the theory of MESFET noise was provided in mid-1975 by the comprehensive analyses of Statz et al. [4] and Pucel et al. [5], who included the effects of velocity saturation both in the small-signal parameters and in the noise performance. These works represent the basis of modern low-noise MESFET noise theory.

11.3.2 Velocity Saturation Models and High Field Diffusion Noise

The Two-Zone Model and Equivalent Circuit. Based on the work of Turner and Wilson [33], Statz and Pucel based their noise analysis on the two-section MESFET model shown in Fig. 11.7. The idealized MESFET structure in (a) consists of source (D), gate (S), and drain (G) electrodes, all of width W. The length of the gate is L. The conducting channel of thickness is uniformly doped to a density N_d with a low field mobility μ_0. Typical values for these parameters for an X-band low noise MESFET are $N_d = 2 \times 10^{17}$–3×10^{17}, $a = 0.1$–0.3 μm and $\mu_0 = 4000$–5000 cm^2/V-s. As shown in Fig. 11.7b, the GaAs velocity field characteristic is idealized by a piecewise linear approximation with v_s constant for fields greater than E_s. The value of E_s was chosen to be 2.9 kV/cm and the saturated velocity v_s to be 1.3×10^7 cm/s. Based on the piecewise linear approximation to the velocity field, the conducting channel under the gate is divided into two zones

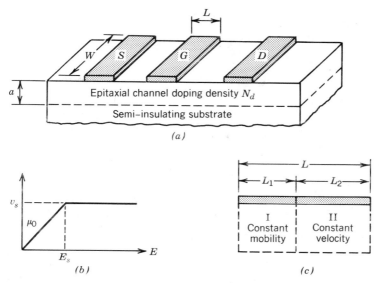

Figure 11.7 Two-section MESFET model used in the noise analysis of Pucel et al. [5]: (*a*) MESFET model, (*b*) velocity-field characteristic assumed in the calculation, (*c*) two-region channel model.

as shown in Fig. 11.7*c* and as originally suggested by Grabene and Ghandi [34]. In the two-zone model, the portion of the conducting channel near the source end is assumed to have constant mobility (region I) while the remaining portion near the drain has a constant (saturated) velocity (region II). The position of the boundary between the two regions is a strong function of source–drain bias and a weaker function of gate–drain bias, and the length L_2 of region II increases with increasing source–drain bias [3].

Based on the two-zone model, it is shown that when the FET is biased into current saturation, the length L_2 is of the order of 2–4 times the active layer thickness *a*. As a consequence, for the majority of contemporary low-noise MESFET designs with gate lengths less than 1 μm, the velocity-saturated zone comprises a large part of the channel length. Despite the approximations implicit in the two-zone model, in comparison with the actual velocity-field characteristic (Fig. 11.2) and the neglect of short channel velocity overshoot effects [35], the model agrees well with measured data for both dc and small-signal MESFET parameters. Figures 11.8 and 11.9 illustrate this agreement. Figure 11.8 shows the agreement between the model and measured *I–V* characteristics of an X-band MESFET. The locus $L_2 = 0$ denotes the bias conditions for which velocity saturation first begins to occur at the drain end of the gate. To the left of this line, below the knee in the *I–V* characteristic, the entire channel is operating in the constant-mobility mode of operation. To the right of the locus line, where the FET is normally operated, a portion of the channel is in velocity saturation. Figure 11.9 shows a comparison of calculated and experimental values of transconductance (measured at the device terminals) and gate–source capacitance C_{gs} as a function of gate–source bias voltage.

Statz et al. [4], like Van der Ziel, treated the noise in region I as thermal but enhanced by field-dependent hot electron effects as described by Baechtold [31, 32]. Region II, however, is not an ohmic conductor. Here the noise contribution must be represented as a high field diffusion noise as shown by Schockley et al. [36] and Van der Ziel [37]. The diffusion noise in region II is proportional to the high field diffusion coefficient and

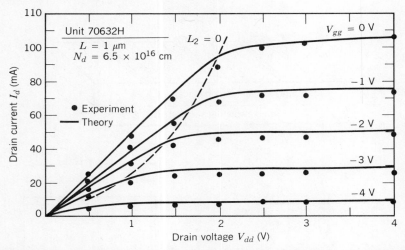

Figure 11.8   Comparison between the theoretical and measured drain current–voltage characteristic for a 1.0-μm-gate-length X-band MESFET (after Pucel et al. [3]).

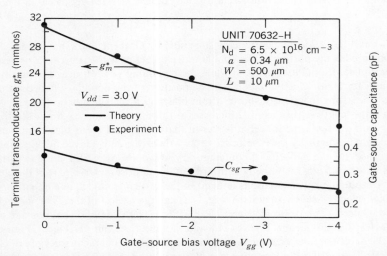

Figure 11.9   Comparison between the modeled and measured values of gate source capacitance and terminal transconductance for an X-band MESFET (after Pucel et al. [3]).

varies linearly with drain current, while the thermal noise generated in region I decreases with increasing drain current. Although the diffusion noise is high, the high degree of correlation existing between the drain noise and the induced gate noise leads to significant noise cancellation at the MESFET output. It is this fact that makes the MESFET a low-noise device.

Figure 11.10 illustrates the high-frequency noise equivalent circuit of the MESFET [3]. The noise generator i_g represents noise induced on the gate of the intrinsic device

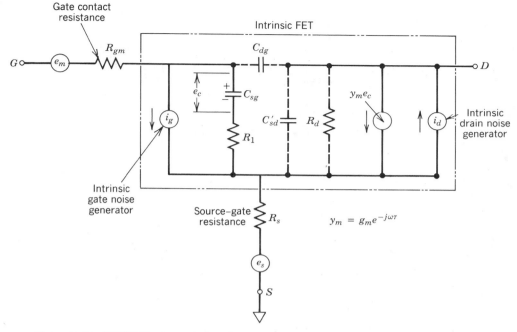

Figure 11.10 MESFET noise equivalent circuit illustrating the intrinsic and extrinsic noise sources (after Pucel et al. [3]).

by drain current fluctuations. The mean-square value of i_g varies as the square of the frequency. The intrinsic drain noise generator i_d, on the other hand, has a flat frequency spectrum. The correlation coefficient C given by

$$jC = \frac{\overline{i_g^* i_d}}{\sqrt{\overline{|i_g^2|}\,\overline{|i_d^2|}}} \tag{11.9}$$

approaches a magnitude of unity for short gate ($L \leqslant 1\ \mu$m) MESFETs. In Eq. (11.9), the asterisk denotes the complex conjugate and the overbars represent statistical averages. A constant-mobility model results in a lower correlation $|C| \sim 0.3$–0.4 [38]. In Fig. 11.10, there are two additional sources of noise. In the gate circuit, the voltage generator e_m is associated with thermal noise originating in the gate resistance R_g. Similarly e_s results from thermal noise generated in the parasitic source resistance R_s. Note that the noise associated with R_i is embedded in the gate noise generator i_g [3].

Noise Figure. Pucel et al. [5] have evaluated MESFET noise figure and minimum noise figure (achieved when the device is matched to an optimum source impedance) using a simplified MESFET equivalent circuit. They show that the equivalent circuit elements C_{dg} (drain–gate capacitance), C_{sd} (source–drain capacitance), and R_d (output drain resistance) have only a small effect (measured in tenths of a decibel) on noise figure. The resulting simplified equivalent circuit is shown in Fig. 11.11. Here the signal source impedance Z_g and the associated thermal noise source have been added to the input circuit.

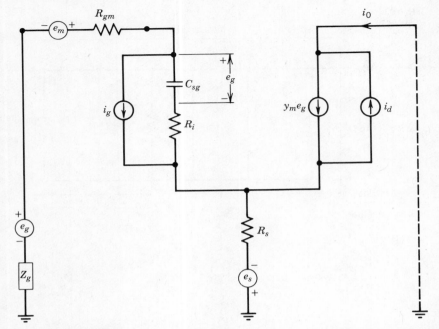

Figure 11.11 Simplified equivalent circuit used in the MESFET noise analysis of Pucel et al. [3].

Using this simplified equivalent circuit and an extended circuit analysis, the device noise figure can be written as

$$F = 1 + \frac{1}{R_g}(r_n + g_n|Z_g + Z_c|^2) \tag{11.10}$$

where R_g is the real part of the source impedance Z_g; r_n and g_n are the noise resistance and noise conductance, respectively [39]; and Z_c is the correlation impedance. The noise properties of the FET are contained in a simple noise network, shown in Fig. 11.12, in front of the FET, which is now noise free. The parameters r_n and g_n represent, respectively, thermal noise voltage and thermal noise current generators at a reference temperature $T_0 = 300$ K; and Z_c is a noiseless impedance (at absolute zero). The noise properties of the combined circuits are the same as that of the noisy FET.

The noise functions are given by the expressions

$$r_n = (R_s + R_{gm})\frac{T_d}{T_0} + K_r\left(\frac{1 + \omega^2 C_{sg}^2 R_i^2}{g_m}\right) \tag{11.11}$$

$$g_n = K_g \frac{\omega^2 C_{sg}^2}{g_m} \tag{11.12}$$

$$Z_c = R_s + R_{gm} + \frac{K_c}{Y_{11}} \tag{11.13}$$

where T_d is the actual device temperature (measured in degrees kelvin) and the param-

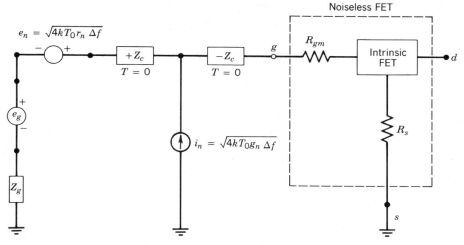

Figure 11.12 Equivalent noise circuit which represents a noisy MESFET by a noise network preceding a noiseless MESFET [3].

eters K_g, K_r, and K_c are noise coefficients representing the properties of the intrinsic noise generators i_g, i_d and their correlation. The reciprocal of Y_{11} is the input impedance of the intrinsic MESFET (Fig. 11.10) and is given by $Y_{11}^{-1} = R_i + 1/j\omega C_{sg}$. At temperatures other than T_0, the noise coefficients and the parameters g_m, R_i, C_{sg}, R_s, R_{sg}, and Y_{11} must be evaluated at T_d.

Minimum Noise Figure. The first stage of a low-noise amplifier is usually designed to have a minimum noise figure. This is obtained by proper matching of the complex source impedance $Z_{g0} = R_g + jX_g$ to the input terminals of the MESFET. Noise match is achieved when

$$R_g = R_{g0} = \sqrt{R_c^2 + \frac{r_n}{g_n}} \tag{11.14a}$$

$$X_g = X_{g0} = -X_c \tag{11.14b}$$

where R_c and X_c are the real and imaginary parts of the correlation impedance [Eq. (11.13)]. Under these conditions, the minimum noise figure is given by

$$F_{\min} = 1 + 2g_n(R_c + R_{g0}) \tag{11.15}$$

which can also be written as a power series expansion in frequency giving (for the first three terms)

$$F_{\min} = 1 + 2\left(\frac{\omega C_{sg}}{g_m}\right)\sqrt{K_g[K_r + g_m(R_s + R_{gm})]}$$
$$+ 2\left(\frac{\omega C_{sg}}{g_m}\right)^2 [K_g g_m(R_{gm} + R_s + K_c R_i)] + \cdots \tag{11.16}$$

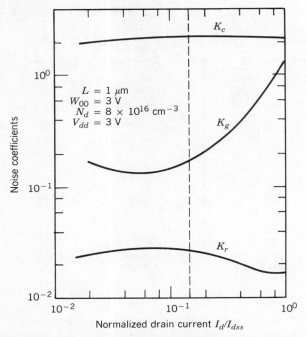

Figure 11.13   Dependence of the GaAs MESFET noise coefficients on the drain current [3].

From this expression, the dependency of F_{min} on the intrinsic device noise sources, as embodied in the noise coefficients K_g, K_c, and K_r as well as from the parasitic resistances R_s and R_g, may be deduced. The gain–bandwidth factor g_m/C_{gs} is an important parameter in determining F_{min}. Here, F_{min} decreases with an increasing gain–bandwidth factor, which is also a function of gate bias voltage and hence the drain current; F_{min} increases with increasing gate capacitance C_{gs} and decreases approximately with g_m^{-1}. These results provide first order information for designing low-noise MESFET devices, in particular that C_{gs}, R_s, and R_g should be minimized.

The three noise coefficients K_g, K_r, and K_c are dependent on both the gate and drain bias voltages and temperature but are independent of frequency. Figure 11.13 illustrates the dependency of the noise coefficients on drain current I_d normalized to I_{dss}, its saturated value at zero gate bias [3]. In addition to the drain current dependency shown, the three noise coefficients depend on gate length, the thickness of the channel, and other device parameters. Since the noise coefficients and the transconductance g_m are functions of the drain current, the minimum noise figure is also expected to exhibit a dependency on drain current.

Comparison of Minimum Noise Figure with Experiment. The predictions of the MESFET two-zone noise model show good agreement with measured device performance. Figure 11.14 shows the variation of the minimum noise figure F_{min} with the normalized drain current for a particular value of gate bias voltage based on the two-zone model [3]. Shown also are experimental data reported by Brehm [40] and by Brehm and Vendelin [41]. Exhibited also is F_{min} in the absence of high field diffusion noise. The two-zone model exhibits the basic features of the experimental data including rapid and near-linear rise in F_{min} for high drain currents, a behavior not exhibited by other noise

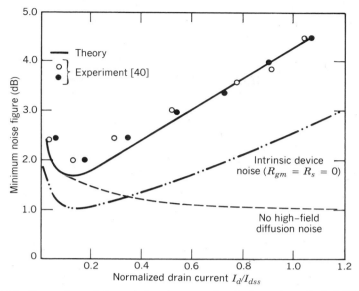

Figure 11.14 Comparison of theoretical and measured noise figure for a GaAs MESFET. The theoretical data are based on the Pucel and Statz noise model [5].

theories. The increase in F_{min} to the left of its minimum is a consequence of the reduced transconductance which occurs at low drain currents. The calculated behavior of the intrinsic device F_{min} (no parasitics) is also shown. The low value of minimum noise figure over the entire range of normalized current results from the strong cancellation of noise due to the correlation of the intrinsic noise sources associated with the gate and drain portions of the equivalent circuit. Such noise cancellation is diminished with the introduction of parasitic elements. For optimum low-noise performance, the parasitic resistances R_g and R_s must be kept small because they increase noise contributions from the noise sources associated with intrinsic device as well as introducing noise of their own. Typically, for a low-noise MESFET operating at microwave frequencies, the parasitic resistances may add 1–2 dB to the minimum noise figure.

Noise Conductance. As discussed in Section 11.5, in relation to optimum noise match, in designing low-noise amplifiers for broadband operation, the input device may not be matched for optimum noise figure, and the mismatch effect on noise figure can be written as

$$F = F_{min} + \frac{g_n}{R_g}\left[(R_g - R_{g0})^2 + (X_g - X_{g0})^2\right] \qquad (11.17)$$

where F_{min} is the minimum noise figure, g_n is the noise conductance, $Z_g = R_g + jX_g$ is the source impedance, and $Z_{g0} = R_{g0} + jX_{g0}$ is the optimum source impedance for minimum noise. The mismatch effect described by Eq. (11.17) is larger for larger values of the noise conductance g_n. The noise conductance can be shown to be inversely proportional to the reciprocal of the square of the cutoff frequency f_c [42]. Thus, achieving a high cutoff frequency not only results in a low value of F_{min} but also in a lower sensitivity of the device noise figure to input mismatch and thus provides low-noise operation over a wide bandwidth. As a result, achieving low-noise conductance is an important objective in designing low-noise MESFET devices.

11.3.3 Fukui Noise Theory

The MESFET noise theory summarized in the preceding section is relatively cumbersome. The behavior of the minimum noise figure as embodied in Eq. (11.16) depends not only on the parasitic elements R_g and R_s but also in a complex way on the three noise coefficients K_g, K_r, and K_c, which are intrinsic and thus depend on both gate and drain bias at a given frequency.

A relatively simple empirical MESFET noise formulation has been developed by Fukui [9, 43] and verified experimentally [44]. An expression for the optimal value of the minimum noise figure (denoted as F_0) was found to be

$$F_0 = 1 + 2\pi K_f f C_{gs} \sqrt{\frac{R_g + R_s}{g_m}} \times 10^{-3} \qquad (11.18)$$

where K_f is an empirical fitting factor of approximately 2.5 which depends on the channel material. It was found that this expression holds regardless of the gate bias voltage required to achieve F_0. In this equation, the equivalent circuit parameters are evaluated at zero gate bias. Thus, the parameters in Eq. (11.18) are fixed and not functions of bias. Apart from this discrepancy, note the similarity in the form of Eqs. (11.18) and (11.16).

The expression for F_0 may be rewritten as

$$F_0 = K_f \frac{f}{f_T} \sqrt{g_m(R_g + R_s)} \qquad (11.19)$$

using the relation that $f_T = g_m \times 10^3/2\pi C_{gs}$, or

$$F_0 = 1 + K_l L f \sqrt{g_m(R_g + R_s)} \qquad (11.20)$$

since f_T is related to the gate length L. In Eq. (11.20), K_l is another fitting factor whose value is approximately 0.27 when L is given in micrometers.

Fukui has also developed empirical expressions for the device transconductance and parasitic resistances [43]. These relationships are shown in Table 11.1. Figure 11.15 shows the geometrical parameters used in this formulation. The figure illustrates a MESFET configuration with a recessed gate. If the gate is not recessed, the geometrical parameters are altered in an obvious way.

Frequently an n^+ layer is applied between the ohmic metal contacts and the n-GaAs channel. This has been shown to be an effective means of reducing the source resistance [45]. In this case, L_2 is approximated by the distance between the source electrode and the edge of the n^+ layer and L_3 the distance along the n layer between the n^+ layer and the effective gate edge.

Although the expressions given by Fukui for g_m, R_s, and R_g are approximate and based on "effective" or average values of the geometrical and material parameters, good agreement was found between the calculated and experimental values of the equivalent circuit elements [9]. In addition, calculated values of F_0 based on the material and geometrical parameters agreed well with measured values. These comparisons were carried out for five device structures, and overall, calculated and experimental noise figures were found to differ by only a few percent at 6 GHz.

TABLE 11.1

	Expression	Parameter Definition
Transconductance	$g_m = K_m Z \left(\dfrac{N}{aL} \right)^{1/3}$ $(\Omega\text{-m})^{-1}$	k_m = constant (~ 0.023) Z = total device width in mm a = effective thickness of the active layer in μm N = effective free carrier concentration in the channel in 10^{16} cm^{-3} L = effective gate length in μm
Gate resistance	$R_g = \dfrac{17z^2}{ZhL_g}$ (Ω)	z = unit gate width in mm h = average gate metal height in μm L_g = average gate metal length in μm
Source resistance	$R_s = R_1 + R_2 + R_3 (\Omega)$	a_1 = effective channel thickness under the source electrode
	$R_1 \simeq \dfrac{2.1}{Z a_1^{0.5} N_1^{0.66}}$	N_1 = effective free carrier concentration in the channel under the source electrode in 10^{16} cm^{-3}
	$R_2 \simeq \dfrac{1.1 L_2}{Z a_2 N_2^{0.82}}$	L_2, L_3 = effective length of each channel section between the source and gate electrodes in μm
	$R_3 \simeq \dfrac{1.1 L_3}{Z a_3 N_3^{0.82}}$	a_2, a_3 = effective thickness of the channel sections in μm N_2, N_3 = effective free carrier concentrations of the channel sections in 10^{16} cm^{-3}

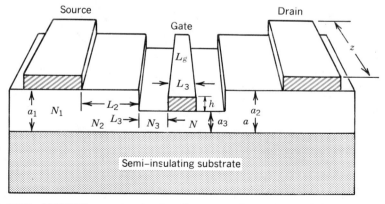

Figure 11.15 MESFET geometry showing the structural parameters used in the Fukui noise model.

Based on the results, design optimization techniques for low-noise devices were formulated. The equation for F_0 may be written as

$$F_0 = 1 + kf\left(\frac{NL^5}{a}\right)^{1/6}\left[\frac{17z^2}{hL_g} + 1.3z^2\left(\frac{f}{hL_g}\right)^{1/2} + \frac{2.1}{a_1^{0.5}N_1^{0.66}} + \frac{1.1L_2}{a_2N_2^{0.82}} + \frac{1.1L_3}{a_3N_3^{0.82}}\right]^{1/2}$$

$$(11.21)$$

where $k = K_k\sqrt{K_m} \simeq 0.040$. The second term under the square root has been added to account for skin effect contributions to the gate resistance.

Equation (11.21) delineates the major contributions of the parasitic elements to the optimal noise figure. The first two terms under the square root correspond to normal gate resistance and skin effect enhancements, respectively. The third term arises from the source contact resistance, while the fourth and fifth terms correspond to the ohmic resistances of the channel between the source contact and the gate.

The expressions for F_0 indicate that shortening the gate length and minimizing the parasitic gate and source resistances are essential to reduce the minimum noise figure. In addition, merely shortening the gate length will not improve F_0 unless the gate width is correspondingly reduced so as to limit the gate resistance. A rough design guideline for the maximum gate width is that width above which the gate metal resistance R_g exceeds the source resistance R_s.

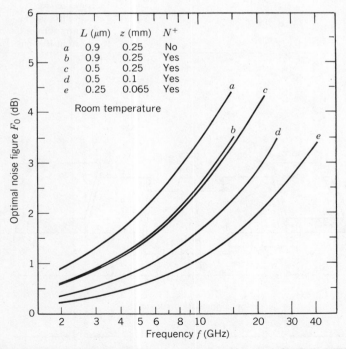

Figure 11.16 Calculated optimal noise figure as a function of frequency for GaAs MESFETs with various design and structural parameters (after Fukui [9]).

The Fukui analysis predicted GaAs MESFETs with excellent noise performance at microwave and low millimeter wave frequencies. Figure 11.16 shows the calculated optimal noise figures for five device structures with the values of gate length and gate width shown. Other design parameters are provided in Table V of Ref. 9. In recent years, devices meeting or exceeding these performance projections have been realized.

11.4 LOW-NOISE MESFET DESIGN AND FABRICATION

11.4.1 General Design Considerations

The work of Pucel et al. [5] and Fukui [9] provides specific low-noise MESFET design guidance that relates material parameters and device structural geometry to the RF performance of the device. Critical to the noise figure and associated gain of the low-noise MESFET are the active channel material quality, particularly at the channel–substrate interface; the gate length; and the source and gate parasitic resistances. These, in turn, are related to a number of material and structural parameters which include the doping characteristics of the channel such as the peak doping density and the doping density and mobility profiles at the lower edge of the channel [46, 47]; the geometry and placement of the gate in the source–drain space [45, 46]; and the design and structure of the source and drain ohmic contacts [48, 49].

At a given operating frequency, with all other material and device parameters optimized, the gate length is the most significant factor determining the device performance. A short gate length results in a high cutoff frequency; however, in order to achieve a sharp pinchoff characteristic and a high value of transconductance, the thickness of the active layer must be sufficiently small compared with the gate length. A higher doping density under the gate results in higher gain [47], but both the minimum active layer thickness and the maximum doping concentration is limited by leakage current and the gate-to-drain breakdown voltage. Control of the doping profile is an issue for very thin (tenths of micrometers) active layers, although MBE mitigates this problem to a great extent [6]. Finally, the electron mobility should be as high as possible, consistent with the channel doping density, and should remain constant in the region of the channel–substrate interface [47].

Variation of MESFET noise performance and associated gain with active layer doping density is shown in Fig. 11.17 [47]. Data are shown for a 0.5-μm gate length FET. The noise figure decreases with increasing doping density while the associated gain increases. Both performance parameters reach optimum values at a doping density of about 2.5×10^{17} cm^{-3}. The slight increase in noise figure for higher doping levels may be a result of a deteriorating doping profile.

The abruptness of the channel doping profile also has an important influence on low-noise MESFET performance. As shown in Fig. 11.14, analysis predicts and experimental results verify that minimum noise figure is achieved at a low drain bias current, typically 0.1–0.3 of I_{DSS}. At drain currents below the value for the minimum noise figure, the noise figure rises due to diminished transconductance at the interface. If, however, the doping profile is less steep, as, for example, is achieved by ion implantation rather than epitaxial growth, the minimum noise figure is observed to be greater and begins to increase at higher values of drain current, as shown in Fig. 11.18 [47].

The parasitic source and gate resistances are the major noise sources for GaAs low-noise MESFETs. Reduction of these parasitic resistances is therefore of comparable importance to reducing the gate length. Techniques are described in the next section.

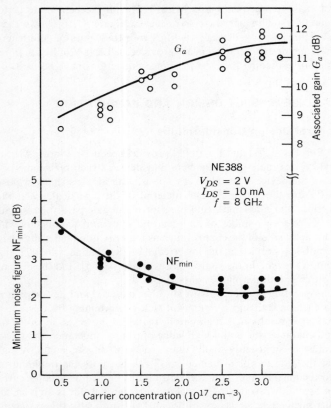

Figure 11.17   Dependence of MESFET minimum noise figure on the doping density of the active layer (after Hasegawa [47]).

11.4.2   Device Optimization for Low-Noise MESFET Performance

Based on the discussions presented in Sections 11.2.2 and 11.2.3, the optimization of GaAs MESFETs for low-noise operation requires the simultaneous optimization of materials and the device structure. Such improvements have been systematically realized over the past one and a half decades, resulting from on-going materials advances as well as significantly improved fabrication technology.

In order to optimize the noise figure and associated gain of a low-noise FET, several critical equivalent circuit elements and device geometrical parameters must also be optimized as follows:

1. Reduce the gate length L.
2. Increase the transconductance g_m.
3. Reduce gate-to-source capacitance C_{gs}.
4. Minimize the parasitic gate and source resistances R_g and R_s.

In some instances, compromises must be reached between these objectives.

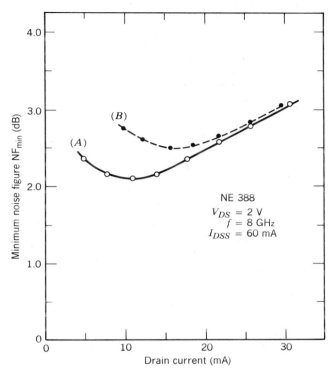

Figure 11.18 Dependence of MESFET minimum noise figure on drain current for two different interface doping profiles at the active layer–substrate interface [47]. Curve *B* shows the performance of a device with a less abrupt interface profile than that represented by *A*.

Gate Length. With current electron beam technology, fabrication of gate lengths as short as 0.1 μm is achievable. Gate lengths ranging from 0.2 to 0.5 μm are routinely fabricated across the industry using this technology. For gate lengths greater than 0.5 μm, contact or stepper lithography is used.

Note that reducing the gate length will effectively increase g_m. The Fukui approximate analysis [43] projects that g_m varies approximately as $L^{-1/3}$. The Pucel analysis [5] provides a more accurate expression in its Eqs. 20 and 21.

Methods of reducing the parasitic gate resistance of short-gate MESFETs are discussed below:

Transconductance. A high value of transconductance will result from a short gate length and from an optimized channel doping profile. Examples of the latter include retrograde [46] and spike doping profiles. These are illustrated schematically in Fig. 11.19. The spike profile provides the basis for the optimized MESFET design discussed in Section 11.4.5.

Gate Source Capacitance. Short gate length is a principal contributor to low gate source capacitance. Some of the techniques used to reduce the source resistance (e.g., offsetting the gate closer to the source than to the drain) may result in a marginal increase in C_{gs}.

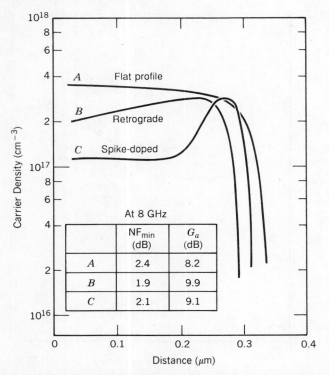

Figure 11.19 Low-noise MESFET device profiles and noise performance of devices employing these carrier profiles. Noise figures and associated gains measured at 8 GHz [46].

Source Resistance. Several techniques are used to minimize the source resistance. These include (1) recessing the gate in order to increase the cross-sectional area of the channel between the source and gate [50] (2), the use of an n^+ contact layer under the ohmic metal to reduce contact resistance [51], (3) offsetting the gate toward the source [46], and (4) the use of a selectively implanted source contact. These techniques are illustrated in Fig. 11.20. Figure 11.21 shows the series source resistance as a function of recess etch depth for a 0.5-μm-gate-length low-noise MESFET with a 280-μm gate width [50].

Gate Resistance. Several techniques are used to keep the parasitic gate resistance as low as possible. As the gate length is reduced, the resistance of the gate metal will increase. Increasing the gate metal thickness will reduce the problem, but lithography ultimately places limits to this approach. A related technique involves the use of special gate metal cross sections [52]. These take the form of a "T" or "mushroom" and permit a realization of a short gate while keeping the gate resistance low. These configurations may experience a small increase in gate source capacitance. Another approach to gate resistance reduction is to restrict the gate width through the use of multiple gate fingers or gate feeds. These techniques are illustrated in Fig. 11.22.

Special gate recess etch profiles have also been used to achieve an effective gate length shorter than the geometrical gate metal length [45].

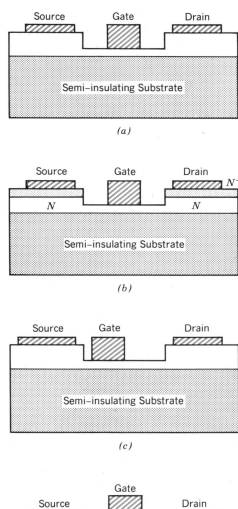

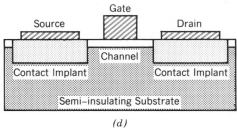

Figure 11.20 Techniques for reducing MESFET source resistance: (*a*) recessing the gate; (*b*) an N$^+$ contact under the ohmic metal reduces the contact resistance and hence the source resistance; (*c*) offsetting the gate toward the source; (*d*) selectively implanting ohmic contact layers.

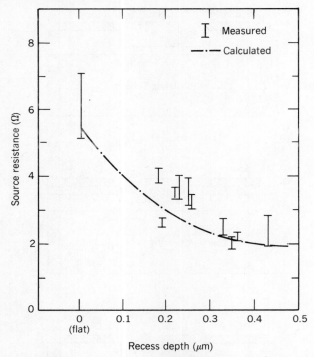

Figure 11.21   MESFET source resistance as a function of gate recess depth (after Ohata et al [50]).

11.4.3   Materials

High-quality, reproducible GaAs material is fundamental to the fabrication of high-performance low-noise MESFET devices. In the context of these devices, "material" encompasses two aspects: (1) the growth and characterization of the insulating substrate and (2) the growth or formation and characterization of the active device region, which consists of one or more thin n-type layers on one surface of the wafer. The active layer is usually formed by one of two methods: (1) epitaxial growth using one of several techniques or (2) ion implantation doping.

Substrates.   Currently most GaAs substrate wafers are produced using the liquid encapsulated Chochralski (LEC) growth process [53, 54]. This growth technique provides high-quality boules of single-crystal GaAs material with maximum impurity levels of approximately 5×10^{15} cm^{-3}. This material is highly resistive (resistivity greater than 10^7 Ω-cm). In 1989, 3-in.-diameter wafers are the industry standard with 4-in. wafers beginning to make an appearance. Substrate material is commercially grown using either high-pressure [55] or low-pressure [56] growth techniques. Currently, most commercial substrate vendors employ the high-pressure growth.

Substrate processing which includes wafer sawing, lapping, and polishing may impact subsequent device processing and performance, especially if not under adequate control. A major issue is residual polishing damage, which may be present in a surface layer ranging in thickness from hundreds of angstroms to tens of micrometers or greater.

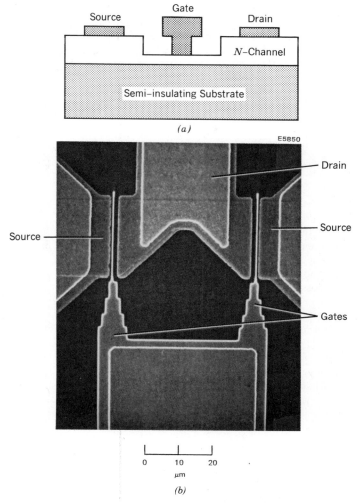

Figure 11.22 Techniques for minimizing MESFET gate resistance: (*a*) T- or mushroom-shaped gate metal cross sections; (*b*) multiple gate fingers and gate feeds.

Such residual surface damage may have a negative impact on subsequent epitaxial layer growth or the formation of conductive layers by ion implantation.

Substrate wafer qualification consists of a combination of physical and chemical measurements of quantities such as bulk impurity levels, bulk and surface defects, resistivity, and wafer flatness [57].

Active Layer Formation. Formation of active device layers for MESFET devices is carried out by either epitaxial growth or by ion implantation. Epitaxial techniques include vapor phase epitaxy (VPE) [58], molecular beam epitaxy (MBE) [59], metal organic chemical vapor deposition (MOCVD) [60], and several hybrid versions of MBE and MOCVD, such as metal organic molecular beam epitaxy (MOMBE) [61]. During the

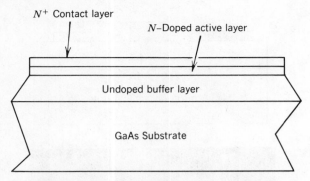

Figure 11.23   Epitaxial MESFET device structure.

1970s, VPE was the most important epitaxial technique for low-noise MESFET fabrication. More recently, MBE has become of increasing importance, particularly for devices operating at high frequencies. The ability of MBE to grow device structures with special profiles and near-atomic-layer precision has greatly enhanced device performance and has made possible the extension of low-noise MESFET devices well into the millimeter wave frequency range [62].

Active Layer Formation by Epitaxy.   Figure 11.23 illustrates an epitaxial device structure. The multiple GaAs layers include (1) an optional N^+-doped contact layer used to reduce the contact resistance at the interfaces with the source and drain ohmic metal contacts, (2) the active N-doped channel layer, and (3) an undoped buffer layer which separates the active device from the substrate. Typical device layers are tenths of micrometers in thickness, while the buffer layer may be a micrometer or greater in thickness.

Molecular beam epitaxy growth takes place when atomic beams of the principal constituents (gallium and arsenic) together with appropriate n-type dopant impurities such as silicon or selenium impinge on a heated substrate ($\sim 700°C$) under ultrahigh vacuum conditions. Molecular beam epitaxy growth rates are slow (approximately tens of angstroms per minute). As a consequence, MBE production rates are low, particularly when thick buffer layers are needed. Currently, despite its slow growth rate, MBE offers the greatest degree of precision and control in growing multilayer epitaxial GaAs films. The MOCVD growth systems, on the other hand, offer faster growth rates (10–100 times) at the expense of some precision. Metal organic epitaxial growth is also relatively free of major growth defects such as oval defects [60], which are associated with MBE growth. Hybrid systems such as MOMBE are also relatively defect free.

For all active layer growth systems, the major uniformity requirements (both over the wafer and wafer to wafer) are related to the layer thickness a and doping density N. The product $N \times a$ is related to the total charge in the device channel and has a major influence on device operation [44]. Uniformity of the $N \times a$ product thus has a large influence on device yield. Figure 11.24 illustrates the uniformity in the $N \times a$ product over a 3-in. GaAs wafer grown by MBE.

Another important material parameter which has a major impact on low-noise MESFET device performance is the abruptness of the transition between the doped channel layer and the buffer layer or substrate. The more abrupt this transition, the more effectively the current is confined in the channel and the higher the transconductance. Figure 11.25 shows the structure of a low-noise FET grown by MBE.

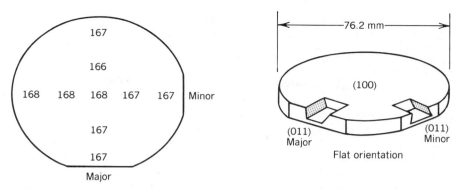

Figure 11.24   Uniformity of the doping density thickness product for a 3-in. GaAs wafer grown by MBE.

STRUCTURE

	THICKNESS (Å)	FREE CARRIER CONC. (cm-3)
GaAs : Si	500	3.0 E18
GaAs : Si	500	2.0 E17
GaAs : Si	500	8.0 E17
GaAs undoped Superlattice buffer	4500	--
Semi-insulating substrate		

SUBSTRATE

Crystal : LEC undoped single-crystal

Diameter (mm) :   **76.2 ± 0.6**

Thickness (μm) :   **635 ± 25**

Orientation :   (1 0 0) **± 0.5°**

Surface finish : polished one side

Edge :   rounded

Flat length (mm) :   major   22 ± 1
                       minor   9 - 14

Figure 11.25   Low-noise MESFET device structure. Material grown by MBE.

A number of low-noise device properties are related to the quality of the buffer layer separating the active channel region from the substrate. These include output resistance, source–drain breakdown voltage, light sensitivity, and backgating [63]. Different types of buffer layers which have been employed include undoped GaAs, AlGaAs, and superlattice (GaAs–AlGaAs) structures [64]. The superlattice consists of alternating layers of undoped GaAs and AlGaAs.

Recently a new type of buffer layer has been reported [65]. This GaAs buffer layer is grown by MBE at a substrate temperature of 200–300°C, compared with normal growth of MBE films at temperatures near 600°C. Growth at the lower temperature results in an excess arsenic mole fraction which generates a high level of electron traps, rendering the material both electrically and optically inactive. The buffer layers are highly resistive, and high-quality GaAs layers can be grown on top of the low-temperature buffer. The MESFET device improvements resulting from these buffer layers include substantial increases in output resistance and drain–source breakdown voltages. In addition, backgating is eliminated.

Active Layer Formation by Ion Implantation. The second important method of active layer formation for low-noise MESFET devices is ion implantation doping. This technique is currently the basis of most active device fabrication in monolithic microwave integrated circuit (MMIC) technology. At higher millimeter wave frequencies, where greater demands are placed on highly precise profiles and sharply defined doping boundaries, epitaxial techniques are more frequently used. In spite of this, ion-implanted MESFET devices operating at frequencies near 60 GHz have been reported [66].

Ion implantation doping has several advantages over epitaxy and at the same time some disadvantages. The advantages include excellent over-the-wafer and wafer-to-wafer uniformity of the active layer ($N \times a$ product), the ability to dope selectively over the wafer using masks, and the ability to scale up to large production rates using automated equipment already developed for silicon manufacture. The major disadvantage of ion implantation doping is a limitation in the degree of interface abruptness of resulting profiles. The latter is a consequence of the stopping physics of high-energy particles in solids [67].

Because the production advantages of ion implantation doping are less important for discrete MESFETs than for MMIC devices (because of the smaller chip areas of the former) and because of the higher performance achievable with epitaxially grown active layers, the latter will continue to be an important source of material for low-noise MESFETs.

The ion implantation doping process involves the direction of a high-energy beam of atomically pure ions onto the substrate wafer. The ion beam voltage may vary from tens to hundreds of kilovolts. The dopant atoms (silicon is most commonly used for low-noise MESFETs) are brought to rest in a thin surface region of the substrate. The thickness of the doped layer depends on the beam energy. In addition to ion beam energy, the other major implant parameter is dose, which is the total number of impurities deposited per unit area of the substrate. The distribution of the dopant atoms is approximately Gaussian in shape and is characterized by Lindhard, Scharff, and Schiott (LSS) statistics [67].

The details of the profiles are illustrated schematically in Fig. 11.26. For higher energy implants, not only is the peak in doping density deeper in the wafer, but also the width of the profile is greater. The result is that the interface profiles of high-energy deep implants are less abrupt than shallower ones. For this reason, low-noise ion-implanted MESFETs are frequently fabricated with shallow implants [68].

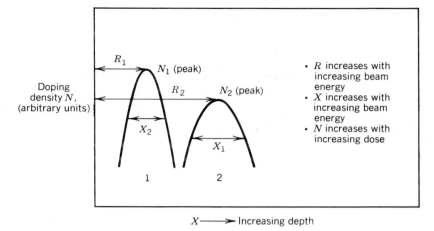

Figure 11.26 Ion implantation doping profiles. Higher-energy implanted layers are deeper with less abrupt interface profiles than low-energy implants.

The properties of MESFET active layers formed by ion implantation directly into semi-insulating substrates may be strongly substrate dependent, particularly in the low dose ranges (1×10^{12}–3×10^{12} cm^{-2}) typically used to form FET channel layers. When GaAs substrates of poor quality are used for ion implantation, variations in carrier concentration and effective depth of the implanted layer are often seen. In order to mitigate this problem, ion implantation into undoped GaAs buffer layers may be employed. Alternatively, if direct implantation into the substrate is used, careful substrate qualification is required [69].

Following ion implantation of GaAs, annealing is required to remove the damage induced by the implant and to activate the dopant atoms introduced. For doses used in MESFET fabrication, typical anneal cycles involve temperatures from 800 to 900°C. Because GaAs is subject to disassociation at temperatures above 600°C, the semiconductor surface must be protected by an encapsulating layer [70] or an arsenic-rich ambient [71] to prevent decomposition at the surface and outdiffusion of the implanted impurities.

Active Layer Qualification. Following formation of the active device structure, diagnostics are performed to verify its properties for low-noise MESFET fabrication. Capacitance–voltage measurements are used to determine the doping profile, which is the variation in doping density with respect to distance measured normal to the plane of the gate. In addition, mobility depth profiles are obtained using a standard Van der Pauw cloverleaf structure with a Schottky diode over the active area. By reverse biasing the Schottky diode, the active layer is depleted and Hall measurements are taken for increasing levels of depletion. From this data, mobility depth profiles may be determined [57]. An additional test which is performed when epitaxial buffer layers are employed is to measure buffer resistivity under standard illumination and in darkness. Light sensitivity here usually correlates with later light sensitivity at the device level, is most likely a result of deep charge trapping centers at the substrate–buffer interface [72], and is related to backgating phenomena [63].

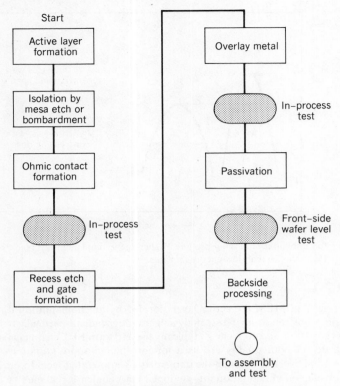

Figure 11.27   MESFET process flow sequence.

11.4.4   Device Fabrication

The fabrication of low-noise GaAs MESFET devices involves a series of operations which include (1) the growth and diagnostic analysis of active device layers on semi-insulating substrates, (2) the definition of active device islands by mesa etching or ion bombardment, (3) the formation of source and drain ohmic metal contacts, (4) the definition by appropriate lithography and formation of the Schottky barrier gate, (5) the application of a passivating layer over the active portion of the device, and (6) a series of substrate-oriented backside processes including substrate thinning, back metalization, and finally the separation of the wafer into individual device chips. Details of the material growth and analysis have been presented in the previous section; the remaining fabrication steps are described here. Figure 11.27 illustrates a typical low-noise MESFET processing sequence.

Mesa Isolation. Mesas which contain the active portions of the FET device are formed by chemical etch or ion milling using photoresist as a mask. Alternatively, the area surrounding the devices may be made inactive by ion bombardment [73]. The latter approach results in a completely planar transistor except for the gate recess.

Ohmic Contact Definition. Source and drain ohmic contacts are formed at device sites over the wafer by (1) photolithographic patterning, (2) the deposition of a multilayer

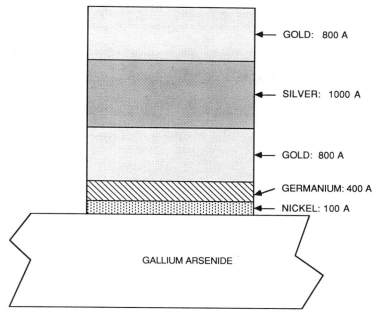

GOLD: 800 A

SILVER: 1000 A

GOLD: 800 A

GERMANIUM: 400 A

NICKEL: 100 A

GALLIUM ARSENIDE

Figure 11.28   MESFET ohmic contact metallurgy.

contact metal system uniformly over the wafer, and (3) the removal of unwanted metal by liftoff and finally alloying of the contacts using either furnace [74] or rapid thermal anneal [75] techniques. For GaAs ohmic contacts, a metallurgical system based on gold–germanium/nickel/gold is commonly used. This operation is usually performed using E-beam-assisted evaporation. Figure 11.28 illustrates a typical metallurgy. Contact delineation, like many MESFET processes, is performed using liftoff. In this procedure, regions of the wafer on which metal is not desired is covered with a masking layer of photoresist and metal deposited over the entire wafer. The masking photoresist is dissolved away in solvent which also removes the overlying metal.

Contact alloying is carried out in a furnace with controlled atmosphere at temperatures of typically 830–880°C for approximately 30 min. Alternatively, rapid thermal anneal is carried out at a peak temperature of 450°C. In this procedure, the wafer is heated by radiation from an incandescent source, also in a controlled atmosphere. In this case the heat-up and cool-down cycle is more rapid so that the entire procedure can be performed in approximately 20–30 s. For either process, smooth morphology of the ohmic metal after alloy is critical to high-resolution edge definition of the ohmic contacts.

Gate Lithography and Formation. This portion of the process is the most critical to the performance of the low-noise MESFET. For high-performance microwave and millimeter wave devices, gate lengths of substantially less than 1.0 μm are required. In addition, the gate must be accurately positioned at the proper depth, if recessed, in the channel between the source and drain contacts.

Several aspects of the gate fabrication process are critical in controlling the device parasitics. As discussed in Section 11.4.2, keeping the parasitic resistances of the source (R_s) and gate (R_g) to a minimum is essential to achieving low-noise performance.

Gate recessing is used as a means of reducing the source resistance. Gate recessing has the effect of increasing the cross-sectional area between the source contact and the channel region under the gate. Gate recessing is also used as a means of adjusting the channel current and pinchoff voltage to their desired values.

Recessing the channel prior to gate formation is a particularly critical process in MESFET fabrication. Location of the recess within the channel and particularly the recess depth must be accurately controlled uniformly over the wafer. The recess etch is usually carried out using an iterative wet-etch and measure technique where channel depth is determined by a measurement of the ungated channel current I_{DS}. As the recess depth increases, the channel current measured at a given drain voltage decreases. This process has recently been automated using computer algorithms to determine subsequent etch times based on the known etch rates, the measured doping profile, and the measured drain current following the first or successive etch cycles. These techniques have substantially increased the precision and reproducibility of the gate recess process.

In addition to recessing the gate, another means of reducing the source resistance is to offset the gate in the channel so that it is situated closer to the source than to the drain. This reduces the effective length of the channel between the source contact and the gate.

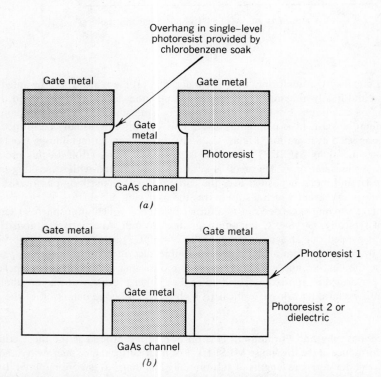

Figure 11.29   Techniques employed to obtain overhanging photoresist profiles required for high-yield liftoff for gate metal patterning: (*a*) chlorobenzine soak; (*b*) multiple-level photoresist or dielectric assisted liftoff.

Gates for low-noise MESFETs are fabricated from aluminum [76] or a multilayer gold-based metallurgy such as titanium platinum gold [77] or a gold refractory metallurgy [78]. Aluminum gates were most common on early devices; during the 1980s, the gold-based metallurgies have become more in favor. Lift-off techniques are most commonly used to define the gate metallurgy [79]. Special techniques employed to achieve the lift-off of short gates with maximum thickness (to minimize gate resistance) include the use of a chlorobenzene resist soak, multiple levels of resist, and dielectric assisted resist. These techniques provide resist layers with undercut sidewalls to facilitate clean lift-off for short gates, as shown in Fig. 11.29.

The above procedures result in gates with an approximate square cross section. In order to further reduce gate resistance, other, in many cases proprietary, fabrication methods are used to fabricate gates with the T-shaped cross section [52] shown in Fig. 11.30. These techniques are based on multiple-resist exposures, electroplating, or other methods. The T-gate structure preserves the short gate length where contact is made to the GaAs but adds more metal to the top portion of the gate finger to reduce resistive loss.

Passivation. Passivation of MESFET devices involves the deposition of a protective insulating film of dielectric material while still in wafer form. Materials used for passivation include silicon dioxide (SiO_2) [80], silicon nitride (S_3N_4) [81], or polyimide [82].

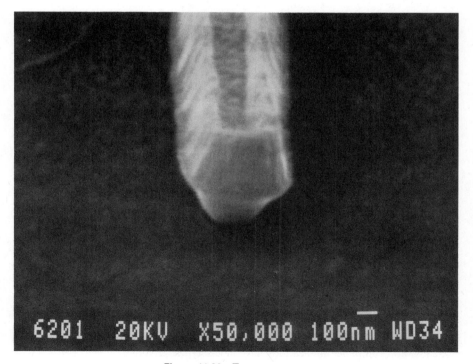

Figure 11.30 T-gate structure.

Composition, thickness, and deposition conditions are important parameters of the passivating medium which may effect the performance and stability of the finished device. The deposition of silicon nitride is most frequently performed by plasma-enhanced chemical vapor deposition (PECVD) [83]. The material is usually deposited uniformly over the surface of the wafer followed by a lithography and etch process to open contact windows. The thickness of the passivating layer is typically on the order of 1000–3000 Å.

Wafer Backside Processing. Following the completion of front-side processing and testing, the wafer is thinned by lapping or grinding from approximately 0.025 in. (for 3.0-in.-diameter substrates) to approximately 0.005–0.008 in. Low-noise device substrates are usually left thicker than power MESFET wafers because achieving minimal thermal impedance is not required for the former. Following back metalization, with a multilayer system such as titanium platimum–gold, the wafer is separated into chips by a scribe-and-break technique along crystal cleavage planes or by sawing.

11.4.5   Design Example—Optimized 0.10-μm Gate Length Low-Noise MESFET

This section describes the design and performance of a highly optimized low-noise MESFET. The devices described were developed at the Hughes Research Laboratories during the period from 1986 to 1988. Most of the techniques described in Section 11.4.2 to enhance the performance of low-noise MESFETs were implemented on these devices. Low-noise operation at frequencies above 60 GHz was achieved. The resulting design represents the state of the art in performance and material and fabrication technology for MESFETs during the latter portion of the 1980s.

Figure 11.31 illustrates the device structure. A spike-doped GaAs profile is employed to obtain a high current density in the channel. The spike profile allows the recessed gate to be placed in close proximity to the channel, resulting in a high transconductance (g_m) together with a short gate length L. This results in a high value of f_T. The device design employs a narrow doping spike (less than 10 nm) and a short gate length ($\simeq 0.1$ μm). Gate structures include cross sections which are approximately square and T-shaped.

Materials. The GaAs MESFET active layers were grown by MBE in order to achieve high purity and precise control of the active layer doping profile. The details of the spike-doped material structure are shown in Fig. 11.32. The peak doping density in the spike is 6×10^{18} cm^{-3} while the doping of the Schottky layer (contacted by the gate) is 5×10^{17} cm^{-3}. The spike is 83 Å thick and contains a total charge of 5×10^{12} cm^{-2}. The spike-doping profile results in a highly localized channel current and high device transconductance. The AlGaAs–GaAs heterojunction buffer confines the carriers to the channel and reduces the output conductance due to the larger bandgap in the AlGaAs layer.

Device Design. The spike-doped GaAs MESFET devices were designed for millimeter wave low-noise amplifier applications. Two device types were investigated; a 50-μm gate

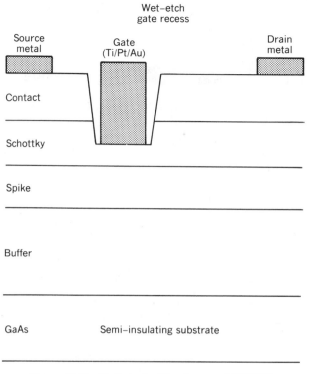

Figure 11.31 Optimized millimeter wave MESFET.

width structure with a single gate feed shown in Fig. 11.33a and a 75-μm gate width design with three gate feeds shown in Fig. 11.33b. For both designs, the gate feeds through a break in the source, thus requiring dual source pads. The 75-μm-wide gate device uses three gate feeds to reduce gate resistance by reducing the effective gate width. The gate manifold effectively feeds six elements of 12.5 μm width each with the outer

GaAs	Contact	6×10^{18} cm^{-3}	500 Å
GaAs	Schottky	1×10^{17} cm^{-3}	200 Å
GaAs	Spike	6×10^{18} cm^{-3}	83 Å
GaAs	Buffer		1000 Å
Al 0.3 Ga 0.7 As	Buffer		2000 Å
GaAs	Buffer		500 Å
GaAs	Semi-insulating substrate		

Figure 11.32 Material structure for optimized millimeter wave MESFET.

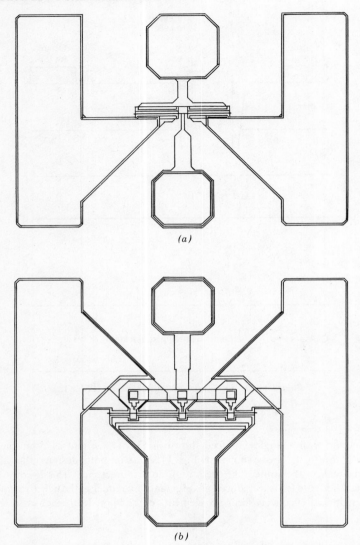

Figure 11.33 Device layout for optimized millimeter wave MESFET: (*a*) 50-μm-gate-width structure with single gate feed; (*b*) 75-μm-width structure with three gate feeds.

gate feeds connected to the center feed by a low-resistance air bridge. The source and drain designs are the same as the 50-μm design. For both device types a single gate stripe is used to allow accurate positioning of the gate in close proximity to the source. The active device area in both designs is on an epitaxially grown mesa while the pads are on semi-insulating buffer layers. Photos of the 50-μm gate-width device are shown in Fig. 11.34.

Device Performance. The spike-doped GaAs MESFETs fabricated have shown excellent dc and RF performance. Average device g_m values of 600–700 mS/mm were ob-

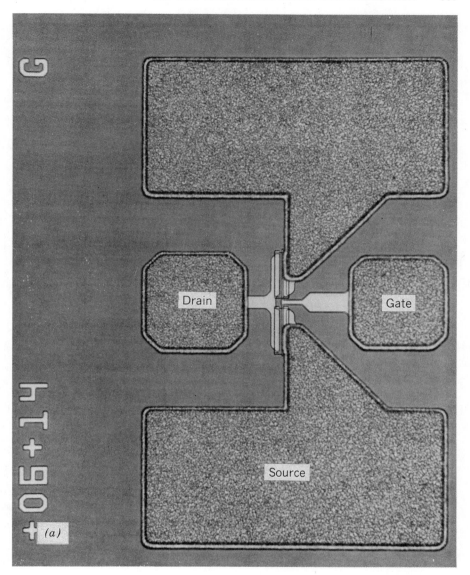

Figure 11.34 Photomicrographs of optimized millimeter wave MESFETs: (*a*) 75-μm device layout; (*b*) T-gate structure with a gate length of approximately 0.2 μm.

tained with high values of up to 800 mS/mm, which are among the highest reported transconductances for GaAs MESFETs.

Wafer-level scattering parameter measurements were made with on-wafer RF probes from 50 MHz to 26.5 GHz. Device current gain (h_{21}) was calculated using commercial simulation software. The projection of the current gain (at 6 dB/octave) was then used

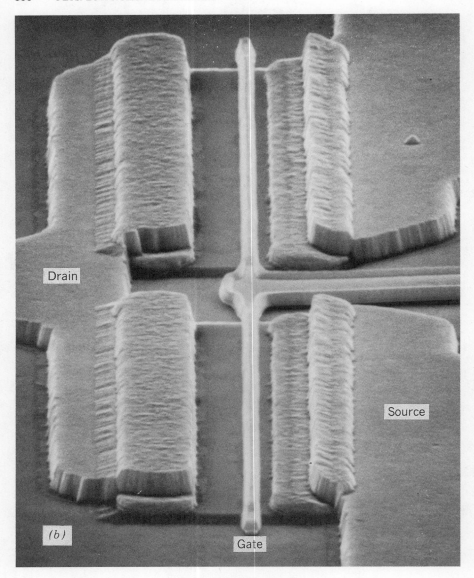

Figure 11.34   (*Continued*)

to determine the f_T of the device under test. Table 11.2 illustrates g_m and f_T performance for a number of device wafers with 0.1-μm gate lengths.

At V-band frequencies (50–75 GHz), these devices produced state-of-the-art low-noise performance. At the device level, noise figures of 2.5 dB with 8 dB of associated gain were measured making appropriate corrections for fixture loss. Single-stage amplifiers were fabricated with noise figures of 3.5 dB measured with respect to the input and output waveguide terminals.

TABLE 11.2

Gate Design	g_m (mS/mm)	f_T (GHz)
Square, 0.1 mm	800	110
Square, 0.1 mm	600	116
T, 0.1 μm	750	80
T, 0.1 μm	700	
T, 0.1 μm	500–600	95
T, 0.1 μm	500–700	110
T, 0.1 μm	500–600	94
T, 0.1 μm	500–700	101
T, 0.1 μm	500–600	92

11.5 LOW-NOISE AMPLIFIER DESIGN

11.5.1 General Principles of Low-Noise Amplifier Design

Amplifier Configuration. Low-noise amplifier configurations include balanced, unbalanced, or single ended and distributed. The noise properties of the distributed amplifier [84] limit the use of this amplifier to very broadband or gain applications. It is not suitable for very low noise applications and will not be considered further.

Low-noise MESFET amplifiers are typically configured in multistage designs whose stages are either balanced or unbalanced. Balanced circuits consist of a pair of amplifier stages whose input and output terminals are connected to the conjugate ports of two 3-dB hybrid couplers. This arrangement, illustrated in Fig. 11.35a, results in the

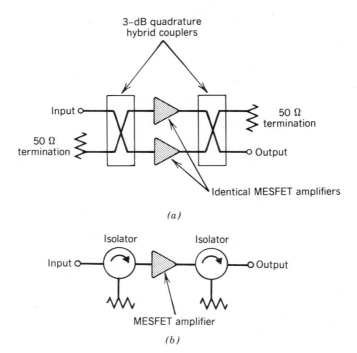

Figure 11.35 MESFET amplifier configurations: (*a*) balanced; (*b*) unbalanced or single ended. Isolators are not always used with the single-ended amplifier.

TABLE 11.3 Comparison between Balanced and Unbalanced Amplifiers

Balanced	Unbalanced
Advantages	Advantages
1. Greater flexibility to obtain input impedance match simultaneously with optimum noise figure 2. Greater stability for high-gain and broadband designs 3. Reduced sensitivity to device variation	1. Smaller number of devices and other components 2. Lower dc power requirements
Disadvantages	Disadvantages
1. Devices must be matched in similar pairs. 2. Loss and noise figure degradation in quadrature couplers	1. More difficulty in optimizing input noise match 2. More prone to instability than balanced amplifiers 3. Often requires isolators or other ferrite components which introduce loss

splitting of the signal applied to the input port of the first coupler into two equal parts which are 90° out of phase. These quadrature signals are fed into the two amplifier channels and recombined at the output coupler. Balanced stages may be cascaded to form multistage amplifiers. An unbalanced, or single-ended, amplifier is shown in Fig. 11.35b. Multiple-stage unbalanced amplifiers may exhibit instabilities which limit the number of stages which may be cascaded. Balanced amplifiers are employed where large bandwidth and high gain is required [85].

Table 11.3 shows the relative advantages and disadvantages of the balanced and unbalanced amplifier configurations. The remainder of this section will focus on the design and optimization of single-ended amplifiers.

Small-signal amplifiers may be further grouped according to the type of input, output, and interstage impedance matching employed. Types of matching include feedback [86], lossy [87], active [88], and reactive [89]. Table 11.4 qualitatively summarizes the char-

TABLE 11.4 Characteristics of Unbalanced Small-Signal MESFET Amplifiers

Type of matching	Feedback	Lossy	Active	Reactive
Noise figure	Moderate to high	High	Moderate to high	Very low to moderate, increases with bandwidth
Gain per stage	Low	Low	Very low	Low to high, decreases with bandwidth and increasing frequency
Bandwidth	Multi-octave	—	—	Narrow to wide, may exceed 2 octaves
Range of frequency (GHz)	0.1–20	0.1–20	0.1–10	1–75 or greater
Input VSWR	Low to moderate	Low to moderate	Very low for common gate	High
Output VSWR	Low	Low	Very low for common drain	Moderate to high

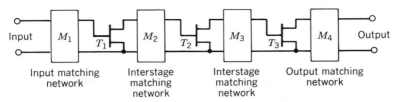

Figure 11.36 Three-stage single-ended amplifier.

acteristics of these amplifiers. In a given amplifier, these matching techniques may be combined; thus feedback or lossy matched stages may be employed for improved input and output VSWR (gain) in the latter stages of a multistage low-noise amplifier which frequently uses reactive matching techniques in the first one or two stages where noise figure is of paramount importance.

Techniques of Design. Figure 11.36 schematically illustrates a three-stage unbalanced amplifier. The matching networks of the amplifier determine the bandwidth, minimize the noise figure over the band, control the gain over the bandwidth, and determine the input and output matches. The amplifier noise figure is almost entirely determined by the input matching network and the noise figure of the first stage transistor. In general, the matching requirements for minimum noise, minimum input VSWR and maximum gain at the input to the amplifier are not compatible, and a compromise must be made between these performance parameters.

In order to carry out the design of a low-noise amplifier, both the device small-signal equivalent circuit model and the noise model are required. The equivalent circuit model used for amplifier design is derived from S-parameter techniques to extract the equivalent circuit parameters from the S-parameter data. The S-parameters may be measured from packaged or carrier-mounted samples or from on-wafer probe measurements. Noise modeling is performed in order to evaluate the conditions for minimum noise match. Characterization techniques are described in the next section. If different MESFET devices are used for the amplifier stages, characterization data is required for each device. Characterization must be carried out at or near the dc bias levels which are optimum for intended performance. Thus for the first-stage device, S-parameters and noise parameters must be determined at the bias current for minimum noise.

Once device characterization and noise data are obtained, the amplifier is designed using computer-aided design (CAD) techniques [90]. A logical design sequence (which is not unique) proceeds as follows: (1) an approximate first-stage input network topology is developed based on a knowledge of the noise parameters and the maximum degree of mismatch permissible while still meeting the amplifier noise specifications; (2) the input network is optimized using CAD; (3) the first-stage output is modeled; (4) the interstage matching network between the first two stages is developed to minimize noise measure*; (5) the subsequent interstage matching networks are developed to maximize and level the gain; (6) the output stage matching network is optimized for gain and gain flatness as well as output VSWR; and (7) the entire amplifier is optimized using CAD.

Frequently, for initial matching network development, the MESFET equivalent circuit is approximated by a unilateral approximation [91]. With this approximation, designs of the input and output matching networks of a stage are treated independently. In the final circuit CAD optimization, the complete MESFET device model is used.

---

* The noise measure M of an amplifier stage with noise figure $F(G)$ and gain G is given by $M(G) = \dfrac{F(G) - l}{l - \dfrac{1}{G}}$

The noise measure of n identical cascaded stages is equal to the noise measure of an individual stage.

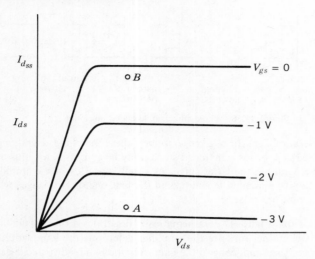

Figure 11.37 MESFET dc drain current versus drain voltage characteristics for various values of gate voltage. Typical operating regions for low noise and gain applications are indicated by A and B, respectively.

11.5.2   dc and RF Characterization of Low-Noise MESFETs

Characterization of low-noise MESFETs involves three different types of measurements. These are (1) dc characterization based on the $I-V$ curves shown in Fig. 11.37; (2) RF characterization, usually carried out by measuring two-port S-parameters over a significant bandwidth; and (3) noise parameter characterization.

These measurements are typically automated to a greater or lesser degree. The dc measurements are, in most manufacturing facilities, performed at the wafer level before die separation using automated probing equipment. These computer-controlled measurements are performed with minimal labor. In recent years, wafer probing has been extended to include the automated on-wafer measurement of microwave S-parameters. Commercially available on-wafer probing systems operating to frequencies in excess of 50 GHz are presently available [8]. Noise parameter characterization has traditionally been performed with packaged or carrier-mounted samples [42], although recently several methods of wafer level noise characterization have been reported [92].

dc Characteristics of Low-Noise MESFETs. Figure 11.37 shows the dc $I-V$ characteristics of a low-noise FET. Such a curve tracer display illustrates the important dc parameters for this device. These include I_{dss}, which is usually measured at a $V_{gs} = 0$ and $V_{ds} = 3-5$ V; the transconductance g_m, which is the ratio of $\Delta I_{dss}/\Delta V_{gs}$ in the saturated region of the transistor ($V_{ds} > V_{dssat}$); and the pinchoff voltage V_p, which is the values of negative gate voltage required to reduce the drain current to a small value (typically $1-10$ μA).

Also shown in Fig. 11.37 are the regions of operation for both low-noise (A) and high-gain (B) devices. Both operate in the saturated region of the transistor. The low-noise bias region occurs at low current (typically $0.1-0.2I_{dss}$) while bias for high gain is at current levels much closer to I_{dss}. Both modes of operation are employed in multistage low-noise amplifiers.

The dc FET parameters are easily measured at the wafer level using autoprobe techniques in conjunction with computer-aided data reduction. Since low-noise MESFET chips are usually small, a 2- or 3-in.-diameter wafer may contain several thousand devices. Thus, because of the sheer numbers, automated measurements are necessary.

RF Characteristics of Low-Noise MESFETs. Microwave GaAs MESFETs intended for low-noise applications are usually characterized by two-port, small-signal S-parameters [91]. This characterization is preferred because (1) a standard impedance (50 Ω) is easily provided at the input and output; (2) such terminations usually result in stable operation at all frequencies; (3) extensive computer software packages for device modeling and amplifier design are readily available; and (4) instrumentation and on-wafer probing systems are available for rapid RF characterization of devices from their two-port S-parameters.

Two-port common source S-parameters are usually measured for low-noise amplifier applications. This configuration, shown in Fig. 11.38, also provides the maximum small-signal gain. Multistage low-noise amplifiers usually employ gain stages after the first or second stage of low-noise amplification.

With the exception of on-wafer measurements, S-parameters are normally measured in a 50-Ω transmission line, usually microstrip. This configuration is shown in Fig. 11.38. In this geometry, the position of the reference planes which define input gate inductance and the output drain inductance must be accurately determined by through-line or short circuit measurements. In this way, the measured S-parameters will include the parasitic inductances of the bonding leads which will also be present when the chip is bonded into an amplifier. Bonding leads are kept as short as possible to minimize the inductances so as to retain circuit bandwidth [91].

Recently, on-wafer probing systems capable of accurate, small-signal S-parameter characterization have become available [8]. These probing systems are based on a tapered coplanar transmission line system which is used to make signal and ground contact to transistors at the wafer level.

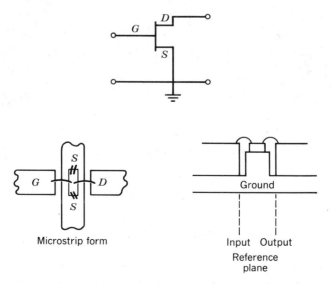

Figure 11.38 Common source MESFET configuration.

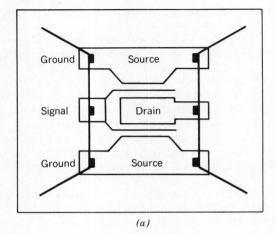

(a)

(b)

Figure 11.39   On-wafer RF probe: (a) ground–signal–ground probe footprint; (b) probe hardware.

This technique offers a number of advantages:

1. It allows device evaluation before backside wafer processing.
2. It eliminates the need for device mounting.
3. It allows nondestructive device evaluation.
4. It greatly improves device evaluation throughput so that 100% RF evaluation is possible.

Figure 11.39 shows an RF wafer probe together with a typical probe footprint layout. In order to ensure accurate measurements, the device pad layout must be matched to the probe. Shown in the figure is a ground–signal–ground layout which requires three contacts at both the input and output of the transistor. This configuration is favored in most applications.

To account for wafer probe loss, an accurate S-parameter characterization of the probe must be obtained. The wafer probe is a noninsertable device which has a miniature coaxial connector at one end and a coplanar footprint on the other. The probe S-parameters are obtained using standard spectrum analysis two-port calibration procedures in conjunction with sets of on-wafer coplanar and coaxial standards (open, short, and load).

Although the setup and calibration time for these measurements is significant, once completed, large numbers of devices can be characterized quickly, with little human labor and with computer-formatted output data.

Noise Characterization

Parameters Influencing the MESFET Amplifier Noise Performance. From the earlier portions of this chapter, it is apparent that a number of MESFET design variables as well as other parameters directly influence the noise figure F of a low-noise microwave amplifier. Following Gupta et al. [92], the "other variables" include six parameters in the case of a single-stage common source stage without feedback. Three of these are related to the circuit: the operating frequency f and the real and imaginary parts of the source or generator admittance $Y_g = G_g + jB_g$. The other three device-related parameters include the temperature T, the dc drain-to-source voltage V_{DS}, and the dc drain current I_{DS}. The dependency of F on these parameters has been extensively investigated by means of theoretical models [4, 5], empirical formulas [9, 43], and experimental measurements [93]. Figure 11.40 illustrates the dependency of F on these six parameters.

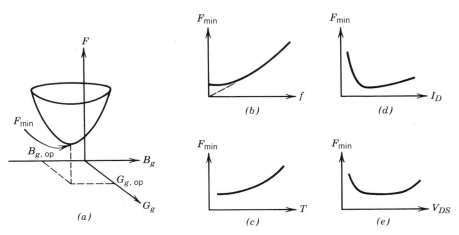

Figure 11.40 Parameters influencing the noise figure of a submicrometer gate MESFET amplifier (after Gupta [92]: (a) generator admittance $G_g + jB_g$, (b) operating frequency f_0, (c) temperature T, (d) dc drain current I_D, (e) dc drain–source voltage V_{DS}.

The admittance Y_g of the generator which drives the low-noise amplifier influences the noise figure $F(Y_g)$ as for any linear two-port network [39] so that

$$F(Y_g) = F_{\min} + \frac{r_n}{G_g} |Y_g - Y_{g0}|^2 \qquad (11.22)$$

where $Y_{g0} = G_{g0} + jB_{g0}$ is the optimum source admittance required for F to be equal to $F_{\min}$ and r_n is the equivalent noise resistance which determines the increase in noise figure resulting from a nonoptimum noise match. Equation (11.22) is the admittance counterpart of Eq. (11.17). Equation (11.22) defines a set of four noise parameters: $F_{\min}$, G_{g0}, B_{g0}, and r_n. Note that each is a function of frequency. Figure 11.40a illustrates the dependence of $F(Y_g)$ on Y_g.

The dependency of $F_{\min}$ on the other parameters—the operating frequency f, the drain current I_D, the temperature T, and the drain source voltage V_{DS}—is also shown in Fig. 11.40. The monotonic decrease of $F_{\min}$ (to 0 dB), as shown by the dashed line in Fig. 40b, results from the decreasing input conductance of the MESFET device with decreasing frequency [94, 95]. In practical low-noise amplifiers, circuit loss results in $F_{\min}$ approaching a value (>0 dB) as shown by the solid line in the figure. The behavior of $F_{\min}$ with drain current I_D (Fig. 11.40d) is discussed in Section 11.3.2. As shown in Fig. 11.40c, $F_{\min}$ increases monotonically with increasing temperature, as described by Weinreb [96]. Finally, as shown in Fig. 11.40e, the drain voltage V_{DS} has little effect on $F_{\min}$ when the device is biased in the current saturation region. In the linear region below the knee of the $I-V$ characteristic, the minimum noise figure is larger because of lower gain. At higher drain voltages, the minimum noise figure also increases because of increasing I_D and the possibility of Gunn domain formation and avalanching in the channel [97].

Noise Characterization Methods. An accurate determination of MESFET device noise parameters is required to perform low-noise amplifier design as well as to evaluate device design and compare relative device performance. Historically, two methods have been used.

The first method, shown schematically in Fig. 11.41, illustrates the conventional approach [42]. The automatic noise figure meter determines the noise figure F_m and gain G_m of the complete amplifier test setup, which includes, in addition to the device under test (with noise figure F), input and output tuners and bias networks. Treating these elements as passive, noisy two ports, the expression for F_m is

$$F_m = F_1 + \frac{F-1}{G_1} + \frac{F_2 - 1}{G_1 G_2} \qquad (11.23)$$

where F_1 and G_1 are the noise figure and available gain of the input matching circuit which includes the input tuner, bias circuit, and half the test fixture containing the test chips, while F_2 denotes similar components in the output circuit. Because the input and output matching networks are passive, $F_1 = 1/G_1$ and $F_2 = 1/G_2$, so that Eq. (11.23) can be rewritten as

$$F = G_1 \left(F_m - \frac{1-G_2}{G_m} \right) \qquad (11.24)$$

where $G_m = GG_1 G_2$ is the total measured gain. An accurate determination of the device noise figure thus requires an accurate measurement of G_1 which can be most accurately determined from the S-parameters of the input network [98] and the test fixture [99].

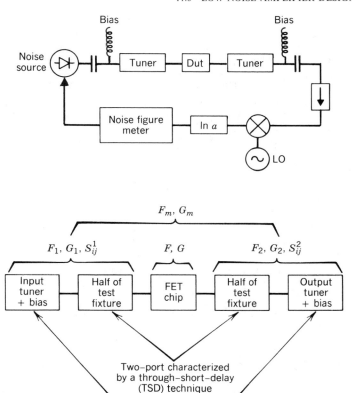

Figure 11.41 Conventional method of MESFET noise characterization [42].

The disadvantages of this technique include: (1) an extensive amount of time is required to find the minimum noise (including the correct bias); (2) a minimum values of F_m may not always provide a minimum device noise figure after loss corrections; and (3) an additional measurement is required to determine the noise resistance (r_n) or conductance (g_n).

A more systematic method of determining noise parameters has been proposed by Lane [100]. In this approach, an alternate method is used to determine the noise figure for different input matching conditions. Equation (11.22) may be written alternatively in terms of input reflection coefficients:

$$F = F_{\min} + \frac{4r_n}{|1 + \Gamma_{\text{opt}}|^2} \frac{|\Gamma_0 - \Gamma_{\text{opt}}|^2}{1 - |\Gamma_0|^2} \qquad (11.25)$$

where Γ_0 is the input reflection coefficient and Γ_{opt} is the reflection coefficient corresponding to the optimum source impedance. At least four measurements are required to determine the four unknowns $F_{\min}$, r_n, and the real and imaginary parats of Γ_{opt}. In practice, more measurements are performed and the four noise coefficients determined by a least-squares fit of Eq. (11.25) [101, 102]. The measurements may in addition be performed on the wafer.

The major advantage of this technique is the possibility of fully automated device noise and gain characterization. Disadvantages are primarily related to accuracy [42] and include: (1) following computation, the results are very sensitive to measurement errors if r_n is large; (2) the measurements are sensitive to low-frequency oscillations to which short-gate-length, high-f_T devices are prone; (3) the matrix of the four-equation system can become singular for some values of input termination; and (4) there is a problem in dealing with input network losses in the case of very low noise figures.

Because of the increasing availability of automated instrumentation, on-wafer RF probing systems operating at higher frequencies and with improved accuracy, and the

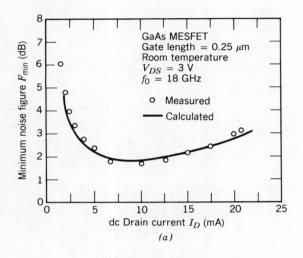

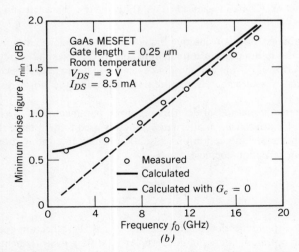

Figure 11.42 Comparison of calculated and measured values of minimum MESFET noise figure at room temperature (after Gupta et al. [92]: (a) variation of minimum noise figure with drain current at 18 GHz, (b) dependence of minimum noise figure on operating frequency at a drain current of 8.5 mA.

desire to reduce testing and characterization costs, on-wafer noise characterization is of increasing interest for both production and research requirements. As discussed above, the least-squares techniques is suited to such measurements subject to the limitations and caveats mentioned.

Gupta et al. [92] have proposed another approach valid for noise characterization of devices with submicrometer gates in the common source configuration. Their method consists of using a simplified noise model based on five linear circuit elements: the gate-to-source capacitance C_{gs}, the total input resistance R_T, the transconductance G_m, the output resistance R_0, and a noise current source of spectral density S_{10} at the transistor output port. These parameters can all be determined by on-wafer measurements while the noise current can be measured at low frequency (30 MHz to 1 GHz). The minimum noise figure of MESFET devices determined by this model as well as the bias and frequency dependence of the noise figure are shown to be in good agreement with microwave noise figure measurements. Figure 11.42 illustrates this agreement for a 0.25-μm-gate-length, low-noise MESFET measured at room temperature. In the figure, the parameter G_c is the conductance associated with the input matching circuit which can be determined by methods described above. In this manner, G_c is found to be nearly frequency independent. If G_c is assumed to be zero, the minimum noise figure in the simplified model approaches zero as shown by the dotted line in Fig. 11.42b. Including G_c causes the minimum noise figure to be frequency independent at low frequency, which agrees with measured results.

The advantages of the Gupta technique include: (1) no device tuning is required to obtain the minimum noise figure, (2) measurements can be performed at the wafer level and can be automated, (3) measurements are simple and fast and less subject (than the least-squares method) to uncertainty as to whether a minimum value of noise figure has been obtained, and (4) the parameters determined in the measurement provide diagnostic aid in isolating the cause of abnormally high or low values of F_{min}.

11.5.3 Design Example: K-Band Three-Stage Low-Noise Amplifier

As a means of illustrating some of the principles of low-noise amplifier design and fabrication, this section describes the details of a three-stage hybrid low-noise amplifier designed to operate in the K-band over a frequency range from 21.5 to 26.5 GHz.

MESFET Devices. The amplifier uses two device types which have the same geometrical layout and similar epitaxial material structures and are designated as DFET-5 and DFET-6. These devices both have a 0.25-μm gate length and a gate width of 60 μm. The device layout is shown in Fig. 11.43. Two 30-μm gate fingers are employed. It was

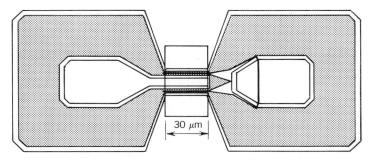

Figure 11.43 Layout of the MESFET used in the K-band amplifier.

determined through earlier evaluation that the DFET-5 lot has a lower noise figure, so this device was used in the first stage of the amplifier, while DFET-6 devices were used in the second and third stages.

Device Characterization. Both device types were characterized by S-parameter measurements from 2 to 26.5 GHz and 14-element equivalent circuit models developed from the measured data. These are shown for both device types in Figs. 11.44a and

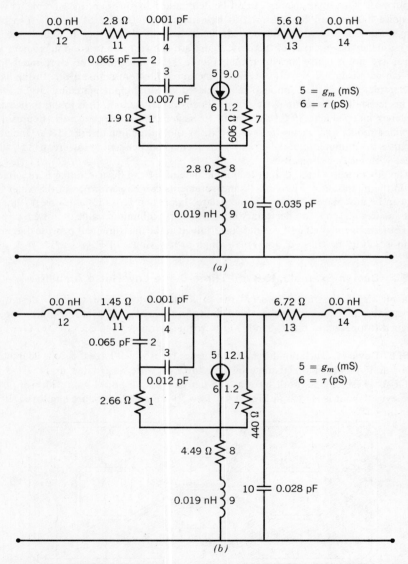

(a)

(b)

Figure 11.44   Small-signal equivalent circuit models for the MESFETs used in the K-band amplifier:(a) model for DFET-5, (b) model for DFET-6.

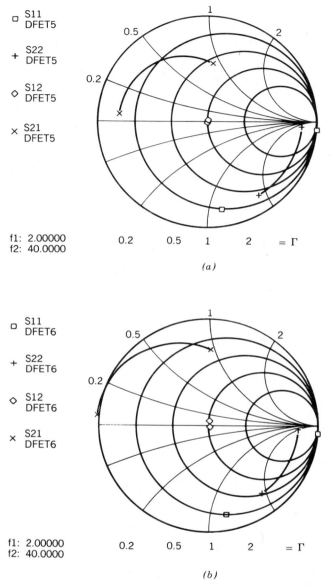

Figure 11.45 Simulated *S*-parameters for the two MESFETs used in the K-band amplifier: (*a*) DFET-5, (*b*) DFET-6.

11.44*b*. The simulated *S*-parameters for these two models are shown in Figs. 11.45*a* and 11.45*b*.

Noise modeling was next performed. The noise parameters $F_{\min}$, R_n and the real and imaginary parts of Γ_{opt} were determined and are shown in Table 11.5 for both devices. A simplified noise model developed by Podel et al. [103] was used.

TABLE 11.5 Noise Parameters for the MESFETS Used in the K-Band Amplifier

Frequency (GHz)	Minimum Noise Figure	Magnitude Γ_{opt}	Angle	R_N (Normalized)
				R_N (Normalized to 50 Ω)
		DFET-5		
18.000	1.800	0.818	43.510	1.835
20.000	1.996	0.804	47.893	1.835
21.000	2.093	0.798	50.037	1.835
22.000	2.191	0.792	52.148	1.835
23.000	2.287	0.787	54.226	1.835
24.000	2.384	0.781	56.269	1.835
25.000	2.480	0.776	58.277	1.835
26.000	2.576	0.772	60.251	1.835
27.000	2.671	0.767	62.189	1.835
28.000	2.766	0.763	64.091	1.835
29.000	2.860	0.760	65.958	1.835
30.000	2.955	0.756	67.789	1.835
		DFET-6		
18.000	2.146	0.800	47.119	1.951
19.000	2.262	0.793	49.468	1.951
20.000	2.377	0.786	51.779	1.951
21.000	2.492	0.780	54.050	1.951
22.000	2.607	0.774	56.280	1.951
23.000	2.720	0.768	58.469	1.951
24.000	2.834	0.763	60.616	1.951
25.000	2.946	0.758	62.721	1.951
26.000	3.058	0.753	64.783	1.951
27.000	3.170	0.749	66.803	1.951
28.000	3.281	0.745	68.780	1.951
29.000	3.391	0.741	70.714	1.951
30.000	3.500	0.738	72.605	1.951

Amplifier Design. Using the equivalent circuit and noise models, the amplifier design was developed following a procedure similar to that outlined in Section 11.5.1 using a commercial CAD software package. Several iterations were carried out to optimize the final design, which is shown in Fig. 11.46.

Amplifier Construction. The three-stage amplifier was fabricated in microstrip on multiple 25-mil-thick quartz substrates with TiWAu metallization. WR-4 waveguide input and output ports are used in the amplifier. To effectively couple the microstrip amplifier circuits to the waveguide ports, a microstrip-to-waveguide transition is employed. The transitions are also fabricated on 25-mil-thick substrates. The insertion loss of a single transition is approximately 0.3 dB over a full 18–26.5-GHz waveguide bandwidth.

The amplifier consists of three cascaded stages. Each stage contains input and output matching circuits. Between the stages are dc blocking circuits. Input matching circuits consist of impedance transformers, high-impedance inductances, and shunt resistors, which serve as out-of-band terminations.

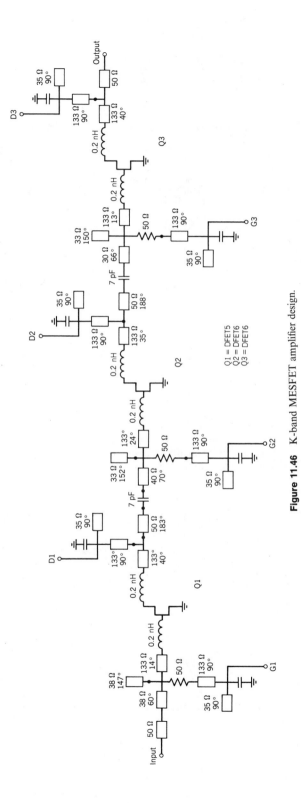

Figure 11.46 K-band MESFET amplifier design.

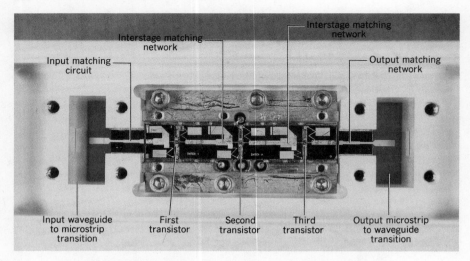

Figure 11.47   Photograph of the K-band MESFET amplifier.

The entire set of cascaded amplifier circuits is mounted in a carrier located in a channel operating below waveguide cutoff. The carrier is sandwiched between the microstrip-to-waveguide transitions. A photograph of an amplifier circuit is shown in Fig. 11.47.

Amplifier Performance. The measured amplifier noise figure and gain are shown in Fig. 11.48. Shown also is the simulated performance based on the final design. Agreement is good. Note that the experimental data have not been corrected for approximately 0.6 dB of input circulator and waveguide-to-microstrip transition loss.

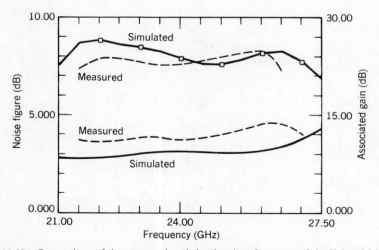

Figure 11.48   Comparison of the measured and simulated performance of the K-band MESFET amplifier.

11.6 CONCLUSIONS AND OUTLOOK

Currently the discrete GaAs MESFET is the most widely used solid-state low-noise amplifying device. It is used throughout the microwave frequency range from below 1 to above 30 GHz. As discussed earlier in this chapter, these devices are capable of excellent performance at twice this frequency using current state-of-the-art material and processing technologies. The low-noise MESFET has been in the field in a wide variety of applications including both military and commercial sockets for over 15 years.

During the past 10 years, MESFET processes have become increasingly mature, although not yet comparable with silicon technology. The low-noise GaAs MESFET is a highly reliable device [104, 105]. This contention is supported both by data from accelerated testing programs as well as field service data. As it is a majority carrier device, the MESFET is also radiation hard in comparison with silicon devices [106], an important property for space and certain military applications. It is expected that because of these attributes as well as its excellent performance, the MESFET will continue to play a major role in low-noise applications.

Currently, there are two important technologies which will impact the future of low-noise discrete MESFET devices: (1) the strong trend toward monolithic integration and (2) the increasing importance and availability of heterojunction field-effect transistors such as the HEMT and MODFET.

Low-noise MESFETs provide the active device for most low-noise MMIC devices. Low-noise HEMT MMICs are beginning to appear, and it is expected that this technology will steadily increase in importance [107]. In general, low-noise HEMT performance is superior to that of the low-noise MESFET because of the higher cutoff frequency and a higher degree of correlation between the gate and drain noise sources of the former device. A detailed comparison of the noise performance of the HEMT and MESFET may be found in a recent paper by Capy [42]. The performance advantages of the HEMT will carry over to the monolithic applications.

Ultimate MESFET Performance. In concluding this survey of low-noise MESFETs, it is appropriate to examine briefly the ultimate performance limits of this device.

As has been shown throughout this chapter, the single most important device parameter which influences low-noise MESFET performance is gate length. Because of the technological advances that have been documented, the gate lengths of low-noise MESFETs have decreased steadily over the past 15–20 years, and as a result, steady improvements in noise performance have resulted. Figure 11.49 illustrates the historical reduction in MESFET gate length over this period. An obvious question is: Will further reductions in gate length result in devices with lower noise figures and higher associated gains at ever higher frequencies? The answer is no. In addition, first order indications of where ultimate performance limitations will occur can be determined on the basis of straightforward physical considerations.

Three physical effects ultimately limit MESFET performance as the gate length is reduced to ever smaller values:

1. Channel pinch-off by the Schottky barrier gate
2. Gate–drain breakdown
3. Tunneling current at the Schottky contact

In a recent paper Golio [108] has examined these physical limitations in conjunction with device scaling to smaller geometries. The results of the analysis may be summarized

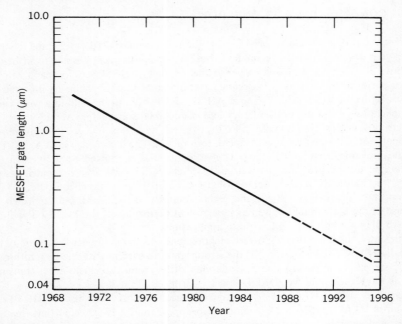

Figure 11.49   Gate Length of MESFETs reported in the literature from 1969 through 1987. Data from Ref. 108.

as follows. Pinchoff voltage considerations limit device scaling to minimum gate lengths of approximately 0.05–0.15 μm. Optimized structures in this range will require special doping profiles such as spike doping or very thin epitaxial layers. Breakdown voltage effects also limit scaling to gate lengths of near 0.1 μm or slightly less. Finally, tunneling at the Schottky barrier will limit scaling to gate lengths of 0.025 to 0.045 μm. The criteria here is that tunneling current densities should remain below 10^4–10^5 A/cm^2, which is generally regarded as the threshold for significant degradation by electromigration [109]. Based on these analyses, GaAs MESFETs with gate lengths below 0.05–0.1 μm will not exhibit substantially improved performance. MESFETs with gate lengths of 0.1 μm are presently being fabricated as laboratory devices. Devices with gate lengths of 0.05 μm are achievable with current technology or with moderate extensions of it. Such devices will have limited dynamic range because of pinchoff considerations and will require extremely thin and carefully controlled epitaxial layers. Assuming the trend line in gate length reduction shown in Fig. 11.46 continues at the historical rate, devices with these ultimate limits should become a reality before the year 2000.

Acknowledgement. The author wishes to acknowledge the contributions of his colleagues at Hughes in the preparation of this chapter. Special thanks are due to Dr. M. Delaney of the Hughes Research Laboratories and to M. Mayer of the Microwave Products Division for supplying unpublished material. F. McCaskill and C. Mirich typed the manuscript and M. Spaid assisted with the references.

REFERENCES

1. C. A. Liechti, "Microwave Field-Effect Transistors—1976," *IEEE Trans. Microwave Theory Tech.*, Vol. MTT-24, pp. 279–300, June 1976.

2. K. H. G. Duh, P. C. Chao, P. M. Smith, L. F. Lester, and B. R. Lee, "60 GHz Low-Noise High Electron Mobility Transistors," *Electron. Lett.*, Vol. 22, pp. 647–649, 5 June 1987.

3. R. A. Pucel, D. J. Masse, and C. F. Krumm, "Noise Performance of Gallium Arsenide Field-Effect Transistors," *IEEE J. Solid-State Circuits*, Vol. SC-11, pp. 243–255, April 1976.

4. H. Statz, H. A. Haus, and R. A. Pucel, "Noise Characteristics of Gallium Arsenide Field Effect Transistors," *IEEE Trans. Electron Devices*, Vol. ED-21, pp. 549–562, September 1974.

5. R. A. Pucel, H. A. Haus, and H. Statz, "Signal and Noise Properties of Gallium Arsenide Microwave Field-Effect Transistors," in *Advances in Electronics and Electron Physics*, Vol. 38, Academic Press, New York, 1975, pp. 195–265.

6. C. E. C. Wood, "Molecular Beam Epitaxy for Microwave Field Effect Transistors," in J. V. Di Lorenzo, Ed., *GaAs FET Principles and Technology*, Artech House, Dedham, 1982, pp. 104–114.

7. J. A. Turner, R. S. Butlin, D. Parker, R. Bennett, A. Peske, and A. Hughes, "The Noise and Gain Performance of Submicron Gate Length GaAs FETs," in J. V. Di Lorenzo, Ed., *GaAs FET Principles and Technology*, Artech House, Dedham, 1982, pp. 151–175.

8. K. E. Jones, E. W. Strid, and K. R. Gleason, "Mm-wave Wafer Probes Span 0–50 GHz," *Microwave J.*, Vol. 30, pp. 177–183, April 1987.

9. H. Fukui, "Optimal Noise Figure of Microwave GaAs MESFET's, *IEEE Trans. Electron Devices*, Vol. ED-26, pp. 1032–1037, July 1979.

10. J. Ruch, "Electron Dynamics in Short Channel Field-Effect Transistors," *IEEE Trans. Electron Devices*, Vol. ED-19, pp. 652–654, May 1972.

11. J. Ruch and W. Fawcett, "Temperature Dependence of the Transport Properties of Gallium Arsenide Determined by a Monte Carlo Method," *J. Appl. Phys.*, Vol. 41, pp. 3843–3849, August 1970.

12. T. Maloney and J. Frey, "Frequency Limits of GaAs and InP Field-Effect Transistors," *IEEE Trans. Electron Devices*, Vol. ED-22, pp. 357–358, June 1975; also corrections in *IEEE Trans. Electron Devices*, Vol. ED-22, p. 620, August 1975.

13. B. Himsworth, "A Two-Dimensional Analysis of Gallium Arsenide Junction Field Effect Transistors with Long and Short Channels," *Solid-State Electron.*, Vol. 15, pp. 1353–1361, December 1972.

14. J. Barnes and R. Lomax, "Two-Dimensional Finite Element Simulation of Semiconductor Devices," *Electron. Lett.*, Vol. 10, pp. 341–343, 8 August 1974.

15. R. Hockney, R. Warriner, and M. Reiser, "Two-Dimensional Particle Models in Semiconductor-Device Analysis," *Electron. Lett.*, Vol. 10, pp. 484–486, 14 November 1974.

16. C. Liechti, E. Gowen, and J. Cohen, "GaAs Microwave Schottky-Gate FET," in *1972 Int. Solid-State Circuits Conf. Dig. Tech. Papers*, pp. 158–159.

17. P. Wolf, "Microwave Properties of Schottky-barrier Field-Effect Transistors, *IBM J. Res. Develop.* Vol. 14, pp. 125–141, March 1970.

18. R. Dawson, "Equivalent Circuit of the Schottky-Barrier Field-Effect Transistor at Microwave Frequencies," *IEEE Trans. Microwave Theory Tech.*, Vol. MTT-23, pp. 499–501, June 1975.

19. G. Vendelin and M. Omori, "Circuit Model for the GaAs M.E.S.F.E.T. Valid to 12 GHz," *Electron. Lett.*, Vol. 11, pp. 60–61, 6 February 1975.

20. G. Vendelin and M. Omori, "Try CAD for Accurate GaAs MESFET Models," *Microwaves*, Vol. 14, pp. 58–70, June 1975.

21. G. D. Vendelin, "Feedback Effects in the GaAs MESFET Model," *IEEE Trans. Microwave Theory Tech.*, Vol. MTT-24, pp. 383–385, June 1976.

22. S. J. Mason, "Power Gain in Feedback Amplifiers," *IRE Trans. Circuit Theory*, Vol. CT-1, pp. 20–25, June 1954.

23. P. Hower and G. Bechtel, "Current Saturation and Small-Signal Characteristics of GaAs Field-Effect Transistors," *IEEE Trans. Electron Devices*, Vol. ED-20, pp. 213–220, March 1973.

24. K. E. Drangeid and R. Sommerhalder, "Dynamic Performance of Schottky-Barrier Field-Effect Transistors," *IBM J. Res. Develop.*, pp. 82–94, March 1970.

25. H. Daembkes, W. Brockerhoff, K. Heime, and A. Cappy, "Improved Short-Channel GaAs MESFET's by Use of Higher Doping Concentration," *IEEE Trans. Electron Devices*, Vol. ED-31, pp. 1032–1037, August 1984.

26. W. Shockley, "A Unipolar 'Field Effect' Transistor," *Proc. IRE*, Vol. 40, pp. 1365–1376, November 1952.

27. A. Van der Ziel, and J. W. Enro, "Small Signal, High-Frequency Theory of Field-Effect Transistors," *IEEE Trans. Electron Devices*, Vol. ED-11, pp. 128–135, April 1969.

28. A. Van der Ziel, "Thermal Noise in Field Effect Transistors," *Proc. IRE* Vol. 50, pp. 1808–1812, August 1962.

29. A. Van der Ziel, "Gate Noise in Field Effect Transistors at Moderately High Frequencies," *Proc. IEEE*, Vol. 51, pp. 461–467, March 1963.

30. H. Rothe and W. Dahlke, "Theory of Noisy Fourpoles," *Proc. IRE*, Vol. 44, pp. 811–818, June 1956.

31. W. Baechtold, "Noise Bahavior of Schottky Barrier Gate Field Effect Transistors at Microwave Frequencies," *IEEE Trans. Electron Devices*, Vol. ED-18, pp. 97–104, February 1971.

32. W. Baechtold, "Noise Behavior of GaAs Field-Effect Transistors with Short Gate Lengths," *IEEE Trans. Electron Devices*, Vol. ED-19, pp. 674–680, May 1972.

33. J. A. Turner and B. L. H. Wilson, "Implications of Carrier Velocity Saturation in a Gallium Arsenide Field-Effect Transistor," in *Proc. 2nd Int. Symp. Gallium Arsenide*, pp. 195–204, 1968.

34. A. B. Grebene and S. K. Ghandi, "General Theory for Pinched Operation of the Junction Gate FET," *Solid-State Electron.*, Vol. 12, pp. 573–589, July 1969.

35. T. J. Maloney and J. Frey, "Effects of Nonequilibrium Velocity-Field Characteristics on the Performance of GaAs and InP Field-Effect Transistors," in *1974 Dig. Tech. Papers, Int. Electron Devices Meeting*, pp. 296–298.

36. W. Schockley, J. A. Copeland and R. P. James, "The Impedance Field Method of Noise Calculations in Active Semiconductor Devices," in P. O. Lowdin, Ed., *Quantum Theory of Atoms, Molecules and the Solid State*, Academic Press, New York, 1966.

37. A. Van der Ziel, "Thermal Noise in the Hot Electron Regime in FETs," *IEEE Trans. Electron Devices (Corresp.)*, Vol. ED-18, p. 977, October 1971.

38. A. Van der Ziel, "Gate Noise in Field Effect Transistors at Moderately High Frequencies," *Proc. IEEE*, Vol. 51, pp. 461–467, March 1963.

39. H. Rothe and W. Dahlke, "Theory of Noisy Fourpoles," *Proc. IRE*, Vol. 44, pp. 811–818, June 1956.

40. G. E. Brehm, "Variation of Microwave Gain and Noise Figure with Bias of GaAs FETs," *Proc. 4th. Biennial Cornell Elec. Eng. Conf.*, pp. 77–85 1973.

41. G. E. Brehm and G. D. Vendelin, "Biasing FETs for Optimum Performance," *Microwaves*, Vol. 13, pp. 38–44, February 1974.

42. A. Capy, "Noise Modeling and Measurement Techniques," *IEEE Trans. Microwave Theory Tech.*, Vol. 36, pp. 1–10, January 1988.

43. H. Fukui, "Design of Microwave GaAs MESFETs for Broad-Band Low-Noise Amplifiers," *IEEE Trans. Microwave Theory Tech.*, Vol. MTT-27, pp. 643–650, July 1979.

44. H. Fukui, "Determination of the Basic Device Parameters of a GaAs MESFET," *Bell System Tech. J.*, Vol. 58, pp. 771–797, March 1979.

45. H. Fukui, J. V. Di Lorenzo, B. S. Hewitt, J. R. Velebir, Jr., H. M. Cox, L. C. Luther, and J. A.

Seman, "Optimization of Low Noise GaAs MESFETs," *IEEE Trans. Electron Devices*, Vol. ED-27, pp. 1034–1037, June 1980.

46. T. Suzuki, A. Nara, M. Nakatani, and T. Ishii, "Highly Reliable GaAs MESFET's with a Statistic Mean NF_{min} of 0.89 dB and a Standard Deviation of 0.07 dB at 4 GHz," *IEEE Trans. Microwave Theory Tech.*, Vol. MTT-27, pp. 1070–1074, December 1979.

47. F. Hasegawa, "Low Noise GaAs FETs," in J. V. Di Lorenzo, Ed., *GaAs FET Principles and Technology*, Artech House, Dedham, 1982, pp. 177–193.

48. H. H. Berger, "Contact Resistance and Contact Resistivity," *J. Electrochem Soc.*, Vol. 119, pp. 507–514, April 1972.

49. Mead, C. A., *Ohmic Contacts to Semiconductors*, B. Schwartz, Ed., Electrochemical Society, New York, pp. 3–16, 1969.

50. K. Ohata, H. Itoh, F. Hasegawa, and Y. Fujiki, "Super Low Noise GaAs MESFET's with a Deep Recess Structure," *IEEE Trans. Electron Devices*, Vol. ED-27, pp. 1029–1034, June 1980.

51. K. Ohata, T. Nozaki, and N. Kawamura, "Improved Noise Performance of GaAs MESFETs with Selectively Implanted n^+ Source Regions," *IEEE Trans. Electron Devices*, Vol. ED-24, pp. 1129–1130, August 1977.

52. P. C. Chao, P. M. Smith, S. C. Palmateer, and J. C. M. Hwang, "Electron-Beam Fabrication of Low Noise GaAs MESFET's Using a New Trilayer Resist Technique," *IEEE Trans. Electron Devices*, Vol. ED-32, pp. 1042–1046, June 1985.

53. E. P. A. Metz, R. C. Miller and J. Mazelsky, "A Technique for Pulling Single Crystals of Volatile Materials," *J. Appl. Phys.*, Vol. 33, pp. 2016–2017, June 1962.

54. M. E. Weiner, D. T. Lassota, and B. Schwartz, "Liquid Encapsulated Czochralski Growth of GaAs," *J. Electrochemical Soc.*, Vol. 118, pp. 301–304, February 1971.

55. T. R. AuCoin, R. L. Roso, M. J. Wade, and R. O. Savage, *Solid-State Tech.*, Vol. 22, p. 59, 1959.

56. R. C. Puttback, G. Elliot, and W. Ford, *5th Amer. Conf. Crystal Growth*, Coronado, CA, July 1981.

57. P. F. Lindquist and W. M. Ford, "Semi-Insulating GaAs Substrates," in J. V. Di Lorenzo, Ed., *GaAs FET Principles and Technology*, Artech House, Dedham, 1982, pp. 1–60.

58. L. Hollan, J. C. Brice, J. P. Hallais, "The Preparation of Gallium Arsenide," in E. Kaldis, Ed., *Current Topics in Materials Science*, Vol. 5, North-Holland, Amsterdam, 1980, pp. 1–217.

59. A. Y. Cho, "Growth of III-V Semiconductors by Molecular Beam Epitazy and Their Properties," *Thin Solid Films*, Vol. 100, pp. 291–317, 1983.

60. J. J. Coleman and P. Daniel Dapkus, "Metalorganic Chemical Vapor Deposition," in D. K. Ferry, Ed., *Gallium Arsenide Technology*, Howard W. Sams, Indianapolis, 1985.

61. Y. Tanaka, Y. Kunitsugu, I. Suemune, Y. Honda, Y. Kan, and M. Yamanishi, "Low Temperature GaAs Epitaxial Growth Using Electron-Cyclotron Resonance/Metalorganic-Molecular-Beam Epitazy," *J. Appl. Phys.*, Vol. 64, pp. 2778–2780, 1 September 1988.

62. B. Kim, H. Q. Tserng, and H. D. Shih, "Millimeter-Wave GaAs FET Prepared by MBE," *IEEE Electron Device Lett.*, Vol. EDL-6, pp. 1–2, January 1985.

63. L. F. Eastman and M. S. Shur, "Substrate Current in GaAs MESFET's," *IEEE Trans. Electron. Devices*, Vol. ED-26, pp. 1359–1361, September 1979.

64. W. J. Schaff, L. F. Eastman, B. Van Rees, and B. Liles, "Superlattice Buffers for GaAs Power MESFET's Grown by MBE," *J. Vac. Sci. Technol. B*, Vol. 2, pp. 265–268, April–June 1984.

65. F. W. Smith, A. R. Calawa, C. L. Chen, M. J. Manfra and L. J. Mahoney, "New Buffer Layer to Eliminate Backgating in GaAs MESFETs," *IEEE Electron Device Lett.*, Vol. EDL-9, pp. 77–80, February 1988.

66. M. Feng, H. Kanber, V. K. Eu, E. T. Watkins, and L. R. Hackett, "Ultrahigh Frequency Operation of Ion-Implanted GaAs Metal-Semiconductor Field-Effect Transistors." *Appl. Phys. Lett.*, Vol. 44, pp. 231–233, 15 January 1984.

67. J. F. Gibbons, W. S. Johnson, and S. W. Mylroie, *Projected Range Statistics*, Dowden, Hutchinson and Ross, Stroudsburg, 1975.

68. G. W. Wang, M. Feng, C. L. Lau, C. Itoh, and T. R. Lepkowski "Ultrahigh-Frequency Performance of Submicrometer-Gate Ion-Implanted GaAs MESFET's," *IEEE Electron Device Lett.*, Vol. 10, pp. 206–208, May 1989.

69. F. Eisen, C. Kirkpatrick, and P. Asbeck, "Implantation into GaAs," in J. V. DiLorenzo, Ed., *GaAs FET Principles and Technology*, Artech House, Dedham, 1982, pp. 117–144.

70. F. H. Eisen and B. M. Welch, in F. Chernow, J. A. Borders, and D. K. Brice, Eds., *Ion Implantation in Semiconduuctors*, New York: Plenum Press, New York, 1971, p. 97.

71. J. Kasahara, M. Arai, and N. Watanabe, "Capless Anneal of Ion Implanted GaAS in Controlled Arsenic Vapor," *J. Appl. Phys.*, Vol. 50, pp. 541–545, January 1979.

72. I. Crossley, I. H. Goodridge, M. J. Cardwell, and R. S. Butlin, "Growth and Characterization of High Quality Epitaxial GaAs for Microwave FETs," in L. F. Eastman, Ed., *Gallium Arsenide and Related Compounds (St. Louis) 1976*, The Institute of Physics, London, 1977, pp. 289–296.

73. D. D'Avanzo, "Proton Isolation for GaAs Integrated Circuits," *IEEE Trans. Electron Devices*, Vol. ED-29, pp. 1051–1059, July 1982.

74. M. Heiblum, M. I. Nathan, and G. A. Chang, "Characteristics of AuGeNi Ohmic Contacts to GaAs," *Solid-State Electron.*, Vol. 25, pp. 185–195, 1982.

75. C. L. Chen, L. J. Mahoney, J. D. Woodhouse, M. C. Finn, and P. M. Nitishin, "Ohmic Contacts to *n*-Type GaAs Using High-Temperature Rapid Thermal Annealing for Self-Aligned Processing, *Appl. Phys. Lett.*, Vol. 50, pp. 1179–1181, 27 April 1987.

76. B. S. Hewitt, H. M. Cox, H. Fukui, J. V. DiLorenzo, W. O. Schlosser, and D. E. Iglasias, "Low Noise GaAs MESFETs: Fabrication and Performance," in C. Hilsum, Ed., *Gallium Arsenide and Related Compounds (Edinburgh, Scotland) 1976*, The Institute of Physics, London, 1977, pp. 246–254.

77. R. Lundgren, *Final Technical Report*, RADC-TR-78-213, October 1978.

78. K. Mizuishi, H. Kurono, H. Sato, and H. Kodera, "Degradation Mechanism of GaAs MESFET's," *IEEE Trans. Electron Devices*, Vol. ED-26, pp. 1008–1014, July 1979.

79. Y. Todokoro, "Double Layer Resist Films for Submicrometer Electrom Beam Lithography," *IEEE Trans. Electron Devices*, Vol. ED-27, pp. 1443–1448, August 1980.

80. P. A. Kinkley and P. R. Selway, "Photoelastic Waveguides and Their Effect on Strip Geometry Lasers," *J. Appl. Phys.*, Vol. 50, pp. 4567–4579, July 1979.

81. R. L. Van Tuyl, V. Kumar, D. C. D'Avanzo, T. W. Taylor, V. E. Peterson, D. P. Hornbuckle, R. A. Fisher, and D. B. Estreich, "A Manufacturing Process for Analog and Digital Gallium Arsenide Integrated Circuits," *IEEE Trans. Electron Devices*, Vol. ED-29, pp. 1031–1038, July 1982.

82. J. G. Tenedorio and P. A. Terzian, "Effects of Si_3N_4, SiO and Polymide Surface Passivations on GaAs MESFET Amplifier RF Stability," *IEEE Electron Device Lett.*, Vol. EDL-5, pp. 199–202, June 1984.

83. E. Y. Chang, G. T. Cibuzar, and K. P. Pande, "Passivation of GaAs FETs with Silicon Nitride Films of Different Stress States," *IEEE Trans. Electron Devices*, Vol. ED-35, pp. 1412–1418, September 1988.

84. K. B. Niclas and B. A. Tucker, "On Noise in Distributed Amplifiers at Microwave Frequencies," *IEEE Trans. Microwave Theory Tech.*, Vol. MTT-31, pp. 661–668, August 1983.

85. K. Kurokawa, "Design Theory of Balanced Transistor Amplifiers," *Bell System Tech. J.*, Vol. 44, pp. 1675–1698, October 1965.

86. R. E. Lehmann, "X-Band Monolithic Three-Stage Low Noise Amplifier Employing Series Feedback," *IEEE Trans. Microwave Theory Tech.*, Vol. MTT-33, pp. 1560–1566, December 1985.

87. K. B. Niclas, "Lossy-Match GaAs MESFET Amplifiers; Design and Performance," *IEEE Trans. Microwave Theory Tech.*, Vol. MTT-30, pp. 2017–2020, November 1982.

88. D. Hornbuckle and R. L. Van Tuyl, "Monnolithic GaAs Direct Coupled Amplifiers," *IEEE Trans. Electron Devices*, Vol. ED-28. pp. 175–182, February 1981.

89. K. B. Niclas, "Multioctave Performance of Single Ended Microwave Solid State Amplifiers," *IEEE Trans. Microwave Theory Tech.*, Vol. MTT-32, pp. 896–908, August 1984.

90. D. J. Mellor, "CAD Synthesis Algorithms for Multistage Amplifier Interstage Network of Arbitrary Topologies," *1988 IEEE Int. Microwave Theory Tech. Symp. Dig.*, Vol. 1., pp. 323–326.

91. G. D. Vendelin, "Small Signal and Nonlinear Applications of GaAs FETs," in J. V. DiLorenzo, Ed., *GaAs FET Principles and Technology*, Artech House, Dedham, 1982, pp. 407–474.

92. M. S. Gupta, O. Pitzalis, Jr., E. G. Rosenbaum, and P. T. Greiling, "Microwave Noise Characterization of GaAs MESFET's: Evaluation by On-Wafer Low-Frequency Output Noise Measurement," *IEEE Trans. Microwave Theory Tech.*, Vol. MTT-35, pp. 1208–1217, December 1987.

93. H. Fukui, Ed., *Low-Noise Microwave Transistors & Amplifiers*, IEEE Press, New York, 1982.

94. C. H. Oxly and A. J. Holden, "Modified Fukui Model for High-Frequency MESFETs", *Electron. Lett.*, Vol. 22, pp. 690–692, 19 June 1986.

95. A. F. Podell, "A Functional GaAs FET Noise Model," *IEEE Trans. Electron Devices*," Vol. ED-28, pp. 511–517, May 1981.

96. S. Weinreb, "Low-Noise Cooled GASFET Amplifiers," *IEEE Trans. Microwave Theory Tech.*, Vol. MTT-28, pp. 1041–1054, October 1980.

97. M. S. Gupta, "Detection of Avalanching in Submicrometer Field-Effect Devices," *IEEE Electron Device Lett.*, Vol. EDL-8, pp. 469–471, October 1987.

98. E. W. Strid, "Measurement of Losses in Noise Matching Networks," *IEEE Trans. Microwave Theory Tech.*, Vol. MTT-29, pp. 247–252, March 1981.

99. R. A. Speciale, "A Generalization of the TSD Network-Analyzer Calibration Procedure, Covering *n*-Port Scattering-Parameter Measurements, Affected by Leakage Errors," *IEEE Trans. Microwave Theory Tech.*, Vol. MTT-25, pp. 1100–1115, December 1977.

100. R. Q. Lane, "The Determination of Device Noise Parameters," *Proc. IEEE*, Vol. 57, pp. 1461–1462, August 1969.

101. G. Garuso and M. Sannio, "Computer-Aided Determination of Microwave Two-Port Noise Parameters," *IEEE Trans. Microwave Theory Tech.*, Vol. MTT-26, pp. 639–642, September 1978.

102. M. Mitama and H. Katoh, "An Improved Computational Method for Noise Parameter Measurement," *IEEE Trans. Microwave Theory Tech.*, Vol. MTT-27, pp. 612–615, June 1979.

103. A. Podell, W. Ku, and L. C. T. Liu, "Simplified Noise Model and Design of Broadband Low-Noise MESFET Amplifiers," *Proc. 7th Biennial Cornell Elec. Eng. Conf.*, Vol. 7, pp. 429—443, 1979.

104. M. Omori, J. N. Wholey, and J. F. Gibbons, "Accelerated Active Lifetest of GaAs FETs and a New Failure Mode," *18th Ann. Proc. Int. Rel. Phys. Symp.*, pp. 134–139, 1980.

105. J. C. Irvin and W. O. Schlosser, "Failure Mechanisms and Reliability of Low Noise GaAs FETs," *Proc. 8th European Microwave Con.*, pp. 401–409, 1978.

106. R. J. Gutman and J. M. Borrego, "Degradation of GaAs MESFETs in Radiation Environments," *IEEE Trans. Rel.*, Vol. 29, pp. 232–236, 1980.

107. W. H. Perkins and T. A. Midford, "MIMIC Technology: Better Performance at Affordable Cost," *Microwave J.*, Vol. 31, pp. 135–143, April 1988.

108. J. M. Golio, "Ultimate Scaling Limits for High Frequency GaAs MESFETs," *IEEE Trans. Electron Devices*, Vol. ED-35, pp. 839–848, July 1988.

109. C. Canali, F. Fantini, A. Scorzoni, L. Umena, and E. Zanoni, "Degradation Mechanism Induced by High Current Density in Al-Gate MESFET's," *IEEE Trans. Electron Devices*, Vol. ED-34, pp. 205–211, February 1987.

INDEX